A Guide to the HPLC Literature

A Guide to the HPLC Literature

VOLUME 3: 1983

HENRI COLIN
ANTE M. KRSTULOVIĆ
JEAN-LOUIS EXCOFFIER
GEORGES GUIOCHON

A Wiley-Interscience Publication
JOHN WILEY & SONS
New York • Chichester • Brisbane • Toronto • Singapore

Library of Congress Cataloging in Publication Data:

Main entry under title:

A Guide to the HPLC literature.

"A Wiley-Interscience publication."
Includes index.
Contents: v.1. 1966–1979—v.2. 1980–1981—v.3. 1982.
1. High performance liquid chromatography—Bibliography.
I. Colin, Henri. II. Title: Guide to the H.P.L.C. literature.

Z5524.C55G84 1984 [QD79.C454] 016.543′0894 84-11967
0-471-80687-0 (v.3)

Printed in the United States of America

10 9 8 7 6 5 4 3 2 1

Preface

During the past decade, high-performance or high-pressure liquid chromatography (HPLC) has been experiencing a phenomenal growth in popularity, as evidenced by the rapidly increasing number of publications dealing with the various aspects of this technique.

HPLC has largely contributed to the development of pharmacology, biology, food research, and the biomedical sciences, as demonstrated by the growing number of meetings dedicated to this topic and by the proliferation of companies offering equipment, products, or services for HPLC users. It is becoming highly difficult to follow the current literature, particularly in the area of applications.

After having performed several tedious, lengthy, and not totally successful literature searches for our own needs, or for those of our colleagues mainly in the biological sciences, we decided to make this effort once and for all and have the whole literature organized in a way that would allow easy retrieval of the papers dealing with a particular area of interest.

This collection of references is dedicated to our 10,000 co-workers without whose contributions it could not have been written. We hope it will prove helpful to those who are still thinking, in spite of the examples to the contrary, that it is easier and faster to find an application already described in the literature, rather than to repeat the work that has already been done.

The first three volumes of this series include references covering the period between 1966 and 1982 inclusive. The citations for the above period are assembled in three volumes: Volume 1 covers the period 1966–1979, Volume 2, 1980–1981, and Volume 3, 1982.

Because of the incredibly high total number of publications dealing with all aspects of HPLC, we found it necessary to limit our choice only to those HPLC modes most commonly used; thus the articles on reversed- and normal-phase chromatography represent the essential part of the references cited. Articles dealing with gel permeation, ion exchange, and affinity chromatography are not included. Likewise, thin-layer chromatography (TLC) is not covered completely: included are only those articles that deal with the correlation between TLC and HPLC, as well as those of fundamental importance. Articles on ion chromatography (IC), very fast chromatographic analyses, and supercritical fluid chromatography will appear in Volume 4. This is justified by the fact that these topics are of rather recent interest and there were practically no references prior to 1980–1981.

PREFACE

The majority of articles were classified only on the basis of the key words as they appear in the titles. A certain number of fundamental articles which we were familiar with are also classified on the basis of the actual content. In the course of this project we were stricken by the tremendous importance of the titles. Accordingly, we want to stress that they should be as informative as possible, while still being reasonably succinct. Because of the incompleteness of the titles of certain publications and, either the impossibility to obtain the articles, or more importantly, to read all of them, it is possible that some citations have escaped our search. This is inevitable when the titles are not complete and explicit enough, regardless of the sophistication of the search system.

Based on the subject topic, the references are presented in 7 sections, 56 chapters, and 87 categories. The format of the references is illustrated with the following example:

> 03-81-0009(0133)
> RECYCLING IN LIQUID CHROMATOGRAPHY USING MICROBORE COLUMNS
> Kucera, Paul; Manius, Gerald
> Qual. Control Dep. Hoffmann-La Roche Inc. Nutley, USA NJ 07110
> J. Chromatogr.; (81) P 1–12; Vol 219; No 1

The numbers contain the following information:

- 03 : chapter number
- 81 : year of publication
- 0009 : reference number in a particular year (1981 in this case)
- 0133 : reference number in a particular chapter

Each citation contains the *title*, *author(s) name(s)*, *affiliation*, *journal title*, *year of publication*, the *first* and *last pages*, and journal *volume* and *number* (when indicated). Only languages other than English are indicated.

Sections containing a large number of references (Amino Acids, Biogenic Amines, Food and Food Additives, Metals and Metalloids, Lipids, Nucleic Acid Components, Organic Acids, Peptides, Steroids, Antibiotics, Various Organics, Various Pharmaceuticals, and Vitamins) are indexed. Indexes for these categories appear at the end of each volume and contain the alphabetically listed key words, as they appear in the titles, or abbreviated words (when the original key words were too long). Compounds containing a large number of substituents were listed as the parent compound "deriv." The meaning of the numbers appearing on the right-hand side of each key word in the index is illustrated with the following example:

CREATININE (80), 32, 102
(80) indicates the year of publication

32 and 102 are reference numbers in a particular year (1980 in this case).

The chapter titles are usually self-explanatory. Several chapter headings with broader, unspecific titles deserve some comments. For example, the section on Hydrodynamics contains articles dealing with the different plate height models, the role of column length, the particle diameter, as well as the hydrodynamic optimization of these parameters. The section Other Detectors includes the articles concerned with detection devices other than those used in UV/VIS spectrophotometry, fluorimetry, electrochemistry, and mass spectrometry. It should be pointed out, however, that the chapters pertaining to the separations of various classes of compounds also contain "related compounds." For example, the chapter on Nucleic Acid Components contains articles dealing not only with nucleotides, poly- and oligonucleotides, nucleosides, and bases, but also nucleoside analogs and various drugs with the purine and pyrimidine structures. Likewise, the section on Biogenic Amines (not including polyamines) also contains articles dealing with the metabolites of these compounds; the chapter on Steroids contains bile acids, and the chapter on Lipids includes prostaglandins.

The authors thank Mrs. Josiane Coyac of the Central Library at École Polytechnique for her helpful assistance in the course of this project.

HENRI COLIN
ANTE M. KRSTULOVIĆ
JEAN-LOUIS EXCOFFIER
GEORGES GUIOCHON

Palaiseau, France
June 1984

CONTENTS

SECTION 1

SECTION 2

SECTION 3

CONTENTS

CONTENTS

SECTION 7

A Guide to the HPLC Literature

SECTION 1

CHAPTER 1.1
Books

--- 1982 ---

01-82-0001 (0001)
FIRST WORKSHOP ON LIQUID CHROMATOGRAPHY-MASS SPECTROMETRY. (1981: MONTREUX. IN: J. CHROMATOGR., 1982; 251(2))
Frei, R. W.; Editor Neth. NL
BOOK ; (82) P 137 pp.; Publisher: Elsevier, Amsterdam, Neth.

01-82-0002 (0002)
INTRODUCTION TO HIGH PERFORMANCE LIQUID CHROMATOGRAPHY. 2ND ED
Hamilton, R. J.; Sewall, P. A.
UK GB
BOOK ; (82) P 240 pp.; Publisher: Chapman and Hall, London, UK

01-82-0003 (0003)
ANALYTICAL METHODS FOR PESTICIDES AND PLANT GROWTH REGULATORS, VOL. 12
HIGH-PERFORMANCE LIQUID CHROMATOGRAPHY OF PESTICIDES
Lawrence, James F.
BOOK ; (82) P 246 pp.; Publisher: Academic Press, New York, N. Y.

01-82-0004 (0004)
REVERSED-PHASE HIGH-PERFORMANCE LIQUID CHROMATOGRAPHY. THEORY, PRACTICE, AND BIOMEDICAL APPLICATIONS
Krstulovic, Ante M.; Brown, Phyllis R.
USA
BOOK ; (82) P 296 pp.; Publisher: Wiley-Interscience, New York, N. Y.

01-82-0005 (0005)
CHROMATOGRAPHIC SCIENCE, VOL. 20: BIOLOGICAL/BIOMEDICAL APPLICATIONS OF LIQUID CHROMATOGRAPHY IV
Hawk, Gerald L.; Editor USA US
BOOK ; (82) P 367 pp; Publisher: Dekker , New York, N. Y.

01-82-0006 (0006)
MANUEL PRATIQUE DE CHROMATOGRAPHIE EN PHASE LIQUIDE
PRACTICAL HANDBOOK OF LIQUID-PHASE CHROMATOGRAPHY. 2ND ED
Rosset, Robert; Caude, Marcel; Jardy, Alain
Fr. FR
BOOK ; (82) P 374 pp.; In Fr; Publisher: Masson, Paris, Fr.

CHAPTER 1.2
Miscellaneous

--- 1982 ---

02-82-0001 (0001)
HIGH-PERFORMANCE LIQUID CHROMATOGRAPHY AND ITS APPLICATION TO PROTEIN CHEMISTRY
Hearn, Milton T. W.
St. Vincent's Sch. Med. Res. Fitzroy Australia A
Adv. Chromatogr. (N. Y.); (82) P 1-82; Vol 20

02-82-0002 (0002)
HIGH-PERFORMANCE LIQUID CHROMATOGRAPHY: APPLICATIONS IN A CHILDREN'S HOSPITAL
Soldin, Steven J.
Hosp Sick Children Univ. Toronto Toronto Can. CA ON
Adv. Chromatogr. (N. Y.); (82) P 139-66; Vol 20

02-82-0003 (0003)
THE USE OF HIGH PRESSURE LIQUID CHROMATOGRAPHY IN UNDERGRADUATE PHARMACY EXPERIMENTS
Houglum, Joel E.; Cascella, Peter J.
Coll. Pharm. South Dakota State Univ. Brookings USA US SD 57007
Am. J. Pharm. Educ.; (82) P 32-4; Vol 46; No 1

02-82-0004 (0004)
PURIFICATION BY REVERSE-PHASE HIGH-PERFORMANCE LIQUID CHROMATOGRAPHY OF AN EPIDERMAL GROWTH FACTOR-DEPENDENT TRANSFORMING GROWTH FACTOR
Anzano, Mario A.; Roberts, Anita B.; Smith, Joseph M.; Lamb, Lois C.; Sporn, Michael B.
Lab. Chemoprev. Natl. Cancer Inst. Bethesda USA US MD 20205
Anal. Biochem.; (82) P 217-24; Vol 125; No 1

02-82-0005 (0005)
FURTHER PURIFICATION OF EPIDERMAL GROWTH FACTOR BY HIGH-PERFORMANCE LIQUID CHROMATOGRAPHY
Matrisian, Lynn M.; Larsen, Brent R.; Finch, Joanne S.; Magun, Bruce E.
Arizona Health Sci. Cent. Univ. Arizona Tucson USA US AZ 85724
Anal. Biochem.; (82) P 339-51; Vol 125; No 2

02-82-0006 (0006)
APPLICATION OF HIGH-PERFORMANCE LIQUID CHROMATOGRAPHY TO THE SEPARATION AND ISOLATION OF BIOPOLYMERS
Unger, K. K.; Roumeliotis, P.
Inst. Anorg. Chem. Anal. Chem. Johannes Gutenberg-Univ. Mainz Fed. Rep. Ger. DE 6500
Anal. Chem. Symp. Ser.; (82) P 455-70; Vol 9; No Affinity Chromatogr. Relat. Tech.

02-82-0007 (0007)
COLUMN LIQUID CHROMATOGRAPHY
Majors, Ronald E.; Barth, Howard G.; Lochmueller, Charles H.
Walnut Creek Instrum. Div. Varian Assoc. Walnut Creek USA US CA 94598
Anal. Chem.; (82) P 323R-363R; Vol 54; No 5

02-82-0008 (0008)
SOME APPLICATIONS OF LIQUID CHROMATOGRAPHY TO THE ANALYSIS OF MEDICINAL PRODUCTS
Hopkins, R. G.
Med. Prod. Sub-Div. Lab. Gov. Chem. London UK GB SE1 9NQ
Anal. Proc. (London); (82) P 193-5; Vol 19; No 4

02-82-0009 (0009)
APPLICATION OF DIRECT COUPLING OF LIQUID PHASE CHROMATOGRAPHY AND MASS SPECTROMETRY TO BIOCHEMICAL PROBLEMS
Dixon, David J.
Hewlett-Packard G.m.b.H. Waldbronn Fed. Rep. Ger. DE 7517
Analusis; (82) P 343-8; Vol 10; No 7; In Fr

02-82-0010 (0010)
PSEUDOPHASE LIQUID CHROMATOGRAPHY
Saitoh, Koichi
Coll. Sci. Tohoku Univ. Sendai Japan JP
Bunseki; (82) P 594-5; No 8; In Japan

02-82-0011 (0011)
HIGH-PERFORMANCE LIQUID CHROMATOGRAPHY. A DRAFT FOR THE FOURTH EDITION OF THE CZECHOSLOVAKIAN PHARMACOPEIA
Subert, J.
Lab. Kontrnlu Leciv KUNZ Brno Czech. CS
Cesk. Farm.; (82) P 295-6; Vol 31; No 7; In Czech

02-82-0012 (0012)
POTENTIAL AND TRENDS OF HPLC
Meyer, Veronika R.
Inst. Org. Chem. Univ. Bern Bern Switz. CH CH-3012
Chimia; (82) P 475-80; Vol 36; No 12; In Ger

02-82-0013 (0013)
PREPARATIVE LC EXTENDS THE POSSIBILITIES OF SYNTHETIC ORGANIC CHEMISTRY
Olieman, C.; Voskamp, D.; Verhage, M.
Lab. Org. Chem. Delft Univ. Technol. Delft Neth. NL
Chromatogr. Sci.; (82) P 1-11; Vol 20; No Biol./Biomed. Appl. Liq. Chromatogr. 4

02-82-0014 (0014)
SOLUBILITY PARAMETER AS A TOOL IN UNDERSTANDING LIQUID CHROMATOGRAPHY
Schoenmakers, P. J.; Billiet, H. A. H.; De Galan, L.
Lab. Anal. Scheikd. Tech. Hogesch. Delft Delft Neth. NL NL-2628
Chromatographia; (82) P 387-98; Vol 15; No 6; In Ger

02-82-0015 (0015)
INFLUENCE OF INJECTOR BYPASS ON LIFETIME OF SMALL-PARTICLE LIQUID CHROMATOGRAPHIC COLUMNS
DiCesare, J. L.; Dong, M. W.; Gant, J. R.
Chromatogr. Div. Perkin-Elmer Corp. Norwalk USA US CT 06856
Chromatographia; (82) P 595-8; Vol 15; No 9

02-82-0016 (0016)
FROM CAPILLARY ANALYSIS TO HIGH-PRESSURE LIQUID CHROMATOGRAPHY - DEVELOPMENT OF A SEPARATION METHOD
Schwedt, Georg
Univ. Goettingen Goettingen Fed. Rep. Ger. DE
CLB, Chem. Labor Betr.; (82) P 99-100, 102-4, 107-8; Vol 33; No 3; In Ger

02-82-0017 (0017)
USE OF MICELLAR AND CYCLODEXTRIN SOLUTIONS IN LIQUID CHROMATOGRAPHIC SEPARATIONS
Armstrong, Daniel W.
Dep. Chem. Georgetown Univ. Washington USA US DC 20057
CONFERENCE; (82); Solution Behav. Surfactants: Theor. Appl. Aspects, (Proc. Int. Symp.); P 1273-82; Vol 2,; Avail.: Mittal, K. L.; Fendler, Eleanor J;

Date: 80 ; Publisher: Plenum, New York, N.Y.

02-82-0018 (0018)
STATE OF ART OF SEPARATION TECHNOLOGY IN GLC AND HPLC
Guenther, W.; Hecker, D.; Schlegelmilch, F.
WGA Duesseldorf Fed. Rep. Ger. DE D-4000
CONFERENCE; Aetherische Oele, Ergeb. Int. Arbeitstag.; (82) P 82-98; In Ger; Avail.: Kubeczka, Karl-Heinz; Date: 1980; Publisher: Thieme, Stuttgart, Fed. Rep. Ger.

02-82-0019 (0019)
HIGH-PRESSURE LIQUID CHROMATOGRAPHY
Decourt, Sylvie
Fr. FR
Gaz. Med. Fr.; (82) P 1713, 1715-16; Vol 89; No 15; In Fr

02-82-0020 (0020)
LIQUID CHROMATOGRAPHY USING FOUR SOLVENTS
Krusche, J. U.; Gottmann, G.
Abt. Anal. Instrum. Du Pont de Nemours G.m.b.H. Bad Nauheim Fed. Rep. Ger. DE D-6350
GIT Fachz. Lab.; (82) P 345-6, 348-9; Vol 26; No 4; In Ger

02-82-0021 (0021)
SCREENING FOR UNKNOWN COMPOUNDS BY HPLC USING A MULTICHANNEL DIODE ARRAY UV-VIS SPECTROPHOTOMETER: APPLICATION TO DRUG METABOLISM
Ghijsen, R. T.; Drenth, B. H. F.; Overzet, F.; De Zeeuw, R. A.
Dep. Toxicol. State Univ. Groningen Groningen Neth. NL 9713 AW
HRC CC, J. High Resolut. Chromatogr. Chromatogr. Commun.; (82) P 192-8; Vol 5; No 4

02-82-0022 (0022)
MODERN LIQUID CHROMATOGRAPHY
Nurok, David
Dep. Chem. Indiana Univ. Indianapolis USA US IN 46223
Int. Sugar J.; (82) P 99-104; Vol 84; No 1000

02-82-0023 (0023)
PROCESS LIQUID CHROMATOGRAPHY: PROMISES AND PROBLEMS
Guillemin, C. L.
Rhone Poulenc Rech. Fr. FR
InTech; (82) P 47-9; Vol 29; No 8

02-82-0024 (0024)
HIGH-PERFORMANCE LIQUID CHROMATOGRAPHY AND ITS APPLICATIONS IN THE BIOLOGICAL FIELD
Cancelliere, G.; Gasparrini, F.; Misiti, D.
Ist. Chim. Org. Univ. Roma Rome Italy IT
Ital. J. Biochem.; (82) P 69-74; Vol 31; No 1

02-82-0025 (0025)
ANOMALIES IN HPLC
Kirschbaum, Joel; Perlman, Solomon; Poet, Raymond B.
Squibb Inst. Med. Res. New Brunswick USA US NJ 08903
J. Chromatogr. Sci.; (82) P 336-40; Vol 20; No 7

02-82-0026 (0026)
OFF-LINE COMBINATION OF LIQUID CHROMATOGRAPHY AND FIELD DESORPTION MASS SPECTROMETRY: PRINCIPLES AND ENVIRONMENTAL, MEDICAL, AND PHARMACEUTICAL APPLICATIONS

Schulten, H. R.
Inst. Chem. Univ. Bonn Bonn Fed. Rep. Ger. DE 5300/1
J. Chromatogr.; (82) P 105-28; Vol 251; No 2

02-82-0027 (0027)
LIQUID CHROMATOGRAPHY SYSTEM FOR FAST, ACCURATE ANALYSIS
Katz, E.; Scott, R. P. W.
Perkin-Elmer Corp. Norwalk USA US CT 06856
J. Chromatogr.; (82) P 159-78; Vol 253; No 2

02-82-0028 (0028)
UNIVERSAL LIQUID CHROMATOGRAPHY METHODS. II. SENSITIVE, LOW-WAVELENGTH, GRADIENT REVERSED-PHASE METHODS
Berry, Vern V.
Cent. Anal. Res. Polaroid Corp. Cambridge USA US MA 02139
J. Chromatogr.; (82) P 279-96; Vol 236; No 2

02-82-0029 (0029)
ANALYSIS OF BODY FUNCTIONS USING A CLINICAL LIQUID CHROMATOGRAPH
Miyagi, Hiroyuki; Miura, Junkichi; Takata, Yoshinori; Kamitake, Seigo; Ganno, Shigetake; Yamagata, Yoh
Hitachi Res. Lab. Hitachi Ltd. Hitachi Japan JP 319-12
J. Chromatogr.; (82) P 733-45; Vol 239

02-82-0030 (0030)
SEPARATIONS IN THIN-LAYER AND HIGH-PERFORMANCE LIQUID CHROMATOGRAPHY USING ALKYL SILICA GEL BONDED PHASES
Issaq, Haleem J.; Klose, John R.; Reitz, Gilbert A.
Frederick Cancer Res. Facil. Natl. Cancer Cent. Frederick USA US MD 21701
J. Liq. Chromatogr.; (82) P 1069-80; Vol 5; No 6

02-82-0031 (0031)
USE OF HIGH-PERFORMANCE LIQUID CHROMATOGRAPHY IN COSMETICS
Mokujima, Keiji
Kokuritsu Eisei Shikensho Tokyo Japan JP
J. SCBJ; (82) P 3-9; Vol 16; No 1; In Japan

02-82-0032 (0032)
ADSORPTION CHROMATOGRAPHY
Shinohara, Isao; Negishi, Naoki
Dep. Polym. Chem. Waseda Univ. Tokyo Japan JP
Kagaku no Ryoiki, Kokan; (82) P 129-40; No 135; In Japan

02-82-0033 (0033)
ANALYTICAL HIGH-PERFORMANCE LIQUID CHROMATOGRAPHY - THE CURRENT STATE AND NEW DIRECTIONS
Novotny, Milos
Dep. Chem. Indiana Univ. Bloomington USA US IN
Kem. - Kemi; (82) P 667-71; Vol 9; No 10

02-82-0034 (0034)
HIGH-PERFORMANCE LIQUID-PHASE CHROMATOGRAPHY
Latour, J. F.
Fr. FR
Lyon Pharm.; (82) P 333-51; Vol 3; No 5; In Fr

02-82-0035 (0035)
SEPARATION OF MIXTURES BY HPLC - FUNDAMENTALS, SEPARATION MECHANISMS, SEPARATING COLUMNS
Ditz, R.

Fa. Waters G.m.b.H. Koenigstein/Taunus Fed. Rep. Ger. DE D-6240
Monatsschr. Brau.; (82) P 50-3; Vol 35; No 2; In Ger

02-82-0036 (0036)
STRATEGY FOR THE CHOICE OF MOBILE AND STATIONARY PHASES IN HPLC
Montalbetti, N.
Clin. Chem. Lab. St. Anna Hosp. Como Italy IT I-22100
Not. - Soc. Ital. Biochim. Clin.; (82) P 577-81; Vol 6; No 10

02-82-0037 (0037)
OPTIMAL CONDITIONS IN ISOCRATIC COLUMN LIQUID CHROMATOGRAPHY
Huber, J. F. K.
Inst. Anal. Chem. Univ. Wien Vienna Austria AT
Oesterr. Chem. Z.; (82) P 159-61; Vol 83; No 6; In Ger

02-82-0038 (0038)
DETECTION LIMITS IN LIQUID CHROMATOGRAPHY
Guiochon, G.; Colin, H.
Lab. Chim. Anal. Ec. Polytech. Palaiseau Fr. FR
Pergamon Ser. Environ. Sci.; (82) P 169-76; Vol 7; No Anal. Tech. Environ. Chem. 2

02-82-0039 (0039)
SUBMITTING HPLC METHODS TO THE COMPENDIA AND REGULATORY AGENCIES
Debesis, Eileen; Boehlert, Judy P.; Givand, Theodore E.; Sheridan, Jane C.
USA US
Pharm. Technol.; (82) P 120, 122, 125, 127, 131-3, 136-7; Vol 6; No 9

02-82-0040 (0040)
LIQUID CHROMATOGRAPHY IN 1982
Freeman, David H.
Dep. Chem. Univ. Maryland College Park USA USP MD 20742
Science (Washington, D. C., 1883); (82) P 235-41; Vol 218; No 4569

02-82-0041 (0041)
HIGH PERFORMANCE LIQUID CHROMATOGRAPHY OF BIOLOGICAL FLUIDS
Takai, Nobuharu
Inst. Ind. Sci. Tokyo Univ. Tokyo Japan JP
Seimitsu Kikai; (82) P 562-5; Vol 48; No 5; In Japan

02-82-0042 (0042)
USE OF LIQUID CHROMATOGRAPHY FOR STUDYING REVERSE OSMOSIS AND ULTRAFILTRATION
Taketani, Yutaka; Matsuura, Takeshi; Sourirajan, S.
Div. Chem. Natl. Res. Counc. Canada Ottawa Can. CA ON K1A 0R9
Sep. Sci. Technol.; (82) P 821-38; Vol 17; No 6

02-82-0043 (0043)
APPLICATIONS OF HIGH PERFORMANCE LIQUID CHROMATOGRAPHY WITH ELECTROCHEMICAL DETECTION IN CLINICAL CHEMISTRY
Richards, Douglas A.
Dep. Biochem. R. Sussex County Hosp. Brighton/Sussex UK GB
Surf. Technol.; (82) P 113-20; Vol 15; No 2

02-82-0044 (0044)
A BRIEF REVIEW ON THE APPLICATION OF CLASSICAL LIQUID CHROMATOGRAPHY IN THE SEPARATION OF NATURAL PRODUCTS
Yao, Qianyuan; Ji,PDezhen
Shandong Inst. Trad. Chinese Med. Peop. Rep. China CN
Zhongcaoyao; (82) P 41-7; Vol 13; No 5; In Ch

SECTION 2

CHAPTER 2.1
Hydrodynamics

--- 1982 ---

01-82-0001 (0001)
FOURIER TRANSFORM DECONVOLUTION OF INSTRUMENT AND COLUMN BAND BROADENING IN LIQUID CHROMATOGRAPHY
Wriaht, Norman A.; Villalanti, Dan C.; Burke, Michael F.
Dep. Chem. Univ. Arizona Tucson USA US AZ 85721
Anal. Chem.; (82) P 1735-8; Vol 54; No 11

01-82-0002 (0002)
COMPARISON OF THREE- AND FIVE-MICROMETER COLUMN PACKINGS FOR REVERSED-PHASE LIQUID CHROMATOGRAPHY
Cooke, Nelson H. C.; Archer, Benedict G.; Olsev, Kristine; Berick, Alan
Beckman Instrum., Inc. Berkelev USA US CA P94710
Anal. Chem.; (82) P 2277-83; Vol 54; No 13

01-82-0003 (0003)
OPTIMAL PARAMETERS IN HIGH-PERFORMQNCE LIQUID CHROMATOGRAPHY (HPLC)
Halasz, Istvan; Goerlitz, Gerhard
Univ. Saarbruecken Saarbruecken Fed. Rep. Ger. DE D-6600
Anaew. Chem.; (82) P 50-62; Vol 94; No 1; In Ger

01-82-0004 (0004)
LIQUID CHROMATOGRAPHY IN RECTANGULAR CHANNELS
Martin, Michel; Jurado-Baizaval, Jose Luis; Guiochon, Georaes
Lab. Chim. Anal. Phvs. Ec. Polvtech. Palaiseau Fr. FR 91128
C. R. Seances Acad. Sci., Ser. 2; (82) P 579-81; Vol 295; No 5; In Fr

01-82-0005 (0005)
THE SELECTION OF OPTIMAL CONDITIONS IN HPLC. II. THE INFLUENCE OF COLUMN DIMENSIONS AND SAMPLE SIZE ON SOLUTE DETECTION
Lauer, H. H.; Rozina, G. P.
Hewlett-Packard G.m.b.H. Waldbronn Fed. Rep. Ger. DE D-7517/2
Chromatoaraphia; (82) P 409-13; Vol 15; No 7

01-82-0006 (0006)
INFLUENCE OF INJECTOR BYPASS ON LIFETIME OF SMALL-PARTICLE LIQUID CHROMATOGRAPHIC COLUMNS
DiCesare, J. L.; Dona, M. W.; Gant, J. R.
Chromatoar. Div. Perkin-Elmer Corp. Norwalk USA US CT 06856
Chromatoaraphia; (82) P 595-8; Vol 15; No 9

01-82-0007 (0007)
LIQUID CHROMATOGRAPHY SYSTEM FOR FAST, ACCURATE ANALYSIS
Katz, E.; Scott, R. P. W.
Perkin-Elmer Corp. Norwalk USA US CT 06856
J. Chromatoar.; (82) P 159-78; Vol 253; No 2

01-82-0008 (0008)
NONEQUILIBRIUM MODEL OF LIQUID COLUMN CHROMATOGRAPHY. I. EXACT EXPRESSIONS FOR ELUTION PROFILE MOMENTS AND RELATION TO PLATE HEIGHT THEORY
Hethcote, Herbert W.; DeLisi, Charles
Lab. Math. Biol. Natl. Inst. Health Bethesda USA US MD 20205
J. Chromatoar.; (82) P 269-81; Vol 240; No 2

01-82-0009 (0009)
NONEQUILIBRIUM MODEL OF LIQUID COLUMN CHROMATOGRAPHY. II. EXPLICIT SOLUTIONS AND NONIDEAL CONDITIONS
DeLisi, Charles; Hethcote, Herbert W.; Brettler, Jeffrev W.
Lab. Math. Biol. Natl. Inst. Health Bethesda USA US MD 20205

J. Chromatogr.; (82) P 283-95; Vol 240; No 2

01-82-0010 (0010)
QUANTITATIVE COMPUTER RESOLUTION OF SEVERELY OVERLAPPING LIQUID CHROMATOGRAPHIC PEAKS
D'Allura, Norman J.; Juvet, Richard S., Jr.
Dep. Chem. Arizona State Univ. Tempe USA US AZ 85287
J. Chromatogr.; (82) P 439-49; Vol 239

01-82-0011 (0011)
DETECTION LIMITS IN LIQUID CHROMATOGRAPHY
Guiochon, G.; Colin, H.
Lab. Chim. Anal. Ec. Polytech. Palaiseau Fr. FR
Pergamon Ser. Environ. Sci.; (82) P 169-76; Vol 7; No Anal. Tech. Environ. Chem. 2

CHAPTER 2.2
Extra-Column Effects

--- 1982 ---

02-82-0001 (0001)
FOURIER TRANSFORM DECONVOLUTION OF INSTRUMENT AND COLUMN BAND BROADENING IN LIQUID CHROMATOGRAPHY
Wright, Norman A.; Villalanti, Dan C.; Burke, Michael F.
Dep. Chem. Univ. Arizona Tucson USA US AZ 85721
Anal. Chem.; (82) P 1735-8; Vol 54; No 11

02-82-0002 (0002)
COMPARISON OF LIQUID SEGMENTED WITH NONSEGMENTED FLOW SYSTEMS IN POST-COLUMN REACTORS FOR LIQUID CHROMATOGRAPHY
Scholten, A. H. M. T.; Brinkman, U. A. Th.; Frei, R. W.
Dep. Anal. Chem. Free Univ. Amsterdam Neth. NL 1081 HV
Anal. Chem.; (82) P 1932-8; Vol 54; No 12

02-82-0003 (0003)
OPTIMIZATION OF FLOW CELLS FOR FLUORESCENCE DETECTION IN LIQUID CHROMATOGRAPHY
Lyons, John W.; Faulkner, Larry R.
Dep. Chem. Univ. Illinois Urbana USA US IL 61801
Anal. Chem.; (82) P 1960-4; Vol 54; No 12

02-82-0004 (0004)
COMPARISON OF THREE- AND FIVE-MICROMETER COLUMN PACKINGS FOR REVERSED-PHASE LIQUID CHROMATOGRAPHY
Cooke, Nelson H. C.; Archer, Benedict G.; Olsev, Kristine; Berick, Alan
Beckman Instrum., Inc. Berkeley USA US CA P94710
Anal. Chem.; (82) P 2277-83; Vol 54; No 13

02-82-0005 (0005)
ISOLATION OF DENATURED PROTEINS AND PEPTIDES BY HIGH-PERFORMANCE LIQUID CHROMATOGRAPHY. EFFECT OF DIFFERENT PERFLUORINATED ACIDS, COLUMN LENGTH AND LARGE-PORE SUPPORTS
Mahoney, Walter C.
Dep. Genet. Univ. Washington Seattle USA US WA 98195
Biochim. Biophys. Acta; (82) P 284-9; Vol 704; No 2

02-82-0006 (0006)
REACTION DETECTOR WITH THREE DIMENSIONAL COILED OPEN TUBES IN HPLC
Engelhardt, H.; Neue, U. D.
Univ. Saarlandes Saarbruecken Fed. Rep. Ger. DE
Chromatographia; (82) P 403-8; Vol 15; No 7

02-82-0007 (0007)
THE SELECTION OF OPTIMAL CONDITIONS IN HPLC. II. THE INFLUENCE OF COLUMN DIMENSIONS AND SAMPLE SIZE ON SOLUTE DETECTION
Lauer, H. H.; Rozing, G. P.
Hewlett-Packard G.m.b.H. Waldbronn Fed. Rep. Ger. DE D-7517/2
Chromatographia; (82) P 409-13; Vol 15; No 7

02-82-0008 (0008)
INTRA-COLUMN INJECTION SYSTEM FOR LIQUID CHROMATOGRAPHY
Katz, E.; Scott, R. P. W.
Perkin-Elmer Corp. Norwalk USA US CT
J. Chromatogr.; (82) P 191-7; Vol 246; No 2

02-82-0009 (0009)
USE OF CONVENTIONAL INSTRUMENTATION WITH MICROBORE COLUMNS IN HIGH-PERFORMANCE LIQUID CHROMATOGRAPHY
Kok, W. T.; Brinkman, U. A. Th.; Frei, R. W.; Hanekamp, H. B.; Nooitgedacht,

F.; Poppe, H.
Dep. Anal. Chem. Free Univ. Amsterdam Neth. NL 1081 HV
J. Chromatogr.; (82) P 357-69; Vol 237; No 3

02-82-0010 (0010)
DETECTION LIMITS IN LIQUID CHROMATOGRAPHY
Guiochon, G.; Colin, H.
Lab. Chim. Anal. Ec. Polytech. Palaiseau Fr. FR
Pergamon Ser. Environ. Sci.; (82) P 169-76; Vol 7; No Anal. Tech. Environ. Chem. 2

CHAPTER 2.3

Measurement of Thermodynamic Properties

--- 1982 ---

03-82-0001 (0001)
MEASUREMENT OF SORPTION ISOTHERMS BY ON-STREAM PRECOLUMN EQUILIBRATION LIQUID CHROMATOGRAPHY
May, Susan; Hux, Robert A.; Cantwell, Frederick F.
Dep. Chem. Univ. Alberta Edmonton Can. CA AB T6G 2G2
Anal. Chem.; (82) P 1279-82; Vol 54; No 8

03-82-0002 (0002)
THERMODYNAMIC FUNCTIONS IN HIGH-PERFORMANCE LIQUID CHROMATOGRAPHY OF SACCHARIDES
Fukuhara, Kenichi; Nakamura, Asao; Tsuboi, Masamichi
Cent. Res. Lab. Ajinomoto Co., Inc. Kawasaki Japan JP 210
Chem. Pharm. Bull.; (82) P 1121-5; Vol 30; No 4

03-82-0003 (0003)
ESTIMATING THE AQUEOUS SOLUBILITY OF AROMATIC HYDROCARBONS BY HIGH PERFORMANCE LIQUID CHROMATOGRAPHY
Whitehouse, Brian G.
Dep. Oceanogr. Dalhousie Univ. Halifax Can. CA NS B3H 4J1
Chemosphere; (82) P 689-99; Vol 11; No 8

03-82-0004 (0004)
THE SOLVENT ELUOTROPIC STRENGTH ON CARBON ADSORBENTS
Colin, H.; Guiochon, G.; Jandera, P.
Lab. Chim. Anal. Phys. Ec. Polytech. Palaiseau Fr. FR F-91128
Chromatographia; (82) P 133-9; Vol 15; No 2

03-82-0005 (0005)
CALCULATION OF THE REDUCED SURFACE EXCESS FROM CONTINUOUS FLOW FRONTAL ANALYSIS SOLID-LIQUID CHROMATOGRAPHY
Noll, Leo A.; Burchfield, Thomas E.
Dep. Energy Bartlesville Energy Technol. Cent. Bartlesville USA US OK 74005
Colloids Surf.; (82) P 33-42; Vol 5; No 1

03-82-0006 (0006)
ADSORPTION FROM LIQUID MIXTURES AND LIQUID CHROMATOGRAPHY
Riedo, Francois; Kovats, Ervin S.
Lab. Chim. Tech. Ec. Polytech. Fed. Lausanne Lausanne Switz. CH CH-1015
J. Chromatogr.; (82) P 1-28; Vol 239

03-82-0007 (0007)
ACTIVITY COEFFICIENTS, INTERFACIAL TENSIONS AND RETENTION IN REVERSED-PHASE LIQUID CHROMATOGRAPHY ON LICHROSORB RP-18 WITH METHANOL-WATER MIXTURES
Hammers, W. E.; Meurs, G. J.; De Ligny, C. L.
Lab. Anal. Chem. State Univ. Utrecht Utrecht Neth. NL 3522 ADD
J. Chromatogr.; (82) P 169-89; Vol 246; No 2

03-82-0008 (0008)
EFFECT OF MOLECULAR STRUCTURE AND CONFORMATIONAL CHANGE OF PROLINE-CONTAINING DIPEPTIDES IN REVERSED-PHASE CHROMATOGRAPHY
Melander, Wayne R.; Jacobson, Jana; Horvath, Csaba
Dep. Chem. Eng. Yale Univ. New Haven USA US CT 06520
J. Chromatogr.; (82) P 269-76; Vol 234; No 2

03-82-0009 (0009)
STUDY OF ADSORPTION FROM SOLUTIONS BY COLUMN CHROMATOGRAPHY
Busev, S. A.; Zverev, S. I.; Larionov, O. G.; Yakubov, E. S.
Chromatogr. Lab. Inst. Phys. Chem. Moscow USSR SU 117312

J. Chromatogr.; (82) P 287-94; Vol 241; No 2

03-82-0010 (0010)
USE OF LIQUID CHROMATOGRAPHY FOR STUDYING INTERFACIAL PROPERTIES OF INORGANIC SOLUTES RELEVANT TO REVERSE-OSMOSIS SEPARATIONS
TaketaniD Yutaka; Matsuura, Takeshi; Sourirajan, S.
Div. Chem. Natl. Res. Counc. Canada Ottawa Can. CA ON K1A 0R9
J. Electrochem. Soc.; (82) P 1485-92; Vol 129; No 7

03-82-0011 (0011)
ADSORPTION ISOTHERMS OF ION-PAIR FORMING REAGENTS USED IN REVERSED-PHASE CHROMATOGRAPHY
Bartha, Akos; Vigh, Gyula
Anal. Kem. Tansz. Veszpremi Vegyip. Egy; Veszprem Hung. HU
Magy. Kem. Lapja; (82) P 216-19; Vol 37; No 5; In Hung

03-82-0012 (0012)
THEORY OF ADSORPTION AND ITS RELATION TO CHROMATOGRAPHIC SEPARATION OF OLIGOMERS ACCORDING TO MOLECULAR WEIGHTS AND TYPES OF FUNCTIONALITY ON POROUS ADSORBENTS
Gorshkov, A. V.; Evreinov, V. V.; Entelis, S. G.
Inst. Khim. Fiz. Moscow USSR SU
Vysokomol. Soedin., Ser. A; (82) P 524-35; Vol 24; No 3; In Russ

03-82-0013 (0013)
DETERMINATION OF K.ETA. AND .ALPHA. COEFFICIENTS FOR OLIGOMERS FROM LIQUID CHROMATOGRAPHIC DATA.
Shilov, A. D.; Basov, B. K.; El'tekov, Yu. A.
Yarosl. Zavod Sint Kauch. Yaroslavl USSR SU
Vysokomol. Soedin., Ser. A; (82) P 653-6; Vol 24; No 3; In Russ

03-82-0014 (0014)
CHROMATOGRAPHIC MEASUREMENT OF ISOTHERMS OF ADSORPTION
Busev, S. A.; Larionov, O. G.
Inst. Fiz. Khim. Moscow USSR SU
Zh. Fiz. Khim.; (82) P 1238-41; Vol 56; No 5; In Russ

03-82-0015 (0015)
DETERMINATION OF ISOTHERMS OF ADSORPTION BY LIQUID COLUMN CHROMATOGRAPHY
Busev, S. A.; Zverev, S. I.; Larionov, O. G.; Yakubov, E. S.
Inst. Fiz. Khim. Moscow USSR SU
Zh. Fiz. Khim.; (82) P 929-32; Vol 56; No 4; In Russ

CHAPTER 2.4

Role of Temperature

--- 1982 ---

04-82-0001 (0001)
EFFECTS OF PRESSURE, TEMPERATURE, AND SOLVENT COMPOSITION IN ANALYSIS BY DIRECT LIQUID INTRODUCTION LIQUID CHROMATOGRAPHY/MASS SPECTROMETRY
Voyksner, Robert D.; Parker, Carol E.; Hass, Ronald J.; Bursey, Maurice M.
Natl. Inst. Environ. Health Sci. Research Triangle Park USA US NC 27709
Anal. Chem.; (82) P 2583-6; Vol 54; No 14

04-82-0002 (0002)
MOBILE PHASE EFFECTS IN REVERSED-PHASE CHROMATOGRAPHY. IV. RETENTION OF n-ALKYLBENZENES AS A FUNCTION OF COLUMN TEMPERATURE AND THE NATURE AND CONCENTRATION OF ORGANIC CO-SOLVENT
Melander, W. R.; Mannan, C. A.; Horvath, C.
Dep. Chem. Eng. Yale Univ. New Haven USA US CT 06520
Chromatographia; (82) P 611-15; Vol 15; No 9

04-82-0003 (0003)
INVESTIGATION OF THE LOW-TEMPERATURE EFFECT IN NORMAL-PHASE MICRO HIGH-PERFORMANCE LIQUID CHROMATOGRAPHY
Jinno, Kiyokatsu; Hirata, Yukio
Sch. Mater. Sci. Toyohashi Univ. Technol. Toyohashi Japan JP 440
HRC CC, J. High Resolut. Chromatogr. Chromatogr. Commun.; (82) P 85-90; Vol 5; No 2

04-82-0004 (0004)
LOW-TEMPERATURE, NORMAL-PHASE HPLC FOR SEPARATION OF THERMALLY LABILE SPECIES
Henderson, David E.; Novak, Frank P.
Dep. Chem. Trinity Coll. Hartford USA US CT 06106
J. Chromatogr. Sci.; (82) P 256-9; Vol 20; No 6

04-82-0005 (0005)
ROLE OF COLUMN TEMPERATURE IN OPEN-TUBULAR MICROCAPILLARY LIQUID CHROMATOGRAPHY
Takeuchi, Toyohire; Kumaki, Masayoshi; Ishii, Daido
Fac. Eng. Nagoya Univ. Nagoya Japan JP 464
J. Chromatogr.; (82) P 309-22; Vol 235; No 2

04-82-0006 (0006)
TEMPERATURE AND ELUENT EFFECTS ON THE SELECTIVITY OF SOME NITROAROMATIC BONDED PHASES IN HIGH-PERFORMANCE LIQUID CHROMATOGRAPHY
Hammers, W. E.; Theeuwes, A. G. M.; Brederode, W. K.; De Ligny, C. L.
Lab. Anal. Chem. State Univ. Utrecht Neth. NL 3522 AD
J. Chromatogr.; (82) P 321-36; Vol 234; No 2

04-82-0007 (0007)
RETENTION BEHAVIOR OF ALKYLBENZENES AS A FUNCTION OF TEMPERATURE AND MOBILE PHASE COMPOSITION IN REVERSED-PHASE CHROMATOGRAPHY
Grushka, Eli; Colin, Henri; Guiochon, Georges
Dep. Inorg. Anal. Chem. Hebrew Univ. Jerusalem Israel IL
J. Chromatogr.; (82) P 325-39; Vol 248; No 3

04-82-0008 (0008)
INFLUENCE OF TEMPERATURE ON THE STABILITY OF SOLID TETRACYCLINE HYDROCHLORIDE, MEASURED BY HIGH-PERFORMANCE LIQUID CHROMATOGRAPHY
Dihuidi, Kabala; Roets, E.; Hoogmartens, J.; Vanderhaeghe, H.
Lab. Farm. Chem. Katholieke Univ. Leuven Louvain Belg. BE B-3000
J. Chromatogr.; (82) P 350-5; Vol 246; No 2

04-82-0009 (0009)
SEPARATION OF SATURATED, MONOUNSATURATED AND DIUNSATURATED ALDEHYDES AS 2,4-DINITROPHENYLHYDRAZONES USING HIGH-PERFORMANCE LIQUID CHROMATOGRAPHY AT INCREASED TEMPERATURE
Reindl, B.; Stan, H. J.
Inst. Sozialmed. Epidemiol. Bundesgesundheitsamt Berlin Fed. Rep. Ger. E 1000/33
J. Chromatogr.; (82) P 481-8; Vol 235; No 2

CHAPTER 2.5
Gradient Elution

--- 1982 ---

05-82-0001 (0001)
DETERMINATION OF TETRAHYDROISOQUINOLINES BY REVERSED-PHASE LIQUID CHROMATOGRAPHY WITH GRADIENT ELUTION AND AMPEROMETRIC DETECTION
St. Claire, R. L., III; Ansari, G. A. S.; Abell, Creed W.
Med. Branch Univ. Texas Galveston USA US TX 77550
Anal. Chem.; (82) P 186-9; Vol 54; No 2

05-82-0002 (0002)
A NEW LOW COST, FULLY AUTOMATED AMINO ACID ANALYZER USING A GRADIENT HPLC
Klapper, David G.
Dep. Bacteriol. Immunol. Univ. North Carolina Chapel Hill USA US NC
CONFERENCE; Methods Protein Sequence Anal., (Proc. Int. Conf.), 4th; (82) P 509-15; Avail.: Elzinga, Marshall; Date: 81; Publisher: Humana, Clifton, N. J.

05-82-0003 (0003)
HPLC GRADIENT ELUTION OF DINITROPHENYLHYDRAZONES OF ALDEHYDES AND KETONES FROM AQUEOUS SAMPLES
Liebezeit, G.
Sonderforschungsber. 95 Univ. Kiel Kiel Fed. Rep. Ger. DE 2300
HRC CC, J. High Resolut. Chromatogr. Chromatogr. Commun.; (82) P 215-16; Vol 5; No 4

05-82-0004 (0004)
HIGH-PERFORMANCE LIQUID CHROMATOGRAPHIC SCREENING FOR UNKNOWN SUNSET YELLOW FCF AND ORANGE GGN METABOLITES IN RAT FECES USING A CONTINUOUS GRADIENT ELUTION SYSTEM
De Beer, Jacques O.; Dierickx, aul J.
Dep. Farmatoxicol. Inst. Hyg. Epidemiol. Brussels Belg. BE B-1050
J. Chromatogr.; (82) P 239-47; Vol 245; No 2

05-82-0005 (0005)
UNIVERSAL LIQUID CHROMATOGRAPHY METHODS. II. SENSITIVE, LOW-WAVELENGTH, GRADIENT REVERSED-PHASE METHODS
Berry, Vern V.
Cent. Anal. Res. Polaroid Corp. Cambridge USA US MA 02139
J. Chromatogr.; (82) P 279-96; Vol 236; No 2

05-82-0006 (0006)
PREDICTION OF PEPTIDE RETENTION TIMES IN REVERSED-PHASE HIGH-PERFORMANCE LIQUID CHROMATOGRAPHY DURING LINEAR GRADIENT ELUTION
Sasagawa, Tatsuru; Okuyama, Tsuneo; Teller, David C.
Dep. Biochem. Univ. Washington Seattle USA US WA 98195
J. Chromatogr.; (82) P 329-40; Vol 240; No 2

05-82-0007 (0007)
CONTINUOUS GRADIENT ELUTION IN MICRO HIGH-PERFORMANCE LIQUID CHROMATOGRAPHY
Takeuchi, Toyohide; Ishii, Daido
Fac. Eng. Nagoya Univ. Nagoya Japan JP 464
J. Chromatogr.; (82) P 41-7; Vol 253; No 1

05-82-0008 (0008)
DETERMINATION OF UV ABSORBING POLYOLEFIN ADDITIVES BY GRADIENT AND ISOCRATIC NORMAL-PHASE HIGH-PERFORMANCE LIQUID CHROMATOGRAPHY
Schabron, J. F.; Smith, V. J.; Ware, J. L.
Phillips Res. Cent. Phillips Pet. Co. Bartlesville USA US OK 74004
J. Liq. Chromatogr.; (82) P 613-24; Vol 5; No 4

05-82-0009 (0009)
HPLC SOLVENT GRADIENT FORMATION
Riggenmann, Hansjuergen
Gynkotek Ges. Bau Wissenschaftlich-Tech. Gerate m.b.H. Munich Fed. Rep. Ger.
DE 8000/21
LaborPraxis; (82) P 474, 476, 478, 481; Vol 6; No 5

CHAPTER 2.6
Normal-Phase Chromatography

2.6.1
Retention Model

--- 1982 ---

06-82-0001 (0001)
THE SILICA GEL SURFACE AND ITS INTERACTIONS WITH SOLVENT AND SOLUTE IN LIQUID CHROMATOGRAPHY
Scott, Raymond P. W.
Perkin-Elmer Corp. Norwalk USA US CT
Adv. Chromatogr. (N. Y.); (82) P 167-215; Vol 20

06-82-0002 (0002)
MECHANISTIC INTERPRETATIONS AND SIMULATIONS OF INDUCED PEAKS IN LIQUID CHROMATOGRAPHY
Stranahan, John J.; Deming, Stanley N.
Dep. Chem. Univ. Houston Houston USA US TX 77004
Anal. Chem.; (82) P 1540-6; Vol 54; No 9

06-82-0003 (0003)
ADSORPTION ISOTHERM AT THE LIQUID-SOLID INTERFACE AND THE INTERPRETATION OF CHROMATOGRAPHIC DATA
Ngoc Le Ha; Ungvarai, Javos; Kovats, Ervin S.
Lab. Chim. Tech. Ec. Polytech. Fed. Lausanne Lausanne Switz. CH 1015
Anal. Chem.; (82) P 2410-21; Vol 54; No 14

06-82-0004 (0004)
THE SOLUBILITY PARAMETER AS A TOOL IN UNDERSTANDING LIQUID CHROMATOGRAPHY
Schoenmakers, P. J.; Billiet, H. A. H.; De Galan, L.
Lab. Anal. Scheikd. Tech. Hogesch. Delft Delft Neth. NL 2628 BX
Chromatographia; (82) P 205-14; Vol 15; No 3

06-82-0005 (0005)
SOLUBILITY PARAMETER AS A TOOL IN UNDERSTANDING LIQUID CHROMATOGRAPHY
Schoenmakers, P. J.; Billiet, H. A. H.; De Galan, L.
Lab. Anal. Scheikd. Tech. Hogesch. Delft Delft Neth. NL NL-2628
Chromatographia; (82) P 387-98; Vol 15; No 6; In Ger

06-82-0006 (0006)
ADSORPTION FROM BINARY SOLVENT MIXTURES ONTO SILICA GEL BY HPLC FRONTAL ANALYSIS
Koester, F.; Findenegg, G. H.
Phys. Chem. II Ruhr-Univ. Bochum Bochum Fed. Rep. Ger. DE D-4630
Chromatographia; (82) P 743-7; Vol 15; No 12

06-82-0007 (0007)
LIQUID-SOLID CHROMATOGRAPHY. RECENT PROGRESS IN THEORETICAL STUDIES CONCERNING THE DEPENDENCE OF THE CAPACITY RATIO UPON THE MOBILE PHASE COMPOSITION
Jaroniec, M.; Oscik, J.
Inst. Chem. M. Curie-Sklodowska Univ. Lublin Pol. PL 20031
HRC CC, J. High Resolut. Chromatogr. Chromatogr. Commun.; (82) P 3-12; Vol 5; No 1

06-82-0008 (0008)
PHOTOCONTROLLED ADSORPTION CHROMATOGRAPHY FOR LYSOZYME USING AZOAROMATIC POLYMER
Ishihara, Kazuhiko; Negishi, Naoki; Shinohara, Isao
Dep. Polym. Chem. Waseda Univ. Tokyo Japan JP 160
J. Appl. Polym. Sci.; (82) P 1897-902; Vol 27; No 6

06-82-0009 (0009)
ADSORPTION FROM LIQUID MIXTURES AND LIQUID CHROMATOGRAPHY
Riedo, Francois; Kovats, Ervin S.

Lab. Chim. Tech. Ec. Polytech. Fed. Lausanne Lausanne Switz. CH CH-1015
J. Chromatogr.; (82) P 1-28; Vol 239

06-82-0010 (0010)
THEORY FOR MOBILE PHASE EFFECTS IN SEPARATIONS OF ISOMERS BY LIQUID-SOLID CHROMATOGRAPHY. APPLICATION TO THE RELATIVE RETENTION OF CERTAIN DIASTEREOMERS ON SILICA
Snyder, L. R.
Tech. Instrum. Corp. Tarrytown USA US NY 10591
J. Chromatogr.; (82) P 165-76; Vol 245; No 2

06-82-0011 (0011)
SOLVENT STRENGTH OF MULTICOMPONENT MOBILE PHASES IN LIQUID-SOLID CHROMATOGRAPHY. FURTHER STUDY OF DIFFERENT MOBILE PHASES AND SILICA AS ADSORBENT
Snyder, L. R.; Glajch, J. L.
Technicon Instrum. Corp. Tarrytown USA US NY 10591
J. Chromatogr.; (82) P 165-82; Vol 248; No 2

06-82-0012 (0012)
CONTRIBUTION TO THE ELUCIDATION OF THE MECHANISM OF SUGAR RETENTION ON AMINE-MODIFIED SILICA IN LIQUID CHROMATOGRAPHY
Verhaar, L. A. Th.; Kuster, B. F. M.
Dep. Chem. Eng. Eindhoven Univ. Technol. Eindhoven Neth. NL 5600 MB
J. Chromatogr.; (82) P 57-64; Vol 234; No 1

06-82-0013 (0013)
INFLUENCE OF WATER CONTENT OF THE MOBILE PHASE ON CHROMATOGRAPHIC PERFORMANCE IN ADSORPTION CHROMATOGRAPHY
Szepesy, L.; Combellas, C.; Caude, M.; Rosset, R. Lab. Chim. Anal. Ec. Super. Phys. Chim. Ind. Ville de Paris; Paris Fr. FR 75231
J. Chromatogr.; (82) P 65-78; Vol 237; No 1

06-82-0014 (0014)
THEORY OF ADSORPTION AND ITS RELATION TO CHROMATOGRAPHIC SEPARATION OF OLIGOMERS ACCORDING TO MOLECULAR WEIGHTS AND TYPES OF FUNCTIONALITY ON POROUS ADSORBENTS
Gorshkov, A. V.; Evreinov, V. V.; Entelis, S. G.
Inst. Khim. Fiz. Moscow USSR SU
Vysokomol. Soedin., Ser. A; (82) P 524-35; Vol 24; No 3; In Russ

2.6.2
Solvent Optimization

--- 1982 ---

06-82-0001 (0001)
THE SILICA GEL SURFACE AND ITS INTERACTIONS WITH SOLVENT AND SOLUTE IN LIQUID CHROMATOGRAPHY
Scott, Raymond P. W.
Perkin-Elmer Corp. Norwalk USA US CT
Adv. Chromatogr. (N. Y.); (82) P 167-215; Vol 20

06-82-0002 (0002)
1,1,2-TRICHLORO-1,2,2-TRIFLUOROETHANE AS A CARRIER SOLVENT IN LIQUID CHROMATOGRAPHY
Glajch, J. L.; Kirkland, J. J.; Schindel, W. G.
Cent. Res. Dev. Dep. E. I. du Pont de Nemours and Co. Wilmington USA US DE 19898
Anal. Chem.; (82) P 1276-9; Vol 54; No 8

06-82-0003 (0003)
CLASSIFICATION OF MULTISOLVENT MOBILE PHASE SYSTEMS IN LIQUID CHROMATOGRAPHY
Glajch, Joseph L.; Kirkland, J. J.
Cent. Res. Dev. Dep. du Pont de Nemours and Co. Wilmington USA US DE 19898
Anal. Chem.; (82) P 2593-6; Vol 54; No 14

06-82-0004 (0004)
SOLVENT OPTIMIZATION IN NORMAL-PHASE LIQUID CHROMATOGRAPHY OF SELECTED STEROIDS
Antle, P. E.
Anal. Instrum. Div. E. I. du Pont de Nemours and Co., Inc. Wilmington USA US DE 19898
Chromatographia; (82) P 277-81; Vol 1; No 5

06-82-0005 (0005)
SEPARATION OF CARBOHYDRATES BY LIQUID CHROMATOGRAPHY ON SILICA GEL, ADDING ADSORPTION MODIFICATORS TO THE ELUENT
Boumahraz, M.; Davydov, V. Ya.; Kiselev, A. V.
Chem. Dep. M. V. Lomonosov State Univ.-Moscow Moscow USSR SU 117234
Chromatographia; (82) P 751-6; Vol 15; No 12

06-82-0006 (0006)
LOW-TEMPERATURE, NORMAL-PHASE HPLC FOR SEPARATION OF THERMALLY LABILE SPECIES
Henderson, David E.; Novak, Frank P.
Dep. Chem. Trinity Coll. Hartford USA US CT 06106
J. Chromatogr. Sci.; (82) P 256-9; Vol 20; No 6

06-82-0007 (0007)
HIGH-PERFORMANCE LIQUID CHROMATOGRAPHIC ANALYSIS OF BASIC DRUGS ON SILICA COLUMNS USING NONAQUEOUS IONIC ELUENTS
Flanagan, R. J.; Storey, G. C. A.; Bhamra, R. K.; Jane, I.
Poisons Univ. Guy's Hosp. London UK GB SE1 9RT
J. Chromatogr.; (82) P 15-37; Vol 247; No 1

06-82-0008 (0008)
THEORY FOR MOBILE PHASE EFFECTS IN SEPARATIONS OF ISOMERS BY LIQUID-SOLID CHROMATOGRAPHY. APPLICATION TO THE RELATIVE RETENTION OF CERTAIN DIASTEREOMERS ON SILICA
Snyder, L. R.
Tech. Instrum. Corp. Tarrytown USA US NY 10591
J. Chromatogr.; (82) P 165-76; Vol 245; No 2

06-82-0009 (0009)
SOLVENT STRENGTH OF MULTICOMPONENT MOBILE PHASES IN LIQUID-SOLID CHROMATOGRAPHY. FURTHER STUDY OF DIFFERENT MOBILE PHASES AND SILICA AS ADSORBENT
Snyder, L. R.; Glajch, J. L.
Technicon Instrum. Corp. Tarrytown USA US NY 10591
J. Chromatogr.; (82) P 165-82; Vol 248; No 2

06-82-0010 (0010)
PRACTICAL OPTIMIZATION OF SOLVENT SELECTIVITY IN LIQUID-SOLID CHROMATOGRAPHY USING A MIXTURE-DESIGN STATISTICAL TECHNIQUE
Glajch, J. L.; Kirkland, J. J.; Snyder, L. R.
Cent. Res. Dev. Dep. E. I. du Pont de Nemours and Co. Wilmington USA US DE 19898
J. Chromatogr.; (82) P 269-80; Vol 238; No 2

06-82-0011 (0011)
TEMPERATURE AND ELUENT EFFECTS ON THE SELECTIVITY OF SOME NITROAROMATIC BONDED PHASES IN HIGH-PERFORMANCE LIQUID CHROMATOGRAPHY
Hammers, W. E.; Theeuwes, A. G. M.; Brederode, W. K.; De Ligny, C. L.
Lab. Anal. Chem. State Univ. Utrecht Neth. NL 3522 AD
J. Chromatogr.; (82) P 321-36; Vol 234; No 2

06-82-0012 (0012)
HIGH-BACK-PRESSURE LIQUID CHROMATOGRAPHY. III. OPEN-TUBULAR MICROCAPILLARY LIQUID CHROMATOGRAPHY USING LIQUEFIED ALKANES AS THE MOBILE PHASE
Takeuchi, Toyohide; Ishii, Daido
Fac. Eng. Nagoya Univ. Nagoya Japan JP 464
J. Chromatogr.; (82) P 51-60; Vol 240; No 1

06-82-0013 (0013)
CONTRIBUTION TO THE ELUCIDATION OF THE MECHANISM OF SUGAR RETENTION ON AMINE-MODIFIED SILICA IN LIQUID CHROMATOGRAPHY
Verhaar, L. A. Th.; Kuster, B. F. M.
Dep. Chem. Eng. Eindhoven Univ. Technol. Eindhoven Neth. NL 5600 MB
J. Chromatogr.; (82) P 57-64; Vol 234; No 1

06-82-0014 (0014)
INFLUENCE OF WATER CONTENT OF THE MOBILE PHASE ON CHROMATOGRAPHIC PERFORMANCE IN ADSORPTION CHROMATOGRAPHY
Szepesy, L.; Combellas, C.; Caude, M.; Rosset, R.
Lab. Chim. Anal. Ec. Super. Phys. Chim. Ind. Ville de Paris Paris Fr. FR 75231
J. Chromatogr.; (82) P 65-78; Vol 237; No 1

06-82-0015 (0015)
APPLICATIONS OF A MODIFIED "ISOHYDRIC SOLVENT SYSTEM" IN HPLC ON SILICA GEL FOR THE ANALYSIS OF THE MACROLIDE ANTIBIOTICS TURIMYCINS AND SPIRAMYCINS
Bens, G. A.; Crombez, E.; De Moerloose, P.
Dep. Pharm. Chem. Drug Qual. Control State Univ. Ghent Ghent Belg. BE B-9000
J. Liq. Chromatogr.; (82) P 1449-65; Vol 5; No 8

06-82-0016 (0016)
A KINETIC AND QUANTITATIVE APPROACH TO PEAK SPLITTING PHENOMENA WHEN USING 2-PROPANOL AS MODIFIER IN ADSORPTION HPLC
Mertens, J.
Vrije Univ. Brussel Brussels Belg. BE 1050
J. Liq. Chromatogr.; (82) P 1467-80; Vol 5; No 8

06-82-0017 (0017)
DETERMINATION OF UV ABSORBING POLYOLEFIN ADDITIVES BY GRADIENT AND ISOCRATIC

NORMAL-PHASE HIGH-PERFORMANCE LIQUID CHROMATOGRAPHY
Schabron, J. F.; Smith, V. J.; Ware, J. L.
Phillips Res. Cent. Phillips Pet. Co. Bartlesville USA US OK 74004
J. Liq. Chromatogr.; (82) P 613-24; Vol 5; No 4

06-82-0018 (0018)
DETERMINATION OF SULFADIAZINE AND N4-ACETYLSULFADIAZINE IN BIOLOGICAL FLUIDS BY LIQUID CHROMATOGRAPHY ON SILICA GEL WITH AN AQUEOUS BUFFER AS MOBILE PHASE
Westerlund, Douglas; Wijkstroem, Agneta
Res. Dev. Lab. Anal. Chem. Astra Laekemedel A.B. Soedertaelje Swed. SE S-151 85
J. Pharm. Sci.; (82) P 1142-5; Vol 71; No 10

06-82-0019 (0019)
STRATEGY FOR THE CHOICE OF MOBILE AND STATIONARY PHASES IN HPLC
Montalbetti, N.
Clin. Chem. Lab. St. Anna Hosp. Como Italy IT I-22100
Not. - Soc. Ital. Biochim. Clin.; (82) P 577-81; Vol 6; No 10

06-82-0020 (0020)
OPTIMAL CONDITIONS IN ISOCRATIC COLUMN LIQUID CHROMATOGRAPHY
Huber, J. F. K.
Inst. Anal. Chem. Univ. Wien Vienna Austria AT
Oesterr. Chem. Z.; (82) P 159-61; Vol 83; No 6; In Ger

CHAPTER 2.7
Reversed-Phase Chromatography

2.7.1
Retention Model

--- 1982 ---

07-82-0001 (0001)
ADSORPTION OF 1-PENTANOL ON ALKYL-MODIFIED SILICA AND ITS EFFECT ON THE RETENTION MECHANISM IN REVERSED-PHASE LIQUID CHROMATOGRAPHY
Wahlund, Karl Gustav; Beijersten, Ingegerd
Dep. Anal. Pharm. Chem. Univ. Uppsala Uppsala Swed. SE S-751 23
Anal. Chem.; (82) P 128-32; Vol 54; No 1

07-82-0002 (0002)
MECHANISTIC INTERPRETATIONS AND SIMULATIONS OF INDUCED PEAKS IN LIQUID CHROMATOGRAPHY
Stranahan, John J.; Deming, Stanley N.
Dep. Chem. Univ. Houston Houston USA US TX 77004
Anal. Chem.; (82) P 1540-6; Vol 54; No 9

07-82-0003 (0003)
ADSORPTION ISOTHERM AT THE LIQUID-SOLID INTERFACE AND THE INTERPRETATION OF CHROMATOGRAPHIC DATA
Ngoc Le Ha; Ungvarai, Janos; Kovats, Ervin S.
Lab. Chim. Tech. Ec. Polytech. Fed. Lausanne Lausanne Switz. CH 1015
Anal. Chem.; (82) P 2410-21; Vol 54; No 14

07-82-0004 (0004)
INTERACTION INDEXES FOR PREDICTION OF RETENTION IN REVERSED-PHASE LIQUID CHROMATOGRAPHY
Jandera, P.; Colin, H.; Guiochon, G.
Dep. Anal. Chem. Univ. Chem. Technol. Pardubice Czech. CS
Anal. Chem.; (82) P 435-41; Vol 54; No 3

07-82-0005 (0005)
THERMODYNAMIC FUNCTIONS IN HIGH-PERFORMANCE LIQUID CHROMATOGRAPHY OF SACCHARIDES
Fukuhara, Kenichi; Nakamura, Asao; Tsuboi, Masamichi
Cent. Res. Lab. Ajinomoto Co., Inc. Kawasaki Japan JP 210
Chem. Pharm. Bull.; (82) P 1121-5; Vol 30; No 4

07-82-0006 (0006)
THE SOLUBILITY PARAMETER AS A TOOL IN UNDERSTANDING LIQUID CHROMATOGRAPHY
Schoenmakers, P. J.; Billiet, H. A. H.; De Galan, L.
Lab. Anal. Scheikd. Tech. Hogesch. Delft Delft Neth. NL 2628 BX
Chromatographia; (82) P 205-14; Vol 15; No 3

07-82-0007 (0007)
SOLUBILITY PARAMETER AS A TOOL IN UNDERSTANDING LIQUID CHROMATOGRAPHY
Schoenmakers, P. J.; Billiet, H. A. H.; De Galan, L.
Lab. Anal. Scheikd. Tech. Hogesch. Delft Delft Neth. NL NL-2628
Chromatographia; (82) P 387-98; Vol 15; No 6; In Ger

07-82-0008 (0008)
STATIONARY PHASE EFFECTS IN REVERSED-PHASE CHROMATOGRAPHY II. SUBSTITUENT SELECTIVITIES FOR RETENTION ON VARIOUS HYDROCARBONACEOUS BONDED PHASES
Melander, W. R.; Horvath, C.
Dep. Chem. Eng. Yale Univ. New Haven USA US CT 06520
Chromatographia; (82) P 86-90; Vol 15; No 2

07-82-0009 (0009)
ADSORPTION FROM LIQUID MIXTURES AND LIQUID CHROMATOGRAPHY
Riedo, Francois; Kovats, Ervin S.
Lab. Chim. Tech. Ec. Polytech. Fed. Lausanne Lausanne Switz. CH CH-1015

J. Chromatogr.; (82) P 1-28; Vol 239

07-82-0010 (0010)
ACTIVITY COEFFICIENTS, INTERFACIAL TENSIONS AND RETENTION IN REVERSED-PHASE LIQUID CHROMATOGRAPHY ON LICHROSORB RP-18 WITH METHANOL-WATER MIXTURES
Hammers, W. E.; Meurs, G. J.; De Ligny, C. L.
Lab. Anal. Chem. State Univ. Utrecht Utrecht Neth. NL 3522 ADD
J. Chromatogr.; (82) P 169-89; Vol 246; No 2

07-82-0011 (0011)
INVESTIGATION OF n-ALKYLBENZAMIDES BY REVERSED-PHASE LIQUID CHROMATOGRAPHY. V. CHARACTERISTICS OF SOME TERTIARY ALKYLBENZAMIDES
P Wells, Martha J. M.; Clark, C. Randall
Sch. Pharm. Auburn Univ. USA US AL 36849
J. Chromatogr.; (82) P 231-40; Vol 244; No 2

07-82-0012 (0012)
LIQUID CHROMATOGRAPHIC STUDY OF THE TWO PHYSICAL STATES OF A DENSELY BONDED ALKYL-SILICA AND THE CORRESPONDING RETENTION PROCESSES
Morel, D.; Serpinet, J.
Lab. Chim. Anal. III Univ. Lyon I Villeurbanne Fr. FR 69622
J. Chromatogr.; (82) P 231-40; Vol 248; No 2

07-82-0013 (0013)
INVESTIGATION OF n-ALKYLBENZAMIDES BY REVERSED-PHASE LIQUID CHROMATOGRAPHY. I. THE STUDY OF A HOMOLOGOUS SERIES OF n-ALKYLBENZAMIDES USING THE SOLVOPHOBIC THEORY AND MOLECULAR CONNECTIVITY
Wells, Martha J. M.; Clark C. Randall
Sch. Pharm. Auburn Univ. Auburn USA US AL 36849
J. Chromatogr.; (82) P 263-77; Vol 243; No 2

07-82-0014 (0014)
MECHANISM OF PROTEIN RETENTION IN REVERSED-PHASE HIGH-PERFORMANCE LIQUID CHROMATOGRAPHY
Barford, R. A.; Sliwinski, B. J.; Breyer, A. C.; Rothbart, H. L.
East. Reg. Res. Cent. Philadelphia USA US PA 19118
J. Chromatogr.; (82) P 281-8; Vol 235; No 2

07-82-0015 (0015)
INVESTIGATION OF n-ALKYLBENZAMIDES BY REVERSED-PHASE LIQUID CHROMATOGRAPHY. P I. ISOCRATIC ELUTION CHARACTERISTICS OF THE C1-C5 n-ALKYLBENZAMIDES
Wells, Martha J. M.; Clark, C. Randall
Sch. Pharm. Auburn Univ. Auburn USA AL 36849
J. Chromatogr.; (82) P 31-41; Vol 235; No 1

07-82-0016 (0016)
INVESTIGATION OF n-ALKYLBENZAMIDES BY REVERSED-PHASE LIQUID CHROMATOGRAPHY. II. APPLICATION OF THE SOLVOPHOBIC THEORY TO THE PREDICTION OF RETENTION DATA FOR THE C1-C5 n-ALKYLBENZAMIDES
Wells, Martha J. M.; Clark, C. Randall; Patterson, Richard M.
Sch. Pharm. Res. Data Anal. Auburn Univ. Auburn USA AL 36849
J. Chromatogr.; (82) P 43-59; Vol 235; No 1

07-82-0017 (0017)
COMPARISON OF MOLECULAR CONNECTIVITY AND A CHROMATOGRAPHIC CORRELATION FACTOR IN REVERSED-PHASE HIGH-PERFORMANCE LIQUID CHROMATOGRAPHY FOR POLYCYCLIC AROMATIC HYDROCARBONS
Hurtubise, Robert J.; Allen, Todd W.; Silver, Howard F.
Chem. Dep. Univ. Wyoming Laramie USA US WY 82071
J. Chromatogr.; (82) P 517-22; Vol 235; No 2

07-82-0018 (0018)
INVESTIGATION OF n-ALKYLBENZAMIDES BY REVERSED-PHASE LIQUID CHROMATOGRAPHY. III. CORRELATION OF CHROMATOGRAPHIC PARAMETERS WITH MOLECULAR CONNECTIVITY INDICES FOR THE C1-C5 n-ALKYLBENZAMIDES
Wells, Martha J. M.; Clark, C. Randall; Patterson, Richard M.
Sch. Pharm. Auburn Univ. Auburn USA US AL 36849
J. Chromatogr.; (82) P 61-74; Vol 235; No 1

07-82-0019 (0019)
EFFECT OF STATIONARY-PHASE STRUCTURE ON RETENTION AND SELECTIVITY IN REVERSED-PHASE LIQUID CHROMATOGRAPHY
Tanaka, Nobuo; Tokuda, Yasuyuki; Iwaguchi, Kazufusa; Araki, Mikio
Fac. Text. Sci. Kyoto Tech. Univ. Kyoto Japan JP 606
J. Chromatogr.; (82) P 761-72; Vol 239

07-82-0020 (0020)
RETENTION OF IONIZED SOLUTES IN REVERSED-PHASE HIGH-PERFORMANCE LIQUID CHROMATOGRAPHY
Rudzinski, W. E.; Bennett, D.; Garica, V.
Dep. Chem. Southwest Texas State Univ. San Marcos USA US TX 78666
J. Liq. Chromatogr.; (82) P 1295-312; Vol 5; No 7

07-82-0021 (0021)
HYDROPHOBICITY AND RETENTION IN REVERSED-PHASE LIQUID CHROMATOGRAPHY
D'Amboise, Marius; Hanai, Toshihiko
Dep. Chem. Univ. Montreal Montreal Can. CA PQ H3C 3V1
J. Liq. Chromatogr.; (82) P 229-44; Vol 5; No 2

2.7.2
Solvent Optimization

--- 1982 ---

07-82-0001 (0001)
THE ISOLATION OF PEPTIDES BY HIGH-PERFORMANCE LIQUID CHROMATOGRAPHY USING PREDICTED ELUTION POSITIONS
Browne, C. A.; Bennett, H. P. J.; Solomon, S.
Endocr. Lab. R. Victoria Hosp. Montreal Can. CA PQ H3A 1A1
Anal. Biochem.; (82) P 201-8; Vol 124; No 1

07-82-0002 (0002)
LIQUID CHROMATOGRAPHIC PHOSPHORESCENCE DETECTION WITH MICELLAR CHROMATOGRAPHY AND POSTCOLUMN REACTION MODES
Weinberger, Robert; Yarmchuk, Paul; Love, L. J. Cline
Dep. Chem. Seton Hall Univ. South Orange USA US NJ 07079
Anal. Chem.; (82) P 1552-8; Vol 54; No 9

07-82-0003 (0003)
SELECTIVITY IN LIQUID CHROMATOGRAPHY WITH MICELLAR MOBILE PHASES
Yarmchuk, Paul; Weinberger, Robert; Hirsch, R. F.; Love, L. J. Cline
Dep. Chem. Seton Hall Univ. South Orange USA US NJ 070W9
Anal. Chem.; (82) P 2233-8; Vol 54; No 13

07-82-0004 (0004)
CLASSIFICATION OF MULTISOLVENT MOBILE PHASE SYSTEMS IN LIQUID CHROMATOGRAPHY
Glajch, Joseph L.; Kirkland, J. J.
Cent. Res. Dev. Dep. E. I. du Pont de Nemours and Co. Wilmington USA US DE 19898
Anal. Chem.; (82) P 2593-6; Vol 54; No 14

07-82-0005 (0005)
INTERACTION INDEXES FOR PREDICTION OF RETENTION IN REVERSED-PHASE LIQUID CHROMATOGRAPHY
Jandera, P.; Colin, H.; Guiochon, G.
Dep. Anal. Chem. Univ. Chem. Technol. Pardubice Czech. CS
Anal. Chem.; (82) P 435-41; Vol 54; No 3

07-82-0006 (0006)
NONAQUEOUS REVERSED-PHASE LIQUID CHROMATOGRAPHIC FRACTIONATION OF POLYSTYRENE
Armstrong, Daniel W.; Bui, K. H.
Dep. Chem. Georgetown Univ. Washington USA US DC 20057
Anal. Chem.; (82) P 706-8; Vol 54; No 4

07-82-0007 (0007)
REVERSED PHASE CHROMATOGRAPHY OF IONIC COMPOUNDS USING SURFACTANTS OF THE SAME CHARGE
Dreux, M.; Lafosse, M.
Lab. Chim. Organ. Phys. Chromatogr. Chim. VII Univ. Orleans Orleans Fr. FR 45046
Analusis; (82) P 87-91; Vol 10; No 2; In Fr

07-82-0008 (0008)
THE SOLVENT ELUOTROPIC STRENGTH ON CARBON ADSORBENTS
Colin, H.; Guiochon, G.; Jandera, P.
Lab. Chim. Anal. Phys. Ec. Polytech. Palaiseau Fr. FR F-91128
Chromatographia; (82) P 133-9; Vol 15; No 2

07-82-0009 (0009)
DIMETHYLFORMAMIDE AS A MOBILE PHASE COMPONENT IN REVERSED-PHASE LIQUID COLUMN CHROMATOGRAPHY ON OCTADECYLSILICA
Ryba, M.

Inst. Org. Chem. Biochem. Czech. Acad. Sci. Prague Czech. CS 16610/6
Chromatographia; (82) P 227-30; Vol 15; No 4

07-82-0010 (0010)
HPLC STUDIES OF PEPTIDES. THE EFFECT OF THE MOBILE PHASE SYSTEM IN REVERSED-PHASE CHROMATOGRAPHY
Hu, Zhenyuan; Xia, Lijun; Xu, Jiacheng; Cheng, Linlin
Shanghai Inst. Org. Chem. Acad. Sin. Shanghai Peop. Rep. China CN
Chromatographia; (82) P 5-8; Vol 15; No 1

07-82-0011 (0011)
MOBILE PHASE EFFECTS IN REVERSED-PHASE CHROMATOGRAPHY. IV. RETENTION BY n-ALKYLBENZENES AS A FUNCTION OF COLUMN TEMPERATURE AND THE NATURE AND CONCENTRATION OF ORGANIC CO-SOLVENT
Melander, W. R.; Mannan, C. A.; Horvath, C.
Dep. Chem. Eng. Yale Univ. New Haven USA US CT 06520
Chromatographia; (82) P 611-15; Vol 15; No 9

07-82-0012 (0012)
CONTAMINANTS IN WATER USED FOR "HIGH-PERFORMANCE" LIQUID CHROMATOGRAPHY
Sampson, Eric J.; Culbreth, Paula H.
Cent. Environ. Health Cent. Dis. Control Atlanta USA US GA 30333
Clin. Chem. (Winston-Salem, N. C.); (82) P 1393-4; Vol 28; No 6

07-82-0013 (0013)
LIQUID-CHROMATOGRAPHIC DETERMINATION OF PENTOBARBITAL IN PLASMA WITH USE OF A RESIN COLUMN AND AN ALKALINE MOBILE PHASE
Gupta, Ram N.; Smith, Paul T.; Eng, Francis
Clin. Chem. Lab. St. Joseph's Hosp. Hamilton Can. CA ON L8N 1Y4
Clin. Chem. (Winston-Salem, N. C.); (82) P 1772-4; Vol 28; No 8

07-82-0014 (0014)
MULTIVARIATE AND UNIVARIATE OPTIMIZATION STUDIES OF LIQUID-CHROMATOGRAPHIC SEPARATION OF STEROID MIXTURES
Fast, Douglas M.; Culbreth, Paula H.; Sampson, Eric J.
Clin. Chem. Div. Cent. Environ. Health Atlanta USA US GA 30333
Clin. Chem. (Winston-Salem, N. C.); (82) P 444-8; Vol 28; No 3

07-82-0015 (0015)
USE OF MICELLAR AND CYCLODEXTRIN SOLUTIONS IN LIQUID CHROMATOGRAPHIC SEPARATIONS
Armstrong, Daniel W.
Dep. Chem. Georgetown Univ. Washington USA US DC 20057
CONFERENCE; (82); Solution Behav. Surfactants: Theor. Appl. Aspects, (Proc. Int. Symp.); P 1273-82; Vol 2,; Avail.: Mittal, K. L.; Fendler, Eleanor J; Date: 80 ; Publisher: Plenum, New York, N.Y.

07-82-0016 (0016)
A CHEAP AND SIMPLE METHOD FOR THE PREPARATION OF "HPLC-PURE" WATER ON THE LABORATORY SCALE BY PHOTOOXIDATION
Frischkorn, C. G. B.; Schlimper, H.
Inst. Appl. Phys. Chem. Nucl. Res. Cent. Juelich Fed. Rep. Ger. DE D-5170
Fresenius' Z. Anal. Chem.; (82) P 541-2; Vol 312; No 6

07-82-0017 (0017)
LIQUID CHROMATOGRAPHY USING FOUR SOLVENTS
Krusche, J. U.; Gottmann, G.
Abt. Anal. Instrum. Du Pont de Nemours G.m.b.H. Bad Nauheim Fed. Rep. Ger. DE D-6350
GIT Fachz. Lab.; (82) P 345-6, 348-9; Vol 26; No 4; In Ger

07-82-0018 (0018)
SIMPLE PURIFICATION OF ACETONITRILE FOR HIGH-PERFORMANCE LIQUID CHROMATOGRAPHY (HPLC) OF PEPTIDES AND PROTEINS
Braunitzer, Gerhard; Ruecknagel, Peter; Oberthuer, Walter
Abt. Proteinchem. Max-Planck-Inst. Biochem. Martinsried/Muenchen Fed. Rep. Ger. DE D-8033
Hoppe-Seyler's Z. Physiol. Chem.; (82) P 485-6; Vol 363; No 5; In Ger

07-82-0019 (0019)
CHROMATOGRAPHIC PERFORMANCE OF DEUTERATED SOLVENTS IN REVERSED PHASE MICRO-HIGH-PERFORMANCE LIQUID CHROMATOGRAPHY
Jinno, K.
Sch. Mater. Sci. Toyohashi Univ. Technol. Toyohashi Japan JP 440
HRC CC, J. High Resolut. Chromatogr. Chromatogr. Commun.; (82) P 364-7; Vol 5; No 7

07-82-0020 (0020)
HPLC RESULTS ENHANCED BY TERNARY MIXING OF MOBILE PHASE
Hodgin, John C.
Micro-Meritics Instrum. Corp. Cleveland USA US OH
Ind. Res. Dev.; (82) P 156-7; Vol 24; No 1

07-82-0021 (0021)
APPLICATION OF GEL PERMEATION CHROMATOGRAPHY AND NONAQUEOUS REVERSE PHASE CHROMATOGRAPHY TO HIGH-PERFORMANCE LIQUID CHROMATOGRAPHIC DETERMINATION OF RETINYL PALMITATE AND .ALPHA.-TOCOPHERYL ACETATE IN INFANT FORMULAS
Landen, William O., Jr.
FDA Atlanta USA US GA 30309
J. Assoc. Off. Anal. Chem.; (82) P 810-16; Vol 65; No 4

07-82-0022 (0022)
EVALUATION OF STATIONARY AND MOBILE PHASES FOR REVERSED-PHASE HIGH PERFORMANCE LIQUID CHROMATOGRAPHY OF PEPTIDES
Wehr, C. T.; Correia, L.; Abbott, S. R.
Varian Instrum. Group Walnut Creek USA US CA 94598
J. Chromatogr. Sci.; (82) P 114-19; Vol 20; No 3

07-82-0023 (0023)
REVERSED-PHASE HPLC FROM pH 1 TO 13
Lee, Dan Peter
Hamilton Co. Reno USA US NV 89510
J. Chromatogr. Sci.; (82) P 203-8; Vol 20; No 5

07-82-0024 (0024)
UNATTENDED OPTIMIZATION OF REVERSED-PHASE HIGH-PERFORMANCE LIQUID CHROMATOGRAPHIC SEPARATIONS USING THE MODIFIED SIMPLEX ALGORITHM
Berridge, J. C.
Pfizer Cent. Res. Kent/Sandwich UK GB CT13 9NJ
J. Chromatogr.; (82) P 1-14; Vol 244; No 1

07-82-0025 (0025)
SELECTIVE GROUP SEPARATION OF PRIMARY AND SECONDARY BIOGENIC AMINES BY REVERSED-PHASE HIGH-PERFORMANCE LIQUID CHROMATOGRAPHY WITH 18-CROWN-6 IN THE MOBILE PHASE
Wiechmann, Manfred
Univ. Bayreuth Bayreuth Fed. Rep. Ger. DE D-8580
J. Chromatogr.; (82) P 129-37; Vol 235; No 1; In Ger

07-82-0026 (0026)
RAPID HIGH-PERFORMANCE LIQUID CHROMATOGRAPHIC DETERMINATION OF ACETAMINOPHEN

IN DOSAGE FORMS, BY USING A TOTALLY AQUEOUS MOBILE PHASE
Gilpin, R. K.; Gaudet, M. H.
IBM Instruments, Inc. Danbury USA US CT 06810
J. Chromatogr.; (82) P 160-4; Vol 248; No 1

07-82-0027 (0027)
REVERSED-PHASE LIQUID CHROMATOGRAPHY ON LICHROSORB RP-18 WITH METHANOL-WATER MIXTURES
Hammers, W. E.; Meurs, G. J.; De Ligny, C. L.
Lab. Anal. Chem. State Univ. Utrecht Utrecht Neth. NL 3522 AD
J. Chromatogr.; (82) P 169-89; Vol 246; No 2

07-82-0028 (0028)
REVERSED-PHASE HIGH-PERFORMANCE LIQUID CHROMATOGRAPHY OF PROCYANIDINS AND OTHER PHENOLICS IN FRESH AND OXIDIZING APPLE JUICES USING A pH SHIFT TECHNIQUE
Lea, Andrew G. H.
Long Ashton Res. Stn. Univ. Bristol Long Ashton/Bristol UK GB BS18 9AF
J. Chromatogr.; (82) P 253-7; Vol 238; No 1

07-82-0029 (0029)
COMMENT ON THE PROPOSED RQ TRANSFORMATION METHOD FOR OPTIMIZING MOBILE PHASE COMPOSITION IN HIGH-PERFORMANCE LIQUID CHROMATOGRAPHY
Tomlinson, E.
Subfac. Pharm. Univ. Amsterdam Amsterdam Neth. NL 1018 TV
J. Chromatogr.; (82) P 258-61; Vol 236; No 1

07-82-0030 (0030)
TEMPERATURE AND ELUENT EFFECTS ON THE SELECTIVITY OF SOME NITROAROMATIC BONDED PHASES IN HIGH-PERFORMANCE LIQUID CHROMATOGRAPHY
Hammers, W. E.; Theeuwes, A. G. M.; Brederode, W. K.; De Ligny, C. L.
Lab. Anal. Chem. State Univ. Utrecht Neth. NL 3522 AD
J. Chromatogr.; (82) P 321-36; Vol 234; No 2

07-82-0031 (0031)
RETENTION BEHAVIOR OF ALKYLBENZENES AS A FUNCTION OF TEMPERATURE AND MOBILE PHASE COMPOSITION IN REVERSED-PHASE CHROMATOGRAPHY
Grushka, Eli; Colin, Henri; Guiochon, Georges
Dep. Inorg. Anal. Chem. Hebrew Univ. Jerusalem Israel IL
J. Chromatogr.; (82) P 325-39; Vol 248; No 3

07-82-0032 (0032)
ANOMALOUS SEPARATION OF 2'-AND 3'-RIBONUCLEOTIDES USING ION SUPRESSION REVERSED-PHASE HIGH-PERFORMANCE LIQUID CHROMATOGRAPHY
Whitehouse, R.P.; Greenstock, C.L.
Med. Biophys. Branch, Whiteshell Nucl. Res. Establ., Pinawa, MB Can.ROE 1LO
J. Chromatogr.; (82) P 376-8; Vol 243; No 2

07-82-0033 (0033)
SEPARATION OF CHLORINATED PHENOLS BY REVERSED-PHASE HIGH-PERFORMANCE LIQUID CHROMATOGRAPHY AT AN ALKALINE pH
McLeod, H. A.; Laver, G.
Food Res. Div. Health Prot. Branch Ottawa Can. CA ON K1A 0L2
J. Chromatogr.; (82) P 385-90; Vol 244; No 2

07-82-0034 (0034)
COMPARISON OF AMINE MODIFIERS USED TO REDUCE PEAK TAILING OF 2-PHENYLETHYL-AMINE DRUGS IN REVERSED-PHASE HIGH PERFORMANCE LIQUID CHROMATOGRAPHY
Gill, R.; Alexander, S. P.; Moffat, A. C.
Home Office Cent. Res. Establ. Aldermaston/Reading/Berks. UK GB RG7 4PN
J. Chromatogr.; (82) P 39-45; Vol 247; No 1

07-82-0035 (0035)
FACTORS INFLUENCING THE RETENTION OF INSULINS IN REVERSED-PHASE HIGH-PERFORMANCE LIQUID CHROMATOGRAPHIC SYSTEMS
Vigh, Gyula; Varga-Puchony, Zita; Hlavay, Jozsef; Papp-Hites, E.
Inst. Anal. Chem. Veszprem Univ. Chem. Eng. Veszprem Hung. HU
J. Chromatogr.; (82) P 51-9; Vol 236; No 1

07-82-0036 (0036)
INVESTIGATION OF THE EFFECT OF ADDED ORGANIC AMINE ON THE CHROMATOGRAPHY OF TRICYCLIC ANTIDEPRESSANT DRUGS USING REVERSED-PHASE CHROMATOGRAPHY ON OCTADECYLSILICA WITH SODIUM LAURYL SULFATE AS PAIRING ION
Hung, C. T.; Taylor, R. B.; Paterson, N.
Sch. Pharm. Robert Gordon's Inst. Technol. Aberdeen UK GB
J. Chromatogr.; (82) P 61-73; Vol 240; No 1

07-82-0037 (0037)
LIQUID CHROMATOGRAPHY WITH CROWN ETHER-CONTAINING MOBILE PHASES. II. RETENTION BEHAVIOR OF .BETA.-LACTAM ANTIBIOTICS IN REVERSED-PHASE HIGH-PERFORMANCE LIQUID CHROMATOGRAPHY
Nakagawa, Terumichi; Shibukawa, Akimasa; Uno, Toyozo
Fac. Pharm. Sci. Kyoto Univ. Kyoto Japan JP 606
J. Chromatogr.; (82) P 695-706; Vol 239

07-82-0038 (0038)
SELECTIVE EFFECTS OF MOBILE AND STATIONARY PHASES IN REVERSED-PHASE HIGH-PERFORMANCE LIQUID CHROMATOGRAPHY OF ECDYSTEROIDS
Wilson, I. D.; Bielby, C. R.; Morgan, E. D.
Dep. Chem. Univ. Keele Keele/Staffordshire UK GB ST5 5BG
J. Chromatogr.; (82) P 97-102; Vol 238; No 1

07-82-0039 (0039)
SEPARATION OF SODIUM SALTS OF PHENYLSULFAMIC ACIDS BY REVERSED-PHASE HIGH-PERFORMANCE LIQUID CHROMATOGRAPHY USING STRONG ELECTROLYTES
Benson G. A.
Sch. Sci. Reg. Tech. Coll. Sligo Ire. IE
J. Chromatogr.; P(82) P 285-90; Vol 243; No 2

07-82-0040 (0040)
RETENTION OF IONIZED SOLUTES IN REVERSED-PHASE HIGH-PERFORMANCE LIQUID CHROMATOGRAPHY
Rudzinski, W. E.; Bennett, D.; Garica, V.
Dep. Chem. Southwest Texas State Univ. San Marcos USA US TX 78666
J. Liq. Chromatogr.; (82) P 1295-312; Vol 5; No 7

07-82-0041 (0041)
THE EFFECT OF BINARY SOLVENT COMPOSITION AND POLARITY ON SEPARATIONS IN REVERSED PHASE THIN-LAYER AND HIGH-PERFORMANCE LIQUID CHROMATOGRAPHY
Issaq, Haleem J.; Klose, John R.; Cutchin, William
Frederick Cancer Res. Facility Natl. Cancer Inst. Frederick USA US MD 21701
J. Liq. Chromatogr.; (82) P 625-41; Vol 5; No 4

07-82-0042 (0042)
LIQUID CHROMATOGRAPH USING MULTIPLE ELUENTS
Hitachi, Ltd. Japan JP
Jpn. Kokai Tokkyo Koho; (070682) P 3 pp.; In Japan; Pat. No.: 82108661; Int. Class: G01N-031/08; B01D-015/08; Doc. Code: A2; Appl.: JP 80/183829 801226

07-82-0043 (0043)
HPLC SOLVENT GRADIENT FORMATION
Riggenmann, Hansjuergen

Gynkotek Ges. Bau Wissenschaftlich-Tech. Gerate m.b.H. Munich Fed. Rep. Ger.
DE 8000/21
LaborPraxis; (82) P 474, 476, 478, 481; Vol 6; No 5

07-82-0044 (0044)
STRATEGY FOR THE CHOICE OF MOBILE AND STATIONARY PHASES IN HPLC
Montalbetti, N.
Clin. Chem. Lab. St. Anna Hosp. Como Italy IT I-22100
Not. - Soc. Ital. Biochim. Clin.; (82) P 577-81; Vol 6; No 10

07-82-0045 (0045)
OPTIMAL CONDITIONS IN ISOCRATIC COLUMN LIQUID CHROMATOGRAPHY
Huber, J. F. K.
Inst. Anal. Chem. Univ. Wien Vienna Austria AT
Oesterr. Chem. Z.; (82) P 159-61; Vol 83; No 6; In Ger

2.7.3
Physicochemical Parameters

--- 1982 ---

07-82-0001 (0001)
INTERACTION INDEXES FOR PREDICTION OF RETENTION IN REVERSED-PHASE LIQUID CHROMATOGRAPHY
Jandera, P.; Colin, H.; Guiochon, G.
Dep. Anal. Chem. Univ. Chem. Technol. Pardubice Czech. CS
Anal. Chem.; (82) P 435-41; Vol 54; No 3

07-82-0002 (0002)
ESTIMATING THE AQUEOUS SOLUBILITY OF AROMATIC HYDROCARBONS BY HIGH PERFORMANCE LIQUID CHROMATOGRAPHY
Whitehouse, Brian G.
Dep. Oceanogr. Dalhousie Univ. Halifax Can. CA NS B3H 4J1
Chemosphere; (82) P 689-99; Vol 11; No 8

07-82-0003 (0003)
THE SOLUBILITY PARAMETER AS A TOOL IN UNDERSTANDING LIQUID CHROMATOGRAPHY
Schoenmakers, P. J.; Billiet, H. A. H.; De Galan, L.
Lab. Anal. Scheikd. Tech. Hogesch. Delft Delft Neth. NL 2628 BX
Chromatographia; (82) P 205-14; Vol 15; No 3

07-82-0004 (0004)
SOLUBILITY PARAMETER AS A TOOL IN UNDERSTANDING LIQUID CHROMATOGRAPHY
Schoenmakers, P. J.; Billiet, H. A. H.; De Galan, L.
Lab. Anal. Scheikd. Tech. Hogesch. Delft Delft Neth. NL NL-2628
Chromatographia; (82) P 387-98; Vol 15; No 6; In Ger

07-82-0005 (0005)
CORRELATIONS BETWEEN LIQUID CHROMATOGRAPHIC CAPACITY RATIO DATA ON LICHROSORB RP-18 AND PARTITION COEFFICIENTS IN THE OCTANOL-WATER SYSTEM
Hammers, W. E.; Meurs, G. J.; De Ligny, C. L.
Lab. Anal. Chem. State Univ. Utrecht Utrecht Neth. NL 3522 AD
J. Chromatogr.; (82) P 1-13; Vol 247; No 1

07-82-0006 (0006)
ADSORPTION BEHAVIOR OF SEVERAL SUPPORTS IN REVERSED-PHASE THIN-LAYER CHROMATOGRAPHY AS DEMONSTRATED BY THE DETERMINATION OF RELATIVE PARTITION COEFFICIENTS OF SOME 4-HYDROXYCOUMARIN DERIVATIVES
Van der Giesen, Willem F.; Janssen, Lambert H. M.
Fac. Pharm. State Univ. Utrecht Utrecht Neth. NL 3511 GH
J. Chromatogr.; (82) P 199-213; Vol 237; No 2

07-82-0007 (0007)
INVESTIGATION OF n-ALKYLBENZAMIDES BY REVERSED-PHASE LIQUID CHROMATOGRAPHY. I. THE STUDY OF A HOMOLOGOUS SERIES OF n-ALKYLBENZAMIDES USING THE SOLVOPHOBIC THEORY AND MOLECULAR CONNECTIVITY
Wells, Martha J. M.; Clark C. Randall
Sch. Pharm. Auburn Univ. Auburn USA US AL 36849
J. Chromatogr.; (82) P 263-77; Vol 243; No 2

07-82-0008 (0008)
ALKYL ARYL KETONES AS A RETENTION INDEX SCALE IN LIQUID CHROMATOGRAPHY
Smith, Roger M.
Dep. Chem. Univ. Technol. Loughborough Leicestershire UK GB LE11 3TU
J. Chromatogr.; (82) P 313-20; Vol 236; No 2

07-82-0009 (0009)
REVERSED-PHASE THIN-LAYER CHROMATOGRAPHY OF POLYNUCLEAR AROMATIC HYDROCARBONS AND CHLORINATED BIPHENYLS. RELATIONSHIP WITH HYDROPHOBICITY AS MEASURED BY

AQUEOUS SOLUBILITY AND OCTANOL-WATER PARTITION COEFFICIENT
Bruggeman, W. A.; Van der Steen, J.; Hutzinger, O.
Lab. Environ. Toxicol. Chem. Univ. Amsterdam Amsterdam Neth. NL 1018 WV
J. Chromatogr.; (82) P 335-46; Vol 238; No 2

07-82-0010 (0010)
AMINO ACIDS AND PEPTIDES. CLXXVIII. CORRELATION BETWEEN THE HYDROPHOBICITY OF SUBSTITUENTS IN THE PHENYLALANINE MOIETY OF OXYTOCIN CARBA ANALOGS AND REVERSED-PHASE-CHROMATOGRAPHIC k' VALUES
Lebl, Michel
Inst. Org. Chem. Biochem. Czech. Acad. Sci. Prague Czech. CS 166 10/6
J. Chromatogr.; (82) P 342-5; Vol 242; No 2

07-82-0011 (0011)
PREDICTION OF RETENTION TIMES FOR AROMATIC ACIDS IN LIQUID CHROMATOGRAPHY
Hanai, T.; Tran, K. C.; Hubert, J.
Dep. Chim. Univ. Montreal Montreal Can. CA PQ H3C 3V1
J. Chromatogr.; (82) P 385-95; Vol 239

07-82-0012 (0012)
COMPARISON OF MOLECULAR CONNECTIVITY AND A CHROMATOGRAPHIC CORRELATION FACTOR IN REVERSED-PHASE HIGH-PERFORMANCE LIQUID CHROMATOGRAPHY FOR POLYCYCLIC AROMATIC HYDROCARBONS
Hurtubise, Robert J.; Allen, Todd W.; Silver, Howard F.
Chem. Dep. Univ. Wyoming Laramie USA US WY 82071
J. Chromatogr.; (82) P 517-22; Vol 235; No 2

07-82-0013 (0013)
INVESTIGATION OF n-ALKYLBENZAMIDES BY REVERSED-PHASE LIQUID CHROMATOGRAPHY. III. CORRELATION OF CHROMATOGRAPHIC PARAMETERS WITH MOLECULAR CONNECTIVITY INDICES FOR THE C1-C5 n-ALKYLBENZAMIDES
Wells, Martha J. M.; Clark, C. Randall; Patterson, Richard M.
Sch. Pharm. Auburn Univ. Auburn USA US AL 36849
J. Chromatogr.; (82) P 61-74; Vol 235; No 1

07-82-0014 (0014)
DETERMINATION OF PARTITION COEFFICIENTS OF VERY HYDROPHOBIC COMPOUNDS BY HIGH-PERFORMANCE LIQUID CHROMATOGRAPHY ON GLYCERYL-COATED CONTROLLED-PORE GLASS
Miyake, Keishiro; Terada, Hiroshi
Fac. Pharm. Sci. Mukogawa Women's Univ. Nishinomiya Japan JP 663
J. Chromatogr.; (82) P 9-20; Vol 240; No 1

07-82-0015 (0015)
HYDROPHOBICITY AND RETENTION IN REVERSED-PHASE LIQUID CHROMATOGRAPHY
D'Amboise, Marius; Hanai, Toshihiko
Dep. Chem. Univ. Montreal Montreal Can. CA PQ H3C 3V1
J. Liq. Chromatogr.; (82) P 229-44; Vol 5; No 2

07-82-0016 (0016)
LABORATORY TO LABORATORY REPRODUCIBILITY OF HIGH-PERFORMANCE LIQUID CHROMATOGRAPHIC RETENTION INDEXES
Baker, John K.; Cates, L. A.; Corbett, M. D.; Huber, Joseph W.; Lattin, Danny L.
Sch. Pharm. Univ. Mississippi University USA US MS 38677
J. Liq. Chromatogr.; (82) P 829-39; Vol 5; No 5

07-82-0017 (0017)
DETERMINATION OF OCTANOL-WATER EQUIVALENT PARTITION COEFFICIENTS OF INDOLIZINE AND SUBSTITUTED 2-PHENYLINDOLIZINES BY REVERSED-PHASE HIGH-PRESSURE LIQUID

CHROMATOGRAPHY AND FRAGMENTATION VALUES
Lins, Claudio L. K.; Block, John H.; Doerge, Robert F.; Barnes, Gerard J.
Sch. Pharm. Oregon State Univ. Corvallis USA US OR 97331
J. Pharm. Sci.; (82) P 614-17; Vol 71; No 6

07-82-0018 (0018)
RELATION BETWEEN HPLC PARAMETER log k' AND BIOLOGICAL ACTIVITY OF SULFA DRUGS
Zhang, Shengqiang; An, Dengkui; Wu, Rujin; Sheng, Longsheng
Dep. Pharm. Anal. Nanjing Coll. Pharm. Nanjing Peop. Rep. China CN
Nanjing Yaoxueyuan Xuebao; (82) P 1-9; No 2; In Ch

CHAPTER 2.8
Liquid–Liquid Chromatography

--- 1982 ---

08-82-0001 (0001)
ADSORPTION OF 1-PENTANOL ON ALKYL-MODIFIED SILICA AND ITS EFFECT ON THE RETENTION MECHANISM IN REVERSED-PHASE LIQUID CHROMATOGRAPHY
Wahlund, Karl Gustav; Beijersten, Ingegerd
Dep. Anal. Pharm. Chem. Univ. Uppsala Uppsala Swed. SE S-751 23
Anal. Chem.; (82) P 128-32; Vol 54; No 1

08-82-0002 (0002)
SIZE FRACTIONATION OF DNA FRAGMENTS RANGING FROM 20 TO 30,000 BASE PAIRS BY LIQUID/LIQUID CHROMATOGRAPHY
Mueller, Werner; Kuetemeier, Gertrud
Abt. Viroidforsch. Max-Planck-Inst. Biochem. Martinsried Fed. Rep. Ger. DE
Eur. J. Biochem.; (82) P 231-8; Vol 128; No 1

08-82-0003 (0003)
REVERSED-PHASE EXTRACTION CHROMATOGRAPHY OF MOLYBDENUM(VI) WITH TRIBUTYL PHOSPHATE ON SILICA GEL
Bhosale, S. N.; Khopkar, S. M.
Indian Inst. Technol. Bombay India IN 400 076
Indian J. Chem., Sect. A; (82) P 147-9; Vol 21A; No 2

08-82-0004 (0004)
DETERMINATION OF URINARY HIPPURIC ACID AND o-CRESOL, AS INDICES OF TOLUENE EXPOSURE, BY LIQUID CHROMATOGRAPHY ON DYNAMICALLY MODIFIED SILICA
Hansen, Steen Honore; Doessing, Martin
Dep. Org. Chem. R. Dan. Sch. Pharm. Copenhagen Den. DK DK-2100
J. Chromatogr.; (82) P 141-8; Vol 229; No 1

08-82-0005 (0005)
DETERMINATION OF IMPURITIES IN PROPRANOLOL HYDROCHLORIDE BY HIGH-PERFORMANCE LIQUID CHROMATOGRAPHY ON DYNAMICALLY MODIFIED SILICA
Helboe, Per
Drug Standardization Lab. Natl. Board Health Broenshoej Den. DK DK-2700
J. Chromatogr.; (82) P 229-38; Vol 245; No 2

08-82-0006 (0006)
NORMAL-PHASE HIGH-PERFORMANCE LIQUID CHROMATOGRAPHY OF PORPHYRIN FREE ACIDS ON SILICA MODIFIED WITH TETRAETHYLENEPENTAMINE
Lim, C. K.; Chan, Jenny Y. Y.
Div. Clin. Chem. MRC Clin. Res. Cent. Harrow/Middlesex UK GB HA1 3UJ
J. Chromatogr.; (82) P 305-10; Vol 228

08-82-0007 (0007)
HIGH-PERFORMANCE LIQUID CHROMATOGRAPHY ON DYNAMICALLY MODIFIED SILICA. III. MODIFICATION OF SILICA WITH LONG-CHAIN AND SYMMETRICAL QUATERNARY AMMONIUM COMPOUNDS
Hansen, Steen Honore; Helboe, Per; Lund, Ulla
Dep. Org. Chem. R. Danish Sch. Pharm. Copenhagen Den. DK DK-2100
J. Chromatogr.; (82) P 319-27; Vol 240; No 2

08-82-0008 (0008)
SILICA GEL LIQUID-LIQUID CHROMATOGRAPHY USING AQUEOUS BINARY PHASE SYSTEMS. HIGH-EFFICIENCY EXTRACTION AND RESOLUTION OF PHENOLS AND CARBOXYLIC ACIDS
Hara, Shoji; Dobashi, Yasuo; Oka, Kitaro
Tokyo Coll. Pharm. Hachioji Japan JP 192-03
J. Chromatogr.; (82) P 677-85; Vol 239

08-82-0009 (0009)
DETERMINATION OF PARTITION COEFFICIENTS OF VERY HYDROPHOBIC COMPOUNDS BY HIGH-PERFORMANCE LIQUID CHROMATOGRAPHY ON GLYCERYL-COATED CONTROLLED-PORE GLASS
Miyake, Keishiro; Terada, Hiroshi
Fac. Pharm. Sci. Mukogawa Women's Univ. Nishinomiya Japan JP 663
J. Chromatogr.; (82) P 9-20; Vol 240; No 1

08-82-0010 (0010)
REVERSED-PHASE LIQUID CHROMATOGRAPHY OF CHROMIUM WITH TRIBUTYL PHOSPHATE ON SILICA GEL
Bhosale, S. N.; Khopkar, S. M.
Dep. Chem. Indian Inst. Technol. Bombay India IN 400076
Mikrochim. Acta; (82) P 433-9; Vol 1; No 5-6

CHAPTER 2.9
Structure-Retention Relationships

--- 1982 ---

09-82-0001 (0001)
HPLC IDENTIFICATION AND SEPARATION OF PHENOLIC COMPOUNDS DERIVATIZED WITH DIAZOTIZED SULFANILIC ACID. STRUCTURAL EFFECTS ON RETENTION TIMES
Baiocchi, C.; Gennaro, M. C.; Campi, E.; Mentasti, E.; Aruga, R.
Ist. Chim. Anal. Univ. Torino Turin Italy IT 10125
Anal. Lett.; (82) P 1539-48; Vol 15; No A19

09-82-0002 (0002)
MOBILE PHASE EFFECTS IN REVERSED-PHASE CHROMATOGRAPHY. IV. RETENTION OF n-ALKYLBENZENES AS A FUNCTION OF COLUMN TEMPERATURE AND THE NATURE AND CONCENTRATION OF ORGANIC CO-SOLVENT
Melander, W. R.; Mannan, C. A.; Horvath, C.
Dep. Chem. Eng. Yale Univ. New Haven USA US CT 06520
Chromatographia; (82) P 611-15; Vol 15; No 9

09-82-0003 (0003)
STATIONARY PHASE EFFECTS IN REVERSED-PHASE CHROMATOGRAPHY II. SUBSTITUENT SELECTIVITIES FOR RETENTION ON VARIOUS HYDROCARBONACEOUS BONDED PHASES
Melander, W. R.; Horvath, C.
Dep. Chem. Eng. Yale Univ. New Haven USA US CT 06520
Chromatographia; (82) P 86-90; Vol 15; No 2

09-82-0004 (0004)
HIGH-BACK-PRESSURE LIQUID CHROMATOGRAPHY. IV. STUDIES ON THE RETENTION BEHAVIOR OF AROMATIC COMPOUNDS
Takeuchi, Toyohide; Ishii, Daido
Fac. Eng. Nagoya Univ. Nagoya Japan JP 464
J. Chromatogr.; (82) P 23-32; Vol 244; No 1

09-82-0005 (0005)
INVESTIGATION OF n-ALKYLBENZAMIDES BY REVERSED-PHASE LIQUID CHROMATOGRAPHY. V. CHARACTERISTICS OF SOME TERTIARY ALKYLBENZAMIDES
P Wells, Martha J. M.; Clark, C. Randall
Sch. Pharm. Auburn Univ. USA US AL 36849
J. Chromatogr.; (82) P 231-40; Vol 244; No 2

09-82-0006 (0006)
LIQUID CHROMATOGRAPHY STUDY OF BROMINATED ANILINES AND INVESTIGATION OF PRODUCT FORMATION IN THE BROMINATION REACTION. I. ANILINES WITHOUT RING SUBSTITUENTS OR WITH ALKYL GROUPS IN THE ORTHO- AND PARA-POSITIONS
Truedsson, L. A.; Smith, B. E. F.
Chem. Cent. Lund Inst. Technol. Lund Swed. SE S-220 07/7
J. Chromatogr.; (82) P 25-46; Vol 234; No 1

09-82-0007 (0007)
INVESTIGATION OF n-ALKYLBENZAMIDES BY REVERSED-PHASE LIQUID CHROMATOGRAPHY. I. THE STUDY OF A HOMOLOGOUS SERIES OF n-ALKYLBENZAMIDES USING THE SOLVOPHOBIC THEORY AND MOLECULAR CONNECTIVITY
Wells, Martha J. M.; Clark C. Randall
Sch. Pharm. Auburn Univ. Auburn USA US AL 36849
J. Chromatogr.; (82) P 263-77; Vol 243; No 2

09-82-0008 (0008)
EFFECT OF MOLECULAR STRUCTURE AND CONFORMATIONAL CHANGE OF PROLINE-CONTAINING DIPEPTIDES IN REVERSED-PHASE CHROMATOGRAPHY
Melander, Wayne R.; Jacobson, Jana; Horvath, Csaba
Dep. Chem. Eng. Yale Univ. New Haven USA US CT 06520
J. Chromatogr.; (82) P 269-76; Vol 234; No 2

09-82-0009 (0009)
RETENTION BEHAVIOR OF ALKYLBENZENES AS A FUNCTION OF TEMPERATURE AND MOBILE PHASE COMPOSITION IN REVERSED-PHASE CHROMATOGRAPHY
Grushaka, Eli; Colin, Henri; Guiochon, Georges
Dep. Inorg. Anal. Chem. Hebrew Univ. Jerusalem Israel IL
J. Chromatogr.; (82) P 325-39; Vol 248; No 3

09-82-0010 (0010)
REVERSED-PHASE THIN-LAYER CHROMATOGRAPHY OF POLYNUCLEAR AROMATIC HYDROCARBONS AND CHLORINATED BIPHENYLS. RELATIONSHIP WITH HYDROPHOBICITY AS MEASURED BY AQUEOUS SOLUBILITY AND OCTANOL-WATER PARTITION COEFFICIENT
Bruggeman, W. A.; Van der Steen, J.; Hutzinger, O.
Lab. Environ. Toxicol. Chem. Univ. Amsterdam Amsterdam Neth. NL 1018 WV
J. Chromatogr.; (82) P 335-46; Vol 238; No 2

09-82-0011 (0011)
LIQUID CHROMATOGRAPHY STUDY OF BROMINATED ANILINES AND INVESTIGATION OF PRODUCT FORMATION IN THE BROMINATION REACTION. II. ANILINES WITH ALKYL GROUPS IN THE META-POSITION
Truedsson, L. A.
Chem. Cent. Lund Inst. Technol. Lund Swed. SE S-220 07/7
J. Chromatogr.; (82) P 47-56; Vol 234; No 1

09-82-0012 (0012)
CORRELATION BETWEEN THE RETENTION OF CARDIAC GLYCOSIDES IN REVERSED-PHASE HIGH-PERFORMANCE LIQUID CHROMATOGRAPHY WITH DIPHENYLSILYL STATIONARY PHASE, THE STRUCTURE OF THEIR MOLECULES AND THEIR BIOLOGICAL ACTIVITY
Daviidov, V. Ya.; El1zalde Gonzalez, M.; Kiselev, A. V.
Chem. Dep. Lomonosov State Univ. Moscow Moscow USSR SU 117234
J. Chromatogr.; (82) P 49-62; Vol 248; No 1

09-82-0013 (0013)
RETENTION BEHAVIOR OF SOME 4'-ALKANOYLBENZO-15-CROWN-5 DERIVATIVES IN REVERSED-PHASE HIGH-PERFORMANCE LIQUID CHROMATOGRAPHY (HPLC)
Stein, Johann; Meerbote, Michael; Beger, Joerg
Sekt. Chem. Bergakad. Freiberg Freiberg Ger. Dem. Rep. DD DDR-9200
Z. Chem.; (82) P 146-7; Vol 22; No 4; In Ger

09-82-0014 (0014)
RELATION BETWEEN CHROMATOGRAPHIC RETENTION AND MOLECULAR STRUCTURE OF ALKYLPYRIDINES AND POLAR STATIONARY LIQUID PHASES
Nabivach, V. M.; Berlizov, Yu. S.; Dmitrikov, V. P.
Dnepropetr. Khim.-Tekhnol. Inst. Dnepropetrovsk USSR SU
Zh. Anal. Khim.; (82) P 480-7; Vol 37; No 3; In Russ

09-82-0015 (0015)
LIQUID ADSORPTION CHROMATOGRAPHY OF CHELATES. EFFECT OF THE STRUCTURE OF CHELATES ON THEIR CHROMATOGRAPHIC BEHAVIOR IN A THIN LAYER OF SORBENT
Timerbaev, A. R.; Petrukhin, O. M.; Zolotov, Yu. A.
V. I. Vernadskii Inst. Geochem. Anal. Chem. Moscow USSR SU
Zh. Anal. Khim.; (82) P 581-8; Vol 37; No 4; In Russ

09-82-0016 (0016)
QUANTITATIVE EVALUATION OF STERIC EFFECTS OF ALKYL GROUPS BY A LIQUID-CHROMATOGRAPHIC METHOD
Chumakov, Yu. I.; Kabulov, B. D.
Samark. Gos. Univ. Samarkand USSR SU
Zh. Fiz. Khim.; (82) P 1463-6; Vol 56; No 6; In Russ

CHAPTER 2.10
Void Volume Determination

--- 1982 ---

10-82-0001 (0001)
ADSORPTION OF 1-PENTANOL ON ALKYL-MODIFIED SILICA AND ITS EFFECT ON THE RETENTION MECHANISM IN REVERSED-PHASE LIQUID CHROMATOGRAPHY
Wahlund, Karl Gustav; Beijersten, Ingegerd
Dep. Anal. Pharm. Chem. Univ. Uppsala Uppsala Swed. SE S-751 23
Anal. Chem.; (82) P 128-32; Vol 54; No 1

10-82-0002 (0002)
COMPARISON OF METHODS USED FOR THE DETERMINATION OF VOID VOLUME IN REVERSED-PHASE LIQUID CHROMATOGRAPHY
Krstulovic, Ante M.; Colin, Henri; Guiochon, Georges
Lab. Chim. Anal. Phys. Ec. Polytech. Palaiseau Fr. FR 91128
Anal. Chem.; (82) P 2438-43; Vol 54; No 14

10-82-0003 (0003)
ADSORPTION FROM LIQUID MIXTURES AND LIQUID CHROMATOGRAPHY
Riedo, Francois; Kovats, Ervin S.
Lab. Chim. Tech. Ec. Polytech. Fed. Lausanne Lausanne Switz. CH CH-1015
J. Chromatogr.; (82) P 1-28; Vol 239

CHAPTER 2.11
Secondary Equilibria

2.11.1
Ion Pair (Theory)

--- 1982 ---

11-82-0001 (0001)
REVERSE-PHASE ION-PAIR CHROMATOGRAPHY OF POLYVALENT IONS USING OLIGONUCLEOTIDES AS MODEL SUBSTANCES
Jost, W.; Unger, K.; Schill, G.
Inst. Anorg. Chem. Anal. Chem. Johannes-Gutenberg Univ. Mainz Fed. Rep. Ger. DE 6500
Anal. Biochem.; (82) P 214-23; Vol 119; No 2

11-82-0002 (0002)
MECHANISTIC INTERPRETATIONS AND SIMULATIONS OF INDUCED PEAKS IN LIQUID CHROMATOGRAPHY
Stranahan, John J.; Deming, Stanley N.
Dep. Chem. Univ. Houston Houston USA US TX 77004
Anal. Chem.; (82) P 1540-6; Vol 54; No 9

11-82-0003 (0003)
THERMODYNAMIC MODEL FOR REVERSED-PHASE ION-PAIR LIQUID CHROMATOGRAPHY
Stranahan, John J.; Deming, Stanley N.
Dep. Chem. Univ. Houston Houston USA US TX 77004
Anal. Chem.; (82) P 2251-6; Vol 54; No 13

11-82-0004 (0004)
EFFECT OF IONIC STRENGTH ON RETENTION AND DETECTOR RESPONSE IN REVERSED-PHASE ION-PAIR LIQUID CHROMATOGRAPHY WITH ULTRAVIOLET-ABSORBING ION INTERACTION REAGENTS
Bidlingmeyer, Brian A.; Warren, F. Vincent, Jr.
Waters Assoc., Inc. Milford USA US MA 01757
Anal. Chem.; (82) P 2351-6; Vol 54; No 13

11-82-0005 (0005)
PEAK-SPLITTING IN REVERSED-PHASE, ION-PAIR HIGH-PERFORMANCE LIQUID CHROMATOGRAPHY OF SYMPATHOMIMETIC DRUGS AND ITS PROBABLE MECHANISM
Low, G. K. C.; Duffield, A. M.; Haddad, P. R.
Dep. Anal. Chem. Univ. New South Wales Kensington Australia AU 2033
Chromatographia; (82) P 289-96; Vol 15; No 5

11-82-0006 (0006)
FEMTOMOLAR ION-PAIR HIGH-PERFORMANCE LIQUID CHROMATOGRAPHIC METHOD FOR DETERMINING DNS-POLYAMINE DERIVATIVES OF RED BLOOD CELL EXTRACTS UTILIZING AN AUTOMATED POLYAMINE ANALYZER
Brown, Nesbitt D.; Strickler, M. Patricia; Whaun, June M.
Div. Biochem. Walter Reed Army Inst. Res. Washington USA US DC 20012
J. Chromatogr.; (82) P 101-8; Vol 245; No 1

11-82-0007 (0007)
NEW APPROACHES TO ON-LINE ION-PAIR EXTRACTION AND DERIVATIZATION FOR HIGH-PERFORMANCE LIQUID CHROMATOGRAPHY-MASS SPECTROMETRY
Vouros, Paul; Lankmayr, E. P.; Hayes, M. J.; Karger, B. L.; McGuire, John M.
Inst. Chem. Anal. Northeastern Univ. Boston USA US MA 02115
J. Chromatogr.; (82) P 175-88; Vol 251; No 2

11-82-0008 (0008)
REGULATION OF SELECTIVITY IN THE SEPARATION OF PAFENOLOL AND POTENTIAL IMPURITIES BY REVERSED-PHASE ION-PAIR CHROMATOGRAPHY
Jansson, Sven Olof; Johansson, Svante
Dep. Anal. Chem. AB Haessle Moelndal Swed. SE S-431 83
J. Chromatogr.; (82) P 41-50; Vol 242; No 1

11-82-0009 (0009)
INVESTIGATION OF THE EFFECT OF ADDED ORGANIC AMINE ON THE CHROMATOGRAPHY OF TRICYCLIC ANTIDEPRESSANT DRUGS USING REVERSED-PHASE CHROMATOGRAPHY ON OCTADECYLSILICA WITH SODIUM LAURYL SULFATE AS PAIRING ION
Hung, C. T.; Taylor, R. B.; Paterson, N.
Sch. Pharm. Robert Gordon's Inst. Technol. Aberdeen UK GB
J. Chromatogr.; (82) P 61-73; Vol 240; No 1

11-82-0010 (0010)
EFFECTS OF COUNTER IONS IN ION-PAIR LIQUID CHROMATOGRAPHY OF HYDROPHOBIC AMINES ON NONPOLAR BONDED PHASES
Jansson, S. O.
Dep. Anal. Chem. AB Haessle Molndal Swed. SE S-431 83
J. Liq. Chromatogr.; (82) P 677-91; Vol 5; No 4

11-82-0011 (0011)
ADSORPTION ISOTHERMS OF ION-PAIR FORMING REAGENTS USED IN REVERSED-PHASE CHROMATOGRAPHY
Bartha, Akos; Vigh, Gyula
Anal. Kem. Tansz. Veszpremi Vegyip. Egy; Veszprem Hung. HU
Magy. Kem. Lapja; (82) P 216-19; Vol 37; No 5; In Hung

11-82-0012 (0012)
ION-PAIR ADSORPTION THIN-LAYER CHROMATOGRAPHY OF QUATERNARY AMMONIUM IONS. PART 8. PARAMETERS OF AN EQUIDISTRIBUTION
Giebelmann, R.
Inst. Gerichtliche Med. Kriminalist. Bereiches Med.
Ernst-Moritz-Arndt-Univ. Greifswald Ger. Dem. Rep. DD DDR-2200
Pharmazie; (82) P 524; Vol 37; No 7; In Ger

11-82-0013 (0013)
ION-PAIR ADSORPTION THIN-LAYER CHROMATOGRAPHY OF QUATERNARY AMMONIUM IONS. PART 9. SELECTIVE COEFFICIENTS
Giebelmann, R.
Inst. Gerichtliche Med. Kriminalist. Bereiches Med.
Ernst-Moritz-Arndt-Univ. Greifswald Ger. Dem. Rep. DD DDR-2200
Pharmazie; (82) P 525; Vol 37; No 7; In Ger

2.11.2
Ion Pair (Applications)

--- 1982 ---

11-82-0001 (0001)
SEPARATION AND QUANTITATION OF (R)- AND (S)-PROPRANOLOL AS THEIR DIASTEREOMERIC DERIVATIVES IN HUMAN PLASMA BY REVERSED-PHASE ION-PAIR CHROMATOGRAPHY
Hermansson, Joergen
Dep. Drugs Natl. Board Health Welfare Uppsala Swed. SE S-751 25
Acta Pharm. Suec.; (82) P 11-24; Vol 19; No 1

11-82-0002 (0002)
SEPARATION OF BASIC, HYDROPHILIC PEPTIDES BY REVERSED-PHASE ION-PAIR CHROMATOGRAPHY. II. ANALYTICAL APPLICATIONS WITH PARTICULAR REFERENCE TO PHOSPHOSERINE PEPTIDES
Fransson, Bengt; Ragnarsson, Ulf; Zetterqvist, Oerjan
Inst. Biochem. Univ. Uppsala Uppsala Swed. SE S-751 23
Anal. Biochem.; (82) P 174-8; Vol 126; No 1

11-82-0003 (0003)
QUANTITATION OF GUANOSINE 5',3'-BISDIPHOSPHATE IN EXTRACTS FROM BACTERIAL CELLS BY ION-PAIR REVERSE-PHASE HIGH-PERFORMANCE LIQUID CHROMATOGRAPHY
Little, R.; Bremer, H.
Univ. Texas Richardson USA US TX 75080
Anal. Biochem.; (82) P 381-8; Vol 126; No 2

11-82-0004 (0004)
AN IMPROVED METHOD FOR THE DETERMINATION OF OLIGONUCLEOTIDE CHAIN LENGTH USING PHOSPHODIESTERASE HYDROLYSIS AND ION-PAIR HIGH-PERFORMANCE LIQUID CHROMATOGRAPHY
Crowther, Jonathan B.; Caronia, John P.; Hartwick, Richard A.
Dep. Chem. Rutgers, State Univ. New Brunswick USA US NJ 08854
Anal. Biochem.; (82) P 65-73; Vol 124; No 1

11-82-0005 (0005)
SEPARATION OF ALIPHATIC ALCOHOLS BY PAIRED ION LIQUID CHROMATOGRAPHY
Gnanasambandan, T.; Freiser, H.
Dep. Chem. Univ. Arizona Tucson USA US AZ 85721
Anal. Chem.; (82) P 1282-5; Vol 54; No 8

11-82-0006 (0006)
BIOGENIC AMINE RESOLUTION IN TISSUE EXTRACTS OF RAT BRAIN USING ION-PAIR HIGH-PERFORMANCE LIQUID CHROMATOGRAPHY WITH ELECTROCHEMICAL DETECTION
Pleece, S. A.; Redfern, P. H; Riley, C. M.; Tomlinson, E.
Sch. Pharm. Pharmacol. Univ. Bath Bath UK GB BA2 7AY
Analyst (London); (82) P 755-60; Vol 107; No 1276

11-82-0007 (0007)
EVALUATION OF A SIMPLE COLORIMETRIC ASSAY FOR SERUM ANGIOTENSIN-CONVERTING ENZYME: COMPARISON WITH A NEW ION-PAIR LIQUID CHROMATOGRAPHY-ASSISTED ASSAY
Kwarts, Eddy; Beukenveld, Gerard; Gazendam, Jurjen
Cent. Lab. Clin. Chem. Univ. Hosp. Groningen Neth. NL
Ann. Clin. Biochem.; (82) P 227-32; Vol 19; No 4

11-82-0008 (0008)
PREPARATIVE ION PAIR HPLC ISOLATION AND IDENTIFICATION OF THE MAIN CONSTITUENTS OF QUINOLINE YELLOW (C.I. 47005)
Masiala-Tsobo, C.; Gelbcke, M.; Dumont, L.; Parmentier, F.; Grimee, R.
Lab. Chim. Pharm. Org. Bromatol. Univ. Libre Bruxelles Brussels Belg. BE B-1050
Bull. Soc. Chim. Belg.; (82) P 227-35; Vol 91; No 3; In Fr

11-82-0009 (0009)
REVERSED-PHASE ION-PAIR PARTITION CHROMATOGRAPHY OF HALOGENATED ANION COMPLEXES OF PLATINUM METALS
Hayakawa, Kazuichi; Nishijima, Miwako; Miyazaki, Motoichi
Fac. Pharm. Sci. Kanazawa Univ. Kanazawa Japan JP
Bunseki Kagaku; (82) P 160-1; Vol 31; No 3; In Japan

11-82-0010 (0010)
SIMULTANEOUS DETERMINATION OF BERBERINE, PALMATINE AND COPTISINE IN CRUDE DRUGS AND ORIENTAL PHARMACEUTICAL PREPARATIONS BY ION-PAIR HIGH-PERFORMANCE LIQUID CHROMATOGRAPHY
Misaki, Tetsuo; Sagara, Kazuhiko; Ojima, Mitsuharu; Kakizawa, Sadao; Oshima, Toshiyuki; Yoshizawa, Hiroshi
Res. Lab. daisho Pharm. Co., Ltd. Omiya Japan JP 330
Chem. Pharm. Bull.; (82) P 354-7; Vol 30; No 1

11-82-0011 (0011)
THE DETERMINATION OF SOME ADDITIVES BY ION-PAIR LIQUID CHROMATOGRAPHY
Cornet, Joost P.; Van de Haar, Gerrit
Food Insp. Serv. Haarlem Neth. NL
Chromatogr. Sci.; (82) P 343-61; Vol 20; No Biol./Biomed. Appl. Liq. Chromatogr. 4

11-82-0012 (0012)
DETERMINATION OF URINARY OXALATE BY REVERSED-PHASE ION-PAIR "HIGH-PERFORMANCE" LIQUID CHROMATOGRAPHY
Larsson, Lasse; Libert, Bo; Asperud, Monica
Dep. Clin. Chem. Univ. Linkoeping Linkoeping Swed. SE S-581 85
Clin. Chem. (Winston-Salem, N. C.); (82) P 2272-4; Vol 28; No 11

11-82-0013 (0013)
DETERMINATION OF NICOTINIC ACID AND NICOTINAMIDE IN FOODS BY ION-PAIR HIGH-PERFORMANCE LIQUID CHROMATOGRAPHY
Kitayama, Yoshimi; Inoue, Masashige; Tamase, Kikuo; Imou, Masako; Hasuike, Akikazu; Sasaki, Michiko; Tanigawa, Kaoru
Nara Pref. Inst. Pub. Health Nara Japan JP 630
Eiyo to Shokuryo; (82) P 121-4; Vol 35; No 2; In Japan

11-82-0014 (0014)
ION-PAIR LIQUID CHROMATOGRAPHIC AND FLUOROMETRIC DETERMINATIOW OF AMINO ACIDS IN THE GUINEA PIG BRAIN
Gaetani, E.; Laureri, C. F.; Vitto, M.; Bordi, F.
Ist. Chim. Farm. Tossicol. Univ. Studi Parma Parma Italy IT
Farmaco, Ed. Prat.; (82) P 253-7; Vol 37; No 8; In Ital

11-82-0015 (0015)
ASCORBIC ACID IN MEAT PRODUCTS. ANALYSIS BY REVERSED-PHASE ION-PAIR HIGH-PRESSURE LIQUID CHROMATOGRAPHY
Parolari, Giovanni
Stazione Sper. Ind. Conserve Alimentari Parma Italy IT
Ind. Conserve; (82) P 19-22; Vol 57; No 1; In Ital

11-82-0016 (0016)
HIGH-PERFORMANCE LIQUID CHROMATOGRAPHY OF SOME COENZYME M (2-MERCAPTOETHANE-SULFONIC ACID) DERIVATIVES BY ION PAIRING ON REVERSED-PHASE COLUMNS
Apostolides, Z.; Vermeulen, N. M. J.; Potgieter, D. J. J.
Dep. Biochem. Univ. Pretoria Pretoria S. Afr. ZA 0002
J Chromatogr.; (82) P 304-7; Vol 246; No 2

11-82-0017 (0017)
EVALUATION OF THIN-LAYER, PAPER, AND HIGH-PERFORMANCE LIQUID CHROMATOGRAPHY FOR IDENTIFICATION OF DYES EXTRACTED AS ION-PAIRS WITH TRI-n-OCTYLAMINE
Puttemans, Marc L.; Dryon, Louis; Massart, Desire L.
Farm. Inst. Vrije Univ. Brussel Brussels Belg. BE B-1090
J. Assoc. Off. Anal. Chem.; (82) P 730-6; Vol 65; No 3

11-82-0018 (0018)
DETERMINATION OF ADENINE NUCLEOTIDES AND INOSINE IN HUMAN MYOCARDIUM BY ION-PAIR REVERSED-PHASE HIGH-PERFORMANCE LIQUID CHROMATOGRAPHY
Ingebretsen, Ole Christian; Bakken, Anne M.; Segadal, Leidulf; Farstad, Mikael
Lab. Clin. Biochem. Univ. Bergen Haukeland Sykehus Norway NO N-5016
J. Chromatogr.; (82) P 119-26; Vol 242; No 1

11-82-0019 (0019)
SIMULTANEOUS DETERMINATION OF NOREPINEPHRINE, DOPAMINE AND SEROTONIN IN RAT BRAIN REGIONS BY ION-PAIR LIQUID CHROMATOGRAPHY ON OCTYL SILANE COLUMNS AND AMPEROMETRIC DETECTION
Warsh, Jerry J.; Chiu, Andrew; Godse, Damodar D.
Dep. Biochem. Psychiatry Clarke Inst. Psychiatry Toronto Can. CA ON M5T 1R8
J. Chromatogr.; (82) P 131-41; Vol 228

11-82-0020 (0020)
SEPARATION OF BASIC, HYDROPHILIC PEPTIDES BY REVERSED-PHASE ION-PAIR CHROMATOGRAPHY. I. ANALYTICAL APPLICATIONS WITH PARTICULAR REFERENCE TO A CLASS OF SERINE PEPTIDE SUBSTRATES OF CYCLIC AMP-STIMULATED PROTEIN KINASE
Fransson, B.; Ragnarsson, U.; Zetterqvist, O.
Biomed. Cent. Univ. Uppsala Uppsala Swed. SE S-751 23
J. Chromatogr.; (82) P 165-71; Vol 240; No 1

11-82-0021 (0021)
DETERMINATION OF MIMOSINE AND 3-HYDROXY-4(1H)-PYRIDONE IN LEUCAENA, AVIAN EXCRETA AND SERUM USING REVERSED-PHASE ION-PAIR HIGH-PERFORMANCE LIQUID CHROMATOGRAPHY
Acamovic, T.; D'Mello, J. P. F.; Fraser, K. W.
Dep. Agric. Biochem. Edinburgh Sch. Agric. Edinburgh UK GB EH9 3JG
J. Chromatogr.; (82) P 169-79; Vol 236; No 1

11-82-0022 (0022)
SEPARATION OF VITAMIN B6 COMPOUNDS BY PAIRED-ION HIGH-PERFORMANCE LIQUID CHROMATOGRAPHY
Tryfiates, George P.; Sattsangi, Saroj
Sch. Med. West Virginia Univ. Morgantown USA US WV 26506
J. Chromatogr.; (82) P 181-6; Vol 227; No 1

11-82-0023 (0023)
ANALYSIS OF PIPERAZINE ESTRONE SULFATE IN TABLETS BY ION-PAIR HIGH-PERFORMANCE LIQUID CHROMATOGRAPHY
Carignan, G.; Lodge, B. A.; Skakum, W.
Health Protect. Branch Bur. Drug Res. Ottawa Can. CA ON K1A 0L2
J. Chromatogr.; (82) P 240-3; Vol 234; No 1

11-82-0024 (0024)
USE OF ION-PAIRED, REVERSED-PHASE THIN-LAYER CHROMATOGRAPHY FOR THE ANALYSIS OF PEPTIDES. A SIMPLE PROCEDURE FOR THE MONITORING OF PREPARATIVE REVERSED-PHASE HIGH-PERFORMANCE LIQUID CHROMATOGRAPHY
Poll, D. J.; Knighton, D. R.; Harding, D. R. K.; Hancock, W. S.
Dep. Chem. Biochem. Biophys. Massey Univ. Palmerston North N. Z. NZ
J. Chromatogr.; (82) P 244-8; Vol 236; No 1

11-82-0025 (0025)
REVERSED-PHASE ION-PAIR HIGH-PERFORMANCE LIQUID CHROMATOGRAPHIC ASSAY OF 5-FLUOROURACIL, 5'-DEOXY-5-FLUOROURIDINE, THEIR NUCLEOSIDES, MONO-, DI-, AND TRIPHOSPHATE NUCLEOTIDES WITH A MIXTURE OF QUATERNARY AMMONIUM IONS
Au, J. L. S.; Wientjes, M. G.; Luccioni, C. M.; Rustum, Y. M.
Dep. Exp. Therapeut. Roswell Park Mem. Inst. Buffalo USA US NY 14263
J. Chromatogr.; (82) P 245-56; Vol 228

11-82-0026 (0026)
DETERMINATION OF CHLOROQUINE AND ITS DE-ETHYLATED METABOLITES IN HUMAN PLASMA BY ION-PAIR HIGH-PERFORMANCE LIQUID CHROMATOGRAPHY
Brown, Nesbitt D.; Poon, Bing T.; Chulay, Jeffrey D.
Div. Biochem. Walter Reed Army Inst. Res. Washington USA US DC 20012
J. Chromatogr.; (82) P 248-54; Vol 229; No 1

11-82-0027 (0027)
SIMULTANEOUS DETERMINATION OF ASCORBIC ACID AND DEHYDROASCORBIC ACID BY REVERSED-PHASE ION-PAIR HIGH-PERFORMANCE LIQUID CHROMATOGRAPHY WITH PRECOLUMN DERIVATIZATION
Keating, Ross W.; Haddad, Paul R.
Dep. Anal. Chem. Univ. New South Wales Kensington Australia AU 2033
J. Chromatogr.; (82) P 249-55; Vol 245; No 2

11-82-0028 (0028)
RAPID ANALYSIS OF 3-METHYLHISTIDINE IN URINE, PLASMA, MUSCLE AND AMNIOTIC FLUID WITH A SINGLE HIGH-PERFORMANCE LIQUID CHROMATOGRAPHIC SYSTEM BUT WITH DIFFERENT ION-PAIRING REAGENTS
Hancock, W. S.; Harding, D. R. K.; Friedman, Zvi
Dep. Chem. Biochem. Biophys. Massey Univ. Palmerston North N. Z. NZ
J. Chromatogr.; (82) P 273-8; Vol 228

11-82-0029 (0029)
USE OF ION-PAIR, REVERSED-PHASE, HIGH-PERFORMANCE LIQUID CHROMATOGRAPHY FOR THE ANALYSIS OF CYTOKININS
Walker, M. A.; Dumbroff, E. B.
Dep. Biol. Univ. Waterloo Waterloo Can. CA ON N2L 3G1
J. Chromatogr.; (82) P 316-23; Vol 237; No 2

11-82-0030 (0030)
ALTERNATING-CURRENT POLAROGRAPHIC DETECTION FOR REVERSED-PHASE ION-PAIR HIGH-PERFORMANCE LIQUID CHROMATOGRAPHY OF SOME BENZOIC ACIDS
Kutner, Wlodzimierz
Inst. Phys. Chem. Pol. Acad. Sci. Warsaw Pol. PL 01-224
J. Chromatogr.; (82) P 342-6; Vol 247; No 2

11-82-0031 (0031)
REVERSED-PHASE ION-PAIR LIQUID CHROMATOGRAPHIC PROCEDURE FOR THE SIMULTANEOUS ANALYSIS OF S-ADENOSYLMETHIONINE, ITS METABOLITES AND THE NATURAL POLYAMINES
Wagner, Joseph; Danzin, Charles; Mamont, Pierre
Cent. Rech. Merrell Int. Strasbourg Fr. FR 67084
J. Chromatogr.; (82) P 349-68; Vol 227; No 2

11-82-0032 (0032)
ION-PAIR HIGH-PERFORMANCE LIQUID CHROMATOGRAPHIC ASSAY OF 4-AMINOPYRIDINE IN SERUM
Shinohara, Y.; Miller, R. D.; Castagnoli, N., Jr.
Med. Cent. Univ. California San Francisco USA US CA 94143
J. Chromatogr.; (82) P 363-72; Vol 230; No 2

11-82-0033 (0033)
ANALYSIS OF BUFLOMEDIL IN MOUSE, RAT AND RABBIT PLASMA BY REVERSED-PHASE ION-PAIR HIGH-PERFORMANCE LIQUID CHROMATOGRAPHY
Thomas, Elizabeth W.
Dep. Drug Metab. Abbott Lab. North Chicago USA US IL 60064
J. Chromatogr.; (82) P 387-91; Vol 228

11-82-0034 (0034)
REVERSED-PHASE ION-PAIR HIGH-PERFORMANCE LIQUID CHROMATOGRAPHY OF MERCAPTO-ACETATE AND N-ACETYLCYSTEINE AFTER DERIVATIZATION WITH N-(1-PYRENE)MALEIMIDE AND N-(7-DIMETHYLAMINO-4-METHYL-3-COUMARINYL)- MALEIMIDE
Kaagedal, Bertil; Kaellberg, Magnus
Dep. Clin. Chem. Univ. Linkoeping Linkoeping Swed. SE S-581 85
J. Chromatogr.; (82) P 409-15; Vol 229; No 2

11-82-0035 (0035)
ROUTINE DETERMINATION OF PLASMA CATECHOLAMINES USING REVERSED-PHASE, ION-PAIR HIGH-PERFORMANCE LIQUID CHROMATOGRAPHY WITH ELECTROCHEMICAL DETECTION
Davies, Cledwyn L.; Molyneux, Stephen G.
Radcliffe Infirm. Univ. Oxford Oxford UK GB OX2 6HE
J. Chromatogr.; (82) P 41-51; Vol 231; No 1

11-82-0036 (0036)
DETERMINATION OF MORPHINE, MORPHINE-3-GLUCURONIDE AND (TENTATIVELY) MORPHINE-6-GLUCURONIDE IN PLASMA AND URINE USING ION-PAIR HIGH-PERFORMANCE LIQUID CHROMATOGRAPHY
Svensson, Jan Olof; Rane, Anders; Saewe, Juliette; Sjoeqvist, Folke
Karolinska Inst. Huddinge Univ. Hosp. Huddinge Swed. SE S-141 86
J. Chromatogr.; (82) P 427-32; Vol 230; No 2

11-82-0037 (0037)
PAIRED-ION HIGH-PERFORMANCE LIQUID-CHROMATOGRAPHIC DETERMINATION OF THE STABILITY OF NOVOBIOCIN IN MASTITIS PRODUCTS STERILIZED BY COBALT-60 IRRADIATION
Hornish, R. E.
Upjohn Co. Kalamazoo USA US MI 49001
J. Chromatogr.; (82) P 481-8; Vol 236; No 2

11-82-0038 (0038)
RESOLUTION OF RNA BY PAIRED-ION REVERSED-PHASE HIGH-PERFORMANCE LIQUID CHROMATOGRAPHY
Nguyen Phi Nga; Bradley, Janice L.; McGuire, Peter M.
Dep. Biochem. Mol. Biol. Univ. Florida Gainesville USA US FL 32610
J. Chromatogr.; (82) P 508-12; Vol 236; No 2

11-82-0039 (0039)
HIGH-PERFORMANCE LIQUID CHROMATOGRAPHIC ANALYSIS OF HEROIN BY REVERSE-PHASE ION-PAIR CHROMATOGRAPHY
Lurie, I. S.; Sottolano, S. M.; Blasof, Stanley
Special Test. Res. Lab. Drug Enforcement Adm. McLean USA US VA 22102
J. Forensic Sci.; (82) P 519-26; Vol 27; No 3

11-82-0040 (0040)
ION-PAIR HIGH-PRESSURE LIQUID CHROMATOGRAPHY OF CIS-TRANS ISOMERS OF RETINOIC ACID IN TISSUES OF VITAMIN A-SUFFICIENT RATS
Sundaresan, P. Ramnathan; Bhat, Pangala V.
Div. Toxicol. FDA Washington USA US DC 20204
J. Lipid Res.; (82) P 448-55; Vol 23; No 3

11-82-0041 (0041)
DETERMINATION OF METHACHOLINE CHLORIDE BY ION-PAIR HIGH-PRESSURE LIQUID CHROMATOGRAPHY
Woodman, Tracy F.; Johnson, Brian; Marwaha, Raj K.
Med. Sch. Univ. Massachusetts Worcester USA US MA 01605
J. Liq. Chromatogr.; (82) P 1341-8; Vol 5; No 7

11-82-0042 (0042)
ASSAY OF n-PROPYLAJMALINE IN BLOOD PLASMA BY ION-PAIR LIQUID-LIQUID CHROMATOGRAPHY
Grundevik, Inger; Persson, B. A.
AB Haessle Moelndal Swed. SE S-431 83
J. Liq. Chromatogr.; (82) P 141-50; Vol 5; No 1

11-82-0043 (0043)
REVERSED-PHASE ION-PAIRING LIQUID-CHROMATOGRAPHIC SEPARATION AND FLUOROMETRIC DETECTION OF POLYAMINES
Simpson, Richard C.; Mohammed, Hussain Y.; Veening, Hans
Dep. Chem. Bucknell Univ. Lewisburg USA US PA 17837
J. Liq. Chromatogr.; (82) P 245-64; Vol 5; No 2

11-82-0044 (0044)
THE PREPARATIVE SCALE REVERSE-PHASE HPLC SEPARATION OF EPIMERIC ALKALOIDS USING CAMPHORSULFONIC ACID AS AN ION PAIRING REAGENT
Salva, Paul S.; Hite, Gilbert J.; Henkel, James G.
Sch. Pharmacy Univ. Connecticut Storrs USA US CT 06268
J. Liq. Chromatogr.; (82) P 305-12; Vol 5; No 2

11-82-0045 (0045)
METAL CATION/ANION SPECIATION VIA PAIRED-ION, REVERSED PHASE HPLC WITH REFRACTIVE INDEX AND/OR INDUCTIVELY COUPLED PLASMA EMISSION SPECTROSCOPIC DETECTION METHODS
Bushee, D.; Krull, I. S.; Savage, R. N.; Smith, S. B., Jr.
Inst. Chem. Anal. Northeast. Univ. Boston USA US MA 02115
J. Liq. Chromatogr.; (82) P 463-78; Vol 5; No 3

11-82-0046 (0046)
DETERMINATION OF AMINE INGREDIENTS IN COUGH-COLD LIQUIDS BY REVERSED-PHASE ION-PAIR HIGH-PERFORMANCE LIQUID CHROMATOGRAPHY
Halstead, G. W.
Control Anal. Res. Dev. Upjohn Co. Kalamazoo USA US MI 49001
J. Pharm. Sci.; (82) P 1108-12; Vol 71; No 10

11-82-0047 (0047)
REVERSED-PHASE ION-PAIR CHROMATOGRAPHY OF TETRACYCLINE, TETRACYCLINE ANALOGS, AND THEIR POTENTIAL IMPURITIES
Hermansson, Joergen; Andersson, Mats
Div. Pharm. Natl. Board Health Welfare Uppsala Swed.
J. Pharm. Sci.; (82) P 222-9; Vol 71; No 2

11-82-0048 (0048)
SEPARATION OF PENICILLIN AND ITS MAJOR DEGRADATION PRODUCTS BY ION-PAIR REVERSED-PHASE HIGH-PRESSURE LIQUID CHROMATOGRAPHY
Ghebre-Sellassie, Isaac; Hem, Stanley L.; Knevel, Adelbert M.
Sch. Pharm. Pharm. Sci. Purdue Univ. West Lafayette USA US PIN 47907
J. Pharm. Sci.; (82) P 351-3; Vol 71; No 3

11-82-0049 (0049)
ION-PAIR HIGH-PERFORMANCE LIQUID CHROMATOGRAPHY OF TERBUTALINE AND CATECHOLAMINES WITH AMINOPHYLLINE IN INTRAVENOUS SOLUTIONS

Williams, David A.; Fuvg, Esther Y. Y.; Newton, David W.
Dep. Chem. Massachusetts Coll. Pharm. Allied Health Sci. Boston USA US MA 02115
J. Pharm. Sci.; (82) P 956-8; Vol 71; No 8

11-82-0050 (0050)
TOXICOLOGICAL ANALYSIS OF PHENYLETHYLAMINES BY HIGH-PERFORMANCE REVERSED-PHASE ION-PAIR PARTITION THIN-LAYER AND LIQUID CHROMATOGRAPHIES. COMMENT
Donike, Manfred
Inst. Biochem., Dtsch. Sporthochsch. Koeln, 5000 Cologne, 41 Fed. Rep. Ger.
J. Toxicol., Clin. Toxicol. (82) P 225-7; Vol 19; No 2

11-82-0051 (0051)
TOXICOLOGICAL ANALYSIS OF PHENYLETHYLAMINES BY HIGH-PERFORMANCE REVERSED-PHASE ION-PAIR PARTITION THIN-LAYER AND LIQUID CHROMATOGRAPHIES. REPLY TO COMMENTS
Verbiesde, N.
Inst. Pharm. Univ. Brussels Brussels Belg. BE
J. Toxicol., Clin. Toxicol.; (82) P 231-2; Vol 19; No 2

11-82-0052 (0052)
NEW CARDIOACTIVE DRUGS. II. DETECTION AND ISOLATION OF CARDIOTONIC AMINES WITH ION-PAIR HPLC
Wagner, H.; Grevel, J.
Inst. Pharm. Biol. Univ. Muenchen Munich Fed. Rep. Ger. DE
Planta Med.; (82) P 36-40; Vol 44; No 1; In Ger

2.11.3
Complex Formation

--- 1982 ---

11-82-0001 (0001)
LIQUID CHROMATOGRAPHIC SEPARATION OF AROMATIC HYDROCARBONS WITH CHEMICALLY BONDED (2,4-DINITROANILINOPROPYL)SILICA
Grizzle, Patrick L.; Thomson, Jane S.
Bartlesville Energy Technol. Cent. U. S. Dep. Energy Bartlesville USA US OK 74003
Anal. Chem.; (82) P 1071-8; Vol 54; No 7

11-82-0002 (0002)
LIQUID CHROMATOGRAPHIC SEPARATIONS OF DIHYDROXYBENZENES WITH SILICA, DIAMINE, AND COBALT COMPLEX-BONDED PHASE COLUMNS
Chang, C. Allen; Tu, Cheng Fan
Dep. Chem. Univ. Texas El Paso USA US TX 79968
Anal. Chem.; (82) P 1179-82; Vol 54; No 7

11-82-0003 (0003)
SELECTIVITY IN LIQUID CHROMATOGRAPHY WITH MICELLAR MOBILE PHASES
Yarmchuk, Paul; Weinberger, Robert; Hirsch, R. F.; Love, L. J. Cline
Dep. Chem. Seton Hall Univ. South Orange USA US NJ 070W9
Anal. Chem.; (82) P 2233-8; Vol 54; No 13

11-82-0004 (0004)
.PI.-.PI. CHARGE TRANSFER INTERACTIONS IN HIGH-PERFORMANCE LIQUID-SOLID CHROMATOGRAPHY
Burger, J. J.; Tomlinson, E.
Subfac. Pharm. Univ. Amsterdam Amsterdam Neth. NL 1018 TV
Anal. Proc. (London); (82) P 126-8; Vol 19; No 3

11-82-0005 (0005)
DETERMINATION OF METAL IONS BY LIQUID CHROMATOGRAPHY INCORPORATING DITHIOCARBAMATES IN THE ELUENT
Smith, Roger M.; Yankey, Lawrence E.
Dep. Chem. Univ. Technol. Loughborough UK GB LE11 3TU
Analyst (London); (82) P 744-8; Vol 107; No 1276

11-82-0006 (0006)
DONOR-ACCEPTOR COMPLEX CHROMATOGRAPHY. REVERSED-PHASE CHROMATOGRAPHY ON CHEMICALLY-BONDED ACCEPTOR-LIGANDS
Hemetsberger, H.; Ricken, H.
Abt. Chem., Org. Chem. 2 Ruhr-Univ. Bochum Bochum Fed. Rep. Ger. DE D-4630
Chromatographia; (82) P 236-40; Vol 15; No 4

11-82-0007 (0007)
CHROMATOGRAPHY OF METAL CHELATES. IV. HPLC OF THE PLATINUM METALS WITH 1-HYDROXY-2-PYRIDINETHIONE
Steinbrech, B.; Schneeweis, G.; Koenig, K. H.
Inst. Anorg. Chem. Univ. Frankfurt/Main Frankfurt/Main Fed. Rep. Ger. DE D-6000
Fresenius' Z. Anal. Chem.; (82) P 499-502; Vol 311; No 5; In Ger

11-82-0008 (0008)
RESOLUTION OF CHIRAL OLEFINIC HYDROCARBONS AND SULFOXIDES BY HIGH-PERFORMANCE LIQUID CHROMATOGRAPHY VIA DIASTEREOMERIC PLATINUM COMPLEXES
Goldman, M.; Kustanovich, Z.; Weinstein, S.; Tishbee, A.; Gil-Av, E.
Dep. Org. Chem. Weizmann Inst. Sci. Rehovot Israel IL
J. Am. Chem. Soc.; (82) P 1093-5; Vol 104; No 4

11-82-0009 (0009)
HIGH-PERFORMANCE LIQUID CHROMATOGRAPHIC SEPARATION OF AMINO SUGARS AND PEPTIDES WITH METAL ION-MODIFIELD MOBILE PHASES
Dua, Virendra K.; Bush, C.Allen
Dep.Chem., Illinois Inst. Technol., Chicago USA, IL 60 616
J. Chromatogr. (82) P 128-33; Vol 244; No 1

11-82-0010 (0010)
SELECTIVE GROUP SEPARATION OF PRIMARY AND SECONDARY BIOGENIC AMINES BY REVERSED-PHASE HIGH-PERFORMANCE LIQUID CHROMATOGRAPHY WITH 18-CROWN-6 IN THE MOBILE PHASE
Wiechmann, Manfred
Univ. Bayreuth Bayreuth Fed. Rep. Ger. DE D-8580
J. Chromatogr.; (82) P 129-37; Vol 235; No 1; In Ger

11-82-0011 (0011)
HIGH-PERFORMANCE LIQUID CHARGE-TRANSFER CHROMATOGRAPHY
Small, David A. P.; Atkinson, Tony; Lowe, Christopher R.
PHLS Cent. Appl. Microbiol. Res. Porton Down/Nr. Salisbury/Wiltshire UK GB SP4 0JG
J. Chromatogr.; (82) P 271-9; Vol 248; No 2

11-82-0012 (0012)
RESOLUTION OF OPTICAL ISOMERS OF DNS-AMINO ACIDS BY HIGH-PERFORMANCE LIQUID CHROMATOGRAPHY WITH L-HISTIDINE AND ITS DERIVATIVES IN THE MOBILE PHASE
Lam, Stanley; Karmen, Arthur
Albert Einstein Coll Med. Bronx USA US NY 10461
J. Chromatogr.; (82) P 451-62; Vol 239

11-82-0013 (0013)
OPTICAL RESOLUTION OF DL-PROLINE BY REVERSED-PHASE HIGH-PERFORMANCE LIQUID CHROMATOGRAPHY USING N-(p-TOLUENESULFONYL)-L-PHENYLALANINE-COPPER(II) AS A CHIRAL ADDITIVE
Nimura, Noriyuki; Toyama, Atsuko; Kinoshita, Toshio
Sch. Pharm. Sci. Kitasato Univ. Tokyo Japan JP 108
J. Chromatogr.; (82) P 482-4; Vol 234; No 2

11-82-0014 (0014)
LIQUID CHROMATOGRAPHY WITH CROWN ETHER-CONTAINING MOBILE PHASES. II. RETENTION BEHAVIOR OF .BETA.-LACTAM ANTIBIOTICS IN REVERSED-PHASE HIGH-PERFORMANCE LIQUID CHROMATOGRAPHY
Nakagawa, Terumichi; Shibukawa, Akimasa; Uno, Toyozo
Fac. Pharm. Sci. Kyoto Univ. Kyoto Japan JP 606
J. Chromatogr.; (82) P 695-706; Vol 239

11-82-0015 (0015)
SEPARATION OF BACTERIAL MENAQUINONES BY HPLC USING REVERSE PHASE (RP18) AND A SILVER LOADED ION EXCHANGER AS STATIONARY PHASES
Kroppenstedt, Reiner M.
Dtsch. Sammlung von Mikroorg. Ges. Biotechnol. Forsch. m.b.H. Braunschweig Darmstadt Fed. Rep. Ger. DE D-6100
J. Liq. Chromatogr.; (82) P 2359-67; Vol 5; No 12

11-82-0016 (0016)
ARGENTATION-HIGH PRESSURE LIQUID CHROMATOGRAPHY OF PROSTAGLANDINS AND MONOHYDROXYEICOSENOIC ACIDS
Powell, William S.
R. Victoria Hosp. McGill Univ. Montreal Can. CA PQ H3A 1A1
Methods Enzymol.; (82) P 530-43; Vol 86; No Prostaglandins Arachidonate Metab.

2.11.4
Chiral Separations

--- 1982 ---

11-82-0001 (0001)
SEPARATION AND QUANTITATION OF (R)- AND (S)-PROPRANOLOL AS THEIR DIASTEREOMERIC DERIVATIVES IN HUMAN PLASMA BY REVERSED-PHASE ION-PAIR CHROMATOGRAPHY
Hermansson, Joergen
Dep. Drugs Natl. Board Health Welfare Uppsala Swed. SE S-751 25
Acta Pharm. Suec.; (82) P 11-24; Vol 19; No 1

11-82-0002 (0002)
RESOLUTION OF OPTICAL ISOMERS BY CHIRAL HIGH-PERFORMANCE LIQUID CHROMATOGRAPHY: SEPARATION OF DIHYDRODIOLS AND TETRAHYDRODIOLS OF BENZO(A)PYRENE AND BENZ(A)ANTHRACENE
Weems, Henri B.; Yavg, Shen K.
Sch. Med. Uniformed Serv. Univ. of the Health Sci. Bethesda USA US MD 20814
Anal. Biochem.; (82) P 156-61; Vol 125; No 1

11-82-0003 (0003)
THE ENANTIOMERIC ANALYSIS OF A MIXTURE OF ALL COMMON PROTEIN AMINO ACIDS BY HIGH-PERFORMANCE LIQUID CHROMATOGRAPHY USING A NEW CHIRAL MOBILE PHASE
Weinstein, S.; Engel, M. H.; Hare, P. E.
Dep. Org. Chem. Weizmann Inst. Sci. Rehovot Israel IL 76100
Anal. Biochem.; (82) P 370-7; Vol 121; No 2

11-82-0004 (0004)
SPECTROPHOTOMETRIC AND POLARIMETRIC DETECTORS IN LIQUID CHROMATOGRAPHY FOR THE DETERMINATION OF ENANTIOMER RATIOS IN COMPLEX MIXTURES
Boehme, W.; Wagner, G. Oehme, U.; Priesnitz, U.
Bodenseewerk Perkin-Elmer Ueberlingen Fed. Rep. Ger. DE
Anal. Chem.; (82) P 709-11; Vol 54; No 4

11-82-0005 (0005)
SEPARATION OF D- AND L-AMINO ACIDS BY HPLC WITH COPPER COMPLEXES OF n,n-DIALKYL-.ALPHA.-AMINO ACIDS AS NEW CHIRAL ADDITIVES
Weinstein, Shulamith
Dep. Org. Chem. Weizmann Inst. Sci. Rehovot Israel IL 76 100
Angew. Chem.; (82) P 221-2; Vol 94; No 3; In Ger

11-82-0006 (0006)
A USEFUL AND CONVENIENTLY ACCESSIBLE CHIRAL STATIONARY PHASE FOR THE LIQUID CHROMATOGRAPHIC SEPARATION OF ENANTIOMERS
Pirkle, William H.; Finn, John M.; Hamper, Bruce C.; Schreiner, James; Pribish, James R.
Sch. Chem. Sci. Univ. Illinois Urbana USA US IL 61801
ACS Symp. Ser.; (82) P 245-60; Vol 185; No Asymmetric React. Processes Chem.

11-82-0007 (0007)
LIQUID CHROMATOGRAPHIC RESOLUTION OF ENANTIOMERIC .ALPHA.-AMINO ACID DERIVATIVES EMPLOYING A CHIRAL DIAMIDE PHASE
Hara, Shoji; Dobashi, Akira; Eguchi, Masakatzu
Tokyo Coll. Pharm. Hachioji Japan JP 192-03
ACS Symp. Ser.; (82) P 266-7; Vol 185; No Asymmetric React. Processes Chem.

11-82-0008 (0008)
LIQUID CHROMATOGRAPHY ON TRIACETYLCELLULOSE. 4. DETERMINATION OF ENANTIOMERIC PURITY IN SPITE OF INCOMPLETE CHROMATOGRAPHIC SEPARATION
Mannschreck, Albrecht; Eiglsperger, Alfred; Stuehler, Georgine
Inst. Org. Chem. Univ. Regensburg Regensburg Fed. Rep. Ger. DE D-8400
Chem. Ber.; (82) P 1568-75; Vol 115; No 4

11-82-0009 (0009)
SEPARATION OF RACEMIC ALKYLAMINO ACIDS WITH HPLC USING A CHIRAL MOBILE PHASE
Gundlach, G.; Sattler, E. L.; Wagenbach, U.
Zent. Biochem. Univ. Giessen Giessen Fed. Rep. Ger. DE D-6300
Fresenius' Z. Anal. Chem.; (82) P 684-6; Vol 311; No 7; In Ger

11-82-0010 (0010)
RESOLUTION OF CHIRAL OLEFINIC HYDROCARBONS AND SULFOXIDES BY HIGH-PERFORMANCE LIQUID CHROMATOGRAPHY VIA DIASTEREOMERIC PLATINUM COMPLEXES
Goldman, M.; Kustanovich, Z.; Weinstein, S.; Tishbee, A.; Gil-Av, E.
Dep. Org. Chem. Weizmann Inst. Sci. Rehovot Israel IL
J. Am. Chem. Soc.; (82) P 1093-5; Vol 104; No 4

11-82-0011 (0011)
DETERMINATION OF (R)- AND (S)-ALPRENOLOL AND (R)- AND (S)-METOPROLOL AS THEIR DIASTEREOMERIC DERIVATIVES IN HUMAN PLASMA BY REVERSED-PHASE LIQUID CHROMATOGRAPHY
Hermansson, Joergen; Von Bahr, Christer
Dep. Drugs Natl. Board Health and Welfare Uppsala Swed. SE S-751 25
J. Chromatogr.; (82) P 113-27; Vol 227; No 1

11-82-0012 (0012)
THEORY FOR MOBILE PHASE EFFECTS IN SEPARATIONS OF ISOMERS BY LIQUID-SOLID CHROMATOGRAPHY. APPLICATION TO THE RELATIVE RETENTION OF CERTAIN DIASTEREOMERS ON SILICA
Snyder, L. R.
Tech. Instrum. Corp. Tarrytown USA US NY 10591
J. Chromatogr.; (82) P 165-76; Vol 245; No 2

11-82-0013 (0013)
.BETA.-CYCLODEXTRIN AS A CHIRAL COMPONENT OF THE MOBILE PHASE FOR SEPARATION OF MANDELIC ACID INTO ENANTIOMERS IN REVERSED-PHASE SYSTEMS OF HIGH-PERFORMANCE LIQUID CHROMATOGRAPHY
Debowski, Janusz; Sybilska, Danuta; Jurczak, Janusz
Inst. Phys. Chem. Polish Acad. Sci. Warsaw Pol. PL 01-224
J. Chromatogr.; (82) P 303-6; Vol 237; No 2

11-82-0014 (0014)
DIFFERENTIAL DETERMINATION OF L-ASCORBIC ACID AND D-ISOASCORBIC ACID BY REVERSED-PHASE HIGH-PERFORMANCE LIQUID CHROMATOGRAPHY WITH ELECTROCHEMICAL DETECTION
Tsao, C. S.; Salimi, S. L.
Linus Pauling Inst. Sci. Med. Palo Alto USA US CA 94306
J. Chromatogr.; (82) P 355-8; Vol 245; No 3

11-82-0015 (0015)
RESOLUTION OF OPTICAL ISOMERS OF DNS-AMINO ACIDS BY HIGH-PERFORMANCE LIQUID CHROMATOGRAPHY WITH L-HISTIDINE AND ITS DERIVATIVES IN THE MOBILE PHASE
Lam, Stanley; Karmen, Arthur
Albert Einstein Coll Med. Bronx USA US NY 10461
J. Chromatogr.; (82) P 451-62; Vol 239

11-82-0016 (0016)
OPTICAL RESOLUTION OF DL-PROLINE BY REVERSED-PHASE HIGH-PERFORMANCE LIQUID CHROMATOGRAPHY USING N-(p-TOLUENESULFONYL)-L-PHENYLALANINE-COPPER(II) AS A CHIRAL ADDITIVE
Nimura, Noriyuki; Toyama, Atsuko; Kinoshita, Toshio
Sch. Pharm. Sci. Kitasato Univ. Tokyo Japan JP 108
J. Chromatogr.; (82) P 482-4; Vol 234; No 2

11-82-0017 (0017)
REVERSED-PHASE LIQUID CHROMATOGRAPHIC RESOLUTION OF UNDERIVATIZED D,L-AMINO ACIDS USING CHIRAL ELUENTS
Nimura, Noriyuki; Toyama, Atsuko; Kasahara, Yoko; Kinoshita, Toshio
Sch. Pharm. Sci. Kitasato Univ. Tokyo Japan JP 108
J. Chromatogr.; (82) P 671-5; Vol 239

11-82-0018 (0018)
PREPARATIVE RESOLUTION OF RACEMATES ON A CHIRAL LIQUID CHROMATOGRAPHY COLUMN
Pirkle, William H.; Finn, John M.
Sch. Chem. Sci. Univ. Illinois Urbana USA US IL 61801
J. Org. Chem.; (82) P 4037-40; Vol 47; No 21

11-82-0019 (0019)
HPLC DETERMINATION OF D AND L-MOXALACTAM IN HUMAN SERUM AND URINE
Ziemniak, J. A.; Chiarmonte, D. A.; Miner, D. J.; Schentag, J. J.
Sch. Pharm. State Univ. New York Buffalo USA US NY 14209
J. Pharm. Sci.; (82) P 399-402; Vol 71; No 4

SECTION 3

CHAPTER 3.1
Preparative Chromatography

3.1.1
Theory

--- 1982 ---

01-82-0001 (0001)
LOW-PRESSURE PREPARATIVE LIQUID CHROMATOGRAPHY. AMINE BONDED PHASES
Zief, Morris; Crane, Laura J.; Horvath, Joseph
USA US
Am. Lab. (Fairfield, Conn.); (82) P 144, 146, 149-50, 152-3; Vol 14; No 3

01-82-0002 (0002)
PREPARATIVE LC EXTENDS THE POSSIBILITIES OF SYNTHETIC ORGANIC CHEMISTRY
Olieman, C.; Voskamp, D.; Verhage, M.
Lab. Org. Chem. Delft Univ. Technol. Delft Neth. NL
Chromatogr. Sci.; (82) P 1-11; Vol 20; No Biol./Biomed. Appl. Liq. Chromatogr. 4

01-82-0003 (0003)
STUDIES ON THE EFFICACY OF PREPARATIVE AND ANALYTICAL HPLC IN PEPTIDE CHEMISTRY
Goehring, Walter; Gemeiner, Manfred; Moroder, Luis; Nyfelder, Rolf ; Wuensch, Erich
Dep. Pept. Chem. Max-Planck-Inst. Biochem. Martinsried Fed. Rep. Ger. DE 8033
CONFERENCE; Chem. Pept. Proteins, Proc. USSR-FRG Symp., 3rd; (82) P 359-65;
Avail.: Voelter, Wolfgang; Wuensch, Erich; Ovchinnikov, Yu; Date: 80;
Publisher: de Gruyter, Berlin, Fed. Rep. Ger.

01-82-0004 (0004)
PREPARATIVE-SCALE HIGH-PERFORMANCE LIQUID CHROMATOGRAPHY ON ANALYTICAL COLUMNS
Verzele, M.; Dewaele, C.; Van Dijck, J.; Van Haver, D.
Lab. Org. Chem. State Univ. Ghent Ghent Belg. BE
J. Chromatogr.; (82) P 231-8; Vol 249; No 2

01-82-0005 (0005)
COMPLETE COMPUTER AUTOMATION OF PREPARATIVE LIQUID CHROMATOGRAPHY THROUGH INTELLIGENT FRACTION COLLECTION, WITH UNLIMITED INJECTION VOLUME AND REPETITIVE COLLECTION OF SEPARATED SOLUTE PEAKS
Sugnaux, Francois R.; Djerassi, Carl
Dep. Chem. Stanford Univ. Stanford USA US CA 94305
J. Chromatogr.; (82) P 373-89; Vol 248; No 3

01-82-0006 (0006)
AMBERLYTE XAD-4 AS A STATIONARY PHASE FOR PREPARATIVE LIQUID CHROMATOGRAPHY IN A RADIALLY COMPRESSED COLUMN
Pietrzyk, Donald J.; Cahill, William J., Jr.
Chem. Dep. Univ. Iowa Iowa City USA US IA 52242
J. Liq. Chromatogr.; (82) 781-95; Vol 5; No 4

3.1.2
Applications

--- 1982 ---

01-82-0001 (0001)
APPLICATION OF HIGH-PERFORMANCE LIQUID CHROMATOGRAPHY TO THE SEPARATION AND ISOLATION OF BIOPOLYMERS
Unger, K. K.; Roumeliotis, P.
Inst. Anorg. Chem. Anal. Chem. Johannes Gutenberg-Univ. Mainz Fed. Rep. Ger. DE 6500
Anal. Chem. Symp. Ser.; (82) P 455-70; Vol 9; No Affinity Chromatogr. Relat. Tech.

01-82-0002 (0002)
PREPARATIVE ION PAIR HPLC ISOLATION AND IDENTIFICATION OF THE MAIN CONSTITUENTS OF QUINOLINE YELLOW (C.I. 47005)
Masiala-Tsobo, C.; Gelbcke, M.; Dumont, L.; Parmentier, F.; Grimee, R.
Lab. Chim. Pharm. Org. Bromatol. Univ. Libre Bruxelles Brussels Belg. BE B-1050
Bull. Soc. Chim. Belg.; (82) P 227-35; Vol 91; No 3; In Fr

01-82-0003 (0003)
HIGH PERFORMANCE LIQUID CHROMATOGRAPHY (HPLC) OF NATURAL PRODUCTS. V. THE USE OF HPLC IN THE CELL-FREE BIOSYNTHETIC CONVERSION OF .ALPHA.-AMINOADIPYL-CYSTEINYL-VALINE (LLD) INTO ISOPENICILLIN N
Neuss, N.; Berry, D. M.; Kupka, J.; Demain, A. L.; Queener, S. W.; Duckworth, D. C.; Huckstep, L. L.
Dep. Nutr. Food Sci. Massachusetts Inst. Technol Cambridge USA US MA 02139
J. Antibiot.; (82) P 580-4; Vol 35; No 5

01-82-0004 (0004)
PURIFICATION OF SPORIDESMIN BY PREPARATIVE HIGH PRESSURE LIQUID CHROMATOGRAPHY
Ronaldson, John W.
Ruakura Agric. Res. Cent. Hamilton N. Z. NZ
J. Chem. Technol. Biotechnol.; (82) P 556-8; Vol 32; No 4

01-82-0005 (0005)
ANALYTICAL AND SEMI-PREPARATIVE HIGH-PERFORMANCE LIQUID CHROMATOGRAPHY OF OLIGOSACCHARIDES OBTAINED BY HYDRAZINOLYSIS OF HEN OVOMUCOID
Parente, Jose Paz; Strecker, Gerard; Leroy, Yves; Montreuil, Jean; Fournet, Bernard
Lab. Chim. Biol. Uviv. Sci. Tech. Lille I Villeneuve d' Ascq Fr. FR 59655
J. Chromatogr.; (82) P 199-204; Vol 249; No 1

01-82-0006 (0006)
PREPARATIVE-SCALE HIGH-PERFORMANCE LIQUID CHROMATOGRAPHY ON ANALYTICAL COLUMNS
Verzele, M.; Dewaele, C.; Van Dijck, J.; Van Haver, D.
Lab. Org. Chem. State Univ. Ghent Ghent Belg. BE
J. Chromatogr.; (82) P 231-8; Vol 249; No 2

01-82-0007 (0007)
PREPARATIVE-SCALE SEPARATION OF ALKENE GEOMETRIC ISOMERS BY LIQUID CHROMATOGRAPHY
Evershed, R. P.; Morgan, E. D.; Thompson, Lorna D.
Dep. Chem. Univ. Keele Keele/Staffordshire UK GB ST5 5BG
J. Chromatogr.; (82) P 350-4; Vol 237; No 2

01-82-0008 (0008)
ANALYTICAL AND PREPARATIVE SEPARATION OF WITHANOLIDES FROM CRUDE EXTRACTS OF ACNISTUS BREVIFLORUS LEAVES BY HIGH-PERFORMANCE LIQUID CHROMATOGRAPHY
Burton, Gerardo; Veleiro, Adriana S.; Gros, Eduardo G.
Fac. Cienc. Exactas Nat. Univ. Buenos Aires Buenos Aires Argent. AR 1428

J. Chromatogr.; (82) P 472-5; Vol 248; No 3

01-82-0009 (0009)
HIGH-PERFORMANCE SEMIPREPARATIVE LIQUID CHROMATOGRAPHY OF DIESEL ENGINE EMISSION PARTICULATE EXTRACTS
Levine, S. P.; Skewes, L. M.
Anal. Sci. Dep. Ford Motor Co. Dearborn USA US MI 48121
J. Chromatogr.; (82) P 532-5; Vol 235; No 2

01-82-0010 (0010)
THE PREPARATIVE SCALE REVERSE-PHASE HPLC SEPARATION OF EPIMERIC ALKALOIDS USING CAMPHORSULFONIC ACID AS AN ION PAIRING REAGENT
Salva, Paul S.; Hite, Gilbert J.; Henkel, James G.
Sch. Pharmacy Univ. Connecticut Storrs USA US CT 06268
J. Liq. Chromatogr.; (82) P 305-12; Vol 5; No 2

01-82-0011 (0011)
PREPARATIVE LIQUID CHROMATOGRAPHIC SEPARATION OF AMINO ACIDS AND PEPTIDES ON AMBERLITE XAD-4
Pietrzyk, Donald J.; Cahill, W. J., Jr.; Stodola, J. D.
Chem. Dep. Univ. Iowa Iowa City USA US IA 52242
J. Liq. Chromatogr.; (82) P 443-61; Vol 5; No 3

01-82-0012 (0012)
PREPARATIVE RESOLUTION OF RACEMATES ON A CHIRAL LIQUID CHROMATOGRAPHY COLUMN
Pirkle, William H.; Finn, John M.
Sch. Chem. Sci. Univ. Illinois Urbana USA US IL 61801
J. Org. Chem.; (82) P 4037-40; Vol 47; No 21

01-82-0013 (0013)
SEPARATION OF VALTRATE AND ISOVALTRATE BY MEANS OF PREPARATIVE LIQUID CHROMATOGRAPHY
Hazelhoff, B.; Jellema, R.; Grobben, H.; Malingre, T. M.
Lab. Pharmacogn. State Univ. Groningen Neth. NL 9713 AW
Pharm. Weekbl., Sci. Ed.; (82) P 21-4; Vol 4; No 1

CHAPTER 3.2
Multidimensional Chromatography

--- 1982 ---

02-82-0001 (0001)
HPLC SAMPLE INJECTION AND COLUMN SWITCHING USING A 10-PORT MULTIFUNCTIONAL VALVE
Harvey, Mack C.; Stearns, Stanley D.
Valco Instrum. Co. Houston USA US TX
Am. Lab. (Fairfield, Conn.); (82) P 68, 70, 72-4; Vol 14; No 5

02-82-0002 (0002)
HIGH-SENSITIVITY PEPTIDE MAPPING OF TRIOSEPHOSPHATE ISOMERASE: A COMPARISON OF HIGH-PERFORMANCE LIQUID CHROMATOGRAPHY WITH TWO-DIMENSIONAL THIN-LAYER METHODS
Oray, Bedii; Jahani, Minoo; Gracy, Robert W.
Dep. Chem. North Texas State Univ. Denton USA US TX 76203
Anal. Biochem.; (82) P 131-8; Vol 125; No 1

02-82-0003 (0003)
ON-LINE MULTIDIMENSIONAL LIQUID CHROMATOGRAPHIC DETERMINATION OF POLYNUCLEAR AROMATIC HYDROCARBONS IN COMPLEX SAMPLES
Sonnefeld, W. J.; Zoller, W. H.; May, W. E.; Wise, S. A.
Dep. Chem. Univ. Maryland College Park USA MD 20742
Anal. Chem.; (82) 723-7; Vol 54; No 4

02-82-0004 (0004)
AUTOMATIC HPLC OF PTERINS WITH OR WITHOUT COLUMN SWITCHING
Niederwieser, A.; Staudenmann, W.; Wetzel, E.
Dep. Pediatr. Univ. Zurich Zurich Switz. CH CH-8032
Biochem. Clin. Aspects Pteridines; (82) P 81-102; Vol 1; No Cancer, Immunol., Metab. Dis.

02-82-0005 (0005)
DETERMINATION OF CYCLOSPORIN A IN BLOOD AND PLASMA BY COLUMN SWITCHING HIGH-PERFORMANCE LIQUID CHROMATOGRAPHY AFTER RAPID SAMPLE PREPARATION
Nussbaumer, K.; Niederberger, W.; Keller, H. P.
Div. Pharm. Res. Dev. Sandoz Ltd. Basel Switz. CH CH-4002
HRC CC, J. High Resolut. Chromatogr. Chromatogr. Commun.; (82) P 424-7; Vol 5; No 8

02-82-0006 (0006)
HPLC SAMPLE INJECTION AND COLUMN SWITCHING USING A 10-PORT MULTIFUNCTIONAL VALVE
Harvey, Mack C.; Stearns, Stanley D.
Valco Instruments Co. USA US
Int. Lab.; (82) P 112, 114, 116, 118; Vol 12; No 5

02-82-0007 (0007)
HPLC SAMPLE INJECTION AND COLUMN SWITCHING USING A MULTIFUNCTIONAL VALVE
Harvey, Mack C.; Stearns, Stanley D.
Valco Instruments Co. USA US
Int. Lab.; (82) P 42, 44, 46-7; Vol 11; No 4

02-82-0008 (0008)
ON-LINE HPLC/GC TECHNIQUES FOR THE ANALYSES OF BIOMASS DERIVED HYDROCARBONS
Luzbetak, Dori J.; Hoffmann, Joseph J.
Off. Arid Lands Stud. Univ. Arizona Tucson USA US AZ 85719
J. Chromatogr. Sci.; (82) P 132-5; Vol 20; No 3

02-82-0009 (0009)
DETERMINATION OF SALBUTAMOL IN HUMAN PLASMA WITH BIMODAL HIGH-PERFORMANCE LIQUID CHROMATOGRAPHY AND A ROTATED DISK AMPEROMETRIC DETECTOR

Oosterhuis, B.; Van Boxtel, C. J.
Dep. Med. Pharmacol. Univ. Hosp. Amsterdam Amsterdam Neth. NL 1054 EG
J. Chromatogr.; (82) P 327-34; Vol 232; No 2

02-82-0010 (0010)
DETERMINATION OF PLASMA NOREPINEPHRINE AND EPINEPHRINE BY HIGH-PERFORMANCE LIQUID CHROMATOGRAPHY USING A TWO-COLUMN SYSTEM AND AN ELECTROCHEMICAL DETECTOR
Lachatre, G.; Nicot, G.; Magnet, C.; Rocca, J. L.; Merle, L.; Valette, J. P.
Dep. Pharmacol. Clin. Hop. Dupuytren Limoges Fr. FR 87031
J. Liq. Chromatogr.; (82) P 1947-65; Vol 5; No 10

CHAPTER 3.3
Trace Analysis

--- 1982 ---

03-82-0001 (0001)
ULTRAMICRODETERMINATION OF HISTAMINES BY HIGH-PERFORMANCE LIQUID CHROMATOGRAPHY
Ohkura, Y.; Kohashi, K.; Tsuruta, Y.; Ishida, S.
Fac. Pharm. Sci. Kyushu Univ. Fukuoka Japan JP
Adv. Biosci.; (82) P 243-53; Vol 33; No Adv. Histamine Res.

03-82-0002 (0002)
DETERMINATION OF TRACE LEVELS OF QUATERNARY AMMONIUM COMPOUNDS IN RIVER WATER BY LIQUID CHROMATOGRAPHY WITH CONDUCTOMETRIC DETECTION
Wee, Victorio T.; Kennedy, Julie M.
Ivorydale Tech. Cent. Procter and Gamble Co. Cincinnati USA US OH 45217
Anal. Chem.; (82) P 1631-3; Vol 54; No 9

03-82-0003 (0003)
DETERMINATION OF METALS IN GROUNDWATERS BY TRACE ENRICHMENT AND LIQUID CHROMATOGRAPHY
Cassidy, R. M.; Elchuk, Steven; McHugh, J. O.
Environ. Res. Branch At. Energy Canada Ltd. Chalk River Can. CA ON K0J 1J0
Anal. Chem.; (82) P 727-31; Vol 54; No 4

03-82-0004 (0004)
TRACE ENRICHMENT OF DIFENZOQUAT IN WATER AND ITS DETERMINATION BY REVERSED-PHASE HIGH-PERFORMANCE LIQUID CHROMATOGRAPHY
Ahmad, Ijaz
Alberta Environ. Cent. Vegreville Can. CA AB T0B 4L0
Anal. Lett.; (82) P 27-37; Vol 15; No A1

03-82-0005 (0005)
REFRACTOMETRIC DETECTION FOR TRACE ANALYSIS BY LIQUID PHASE CHROMATOGRAPHY. APPLICATION TO THE CASE OF DILUTE AQUEOUS SOLUTIONS OF ISOPROPYL ETHER
Desbene, P. L.; Jehanno, N.
Lab. Chim. Org. Struct. Univ. Pierre et Marie Curie Paris Fr. FR 75230
Analusis; (82) P 292-4; Vol 10; No 6; In Fr

03-82-0006 (0006)
DETERMINATION OF ORGANIC CONTAMINANTS IN ULTRA-PURE WATER BY REVERSED-PHASE HIGH-PERFORMANCE LIQUID CHROMATOGRAPHY WITH ULTRAVIOLET DETECTION
Reust, Juerg B.; Meyer, Veronika R.
Inst. Anorg. Anal. Phys. Chem. Univ. Berne Berne Switz. CH CH-3012
Analyst (London); (82) P 673-9; Vol 107; No 1275

03-82-0007 (0007)
DETERMINATION OF ALDICARB RESIDUES IN WATER BY COMBINED HIGH PERFORMANCE LIQUID CHROMATOGRAPHY/MASS SPECTROMETRY
Wright, L. H.; Jackson, M. D.; Lewis, R. G.
EPA Research Triangle Park USA US NC 27711
Bull. Environ. Contam. Toxicol.; (82) P 740-7; Vol 28; No 6

03-82-0008 (0008)
HIGH PERFORMANCE REVERSED-PHASE LIQUID CHROMATOGRAPHY OF TRACE AMOUNTS OF INORGANIC AND ORGANIC MERCURY AS DIETHYLDITHIOCARBAMATE CHELATE
Inoue, Sadanobu; Hoshi, Suwaru; Sasaki, Mitsuo
Dep. Environ. Eng. Kitami Inst. Technol. Kitami Japan JP 090
Bunseki Kagaku; (82) P E243-E246; Vol 31; No 7

03-82-0009 (0009)
A HIGH-PERFORMANCE LIQUID CHROMATOGRAPHIC DETERMINATION OF FREE AMINO ACIDS IN

NATURAL WATERS IN THE PICOMOLAR (M .TIMES. 10-12) RANGE SUITABLE FOR SHIPBOARD USE
Evens, Roger; Braven, James; Brown, Leslie; Butler, Ian
Dep. Environ. Sci. Plymouth Polytech. Plymouth/Devon UK GB
Chem. Ecol.; (82) P 99-106; Vol 1; No 2

03-82-0010 (0010)
DIRECT ENRICHMENT OF TRYPTOPHAN AND ITS METABOLITES IN PLASMA ONTO A PRE-COLUMN FOLLOWED BY REVERSE PHASE HPLC ANALYSIS
Yoshida, Hisanobu; Morita, Ikue; Masujima, Tsutomu; Imai, Hideo
Sch. Med. Hiroshima Univ. Hiroshima Japan JP 734
Chem. Pharm. Bull.; (82) P 3827-30; Vol 30; No 10

03-82-0011 (0011)
TRACE ENRICHMENT OF FLUORINATED ORGANIC ACIDS USED AS GROUND-WATER TRACERS BY LIQUID CHROMATOGRAPHY
Stetzenbach, Klaus J.; Jensen, Stephen L.; Thompson, Glenn M.
Dep. Hydrol. Water Resour. Univ. Arizona Tucson USA US AZ 85721
Environ. Sci. Technol.; (82) P 250-4; Vol 16; No 5

03-82-0012 (0012)
LIQUID CHROMATOGRAPHY WITH PRECOLUMN SAMPLE PRECONCENTRATION AND ELECTROCHEMICAL DETECTION: DETERMINATION OF AROMATIC AMINES IN ENVIRONMENTAL SAMPLES
Rice, John R.; Kissinger, Peter T.
Dep. Chem. Purdue Univ. West Lafayette USA US IN 47907
Environ. Sci. Technol; (82) P 263-8; Vol 16; No 5

03-82-0013 (0013)
DETERMINATION OF TRACES OF SODIUM ALKYLBENZENESULFONATE BY HIGH-PERFORMANCE LIQUID CHROMATOGRAPHY. APPLICATION TO WATER
Saito, Toshihide; Higashi, Kunishige; Hagiwara, Kazuyoshi
Gov. Ind. Res. Inst. Ikeda Japan JP 563
Fresenius' Z. Anal. Chem.; (82) P 21-3; Vol 313; No 1

03-82-0014 (0014)
METHOD FOR DETERMINATION OF FORMALDEHYDE IN AIR IN THE PPT-RANGE BY HPLC AFTER EXTRACTION AS 2,4-DINITROPHENYLHYDRAZONE
Tuss, H.; Neitzert, V.; Seiler, W.; Neeb, R.
Max-Planck-Inst. Chem. Mainz Fed. Rep. Ger. DE D-6500
Fresenius' Z. Anal. Chem.; (82) P 613-17; Vol 312; No 7

03-82-0015 (0015)
TRACE ANALYSIS OF NATURAL CAPSAICINOIDS IN ANIMAL FEED, HUMAN URINE, AND WASTEWATER BY HIGH-PRESSURE LIQUID CHROMATOGRAPHY
Johnson, Eric K.; Thompson, Harold C., Jr.; Bowman, Malcolm C.
Dep. Health Hum. Serv. Food and Drug Adm. Jefferson USA AR 72079
J. Agric. Food Chem.; (82) P 354-9; Vol 30; No 2

03-82-0016 (0016)
LIQUID CHROMATOGRAPHIC DETERMINATION OF NATAMYCIN IN CHEESE AT RESIDUE LEVELS
Tuinstra, Louis G. M. Th.; Traag, Wim A.
State Inst. Qual. Control Agric. Prod. Wageningen Neth. NL 6708 PD
J. Assoc. Off. Anal. Chem.; (82) P 820-2; Vol 65; No 4

03-82-0017 (0017)
FEMTOMOLAR ION-PAIR HIGH-PERFORMANCE LIQUID CHROMATOGRAPHIC METHOD FOR DETERMINING DNS-POLYAMINE DERIVATIVES OF RED BLOOD CELL EXTRACTS UTILIZING AN AUTOMATED POLYAMINE ANALYZER
Brown, Nesbitt D.; Strickler, M. Patricia; Whaun, June M.

Div. Biochem. Walter Reed Army Inst. Res. Washington USA US DC 20012
J. Chromatogr.; (82) P 101-8; Vol 245; No 1

03-82-0018 (0018)
TRACE ENRICHMENT AND SEPARATION OF CHOLESTEROL OXIDATION PRODUCTS BY ADSORPTION HIGH-PERFORMANCE LIQUID CHROMATOGRAPHY
Csiky, Istvan
Dep. Anal. Chem. Univ. Lund Lund Swed. SE S-220 07
J. Chromatogr.; (82) P 381-9; Vol 241; No 2

03-82-0019 (0019)
DETERMINATION OF CARBOFURAN AND ITS METABOLITES BY HIGH-PERFORMANCE LIQUID CHROMATOGRAPHY USING ON-LINE TRACE ENRICHMENT
Cramer, Paul H.; Drinkwine, Arbor D.; Going, John E.; Carey, Ann E.
Midwest Res. Inst. Kansas City USA US MO 64110
J. Chromatogr.; (82) P 489-500; Vol 235; No 2

03-82-0020 (0020)
RAPID ISOLATION OF NANOGRAM AMOUNTS OF CRUSTACEAN ERYTHROPHORE CONCENTRATING HORMONE FROM INVERTEBRATE NERVE TISSUE BY RP-HPLC
Jaffe, Howard; Loeb, Marcia; Hayes, Dora K.; Holston, Nancy
Agric. Environ. Qual. Inst. ARS Beltsville USA US MD 20705
J. Liq. Chromatogr.; (82) P 1375-90; Vol 5; No 7

03-82-0021 (0021)
HIGH-PERFORMANCE LIQUID CHROMATOGRAPHIC DETERMINATION OF TRACE QUANTITIES OF AZINPHOS METHYL AND AZINPHOS METHYL OXON IN VARIOUS WATER SOURCES BY DIRECT INJECTION AND TRACE ENRICHMENT
Bushway, R. J.
Dep. Food Sci. Univ. Maine Orono USA US ME 04469
J. Liq. Chromatogr.; (82) P 49-62; Vol 5; No 1

03-82-0022 (0022)
A METHOD FOR THE ESTIMATION OF SUBMICROMOLAR CONCENTRATIONS OF ADENOSINE IN PLASMA USING HIGH-PRESSURE LIQUID CHROMATOGRAPHY
Ballard, Heather; Cotterrell, D.; Karim, F.; Reed, Jane
Dep. Physiol. Univ. Leeds Leeds UK GB LS2 9JT
J. Physiol. (London); (82) P 4P-5P; Vol 330

03-82-0023 (0023)
SIMULTANEOUS DETERMINATION OF PICOGRAM LEVELS OF 5HTP, 5HT AND 5HIAA IN DESCRETE BRAIN AREAS OF RAT USING HPLC WITH ELECTROCHEMICAL DETECTION
Oomagari, Kiyoshi; Uchimura, Hideyuki; Matsumoto, Takashi; Matsuo, Hideyasu; Saito, Masashi; Hirano, Makoto; Kin, Choju; Shimamura, Manabu
Dep. Psychiatry Saga Med. Sch. Saga Japan JP
Neurosciences (Kobe, Jpn.); (82) P 64-5; Vol 8; No 1; In Japan

CHAPTER 3.4
TLC

--- 1982 ---

04 82-0001 (0001)
HIGH-SENSITIVITY PEPTIDE MAPPING OF TRIOSEPHOSPHATE ISOMERASE: A COMPARISON OF HIGH-PERFORMANCE LIQUID CHROMATOGRAPHY WITH TWO-DIMENSIONAL THIN-LAYER METHODS
Oray, Bedii; Jahani, Minoo; Gracy, Robert W.
Dep. Chem. North Texas State Univ. Denton USA US TX 76203
Anal. Biochem.; (82) P 131-8; Vol 125; No 1

04-82-0002 (0002)
ADSORPTION BEHAVIOR OF SEVERAL SUPPORTS IN REVERSED-PHASE THIN-LAYER CHROMATOGRAPHY AS DEMONSTRATED BY THE DETERMINATION OF RELATIVE PARTITION COEFFICIENTS OF SOME 4-HYDROXYCOUMARIN DERIVATIVES
Van der Giesen, Willem F.; Janssen, Lambert H. M.
Fac. Pharm. State Univ. Utrecht Utrecht Neth. NL 3511 GH
J. Chromatogr.; (82) P 199-213; Vol 237; No 2

04-82-0003 (0003)
STUDIES ON THE REVERSED-PHASE THIN-LAYER CHROMATOGRAPHY OF ECDYSTEROIDS ON C12 BONDED AND PARAFFIN-COATED SILICA
Wilson, I. D.; Bielby, C. R.; Morgan, E. D.
Dep. Chem. Univ. Keele Keele/Staffordshire UK GB ST5 5BG
J. Chromatogr.; (82) P 202-6; Vol 242; No 1

04-82-0004 (0004)
DETERMINATION OF ANTHRANILATE, 3-HYDROXYANTHRANILATE AND 2,3-DIHYDROXYBENZOATE IN THE ENZYME ASSAY OF ANTHRANILATE HYDROXYLASE BY THIN-LAYER AND HIGH-PERFORMANCE LIQUID CHROMATOGRAPHY
Schott, Hans Henning; Langenbach-Schmidt, Doris; Zinke, Karen
Biochem. Inst. Justus Liebig-Univ. Giessen Giessen Fed. Rep. Ger. DE D-6300
J. Chromatogr.; (82) P 85-91; Vol 245; No 1

04-82-0005 (0005)
SEPARATIONS IN THIN-LAYER AND HIGH-PERFORMANCE LIQUID CHROMATOGRAPHY USING ALKYL SILICA GEL BONDED PHASES
Issaq, Haleem J.; Klose, John R.; Reitz, Gilbert A.
Frederick Cancer Res. Facil. Natl. Cancer Cent. Frederick USA US MD 21701
J. Liq. Chromatogr.; (82) P 1069-80; Vol 5; No 6

04-82-0006 (0006)
THE EFFECT OF BINARY SOLVENT COMPOSITION AND POLARITY ON SEPARATIONS IN REVERSED PHASE THIN-LAYER AND HIGH-PERFORMANCE LIQUID CHROMATOGRAPHY
Issaq, Haleem J.; Klose, John R.; Cutchin, William
Frederick Cancer Res. Facility Natl. Cancer Inst. Frederick USA US MD 21701
J. Liq. Chromatogr.; (82) P 625-41; Vol 5; No 4

04-82-0007 (0007)
SEPARATION OF PIGMENTS BY CENTRIFUGAL LIQUID CHROMATOGRAPHY
Utsugi, Hiroshi; Endo, Atsushi; Suzuki, Noboru; Umemura, Mitsuhiko
Coll. Eng. Utsunomiya Univ. Utsunomiya Japan JP
Kagaku Kogaku; (82) P 241-3; Vol 46; No 4; In Japan

04-82-0008 (0008)
MODERN METHODS OF PLANAR LIQUID CHROMATOGRAPHY
Tyihak, Erno
Gyogynoveny Kut. Int. Budakalasz Hung. HU
Magy. Kem. Lapja; (82) P 289-98; Vol 37; No 7; In Hung

04-82-0009 (0009)
ION-PAIR ADSORPTION THIN-LAYER CHROMATOGRAPHY OF QUATERNARY AMMONIUM IONS.

PART 8. PARAMETERS OF AN EQUIDISTRIBUTION
Giebelmann, R.
Inst. Gerichtliche Med. Kriminalist. Bereiches Med.
Ernst-Moritz-Arndt-Univ. Greifswald Ger. Dem. Rep. DD DDR-2200
Pharmazie; (82) P 524; Vol 37; No 7; In Ger

04-82-0010 (0010)
ION-PAIR ADSORPTION THIN-LAYER CHROMATOGRAPHY OF QUATERNARY AMMONIUM IONS.
PART 9. SELECTIVE COEFFICIENTS
Giebelmann, R.
Inst. Gerichtliche Med. Kriminalist. Bereiches Med.
Ernst-Moritz-Arndt-Univ. Greifswald Ger. Dem. Rep. DD DDR-2200
Pharmazie; (82) P 525; Vol 37; No 7; In Ger

04-82-0011 (0011)
PURIFICATION OF BIOLOGICAL SAMPLES BY HPLC FOR THE SIMULTANEOUS DETERMINATION
OF ANABOLICS BY HPTLC
Smets, Felix; Vandewalle, Marc
Inst. Hyg. Epidemiol. Brussels Belg. BE B-1050
Z. Lebensm.-Unters. Forsch.; (82) P 29-30; Vol 175; No 1

04-82-0012 (0012)
LIQUID ADSORPTION CHROMATOGRAPHY OF CHELATES. EFFECT OF THE STRUCTURE OF
CHELATES ON THEIR CHROMATOGRAPHIC BEHAVIOR IN A THIN LAYER OF SORBENT
Timerbaev, A. R.; Petrukhin, O. M.; Zolotov, Yu. A.
V. I. Vernadskii Inst. Geochem. Anal. Chem. Moscow USSR SU
Zh. Anal. Khim.; (82) P 581-8; Vol 37; No 4; In Russ

SECTION 4

CHAPTER 4.1
Microchromatography

--- 1982 ---

01-82-0001 (0001)
MICROBORE COLUMNS FOR HPLC
Ryall, Ralph R.; Kessler, Howard D., Jr.
Whatman Chem. Sep. Clifton USA US NJ 07014
Am. Lab. (Fairfield, Conn.); (82) P 49-51, 53-4, 56; Vol 14; No 5

01-82-0002 (0002)
LASER-INDUCED FLUORESCENCE DETECTION FOR CONVENTIONAL AND MICROCOLUMN LIQUID CHROMATOGRAPHY
Folestad, Staffan; Johnson, Lars; Josefsson, Bjoern; Galle, Bo
Dep. Anal. Mar. Chem. Chalmers Univ. Technol. Goeteborg Swed. SE S-412 96
Anal. Chem.; (82) P 925-9; Vol 54; No 5

01-82-0003 (0003)
MICRO HIGH-PERFORMANCE LIQUID CHROMATOGRAPHY OF ORGANICALLY BOUND METALS WITH INDUCTIVELY COUPLED PLASMA ATOMIC EMISSION SPECTROMETRIC DETECTION
Jinno, Kiyokatsu; Tsuchida, Hiroyuki
Sch. Mater. Sci. Toyohashi Univ. Technol. Toyohashi Japan JP 440
Anal. Lett.; (82) P 427-37; Vol 15; No A5

01-82-0004 (0004)
MICROCOLUMN HIGH-PERFORMANCE LIQUID CHROMATOGRAPHY
Westwood, Steven A.; Games, David E.; Lant, Michael S.; Woodhall, Brian J.
Dep. Chem. Univ. Coll. Cardiff Cardiff UK GB CF1 1XL
Anal. Proc. (London); (82) P 121-3; Vol 19; No 3

01-82-0005 (0005)
AN INTERFACE FOR THE COMBINATION OF MICRO HIGH-PERFORMANCE LIQUID CHROMATOGRAPHY AND INFRARED SPECTROMETRY
Jinno, K.; Fujimoto, C.; Hirata, Y.
Sch. Mater. Sci. Toyohashi Univ. Technol. Toyohashi Japan JP 440
Appl. Spectrosc.; (82) P 67-9; Vol 36; No 1

01-82-0006 (0006)
MICROBORE HIGH-PERFORMANCE LIQUID CHROMATOGRAPHY MASS SPECTROMETRY
Games, David E.; Lant, Michael S.; Westwood, Steven A.; Cocksedge, Michael J.; Evans, Nigel; Williamson, James; Woodhall, Brian J.
Dep. Chem. Univ. Coll. Cardiff Cardiff UK GB CF1 1XL
Biomed. Mass Spectrom.; (82) P 215-24; Vol 9; No 5

01-82-0007 (0007)
MICRO HIGH-PERFORMANCE LIQUID CHROMATOGRAPHY OF MUTAGEN, ETHYL N-ETHYL-N-NITROSOCARBAMATE, FORMED FROM URETHANE AND NITRITE
Kohashi, Kazuya; Kawazoe, Yuichi; Horiuchi, Tadao; Ohkura, Yosuke
Fac. Pharm. Sci. Kyushu Univ. Fukuoka Japan JP 812
Bunseki Kagaku; (82) P E41-E44; Vol 31; No 1

01-82-0008 (0008)
ON-LINE DIRECT LIQUID INTRODUCTION INTERFACE FOR MICRO-LIQUID CHROMATOGRAPHY/MASS SPECTROMETRY: APPLICATION TO DRUG ANALYSIS
Eckers, Christine; Skrabalak, Dale S.; Henion, Jack
New York State Coll. Vet. Med. Cornell Univ. Ithaca USA US NY 14853
Clin. Chem. (Winston-Salem, N. C.); (82) P 1882-6; Vol 28; No 9

01-82-0009 (0009)
SEPARATION OF ISOTOPICALLY LABELED MOLECULES BY MICRO HIGH-PERFORMANCE LIQUID CHROMATOGRAPHY WITH A PACKED MICRO GLASS CAPILLARY COLUMN
Jinno, Kiyokatsu; Hirata, Yukio; Hiyoshi, Yoshihiko

Sch. Mater. Sci. Toyohashi Univ. Technol. Toyohashi Japan JP 440
HRC CC, J. High Resolut. Chromatogr. Chromatogr. Commun.; (82) P 102-3; Vol 5; No 2

01-82-0010 (0010)
CHROMATOGRAPHIC PERFORMANCE OF DEUTERATED SOLVENTS IN REVERSED PHASE MICRO HIGH-PERFORMANCE LIQUID CHROMATOGRAPHY
Jinno, K.
Sch. Mater. Sci. Toyohashi Univ. Technol. Toyohashi Japan JP 440
HRC CC, J. High Resolut. Chromatogr. Chromatogr. Commun.; (82) P 364-7; Vol 5; No 7

01-82-0011 (0011)
INVESTIGATION OF THE LOW-TEMPERATURE EFFECT IN NORMAL-PHASE MICRO HIGH-PERFORMANCE LIQUID CHROMATOGRAPHY
Jinno, Kiyokatsu; Hirata, Yukio
Sch. Mater. Sci. Toyohashi Univ. Technol. Toyohashi Japan JP 440
HRC CC, J. High Resolut. Chromatogr. Chromatogr. Commun.; (82) P 85-90; Vol 5; No 2

01-82-0012 (0012)
DIRECT COUPLING OF AN ULTRA-MICRO HIGH-PERFORMANCE LIQUID CHROMATOGRAPH AND A MASS SPECTROMETER
Takeuchi, T.; Ishii, D.; Saito, A.; Ohki, T.
Fac. Eng. Nagoya Univ. Nagoya Japan JP 464
HRC CC, J. High Resolut. Chromatogr. Chromatogr. Commun.; (82) P 91-2; Vol 5; No 2

01-82-0013 (0013)
MICROBORE COLUMNS FOR HPLC
Ryall, Ralph R.; Kessler, Howard D., Jr.
Whatman Chem. Separation, Inc. USA US
Int. Lab.; (82) P 68, 70, 72, 74, 76, 78; Vol 12; No 5

01-82-0014 (0014)
SEPARATION OF CHLORO COMPLEXES OF METALS BY MICRO-HIGH-PERFORMANCE LIQUID CHROMATOGRAPHY
Tsuda, Takao; Nozu, Takashi; Nakagawa, Genkichi
Lab. Anal. Chem. Nagoya Inst. Technol. Nagoya Japan JP 466
J Chromatogr.; (82) P 331-6; Vol 242; No 2

01-82-0015 (0015)
OPEN TUBULAR COLUMN LC: THEORY AND PRACTICE
Yang, Frank J.
Walnut Creek Instrum. Div. Varian Assoc. Walnut Creek USA US CA 94598
J. Chromatogr. Sci.; (82) P 241-51; Vol 20; No 6

01-82-0016 (0016)
APPLICATION OF MICROBORE COLUMNS TO LIQUID CHROMATOGRAPHY-MASS SPECTROMETRY
Krien, Paul; Devant, Gerard; Hardy, Michel
Nermag S. A. Rueil-Malmaison Fr. FR 92500
J. Chromatogr.; (82) P 129-39; Vol 251; No 2

01-82-0017 (0017)
HIGH-BACK-PRESSURE LIQUID CHROMATOGRAPHY IV. STUDIES ON THE RETENTION BEHAVIOR OF AROMATIC COMPOUNDS
Takeucti, Toyohide; Ishii, Daido
Fac. Eng. Nagoya Univ. Nagoya Japan JP 464
J. Chromatogr.; (82) P 23-32; Vol 244; No 1

01-82-0018 (0018)
OPEN-TUBULAR MICROCAPILLARY LIQUID CHROMATOGRAPHY WITH ELECTROOSMOSIS FLOW USING A UV DETECTOR
Tsuda, Takao; Nomura, Kazuhiro; Nakagawa, Gdnkichi
Lab. Anal. Chem. Nagoya Inst. Technol. Nagoya Japan JP 466
J. Chromatogr.; (82) P 241-7; Vol 248; No 2

01-82-0019 (0019)
FUSED-SILICA NARROW-BORE MICROPARTICLE-PACKED-COLUMN HIGH-PERFORMANCE LIQUID CHROMATOGRAPHY
Yang, Frank J.
Walnut Creek Div. Varian Associates Walnut Creek USA US CA 94598
J. Chromatogr.; (82) P 265-77; Vol 236; No 2

01-82-0020 (0020)
ROLE OF COLUMN TEMPERATURE IN OPEN-TUBULAR MICROCAPILLARY LIQUID CHROMATOGRAPHY
Takeuchi, Toyohide; Kumaki, Masayoshi; Ishii, Daido
Fac. Eng. Nagoya Univ. Nagoya Japan JP 464
J. Chromatogr.; (82) P 309-22; Vol 235; No 2

01-82-0021 (0021)
DUAL ELECTROCHEMICAL DETECTOR FOR MICRO-HIGH-PERFORMANCE LIQUID CHROMATOGRAPHY AND ITS APPLICATION TO THE SELECTIVE DETECTION OF CATECHOLAMINES
Goto, Masashi; Sakurai, Eizo; Ishii, Daido
Fac. Eng. Nagoya Univ. Nagoya Japan JP 464
J. Chromatogr.; (82) P 357-66; Vol 238; No 2

01-82-0022 (0022)
USE OF CONVENTIONAL INSTRUMENTATION WITH MICROBORE COLUMNS IN HIGH-PERFORMANCE LIQUID CHROMATOGRAPHY
Kok, W. T.; Brinkman, U. A. T.; Frei, R. W.; Hanekamp, H. B.; Nooitgedacht, F.; Poppe, H.
Dep. Anal. Chem. Free Univ. Amsterdam Neth. NL 1081 HV
J. Chromatogr.; (82) P 357-69; Vol 237; No 3

01-82-0023 (0023)
MICRO-HIGH-PERFORMANCE LIQUID CHROMATOGRAPHY WITH LONG MICROPACKED FLEXIBLE FUSED-SILICA COLUMNS
Takeuchi, Toyohide; Ishii, Daido
Fac. Eng. Nagoya Univ. Nagoya Japan JP 464
J. Chromatogr.; (82) P 409-18; Vol 238; No 2

01-82-0024 (0024)
CONTINUOUS GRADIENT ELUTION IN MICRO HIGH-PERFORMANCE LIQUID CHROMATOGRAPHY
Takeuchi, Toyohide; Ishii, Daido
Fac. Eng. Nagoya Univ. Nagoya Japan JP 464
J. Chromatogr.; (82) P 41-7; Vol 253; No 1

01-82-0025 (0025)
PACKED MICROCAPILLARY LIQUID CHROMATOGRAPHY WITH REDUCED I.D. COLUMNS
Tsuda, Takao; Tanaka, Isao; Nakagawa, Genkichi
Lab. Anal. Chem. Nagoya Inst. Technol. Nagoya Japan JP 466
J. Chromatogr.; (82) P 507-13; Vol 239

01-82-0026 (0026)
HIGH-BACK-PRESSURE LIQUID CHROMATOGRAPHY. III. OPEN-TUBULAR MICROCAPILLARY LIQUID CHROMATOGRAPHY USING LIQUEFIED ALKANES AS THE MOBILE PHASE
Takeuchi, Toyohide; Ishii, Daido

Fac. Eng. Nagoya Univ. Nagoya Japan JP 464
J. Chromatogr.; (82) P 51-60; Vol 240; No 1

01-82-0027 (0027)
ANALYSIS OF A MODEL IONIC NITROSAMINE BY MICROBORE HIGH-PERFORMANCE LIQUID CHROMATOGRAPHY USING A THERMAL ENERGY ANALYZER CHEMILUMINESCENCE DETECTOR
Massey, Robert C.; Crews, Colin; McWeeny, David J.; Knowles, Michael E.
Food Lab. Minist. Agric. Fish. Food Norwich UK GB NR2 4SX
J. Chromatogr.; (82) P 527-9; Vol 236; No 2

01-82-0028 (0028)
"BUFFER MEMORY" TECHNIQUE FOR THE COMBINATION OF MICRO-HIGH-PERFORMANCE LIQUID CHROMATOGRAPHY AND INFRARED SPECTROMETRY
Jinno, Kiyokatsu; Fujimoto, Chuzo; Ishii, Daido
Sch. Mater. Sci. Toyohashi Univ. Technol. Toyohashi Japan JP 440
J. Chromatogr.; (82) P 625-32; Vol 239

SECTION 5

CHAPTER 5.1
Sorbents

5.1.1
Generalities

--- 1982 ---

01-82-0001 (0001)
OPTIMAL PARAMETERS IN HIGH-PERFORMANCE LIQUID CHROMATOGRAPHY (HPLC)
Halasz, Istvan; Goerlitz, Gerhard
Univ. Saarbruecken Saarbruecken Fed. Rep. Ger. DE D-6600
Angew. Chem.; (82) P 50-62; Vol 94; No 1; In Ger

01-82-0002 (0002)
HIGH SPEED LIQUID CHROMATOGRAPHIC PACKING
Yanagihara, Yuzo; Noguchi, Kohji; Honda, Makoto
Asahi Chemical Industry Co., Ltd. Japan JP
Eur. Pat. Appl.; (010682) P 39 pp.; Pat. No.: 43074; Int. Class: C08F-008/12; B01D-015/08; Doc. Code: A2; Appl.: EP 81/104826 810623; Priority Appl.: JP 80/85243 800625; JP 80/183703 801226; Des. Countries : CH; DE; FR; GB; IT; NL; SE

01-82-0003 (0003)
INFLUENCE OF THE PACKING MATERIAL AND THE COLUMN FILTERS ON THE RELIABILITY OF A HIGH-PERFORMANCE LIQUID CHROMATOGRAPH - MASS SPECTROMETER INTERFACE BASED ON THE DIRECT LIQUID INLET PRINCIPLE
Mauchamp, Bernard; Krien, Paul
Lab. Phytopharm. Inst. Natl. Rech. Agron. Versailles Fr. FR F-78000
J. Chromatogr.; (82) P 17-24; Vol 236; No 1

01-82-0004 (0004)
SIZE EXCLUSION AT THE EXTERNAL SURFACES OF SPHERICAL PACKINGS USED IN HIGH-PERFORMANCE LIQUID CHROMATOGRAPHY
Squire, Phil G.; Magnus, Arne; Himmel, Michael E.
Dep. Biochem. Colorado State Univ. Fort Collins USA US CO 80523
J. Chromatogr.; (82) P 255-66; Vol 242; No 2

01-82-0005 (0005)
LIQUID CHROMATOGRAPHY PACKING MATERIAL
Showa Denko K. K.
Japan JP
Jpn. Kokai Tokkyo Koho; (052082) P 5 pp.; In Japan; Pat. No.: 82 80557; Int. Class: G01N-031/08; B01D-015/08; B01J-020/26; C08F-002/18; C08F-008/00; C08F-212/10; C08F-257/02; Doc. ; Appl.: JP 80/157059 801110

01-82-0006 (0006)
PARTICLE SIZE MEASUREMENTS FOR POLYMER LATEXES BY LIQUID CHROMATOGRAPHY USING POROUS SILICA PACKINGS
Kato, Tadaya; Takami, Hajime; Takahashi, Akira
Fac. Eng. Mie Univ. Tsu Japan JP 514
Kobunshi Ronbunshu; (82) P 487-92; Vol 39; No 8; In Japan

01-82-0007 (0007)
A COMPARISON OF HPLC COLUMNS. COMMENTS
Waraska, John
Waters Assoc. Milford USA US MA
Pharm. Technol.; (82) P 36-7; Vol 6; No 9

01-82-0008 (0008)
A COMPARISON OF HPLC COLUMNS. REPLY TO COMMENTS
George, R.; Patel, C.
USA US
Pharm. Technol.; (82) P 37, 39; Vol 6; No 9

01-82-0009 (0009)
A COMPARISON OF HPLC COLUMNS
George, Ronald C.; Patel, Chittaranjan
Ames Div. Miles Lab., Inc. Elkhart USA US IN
Pharm. Technol.; (82) P 88, 90, 93-4, 96, 98-9; Vol 6; No 1

01-82-0010 (0010)
FILLING COLUMNS FOR LIQUID CHROMATOGRAPHY
Karapet'yan, Sh. A.
USSR SU
From: Otkrytiya, Izobret., Prom. Obraztsy, Tovarnye Znaki 1982, (11), 183.
U.S.S.R.; (032382) In Russ; Pat. No.: 915004; Int. Class: G01N-031/08; Doc.
Code: A1; Appl.: SU 2965345 800730

5.1.2
Polar Inorganic Phases

--- 1982 ---

01-82-0001 (0001)
THE SILICA GEL SURFACE AND ITS INTERACTIONS WITH SOLVENT AND SOLUTE IN LIQUID CHROMATOGRAPHY
Scott, Ravmond P. W.
Perkin-Elmer Corp. Norwalk USA US CT
Adv. Chromatogr. (N. Y.); (82) P 167-215; Vol 20

01-82-0002 (0002)
SEPARATION OF CARBOHYDRATES BY LIQUID CHROMATOGRAPHY ON SILICA GEL. ADDING ADSORPTION MODIFICATORS TO THE ELUENT
Boumahraz, M.; Davvdov, V. Ya.; Kiselev, A. V.
Chem. Dep. M. V. Lomonosov State Univ.-Moscow Moscow USSR SU 117234
Chromatographia; (82) P 751-6; Vol 15; No 12

01-82-0003 (0003)
EVALUATION OF SILICA AND POLAR BONDED COLUMNS FOR LIQUID CHROMATOGRAPHIC ANALYSIS OF TEMEPHOS FORMULATIONS
Mount, Dwight L.; Miles, James W.
Parasit. Dis. Div. Cent. Infect. Dis. Atlanta USA US GA 30333
J. Assoc. Off. Anal. Chem.; (82) P 575-9; Vol 65; No 3

01-82-0004 (0004)
SOLVENT STRENGTH OF MULTICOMPONENT MOBILE PHASES IN LIQUID-SOLID CHROMATOGRAPHY. FURTHER STUDY OF DIFFERENT MOBILE PHASES AND SILICA AS ADSORBENT
Snvder, L. R.; Glajch, J. L.
Technicon Instrum. Corp. Tarrvtown USA US NY 10591
J. Chromatogr.; (82) P 165-82; Vol 248; No 2

5.1.3
Polar Bonded Phases

--- 1982 ---

01-82-0001 (0001)
LOW-PRESSURE PREPARATIVE LIQUID CHROMATOGRAPHY. AMINE BONDED PHASES
Zief, Morris; Crane, Laura J.; Horvath, Joseph
USA US
Am. Lab. (Fairfield, Conn.); (82) P 144, 146, 149-50, 152-3; Vol 14; No 3

01-82-0002 (0002)
LIQUID CHROMATOGRAPHIC SEPARATION OF AROMATIC HYDROCARBONS WITH CHEMICALLY BONDED (2,4-DINITROANILINOPROPYL)SILICA
Grizzle, Patrick L.; Thomson, Jane S.
Bartlesville Energy Technol. Cent. U. S. Dep. Energy Bartlesville USA US OK 74003
Anal. Chem.; (82) P 1071-8; Vol 54; No 7

01-82-0003 (0003)
LIQUID CHROMATOGRAPHIC SEPARATIONS OF DIHYDROXYBENZENES WITH SILICA, DIAMINE, AND COBALT COMPLEX-BONDED PHASE COLUMNS
Chang, C. Allen; Tu, Cheng Fan
Dep. Chem. Univ. Texas El Paso USA US TX 79968
Anal. Chem.; (82) P 1179-82; Vol 54; No 7

01-82-0004 (0004)
DETERMINATION OF SULFATHIAZOLE IN HONEY FROM MEDICATED HONEY BEE COLONIES BY HIGH-PERFORMANCE LIQUID CHROMATOGRAPHY ON A CYANOAMINO POLAR PHASE
Argauer, Robert J.; Shimanuki, Hachiro; Knox, David A.
ARS USDA Beltsville USA US MD 20705
Environ. Entomol.; (82) P 820-3; Vol 11; No 4

01-82-0005 (0005)
STUDIES ON THE OPTIMIZATION OF THE SEPARATION OF WATER-SOLUBLE VITAMINS BY AMINE-BOUND PHASE HPLC
Calcagno, C.; Lanteri, S.
Ist. Sci. Farm. Univ. Genova Genoa Italy IT
Farmaco, Ed. Prat.; (82) P 152-9; Vol 37; No 5; In Ital

01-82-0006 (0006)
STUDIES ON THE PERFLUOROAMIDE-BONDED PHASE AS HPLC PACKING. III. APPLICATION OF PERFLUOROAMIDE-BONDED SILICA IN THE SEPARATION OF CHOLANIC ACIDS AND ESTERS
Li, Wenhua; Wang, Jinhua
Shanghai Inst. Org. Chem. Acad. Sin. Shanghai Peop. Rep. China CN
Huaxue Xuebao; (82) P 377-80; Vol 40; No 4; In Ch

01-82-0007 (0007)
STUDIES ON THE FLUOROORGANIC BONDED PHASE AS HPLC PACKINGS. II. PREPARATION, PROPERTIES AND APPLICATION OF PERFLUOROAMIDE-BONDED YWG SILICA
Li, Wenhua; Wang, Jinhua; Lin, Zhiping
Shanghai Inst. Org. Chem. Acad. Sin. Shanghai Peop. Rep. China CN
Huaxue Xuebao; (82) P 49-56; Vol 40; No 1; In Ch

01-82-0008 (0008)
PREPARATION AND EVALUATION OF N-(2-AMINOETHYL)-.GAMMA.-AMINOPROPYLTRIMETHOXY-SILANE-TREATED SILICA COLUMNS FOR HPLC
Okamoto, M.; Yamada, F.
Gifu Prefect. Inst. Public Health Gifu Japan JP 500
HRC CC, J. High Resolut. Chromatogr. Chromatogr. Commun.; (82) P 443-4; Vol 5; No 8

01-82-0009 (0009)
HIGH-PERFORMANCE ADSORPTION CHROMATOGRAPHY OF POLYMERS: SEPARATION OF STYRENE-ACRYLONITRILE COPOLYMERS ON CN SILICA
Danielewicz, M.; Kubin, M.; Vozka, S.
Inst. Macromol. Chem. Czech. Acad. Sci. Prague Czech. CS 162 03/6
J. Appl. Polym. Sci.; (82) P 3629-31; Vol 27; No 9

01-82-0010 (0010)
EVALUATION OF SILICA AND POLAR BONDED COLUMNS FOR LIQUID CHROMATOGRAPHIC ANALYSIS OF TEMEPHOS FORMULATIONS
Mount, Dwight L.; Miles, James W.
Parasit. Dis. Div. Cent. Infect. Dis. Atlanta USA US GA 30333
J. Assoc. Off. Anal. Chem.; (82) P 575-9; Vol 65; No 3

01-82-0011 (0011)
SEPARATION OF SORBITOL AND MANNOHEPTULOSE FROM FRUCTOSE, GLUCOSE AND SUCROSE ON REVERSED-PHASE AND AMINE-MODIFIED HPLC COLUMNS
Shaw, Philip E.; Wilson, Charles W., III
U. S. Citrus Subtrop. Prod. Lab. Winter Haven USA US FL 33880
J. Chromatogr. Sci.; (82) P 209-12; Vol 20; No 5

01-82-0012 (0012)
PREPARATION AND EVALUATION OF BENZYLDIMETHYLCHLOROSILANE-TREATED SILICA COLUMN FOR HIGH-PERFORMANCE LIQUID CHROMATOGRAPHY
Okamoto, Mitsuyoshi; Yamada, Fujizo
Gifu Prefect. Inst. Public Health Gifu Japan JP 500
J. Chromatogr.; (82) P 167-70; Vol 247; No 1

01-82-0013 (0013)
HIGH-PERFORMANCE LIQUID CHROMATOGRAPHIC SEPARATION OF POTATO GLYCOALKALOIDS USING A RADIALLY COMPRESSED AMINO COLUMN
Bushway, R. J.
Dep. Food Sci. Univ. Maine Orono USA US ME 04469
J. Chromatogr.; (82) P 180-3; Vol 247; No 1

01-82-0014 (0014)
TEMPERATURE AND ELUENT EFFECTS ON THE SELECTIVITY OF SOME NITROAROMATIC BONDED PHASES IN HIGH-PERFORMANCE LIQUID CHROMATOGRAPHY
Hammers, W. E.; Theeuwes, A. G. M.; Brederode, W. K.; De Ligny, C. L.
Lab. Anal. Chem. State Univ. Utrecht Neth. NL 3522 AD
J. Chromatogr.; (82) P 321-36; Vol 234; No 2

01-82-0015 (0015)
CORRELATION BETWEEN THE RETENTION OF CARDIAC GLYCOSIDES IN REVERSED-PHASE HIGH-PERFORMANCE LIQUID CHROMATOGRAPHY WITH DIPHENYLSILYL STATIONARY PHASE, THE STRUCTURE OF THEIR MOLECULES AND THEIR BIOLOGICAL ACTIVITY
Daviidov, V. Ya.; Elizalde Gonzalez, M.; Kiselev, A. V.
Chem. Dep. Lomonosov State Univ. Moscow Moscow USSR SU 117234
J. Chromatogr.; (82) P 49-62; Vol 248; No 1

01-82-0016 (0016)
HIGH-PERFORMANCE LIQUID CHROMATOGRAPHY OF INORGANIC AND ORGANIC ANIONS USING ULTRAVIOLET DETECTION AND AN AMINO COLUMN
Cortes, Hernan J.
Anal. Lab. Dow Chem. Co. Midland USA US MI 48640
J. Chromatogr.; (82) P 517-20; Vol 234; No 2

01-82-0017 (0017)
CONTRIBUTION TO THE ELUCIDATION OF THE MECHANISM OF SUGAR RETENTION ON AMINE-MODIFIED SILICA IN LIQUID CHROMATOGRAPHY

Verhaar, L. A. Th.; Kuster, B. F. M.
Dep. Chem. Eng. Eindhoven Univ. Technol. Eindhoven Neth. NL 5600 MB
J. Chromatogr.; (82) P 57-64; Vol 234; No 1

01-82-0018 (0018)
SORBENTS MADE OF CHEMICALLY MODIFIED SILICA GEL FOR EXCLUSION AND ADSORPTION LIQUID CHROMATOGRAPHY
Lazaris, A. Ya.; Beloded, L. N.
USSR SU
Khim. Prom-st., Ser.: Metody Anal. Kontrolya Kach. Prod. Khim. Prom-sti.; (82) P 23-6; No 2; In Russ

5.1.4
Nonpolar Bonded Phases

--- 1982 ---

01-82-0001 (0001)
COMPARISON OF THREE- AND FIVE-MICROMETER COLUMN PACKINGS FOR REVERSED-PHASE LIQUID CHROMATOGRAPHY
Cooke, Nelson H. C.; Archer, Benedict G.; Olsev, Kristine; Berick, Alan
Beckman Instrum., Inc. Berkeley USA US CA P94710
Anal. Chem.; (82) P 2277-83; Vol 54; No 13

01-82-0002 (0002)
COMPARISON OF COLUMNS FOR REVERSED-PHASE LIQUID CHROMATOGRAPHY
Goldberg, Alan P.
E. I. du Pont de Nemours and Co., Inc. Wilmington USA US DE 19898
Anal. Chem.; (82) P 342-5; Vol 54; No 2

01-82-0003 (0003)
THE EFFECT OF END-CAPPING REAGENT ON LIQUID CHROMATOGRAPHIC PERFORMANCE
Lochmueller, C. H.; Marshall, David B.
P. M. Gross Chem. Lab. Duke Univ. Durham USA US NC 27706
Anal. Chim. Acta; (82) P 63-72; Vol 142

01-82-0004 (0004)
APPLICATION OF SELECTIVE CHEMICALLY BONDED STATIONARY PHASES IN REVERSE-PHASE LC
Berendsen, Gert E.; Pikaart, K. A.; De Galan, Leo
Lab. Anal. Chem. Delft Univ. Technol. Delft Neth. NL
Chromatogr. Sci.; (82) P 97-110; Vol 20; No Biol./Biomed. Appl. Liq. Chromatogr. 4

01-82-0005 (0005)
STATIONARY PHASE EFFECTS IN REVERSED-PHASE CHROMATOGRAPHY II. SUBSTITUENT SELECTIVITIES FOR RETENTION ON VARIOUS HYDROCARBONACEOUS BONDED PHASES
Melander, W. R.; Horvath, C.
Dep. Chem. Eng. Yale Univ. New Haven USA US CT 06520
Chromatographia; (82) P 86-90; Vol 15; No 2

01-82-0006 (0006)
STUDIES ON ORGANIC FLUORINE-BONDED PHASE AS HPLC PACKING. IV. 1,1,2,2-TETRAHYDRO-7-TRIFLUOROMETHYLDODECAFLUOROOCTYL-BONDED PHASE PACKING
Li, Wenhua; Wang, Jinhua; Chen, Lifo
Shanghai Inst. Org. Chem. Acad. Sin. Shanghai Peop. Rep. China CN
Huaxue Xuebao; (82) P 1159-65; Vol 40; No 12; In Ch

01-82-0007 (0007)
ENDCAPPING OF HIGH PERFORMANCE LIQUID CHROMATOGRAPHY PACKINGS DERIVED FROM SILICA GEL
Dewaele, C.; Mussche, P.; Verzele, M.
Lab. Organ. Chem. State Univ. Ghent Ghent Belg. BE B-9000
HRC CC, J. High Resolut. Chromatogr. Chromatogr. Commun.; (82) P 616-20; Vol 5; No 11

01-82-0008 (0008)
EVALUATION OF STATIONARY AND MOBILE PHASES FOR REVERSED-PHASE HIGH PERFORMANCE LIQUID CHROMATOGRAPHY OF PEPTIDES
Wehr, C. T.; Correia, L.; Abbott, S. R.
Varian Instrum. Group Walnut Creek USA US CA 94598
J. Chromatogr. Sci.; (82) P 114-19; Vol 20; No 3

01-82-0009 (0009)
PREPARATION AND EVALUATION OF BENZYLDIMETHYLCHLOROSILANE-TREATED SILICA COLUMN

FOR HIGH-PERFORMANCE LIQUID CHROMATOGRAPHY
Okamoto, Mitsuyoshi; Yamada, Fujizo
Gifu Prefect. Inst. Public Health Gifu Japan JP 500
J. Chromatogr.; (82) P 167-70; Vol 247; No 1

01-82-0010 (0010)
ADSORPTION BEHAVIOR OF SEVERAL SUPPORTS IN REVERSED-PHASE THIN-LAYER CHROMATOGRAPHY AS DEMONSTRATED BY THE DETERMINATION OF RELATIVE PARTITION COEFFICIENTS OF SOME 4-HYDROXYCOUMARIN DERIVATIVES
Van der Giesen, Willem F.; Janssen, Lambert H. M.
Fac. Pharm. State Univ. Utrecht Utrecht Neth. NL 3511 GH
J. Chromatogr.; (82) P 199-213; Vol 237; No 2

01-82-0011 (0011)
LIQUID CHROMATOGRAPHIC STUDY OF THE TWO PHYSICAL STATES OF A DENSELY BONDED ALKYL-SILICA AND THE CORRESPONDING RETENTION PROCESSES
Morel, D.; Serpinet, J.
Lab. Chim. Anal. III Univ. Lyon I Villeurbanne Fr. FR 69622
J. Chromatogr.; (82) P 231-40; Vol 248; No 2

01-82-0012 (0012)
COMPARISON OF STYRENE-DIVINYLBENZENE COPOLYMER AND C18 SILICA REVERSED-PHASE PACKINGS FOR HIGH-PERFORMANCE LIQUID CHROMATOGRAPHY OF GIBBERELLINS
Greyson, Richard I.; Patçh, Andrea M.
Dep. Plant Sci. Univ. West. Ontario London Can. CA ON N6A 5B7
J. Chromatogr.; (82) P 349-52; Vol 242; No 2

01-82-0013 (0013)
COMPARISON OF THE HIGH-PERFORMANCE LIQUID CHROMATOGRAPHY OF PEPTIDES AND PROTEINS ON 100- AND 300-.ANG. REVERSED-PHASE SUPPORTS
Wilson, Kenneth J.; Van Wieringen, Erika; Klauser, Stefan; Berchtold, Martin W.; Hughes, Graham J.
Biochem. Inst. Univ. Zurich Zurich Switz. CH CH-8028
J. Chromatogr.; (82) P 407-16; Vol 237; No 3

01-82-0014 (0014)
CORRELATION BETWEEN THE RETENTION OF CARDIAC GLYCOSIDES IN REVERSED-PHASE HIGH-PERFORMANCE LIQUID CHROMATOGRAPHY WITH DIPHENYLSILYL STATIONARY PHASE, THE STRUCTURE OF THEIR MOLECULES AND THEIR BIOLOGICAL ACTIVITY
Daviidov, V. Ya.; Elizalde Gonzalez, M.; Kiselev, A. V.
Chem. Dep. Lomonosov State Univ. Moscow Moscow USSR SU 117234
J. Chromatogr.; (82) P 49-62; Vol 248; No 1

01-82-0015 (0015)
ADSORPTION CHROMATOGRAPHY OF PROTEINS ON SILICONIZED POROUS GLASS
Mizutani, Takaharu; Narihara, Takahiko
Fac. Pharm. Sci. Nagoya City Univ. Nagoya Japan JP 467
J. Chromatogr.; (82) P 755-60; Vol 239

01-82-0016 (0016)
EFFECT OF STATIONARY-PHASE STRUCTURE ON RETENTION AND SELECTIVITY IN REVERSED-PHASE LIQUID CHROMATOGRAPHY
Tanaka, Nobuo; Tokuda, Yasuyuki; Iwaguchi, Kazufusa; Araki, Mikio
Fac. Text. Sci. Kyoto Tech. Univ. Kyoto Japan JP 606
J. Chromatogr.; (82) P 761-72; Vol 239

01-82-0017 (0017)
SELECTIVE EFFECTS OF MOBILE AND STATIONARY PHASES IN REVERSED-PHASE HIGH-PERFORMANCE LIQUID CHROMATOGRAPHY OF ECDYSTEROIDS
Wilson, I. D.; Bielby, C. R.; Morgan, E. D.

Dep. Chem. Univ. Keele Keele/Staffordshire UK GB ST5 5BG
J. Chromatogr.; (82) P 97-102; Vol 238; No 1

01-82-0018 (0018)
SEPARATIONS IN THIN-LAYER AND HIGH-PERFORMANCE LIQUID CHROMATOGRAPHY USING ALKYL SILICA GEL BONDED PHASES
Issaq, Haleem J.; Klose, John R.; Reitz, Gilbert A.
Frederick Cancer Res. Facil. Natl. Cancer Cent. Frederick USA US MD 21701
J. Liq. Chromatogr.; (82) P 1069-80; Vol 5; No 6

01-82-0019 (0019)
QUANTITATIVE DETERMINATION OF METHYL RED ADSORPTION ON STATIONARY PHASES USED IN LIQUID CHROMATOGRAPHY
Wouters, I.; Quintens, I.; Roets, E.; Hoogmartens, J.
Inst. Farm. Wetensch. Katholieke Univ. Leuven Louvain Belg. BE 3000
J. Liq. Chromatogr.; (82) P 25-37; Vol 5; No 1

5.1.5
Miscellaneous Phases

--- 1982 ---

01-82-0001 (0001)
PRECOLUMNS OF AMBERLITE XAD-2 FOR DIRECT INJECTION LIQUID CHROMATOGRAPHIC DETERMINATION OF METHAQUALONE IN BLOOD PLASMA
Hux, Robert A.; Mohammed, Hussain Y.; Cantwell, Frederick F.
Dep. Chem. Univ. Alberta Edmonton Can. CA AB T6G 2G2
Anal. Chem.; (82) P 113-17; Vol 54; No 1

01-82-0002 (0002)
DERIVATIZATION OF KEL-F PARTICLES WITH n-BUTYLLITHIUM FOR LIQUID CHROMATOGRAPHY
Huth, Jeffrey A.; Danielson, Neil D.
Dep. Chem. Miami Univ. Oxford USA US OH 45056
Anal. Chem.; (82) P 930-4; Vol 54, No 5

01-82-0003 (0003)
HIGH-PERFORMANCE LIQUID CHROMATOGRAPHY USING POLYMER GELS FOR DIRECT ANALYSES AND CHARACTERIZATIONS OF SOLUBLE ORGANIC POLLUTANTS IN WATER
Urano, Kohei; Furyune, Hiroyuki; Hayashi, Koji
Dep. Saf. Environ. Eng. Yokohama Natl. Univ. Yokohama Japan JP 240
Bull. Chem. Soc. Jpn.; (82) P 2248-53; Vol 55; No 7

01-82-0004 (0004)
THE SOLVENT ELUOTROPIC STRENGTH ON CARBON ADSORBENTS
Colin, H.; Guiochon, G.; Jandera, P.
Lab. Chim. Anal. Phys. Ec. Polytech. Palaiseau Fr. FR F-91128
Chromatographia; (82) P 133-9; Vol 15; No 2

01-82-0005 (0005)
LIQUID-CHROMATOGRAPHIC DETERMINATION OF PENTOBARBITAL IN PLASMA WITH USE OF A RESIN COLUMN AND AN ALKALINE MOBILE PHASE
Gupta, Ram N.; Smith, Paul T.; Eng, Francis
Clin. Chem. Lab. St. Joseph's Hosp. Hamilton Can. CA ON L8N 1Y4
Clin. Chem. (Winston-Salem, N. C.); (82) P 1772-4; Vol 28; No 8

01-82-0006 (0006)
PHOTOCONTROLLED ADSORPTION CHROMATOGRAPHY FOR LYSOZYME USING AZOAROMATIC POLYMER
Ishihara, Kazuhiko; Negishi, Naoki; Shinohara, Isao
Dep. Polym. Chem. Waseda Univ. Tokyo Japan JP 160
J. Appl. Polym. Sci.; (82) P 1897-902; Vol 27; No 6

01-82-0007 (0007)
REVERSED-PHASE HPLC FROM pH 1 TO 13
Lee, Dan Peter
Hamilton Co. Reno USA US NV 89510
J. Chromatogr. Sci.; (82) P 203-8; Vol 20; No 5

01-82-0008 (0008)
SURFACE MODIFIED OPEN-PORE POLYURETHANE PACKINGS FOR LIQUID CHROMATOGRAPHY
Herman, D. P.; Field, L. R.
Dep. Chem. Univ. Washington Seattle USA US WA 98195
J. Chromatogr. Sci.; (82) P 55-61; Vol 20; No 2

01-82-0009 (0009)
SEPARATION OF INDOLE METABOLITES FROM URINE WITH AN ODS-TYPE RESIN BY HIGH-PERFORMANCE LIQUID CHROMATOGRAPHY
Yamaguchi, Tomoko; Yokota, Kazuko; Uematsu, Fumie
Dep. Biochem. Tokyo Women's Med. Coll. Tokyo Japan JP

J. Chromatogr.; (82) P 166-72; Vol 231; No 1

01-82-0010 (0010)
A METHOD FOR HIGH-PERFORMANCE LIQUID CHROMATOGRAPHIC SCREENING OF UV-POSITIVE COMPONENTS IN URINE ELUATE FROM SEPHADEX G-10 AND MODIFICATIONS FOR DETERMINATION OF URINARY SALICYLIC, SALICYLURIC AND GENTISIC ACIDS
Fornstedt, Nermin
Inst. Biochem. Univ. Uppsala Uppsala Swed. SE S-751 23
J. Chromatogr.; (82) P 253-61; Vol 230; No 2

01-82-0011 (0011)
REVERSED-PHASE LIQUID CHROMATOGRAPHY WITH AMPEROMETRIC DETECTION OF LIPOPHILIC DOPAMINE ANALOGS AND DETERMINATION OF BRAIN AND SERUM CONCENTRATIONS AFTER SAMPLE CLEAN-UP ON SMALL SEPHADEX G-10 COLUMNS
Feenstra, Matthijs G. P.; Homan, Jan Willem; Dijkstra, Durk; Mulder, Theo B. A.; Rollema, Hans; Westerink, Ben H. C.; Horn, Alan S.
Dep. Med. Chem. State Univ. Groningen Groningen Neth. NL 9713 AW
J. Chromatogr.; (82) P 271-87; Vol 20; No 2

01-82-0012 (0012)
COMPARISON OF STYRENE-DIVINYLBENZENE COPOLYMER AND C18 SILICA REVERSED-PHASE PACKINGS FOR HIGH-PERFORMANCE LIQUID CHROMATOGRAPHY OF GIBBERELLINS
Greyson, Richard I.; Patch, Andrea M.
Dep. Plant Sci. Univ. West. Ontario London Can. CA ON N6A 5B7
J. Chromatogr.; (82) P 349-52; Vol 242; No 2

01-82-0013 (0013)
PREPARATION AND PROPERTIES OF STATIONARY PHASES CONTAINING IMMOBILIZED, ELECTRICALLY NEUTRAL NONMACROCYCLIC IONOPHORES FOR LIQUID-SOLID CHROMATOGRAPHY
Grossmann, P.; Simon, W.
Dep. Org. Chem. ETH-Zentrum Zurich Switz. CH CH-8092
J. Chromatogr.; (82) P 351-63; Vol 235; No 2

01-82-0014 (0014)
AMBERLYTE XAD-4 AS A STATIONARY PHASE FOR PREPARATIVE LIQUID CHROMATOGRAPHY IN A RADIALLY COMPRESSED COLUMN
Pietrzyk, Donald J.; Cahill, William J., Jr.
Chem. Dep. Univ. Iowa Iowa City USA US IA 52242
J. Liq. Chromatogr.; (82) 781-95; Vol 5; No 4

01-82-0015 (0015)
PREPARATIVE LIQUID CHROMATOGRAPHIC SEPARATION OF AMINO ACIDS AND PEPTIDES ON AMBERLITE XAD-4
Pietrzyk, Donald J.; Cahill, W. J., Jr.; Stodola, J. D.
Chem. Dep. Univ. Iowa Iowa City USA US IA 52242
J. Liq. Chromatogr.; (82) P 443-61; Vol 5; No 3

01-82-0016 (0016)
DIRECT DETERMINATION OF URINARY CATECHOLAMINES BY HIGH-PERFORMANCE LIQUID CHROMATOGRAPHY WITH BORIC ACID COLUMN
Takeshita, Eiko; Noguchi, Ayako; Uchida, Teruko; Matsumoto, Kojiro ; Kano, Shojiro
Div. Endocrinol., Lab. Med. Keio Univ. Hosp. Tokyo Japan JP 160
Rinsho Byori; (82) P 450-4; Vol 30; No 4; In Japan

01-82-0017 (0017)
SYNTHESIS OF POROUS CROSSLINKED COPOLYMERS OF DIVINYL ETHERS OF HYDROQUINONE AND DIETHYLENE GLYCOL AND THEIR TESTING AS SORBENTS FOR LIQUID CHROMATOGRAPHY
Vokina, O. G.; Tsarik, L. Ya.; Kalabina, A. V.; Kuzaev, A. I.
Irkutsk. Gos. Univ. Irkutsk USSR SU
Vysokomol. Soedin., Ser. A; (82) P 1378-85; Vol 24; No 7; In Russ

CHAPTER 5.2
Columns

5.2.1
Column Technology

--- 1982 ---

02-82-0001 (0001)
HIGH-PERFORMANCE LIQUID CHROMATOGRAPHIC ANALYSIS OF PORPHYRINS AND THEIR ISOMERS WITH RADIAL COMPRESSION COLUMNS
Seubert, Andreas; Seubert, Sigrid
Dep. Dermatol. Univ. Goettingen Goettingen Fed. Rep. Ger. DE D-3400
Anal. Biochem.; (82) P 303-7; Vol 124; No 2

02-82-0002 (0002)
DETERMINATION OF BENZODIAZEPINES IN SERUM BY HIGH PERFORMANCE LIQUID CHROMATOGRAPHY USING RADIALLY COMPRESSED COLUMNS AND AN AQUEOUS METHANOLIC MOBILE PHASE
Kinberger, Bengt; Wahrgren, Peter
Dep. Clin. Chem. Cent. Hosp. Halmstad Swed. SE S-301 85
Anal. Lett.; (82) P 549-57; Vol 15; No B6

02-82-0003 (0003)
HIGH PERFORMANCE LIQUID CHROMATOGRAPHY. PART I. MANUFACTURE OF EFFECTIVE AND REPRODUCTIBLES ANALYTICAL COLUMNS
Saint-Yrieix, A.; Amouroux, J.
IRCHA Vert le Petit Fr. FR 91710
Bull. Soc. Chim. Fr.; (82) P I-225/I-240; No 5-6, Pt. 1; In Fr

02-82-0004 (0004)
LIQUID CHROMATOGRAPHY IN RECTANGULAR CHANNELS
Martin, Michel; Jurado-Baizaval, Jose Luis; Guiochon, Georges
Lab. Chim. Anal. Phys. Ec. Polytech. Palaiseau Fr. FR 91128
C. R. Seances Acad. Sci., Ser. 2; (82) P 579-81; Vol 295; No 5; In Fr

02-82-0005 (0005)
INFLUENCE OF INJECTOR BYPASS ON LIFETIME OF SMALL-PARTICLE LIQUID CHROMATOGRAPHIC COLUMNS
DiCesare, J. L.; Dong, M. W.; Gant, J. R.
Chromatogr. Div. Perkin-Elmer Corp. Norwalk USA US CT 06856
Chromatographia; (82) P 595-8; Vol 15; No 9

02-82-0006 (0006)
A SIMPLE REVERSE-PHASE HIGH PRESSURE LIQUID CHROMATOGRAPHIC DETERMINATION OF CONJUGATED BILE ACIDS IN SERUM AND BILE USING A NOVEL RADIAL COMPRESSION SEPARATION SYSTEM
Ruben, Annie T.; Van Berge-Henegouwen, Gerard P.
Lab. Med. Res. Munic. Hosp. Arnhem Neth. NL 6815 AD
Clin. Chim. Acta; (82) P 41-50; Vol 119; No 1-2

02-82-0007 (0007)
GLASS COLUMN FOR HIGH-PRESSURE LIQUID CHROMATOGRAPHY
Svoboda, Vratislav; Kleimann, Imrich Czech. CS
Czech.; (020182) P 5 pp.; In Czech; Pat. No.: 195043; Int. Class: G01N-031/08; Doc. Code: B; Appl.: CS 77/2693 770425

02-82-0008 (0008)
LIQUID CHROMATOGRAPHY WITH PRECOLUMN SAMPLE PRECONCENTRATION AND ELECTROCHEMICAL DETECTION: DETERMINATION OF AROMATIC AMINES IN ENVIRONMENTAL SAMPLES
Rice, John R.; Kissinger, Peter T.
Dep. Chem. Purdue Univ. West Lafayette USA US IN 47907
Environ. Sci. Technol; (82) P 263-8; Vol 16; No 5

02-82-0009 (0009)
COLUMN TIGHTLY CLOSABLE TOWARDS THE EXTERIOR, PARTICULARLY FOR CHROMATOGRAPHIC PURPOSES AND ITS USE IN HPLC CHROMATOGRAPHY
Knauer, Herbert
Knauer, Herbert, und Co. G.m.b.H. Fed. Rep. Ger. DE
Eur. Pat. Appl.; (120282) P 23 pp.; In Ger; Pat. No.: 40663; Int. Class: G01N-031/08; B01D-015/08; Doc. Code: A1; Appl.: EP 80/710015 800517

02-82-0010 (0010)
ANALYSIS OF MULTIRING AROMATIC HYDROCARBONS BY HPLC USING A REVERSE-PHASE RADIAL COMPRESSION COLUMN SYSTEM
Alexander, Robert; Cumbers, Kimberley M.; Kagi, Robert I.
Sch. Appl. Chem. Western Australian Inst. Technol. Bentley Australia AU 6102
Int. J. Environ. Anal. Chem.; (82) P 161-7; Vol 12; No 3-4

02-82-0011 (0011)
HIGH-PRESSURE LIQUID CHROMATOGRAPHIC DETERMINATION OF AFLATOXINS BY USING RADIAL COMPRESSION SEPARATION
Shepherd, Eric C.; Phillips, Timothy D.; Heidelbaugh, Norman D.; Hayes, A. Wallace
Dep. Vet. Public Health Texas A and M Univ. College Station USA US TX 77843
J. Assoc. Off. Anal. Chem.; (82) P 665-71; Vol 65; No 3

02-82-0012 (0012)
HIGH-PERFORMANCE LIQUID CHROMATOGRAPHY OF BILE ACIDS WITH A REVERSED-PHASE RADIAL COMPRESSION COLUMN
Reid, A. D.; Baker, P. R.
Ninewells Hosp. Med. Sch. Univ. Dep. Surg. Dundee UK GB DD1 9SY
J. Chromatogr.; (82) P 149-56; Vol 247; No 1

02-82-0013 (0013)
HIGH-PERFORMANCE LIQUID CHROMATOGRAPHIC SEPARATION OF POTATO GLYCOALKALOIDS USING A RADIALLY COMPRESSED AMINO COLUMN
Bushway, R. J.
Dep. Food Sci. Univ. Maine Orono USA US ME 04469
J. Chromatogr.; (82) P 180-3; Vol 247; No 1

02-82-0014 (0014)
LIQUID CHROMATOGRAPHIC ASSAY OF PHENOTHIAZINE, THIOXANTHENE AND BUTYROPHENONE NEUROLEPTICS AND ANTIHISTAMINES IN BLOOD AND PLASMA WITH CONVENTIONAL AND RADIAL COMPRESSION COLUMNS AND UV AND ELECTROCHEMICAL DETECTION
Curry, Stephen H.; Brown, Edgar A.; Hu, Oliver Y. P.; Perrin, John H.
Coll. Pharm. Univ. Florida Gainesville USA US FL 32610
J. Chromatogr.; (82) P 361-76; Vol 231; No 2

02-82-0015 (0015)
SIMULTANEOUS DETERMINATION OF THIORIDAZINE AND ITS S-OXIDIZED AND N-DEMETHYLATED METABOLITES USING HIGH-PERFORMANCE LIQUID CHROMATOGRAPHY ON RADIALLY COMPRESSED SILICA
Kilts, Clinton D.; Patrick, Kennerly S.; Breese, George R.; Mailman, Richard B.
Med. Cent. Duke Univ. Durham USA US NC 27710
J. Chromatogr.; (82) P 377-91; Vol 231; No 2

02-82-0016 (0016)
AMBERLYTE XAD-4 AS A STATIONARY PHASE FOR PREPARATIVE LIQUID CHROMATOGRAPHY IN A RADIALLY COMPRESSED COLUMN
Pietrzyk, Donald J.; Cahill, William J., Jr.
Chem. Dep. Univ. Iowa Iowa City USA US IA 52242
J. Liq. Chromatogr.; (82) 781-95; Vol 5; No 4

02-82-0017 (0017)
LOBAR PREPACKED COLUMNS IN COMBINATION WITH A CONVENTIONAL LIQUID CHROMATOGRAPH
Kraus, L.; Linnenbrink, N.; Richter, R.
Univ. Hamburg Hamburg Fed. Rep. Ger. DE D-2000/13
Kontakte (Darmstadt); (82) P 20-1; No 1; In Ger

02-82-0018 (0018)
SELECTION OF STRUCTURAL MATERIALS FOR LIQUID CHROMATOGRAPH AND TECHNOLOGICAL FEATURE OF ALLOY 4200 USE
Bailinson, Yu.Ya.; Martirosov, V.A.; Peskin, F.S.; USSR
Pribory dlya Nauch. Issled. i Automatiz. Eksperim., L.(1982) P 71-5; In Russ
From Ref. Zh., Khim. 1982 Abstr. No. 15G53

5.2.2
Column Packing and Testing

--- 1982 ---

02-82-0001 (0001)
HIGH PERFORMANCE LIQUID CHROMATOGRAPHY. PART 1. MANUFACTURE OF EFFECTIVE AND REPRODUCTIBLES ANALYTICAL COLUMNS
Saint-Yrieix, A.; Amouroux, J.
IRCHA Vert le Petit Fr. FR 91710
Bull. Soc. Chim. Fr.; (82) P I-225/I-240; No 5-6, Pt. 1; In Fr

02-82-0002 (0002)
PACKING OF HPLC COLUMN WITH ODS-C16 MICROSPHERES AND ITS PERFORMANCES
Zhao, Deyong
Beijing Municip. Inspect. Stn. Pharm. Beijing Peop. Rep. China CN
Huaxue Shiji; (82) P 306-7, 301; Vol 4; No 5; In Ch

02-82-0003 (0003)
PACKING HIGH PERFORMANCE LIQUID CHROMATOGRAPHY COLUMNS USING ORGANIC ACIDS AS SLURRY SOLVENTS
Liao, John C.; Ponzo, John L.
Ames Anal. Serv. Miles Lab., Inc. Elkhart USA US IN 46515
J. Chromatogr. Sci.; (82) P 14-18; Vol 20; No 1

CHAPTER 5.3
Injection

--- 1982 ---

03-82-0001 (0001)
SELECTING AN HPLC AUTOMATIC INJECTOR
Walcek, E.; Ball, D.
LC Prod. Micromeritics Corp. USA US
Am. Lab. (Fairfield, Conn.); (82) P 118, 120-1, 123; Vol 14; No 9

03-82-0002 (0002)
EVALUATION OF AUTOMATIC LIQUID SAMPLERS FOR HPLC
Winkelbauer, P.
Lord Corp. USA US
Am. Lab. (Fairfield, Conn.); (82) P 44, 46; Vol 14; No 5

03-82-0003 (0003)
HPLC SAMPLE INJECTION AND COLUMN SWITCHING USING A 10-PORT MULTIFUNCTIONAL VALVE
Harvey, Mack C.; Stearns, Stanley D.
Valco Instrum. Co. Houston USA US TX
Am. Lab. (Fairfield, Conn.); (82) P 68, 70, 72-4; Vol 14; No 5

03-82-0004 (0004)
EFFECT OF WATER ADDITION IN THE SAMPLE VALVE IN REVERSED-PHASE HPLC
Yun, Zihou
Dep. Chem. Beijing Norm. Univ. Beijing Peop. Rep. China CN
Beijing Shifan Daxue Xuebao, Ziran Kexueban; (82) P 91-3; No 3; In Ch

03-82-0005 (0005)
INFLUENCE OF INJECTOR BYPASS ON LIFETIME OF SMALL-PARTICLE LIQUID CHROMATOGRAPHIC COLUMNS
DiCesare, J. L.; Dong, M. W.; Gant, J. R.
Chromatogr. Div. Perkin-Elmer Corp. Norwalk USA US CT 06856
Chromatographia; (82) P 595-8; Vol 15; No 9

03-82-0006 (0006)
HPLC ASSAY OF PYRAZINOIC ACID IN HUMAN PLASMA IN THE PRESENCE OF PYRAZINAMIDE AND OTHER ANTITUBERCULOSIS DRUGS USING AN AUTOMATIC SAMPLER
Ratti, B.; Toselli, A.; Beretta, E.; Bernareggi, A.
Lab. Ric. Gruppo Lepetit S.p.A. Milan Italy IT
Farmaco, Ed. Prat.; (82) P 226-34; Vol 37; No 7

03-82-0007 (0007)
UNIT CONSTRUCTION SYSTEM WITH A SIMPLE ELECTRONIC CIRCUIT FOR AUTOMATIC SAMPLE INJECTION IN HPLC
Neidhart, B.; Lippmann, C.; Abendroth, R.
Inst. Arbeitsphysiol. Univ. Dortmund Dortmund Fed. Rep. Ger. DE 4600/1
GIT Fachz. Lab.; (82) P 32-3; No Suppl. Chromatogr.; In Ger

03-82-0008 (0008)
HPLC SAMPLE INJECTION AND COLUMN SWITCHING USING A 10-PORT MULTIFUNCTIONAL VALVE
Harvey, Mack C.; Stearns, Stanley D.
Valco Instruments Co. USA US
Int. Lab.; (82) P 112, 114, 116, 118; Vol 12; No 5

03-82-0009 (0009)
HPLC SAMPLE INJECTION AND COLUMN SWITCHING USING A MULTIFUNCTIONAL VALVE
Harvey, Mack C.; Stearns, Stanley D.
Valco Instruments Co. USA US
Int. Lab.; (82) P 42, 44, 46-7; Vol 11; No 4

03-82-0010 (0010)
INTRA-COLUMN INJECTION SYSTEM FOR LIQUID CHROMATOGRAPHY
Katz, E. Scott, R. P. W.
Perkin-Elmer Corp. Norwalk USA US CT
J. Chromatogr.; (82) P 191-7; Vol 246; No 2

03-82-0011 (0011)
AUTOMATED SAMPLE INJECTION SYSTEM FOR A LIQUID CHROMATOGRAPH
Shimadzu Seisakusho Ltd. Japan JP
Jpn. Kokai Tokkyo Koho; (042282) P 9 pp.; In Japan; Pat. No.: 82 66358; Int. Class: G01N-031/08; B01D-015/08; Doc. Code: A2; Appl: JP 80/141513 801009

03-82-0012 (0012)
REMOTE-CONTROLLED INJECTION OF RADIOACTIVE SOLUTIONS INTO A HIGH-PERFORMANCE LIQUID CHROMATOGRAPH
Vandewalle, T.; Vandecasteele, C.; Van Haver, D.
Inst. Nucl. Sci. Rijksuniv. Ghent Ghent Belg. BE B-9000
Radiochem. Radioanal. Lett.; (82) P 29-32; Vol 51; No 1

CHAPTER 5.4

Pumping Systems

--- 1982 ---

04-82-0001 (0001)
PUMPS FOR LIQUID CHROMATOGRAPHY AND A CHROMATOGRAPH CONTAINING THESE PUMPS
Smith, Sydney Warren
Applied Chromatography Systems Ltd. UK GB
Ger. Offen.; (060982) P 24 pp.; In Ger; Pat. No.: 3144257; Int. Class: G01N-031/08; F04B-001/06; Doc. Code: A1; Appl.: DE 31257 811107; Priority Appl.: GB 80/35838 801107

04-82-0002 (0002)
OPEN-TUBULAR MICROCAPILLARY LIQUID CHROMATOGRAPHY WITH ELECTROOSMOSIS FLOW USING A UV DETECTOR
Tsuda, Takao; Nomura, Kazuhiro; Nakagawa, Gdnkichi
Lab. Anal. Chem. Nagoya Inst. Technol. Nagoya Japan JP 466
J. Chromatogr.; (82) P 241-7; Vol 248; No 2

04-82-0003 (0003)
INCREASED SIGNAL-TO-NOISE RATIO THROUGH GROUNDING AND THE PROPER SELECTION OF A PUMP FOR HIGH-PERFORMANCE LIQUID CHROMATOGRAPHY WITH ELECTROCHEMICAL DETECTION
Helpern, J. A.; Ewing, J. R.; Welch, K. M. A.
Henry Ford Hosp. Detroit USA US MI 48202
J. Chromatogr.; (82) P 491-2; Vol 240; No 2

04-82-0004 (0004)
HIGH-BACK-PRESSURE LIQUID CHROMATOGRAPHY. II. DEVELOPMENT OF A POST-COLUMN-CONTROLLED FLOW SYSTEM
Takeuchi, Toyohide; Ishii, Daido
Fac. Eng. Nagoya Univ. Nagoya Japan JP 464
J. Chromatogr.; (82) P 633-41; Vol 239

04-82-0005 (0005)
COMPUTATION OF HPLC PARAMETERS AND COMPARISON OF NOISE LEVELS OF A DUAL PISTON RECIPROCATING AND A TRIPLE PISTON PUMP
Gopal, N. G. S.; Sharma, G.
ISOMED Bhabha At. Res. Cent. Bombay India IN 400 085
J. Liq. Chromatogr.; (82) P 869-79; Vol 5; No 5

04-82-0006 (0006)
DEVELOPER SOLVENT MIXING APPARATUS FOR LIQUID CHROMATOGRAPHY
Shimadzu Seisakusho Ltd. Japan JP
Jpn. Kokai Tokkyo Koho; (011182) P 3 pp.; In Japan; Pat. No.: 82 04548; Int. Class: G01N-031/08; B01D-015/08; Doc. Code: A2; Appl.: JP 80/78667 800610

04-82-0007 (0007)
HPLC SOLVENT GRADIENT FORMATION
Riggenmann, Hansjuergen
Gynkotek Ges. Bau Wissenschaftlich-Tech. Gerate m.b.H. Munich Fed. Rep. Ger. DE 8000/21
LaborPraxis; (82) P 474, 476, 478, 481; Vol 6; No 5

04-82-0008 (0008)
SOLVENT MIXING IN HPLC USING LOW PRESSURE SOLVENT METERING PUMPS
Baldwin, Lawrence G.; Ehrlich, Burney J.; Dixon, Jack B.
Tracor, Inc.
USA US
U.S.; (011982) P 25 pp.; Pat. No.: 4311586; Int. Class: B01D-015/08; Nat. Class: 210-101; Doc. Code: A; Appl.: US 142679 800422

04-82-0009 (0009)
LIQUID CHROMATOGRAPHIC PUMP MODULE
Brownlee, Robert
USA US
U.S., (083182) P 7 pp.; Pat. No.: 4347131; Int. Class: B01D-015/08; Nat. Class: 210-101; Doc. Code: A; Appl.: US 258404 810428

CHAPTER 5.5
Various Equipment

--- 1982 ---

05-82-0001 (0001)
AN ADVANCED LC SYSTEM
Conlon, R. D.; Ettre, L. S.; Schmid, C. E.; Schwartz, A.
Chromatogr. Div. Perkin-Elmer Corp. Norwalk USA US CT 06856
Am. Lab. (Fairfield, Conn.); (82) P 104-9, 111; Vol 14; No 9

05-82-0002 (0002)
MICROSAMPLE-FILTERING DEVICE FOR LIQUID CHROMATOGRAPHY OR FLOW INJECTION ANALYSIS
Gardner, Wayne S.; Vanderploeg, Henry A.
Great Lakes Environ. Res. Lab. Natl. Oceanic Atmos. Adm. Ann Arbor USA US MI 48104
Anal. Chem.; (82) P 2129-30; Vol 54; No 12

05-82-0003 (0003)
PRESSURE ACTIVATED SHUTDOWN FOR LIQUID CHROMATOGRAPHY SYSTEMS
Lin, Peter Y. T.; Blank, C. LeRoy
Dep. Chem. Univ. Oklahoma Norman USA US OK 73019
Anal. Chem.; (82) P 833-5; Vol 54; No 4

05-82-0004 (0004)
AN INTERFACE FOR THE COMBINATION OF MICRO HIGH-PERFORMANCE LIQUID CHROMATOGRAPHY AND INFRARED SPECTROMETRY
Jinno, K.; Fujimoto, C.; Hirata, Y.
Sch. Mater. Sci. Toyohashi Univ. Technol. Toyohashi Japan JP 440
Appl. Spectrosc.; (82) P 67-9; Vol 36; No 1

05-82-0005 (0005)
APPARATUS FOR SOLVENT DEGASSING AND SOLVENT REPLENISHING IN LIQUID CHROMATOGRAPHS
Schrenker, Helge; Hupe, Peter
Hewlett-Packard G.m.b.H. Fed. Rep. Ger. DE
Ger. Offen.; (011482) P 10 pp.; In Ger; Pat. No.: 3023383; Int. Class: B01D-019/00; B01D-015/08; Doc. Code: A1; Appl.: DE 3023383 800623

05-82-0006 (0006)
THERMOSTAT DEVICE FOR A LIQUID CHROMATOGRAPH
Schrenker, Helge
Hewlett-Packard Co. USA US
Ger. Offen.; (012882) P 9 pp.; In Ger; Pat. No.: 3026267; Int. Class: B01D-015/08; G01N-033/80; Doc. Code: A1; Appl.: DE 3026267 800711

05-82-0007 (0007)
FLOW RESTRICTOR FOR ADJUSTING BACKPRESSURE IN LIQUID CHROMATOGRAPHY DETECTORS
Nikelly, J. G.
Dep. Chem. Philadelphia Coll. Pharm. Sci. Philadelphia USA US PA 19104
HRC CC, J. High Resolut. Chromatogr. Chromatogr. Commun.; (82) P 506-7; Vol 5; No 9

05-82-0008 (0008)
INTEGRATED SYSTEMS MAKE HPLC BEHAVIOR RELIABLE, VERSATILE
Nelson, John
Waters Assoc. Milford USA US MA
Ind. Res. Dev.; (82) P 100-4; Vol 24; No 4

05-82-0009 (0009)
EQUIPMENT FOR HPLC. V
McNair, H. M.

Dep. Chem. Virginia Polytech. Inst. State Univ. Blacksburg USA US VA 24061
J. Chromatogr. Sci.; (82) P 537-50; Vol 20; No 12

05-82-0010 (0010)
INFLUENCE OF THE PACKING MATERIAL AND THE COLUMN FILTERS ON THE RELIABILITY OF A HIGH-PERFORMANCE LIQUID CHROMATOGRAPH - MASS SPECTROMETER INTERFACE BASED ON THE DIRECT LIQUID INLET PRINCIPLE
Mauchamp, Bernard; Krien, Paul
Lab. Phytopharm. Inst. Natl. Rech. Agron. Versailles Fr. FR F-78000
J. Chromatogr.; (82) P 17-24; Vol 236; No 1

05-82-0011 (0011)
USE OF CONVENTIONAL INSTRUMENTATION WITH MICROBORE COLUMNS IN HIGH-PERFORMANCE LIQUID CHROMATOGRAPHY
Kok, W. T.; Brinkman, U. A. T.; Frei, R. W.; Hanekamp, H. B.; Nooitgedacht, F.; Poppe, H.
Dep. Anal. Chem. Free Univ. Amsterdam Neth. NL 1081 HV
J. Chromatogr.; (82) P 357-69; Vol 237; No 3

05-82-0012 (0012)
RAPID SQUARE WAVE INTERFERENCE ON HIGH-PERFORMANCE LIQUID CHROMATOGRAPHIC OUTPUT CAUSED BY A DIGITAL DISPLAY
Bosanquet, Andrew G.; Tooley, Mark A.; Gilby, Edward D.
Dep. Clin. Invest. R. United Hosp. Bath/Avon UK GB BA1 3NG
J. Chromatogr.; (82) P 365-7; Vol 244; No 2

05-82-0013 (0013)
COMPLETE COMPUTER AUTOMATION OF PREPARATIVE LIQUID CHROMATOGRAPHY THROUGH INTELLIGENT FRACTION COLLECTION, WITH UNLIMITED INJECTION VOLUME AND REPETITIVE COLLECTION OF SEPARATED SOLUTE PEAKS
Sugnaux, Francois R.; Djerassi, Carl
Dep. Chem. Stanford Univ. Stanford USA US CA 94305
J. Chromatogr.; (82) P 372-89; Vol 248; No 3

05-82-0014 (0014)
SIMPLE PROGRAMMABLE CONTROLLER ALLOWING THE TIMED COLLECTION OF FRACTIONS IN HIGH-PERFORMANCE LIQUID CHROMATOGRAPHY
Smith, A. l.; McDermott, E. R.; Biggins, J. A.; Boakes, R. J.
MRC Neuroendocrinol. Unit Newcastle Gen. Hosp. Newcastle upon Tyne UK GB NE4 6BE
J. Chromatogr.; (82) P 489-92; Vol 236; No 2

05-82-0015 (0015)
HIGH-BACK-PRESSURE LIQUID CHROMATOGRAPHY. II. DEVELOPMENT OF A POST-COLUMN-CONTROLLED FLOW SYSTEM
Takeuchi, Toyohide; Ishii, Daido
Fac. Eng. Nagoya Univ. Nagoya Japan JP 464
J. Chromatogr.; (82) P 633-41; Vol 239

05-82-0016 (0016)
NUTRITIONAL ANALYSIS OF FOOD MATERIALS AND ITS APPARATUS. HIGH PERFORMANCE LIQUID CHROMATOGRAPHY
Ueda, Suehiro
Chimazu Seisakusho K. K. Japan JP
Jpn. Fudo Saiensu; (82) P 49-57; Vol 21; No 9; In Japan

CHAPTER 5.6
Automation

--- 1982 ---

06-82-0001 (0001)
COMPUTERIZED LIQUID CHROMATOGRAPH
Hitachi. Ltd. Jpn. Kokai Tokyo Koho JP 82 70,458 (Cl. G 01N31/08)
30 Apr. 1982, Appl. 80/146,925, 22 Oct. 1980; 4 pp.;(82)

06-82-0002 (0002)
AUTOMATIC ANALYSIS OF BIOGENIC CATECHOLAMINES BY HIGH-PERFORMANCE LIQUID CHROMATOGRAPHY WITH FLUOROMETRIC REACTION DETECTION
Mori, K.
Natl. Inst. Ind. Health Kawasaki Japan JP 213
Adv. Biosci.; (82) P 275-84; Vol 36; No Synth., Storage Secretion Adrenal Catecholamines

06-82-0003 (0003)
EVALUATION OF AUTOMATIC LIQUID SAMPLERS FOR HPLC
Winkelbauer, P.
Lord Corp. USA US
Am. Lab. (Fairfield, Conn.); (82) P 44, 46; Vol 14; No 5

06-82-0004 (0004)
AN AUTOMATED UV DETECTOR FOR LC
Burce, Gary L.; Klotter, Kevin
USA US
Am. Lab. (Fairfield, Conn.); (82) P 74, 76, 78-9, 81-5; Vol 14; No 3

06-82-0005 (0005)
QUANTITATION OF MULTIPLE STEROIDS IN HUMAN SERUM AND URINE BY COMBINATION OF AUTOMATIC HPLC AND RIA
Kage, A.; Fenner, A.; Weber, B.; Schoeneshoefer, M.
Inst. Clin. Chem. Biochem. FU Berlin Berlin Fed. Rep. Ger. DE D-1000/45
Anal. Chem. Symp. Ser.; (82) P 327-32; Vol 10; No Adv. Steroid Anal.

06-82-0006 (0006)
TRANSIENT ELECTROCHEMICAL TECHNIQUES IN LIQUID CHROMATOGRAPHY WITH MICROPROCESSOR-BASED INSTRUMENTATION
Anderson, J. E.; Bond, A. M.; Heritage, I. D.; Jones, R. D.; Wallace, G. G.
Div. Chem. Phys. Sci. Deakin Univ. Waurn Ponds Australia AU 3217
Anal. Chem.; (82) P 1702-5; Vol 54; No 11

06-82-0007 (0007)
AUTOMATIC DIALYZING-INJECTION SYSTEM FOR LIQUID CHROMATOGRAPHY OF IONS AND SMALL MOLECULES
Nordmeyer, Francis R.; Hansen, Lee D.
Dep. Chem. Brigham Young Univ. Provo USA US UT 84602
Anal. Chem.; (82) P 2605-7; Vol 54; No 14

06-82-0008 (0008)
A RELIABLE SEMIAUTOMATED METHOD FOR THE DETERMINATION OF TOTAL THIAMINE IN WHOLE BLOOD BY THE THIOCHROME METHOD WITH HIGH-PERFORMANCE LIQUID CHROMATOGRAPHY
Schrijver, Jaap; Speek, Andries J.; Klosse, Jan A.; Van Rijn, Herman J. M.; Schreurs, Wil H. P.
Dep. Clin. Biochem. Inst. CIVO-Toxicol. Nutr. Zeist Neth. NL
Ann. Clin. Biochem.; (82) P 52-6; Vol 19; No 1

06-82-0009 (0009)
QUANTITATIVE DETERMINATION OF SUPROFEN IN HUMAN PLASMA AND URINE BY FULLY AUTOMATED HIGH-PERFORMANCE LIQUID CHROMATOGRAPHY

Mueller, H.; Zulliger, Hans W.; Doerig, B. A.; Egli, R. A.
Pharm. Res. Dev. Dep. Cilag A.-G. Ltd. Schaffhausen Switz. CH CH-8201
Arzneim.-Forsch.; (82) P 257-60; Vol 32; No 3

06-82-0010 (0010)
AUTOMATED ROUTINE ESTIMATION OF NEOPTERIN IN HUMAN URINE BY HPLC ON REVERSED PHASE
Fuchs, Dietmar; Hausen, Arno; Reibnegger, Gilbert; Wachter, Helmut
Inst. Med. Chem. Biochem. Univ. Innsbruck Innsbruck Austria AT A-6020
Biochem. Clin. Aspects Pteridines; (82) P 67-79; Vol 1; No Cancer, Immunol., Metab. Dis.

06-82-0011 (0011)
AUTOMATIC HPLC OF PTERINS WITH OR WITHOUT COLUMN SWITCHING
Niederwieser, A.; Staudenmann, W.; Wetzel, E.
Dep. Pediatr. Univ. Zurich Zurich Switz. CH CH-8032
Biochem. Clin. Aspects Pteridines; (82) P 81-102; Vol 1; No Cancer, Immunol., Metab. Dis.

06-82-0012 (0012)
A NEW LOW COST, FULLY AUTOMATED AMINO ACID ANALYZER USING A GRADIENT HPLC
Klapper, David G.
Dep. Bacteriol. Immunol. Univ. North Carolina Chapel Hill USA US NC
CONFERENCE; Methods Protein Sequence Anal., (Proc. Int. Conf.), 4th; (82) P 509-15; Avail.: Elzinga, Marshall; Date: 81; Publisher: Humana, Clifton, N. J.

06-82-0013 (0013)
HPLC ASSAY OF PYRAZINOIC ACID IN HUMAN PLASMA IN THE PRESENCE OF PYRAZINAMIDE AND OTHER ANTITUBERCULOSIS DRUGS USING AN AUTOMATIC SAMPLER
Ratti, B.; Toselli, A.; Beretta, E.; Bernareggi, A.
Lab. Ric. Gruppo Lepetit S.p.A. Milan Italy IT
Farmaco, Ed. Prat.; (82) P 226-34; Vol 37; No 7

06-82-0014 (0014)
AUTOMATED HPLC SEPARATES SAMPLES IN LAB QUANTITIES
DiCesare, J. L.; Vandemark, F. L.
Perkin-Elmer Corp. Norwalk USA US CT
Ind. Res. Dev.; (82) P 138-43; Vol 24; No 2

06-82-0015 (0015)
AN AUTOMATED UV DETECTOR FOR LC
Burce, Gary L.; Klotter, Kevin
Varian Instrument Group USA US
Int. Lab.; (82) P 26, 28-32, 34, 36, 38; Vol 12; No 6

06-82-0016 (0016)
A FULLY AUTOMATED HIGH-PERFORMANCE LIQUID CHROMATOGRAPHIC PROCEDURE FOR ISOLATION AND PURIFICATION OF AMADORI COMPOUNDS
Moll, Nicole; Gross, Bernard; That Vinh; Moll, Manfred
Lab. Chim. Org. III Univ. Nancy I Vandoeuvre les Nancy Fr. FR 54506
J. Agric. Food Chem.; (82) P 782-6; Vol 30; No 4

06-82-0017 (0017)
AUTOMATED HIGH-PERFORMANCE LIQUID CHROMATOGRAPHIC DETERMINATION OF INTERMEDIATES AND SIDE REACTION PRODUCTS IN FD & C RED NO. 3
Goldberg, Allen L.; Calvey, Robert J.
Div. Color Technol. Food and Drug Adm. Washington USA US DC 20204
J. Assoc. Off. Anal. Chem.; (82) P 103-7; Vol 65; No 1

06-82-0018 (0018)
AUTOMATED HIGH-PRESSURE LIQUID CHROMATOGRAPHIC SYSTEM FOR DETERMINATION OF MANNITOL, SORBITOL, AND XYLITOL IN CHEWING GUMS AND CONFECTIONS
Samarco, Eugene C.; Parente, Ernest S.
Res. Dev. Dep. Warner-Lambert Co. Morris Plains USA US NJ 07950
J. Assoc. Off. Anal. Chem.; (82) P 76-8; Vol 65; No 1

06-82-0019 (0019)
AUTOMATED DETERMINATION OF 5-HYDROXYINDOLYLACETIC ACID IN URINE BY HIGH-PERFORMANCE LIQUID CHROMATOGRAPHY
Garnier, J. P.; Bousquet, B.; Dreux, C.
Lab. Biochim. Hop. Saint-Louis Paris Fr. FR 75010
J. Autom. Chem.; (82) P 65-8; Vol 4; No 2

06-82-0020 (0020)
AN AUTOMATED HPLC METHOD FOR THE ASSAY OF PROPRANOLOL AND ITS BASIC METABOLITES IN PLASMA AND URINE
Lo, Man Wai; Silber, Bernie; Riegelman, Sidney
Sch. Pharm. Univ. California San Francisco USA US CA 94143
J. Chromatogr. Sci.; (82) P 126-31; Vol 20; No 3

06-82-0021 (0021)
AN AUTOMATED HPLC ASSAY FOR SIMULTANEOUS QUANTITATION OF METHYLATED XANTHINES AND URIC ACIDS IN URINE
Tang-Liu, D. Dan-Shya; Riegelman, Sidney
Dep. Pharm. Chem. Univ. California San Francisco USA US CA 94143
J. Chromatogr. Sci.; (82) P 155-9; Vol 20; No 4

06-82-0022 (0022)
FEMTOMOLAR ION-PAIR HIGH-PERFORMANCE LIQUID CHROMATOGRAPHIC METHOD FOR DETERMINING DNS-POLYAMINE DERIVATIVES OF RED BLOOD CELL EXTRACTS UTILIZING AN AUTOMATED POLYAMINE ANALYZER
Brown, Nesbitt D.; Strickler, M. Patricia; Whaun, June M.
Div. Biochem. Walter Reed Army Inst. Res. Washington USA US DC 20012
J. Chromatogr.; (82) P 101-8; Vol 245; No 1

06-82-0023 (0023)
AUTOMATED HIGH-PERFORMANCE LIQUID CHROMATOGRAPHIC METHOD FOR THE DETERMINATION OF ANTIPYRINE AND ITS METABOLITES IN URINE. SOME PRELIMINARY RESULTS OBTAINED FROM SMOKERS AND NONSMOKERS
Nakagawa, Akihiko; Nakamura, Kanichi; Ishizaki, Takashi; Chiba, Kan
Anal. Metab. Res. Lab. Sankyo Co., Ltd. Tokyo Japan JP 140
J. Chromatogr.; (82) P 349-60; Vol 231; No 2

06-82-0024 (0024)
COMPLETE COMPUTER AUTOMATION OF PREPARATIVE LIQUID CHROMATOGRAPHY THROUGH INTELLIGENT FRACTION COLLECTION, WITH UNLIMITED INJECTION VOLUME AND REPETITIVE COLLECTION OF SEPARATED SOLUTE PEAKS
Sugnaux, Francois R.; Djerassi, Carl
Dep. Chem. Stanford Univ. Stanford USA US CA 94305
J. Chromatogr.; (82) P 372-89; Vol 248; No 3

06-82-0025 (0025)
AROMATIC HYDROCARBON METABOLITES IN FISH: AUTOMATED EXTRACTION AND HIGH-PERFORMANCE LIQUID-CHROMATOGRAPHIC SEPARATION INTO CONJUGATE AND NONCONJUGATE FRACTIONS
Krahn, Margaret M.; Collier, Tracy K.; Malins, Donald C.
Environ. Conservation Div. Northwest and Alaska Fis. Cent. Seattle USA US WA 98112
J. Chromatogr.; (82) P 441-52; Vol 236; No 2

06-82-0026 (0026)
AUTOMATION AND DATA PROCESSING IN HPLC
Jeuken, S.
Fa. Hewlett Packard G.m.b.H. Frankfurt Fed. Rep. Ger. DE D-6000/56
Monatsschr. Brau.; (82) P 60-2; Vol 35; No 2; In Ger

06-82-0027 (0027)
EVALUATION OF AN AUTOMATED HPLC SYSTEM HAVING MULTIPLE DETECTORS
Klee, M. S.; Rogers, L. B.; Carreira, L. A.; Azarraga, L.
Dep. Chem. Univ. Georgia Athens USA US GA 30602
Pergamon Ser. Environ. Sci.; (82) P 177-92; Vol 7; No Anal. Tech. Environ. Chem. 2

06-82-0028 (0028)
SIMPLIFIED METHOD FOR THE DETERMINATION OF 1,25-DIHYDROXYVITAMIN D USING AUTOMATED HIGH PRESSURE LIQUID CHROMATOGRAPHY: APPLICATION TO THE DIFFERENTIAL DIAGNOSIS OF HYPERCALCEMIA
Endres, D.; Lu, J.; Mueller, J.; Adams, J.; Holick, M.; Broughton, A.
Nichols Inst. San Juan Capistrano USA US CA 92675
Proc. Workshop Vitam. D; (82) P 813-15; Vol 5th; No Vitam. D: Chem., Biochem. Clin. Endocrinol. Calcium Metab.

06-82-0029 (0029)
REMOTE-CONTROLLED INJECTION OF RADIOACTIVE SOLUTIONS INTO A HIGH-PERFORMANCE LIQUID CHROMATOGRAPH
Vandewalle, T.; Vandecasteele, C.; Van Haver, D.
Inst. Nucl. Sci. Rijksuniv. Gent Gent Belg. BE B-9000
Radiochem. Radioanal. Lett.; (82) P 29-32; Vol 51; No 1

CHAPTER 5.7
Detection

5.7.1
Generalities

--- 1982 ---

07-82-0001 (0001)
ADVANCES IN DETECTOR TECHNOLOGY IN HIGH-PERFORMANCE LIQUID CHROMATOGRAPHY
Knox, John H.
Dep. Chem. Univ. Edinburgh Edinburgh UK GB EH9 3JJ
Anal. Proc. (London); (82) P 166-70; Vol 19; No 4

07-82-0002 (0002)
THE SELECTION OF OPTIMAL CONDITIONS IN HPLC. II. THE INFLUENCE OF COLUMN DIMENSIONS AND SAMPLE SIZE ON SOLUTE DETECTION
Lauer, H. H.; Rozing, G. P.
Hewlett-Packard G.m.b.H. Waldbronn Fed. Rep. Ger. DE D-7517/2
Chromatographia; (82) P 409-13; Vol 15; No 7

07-82-0003 (0003)
FLOW RESTRICTOR FOR ADJUSTING BACKPRESSURE IN LIQUID CHROMATOGRAPHY DETECTORS
Nikelly, J. G.
Dep. Chem. Philadelphia Coll. Pharm. Sci. Philadelphia USA US PA 19104
HRC CC, J. High Resolut. Chromatogr. Chromatogr. Commun.; (82) P 506-7; Vol 5; No 9

07-82-0004 (0004)
NEW WAYS TO INCREASE THE SPECIFICITY OF DETECTION IN LIQUID CHROMATOGRAPHY
Di Cesare, J. L.; Ettre, L. S.
Chromatogr. Div. Perkin-Elmer Corp. Norwalk USA US CT 06856
J. Chromatogr.; (82) P 1-16; Vol 251; No 1

07-82-0005 (0005)
ADVERSE EFFECTS OF LAG FILTERS USED FOR DETECTOR BASELINE NOISE SUPPRESSION IN HIGH-PERFORMANCE LIQUID CHROMATOGRAPHY
Haddad, P. R.; Keating, R. W.; Low, G. K. C.
Dep. Anal. Chem. Univ. New South Wales Kensington Australia AU 2033
J. Liq. Chromatogr.; (82) P 853-67; Vol 5; No 5

07-82-0006 (0006)
PEAK HOMOGENEITY DETERMINATION FOR THE VALIDATION OF HIGH-PERFORMANCE LIQUID CHROMATOGRAPHY ASSAY METHODS
Carter, G. T.; Schiesswohl, R. E.; Burke, H.; Yang, R.
Phys. Anal. Chem. Dep. Ciba-Geigy Corp. Suffern USA US NY 10901
J. Pharm. Sci.; (82) P 317-21; Vol 71; No 3

07-82-0007 (0007)
OPTICAL DETECTORS IN HPLC - POSSIBILITIES AND TRENDS
Baumann, Wolfram
Inst. Phys. Chem. Univ. Mainz Mainz Fed. Rep. Ger. DE 6500
LaborPraxis; (82) P 861-3; Vol 6; No 7-8; In Ger

07-82-0008 (0008)
DETECTORS IN HPLC
Heisz, O.
Abt. VWA Philips G.m.b.H. Kassel Fed. Rep. Ger. DE D-3500
Monatsschr. Brau.; (82) P 56-60; Vol 35; No 2; In Ger

07-82-0009 (0009)
DETECTION LIMITS IN LIQUID CHROMATOGRAPHY
Guiochon, G.; Colin, H.
Lab. Chim. Anal. Ec. Polytech. Palaiseau Fr. FR
Pergamon Ser. Environ. Sci.; (82) P 169-76; Vol 7; No Anal. Tech. Environ. Chem. 2

5.7.2
UV/VIS—Generalities

--- 1982 ---

07-82-0001 (0001)
AN AUTOMATED UV DETECTOR FOR LC
Burce, Gary L.; Klotter, Kevin
USA US
Am. Lab. (Fairfield, Conn.); (82) P 74, 76, 78-9, 81-5; Vol 14; No 3

07-82-0002 (0002)
REFRACTIVE INDEX AND ABSORPTION DETECTOR FOR LIQUID CHROMATOGRAPHY BASED ON FABRY-PEROT INTERFEROMETRY
Woodruff, Steven D.; Yeung, Edward S.
Ames Lab. Iowa State Univ. Ames USA US IA 50011
Anal. Chem.; (82) P 1174-8; Vol 54; No 7

07-82-0003 (0003)
EFFECT OF IONIC STRENGTH ON RETENTION AND DETECTOR RESPONSE IN REVERSED-PHASE ION-PAIR LIQUID CHROMATOGRAPHY WITH ULTRAVIOLET-ABSORBING ION INTERACTION REAGENTS
Bidlingmeyer, Brian A.; Warren, F. Vincent, Jr.
Waters Assoc., Inc. Milford USA US MA 01757
Anal. Chem.; (82) P 2351-6; Vol 54; No 13

07-82-0004 (0004)
SPECTROPHOTOMETRIC AND POLARIMETRIC DETECTORS IN LIQUID CHROMATOGRAPHY FOR THE DETERMINATION OF ENANTIOMER RATIOS IN COMPLEX MIXTURES
Boehme, W.; Wagner, G. Oehme, U.; Priesnitz, U.
Bodenseewerk Perkin-Elmer Ueberlingen Fed. Rep. Ger. DE
Anal. Chem.; (82) P 709-11; Vol 54; No 4

07-82-0005 (0005)
LIQUID PHASE CHROMATOGRAPHY OF JASMINE CONCRETES. ELUANT ANALYSIS BY UV-VISIBLE SPECTROPHOTOMETER WITH PARALLEL DETECTION
Linossier, J. L.; Peyron, L.; Rocca, J. L.
Cent. Rech. Saint-Fons Rhone-Poulenc Rech. Saint-Fons Fr. FR 69190
Analusis; (82) P 310-18; Vol 10; No 7; In Fr

07-82-0006 (0006)
CONTINUOUS AND INSTANTANEOUS RECORDING METHOD OF THREE DIMENSIONAL CHROMATOGRAM IN LINEAR PHOTODIODE ARRAY SPECTROMETER SYSTEM FOR LIQUID CHROMATOGRAPHY
Terada, Makoto; Jawamoto, Hiroshia; Ichise, Mitsunojo; Kojima, Tsugio
Fac. Eng. Kyot Univ. Kyoto Japan JP
Bunseki Kagaku; (82) P 6-12; Vol 31; No 1; In Japan

07-82-0007 (0007)
RAPID SCAN UV/VISIBLE DETECTION IN LIQUID CHROMATOGRAPHY . APPLICATION TO ENVIRONMENTAL ANALYSIS
Warwick, C. J.; Bagon, D. A.
Occup. Med. Hyg. Lab. London UK GB NW2 6LN
Chromatographia; (82) P 433-6; Vol 15; No 7

07-82-0008 (0008)
HPLC WITH UV-VIS SPECTROPHOTOMETRIC DETECTION FOR BLOOD AND TISSUE DRUG ASSAYS
Desager, J. P.; Sclavons, M.
Lab. Pharmacother. Univ. Cathol. Louvain Brussels Belg. BE B-1200
Chromatographia; (82) P 451-2; Vol 15; No 7

07-82-0009 (0009)
ANALYSIS OF ANTIEPILEPTIC DRUGS BY HPLC AND UV DETECTION WITH PHOTODIODE

ARRAYS
Dorner, Wolf G.
Inzell Fed. Rep. Ger. DE 8221
GIT Labor-Med.; (82) P 352-3, 356-7; Vol 5; No 5; In Ger

07-82-0010 (0010)
SCREENING FOR UNKNOWN COMPOUNDS BY HPLC USING A MULTICHANNEL DIODE ARRAY UV-VIS SPECTROPHOTOMETER: APPLICATION TO DRUG METABOLISM
Ghijsen, R. T.; Drenth, B. H. F.; Overzet, F.; De Zeeuw, R. A.
Dep. Toxicol. State Univ. Groningen Groningen Neth. NL 9713 AW
HRC CC, J. High Resolut. Chromatogr. Chromatogr. Commun.; (82) P 192-8; Vol 5; No 4

07-82-0011 (0011)
SIMULTANEOUS USE OF FLUORESCENCE, ULTRAVIOLET, AND ELECTROCHEMICAL DETECTORS IN HIGH-PERFORMANCE LIQUID CHROMATOGRAPHY - SEPARATION AND IDENTIFICATION OF PHENOLIC ANTIOXIDANTS AND RELATED COMPOUNDS
Masoud, A. N.; Cha, Young Nam
Dep. Environ. Health Sci. Johns Hopkins Univ. Baltimore USA US MD 21205
HRC CC, J. High Resolut. Chromatogr. Chromatogr. Commun.; (82) P 299-305; Vol 5; No 6

07-82-0012 (0012)
AN AUTOMATED UV DETECTOR FOR LC
Burce, Gary L.; Klotter, Kevin
Varian Instrument Group USA US
Int. Lab.; (82) P 26, 28-32, 34, 36, 38; Vol 12; No 6

07-82-0013 (0013)
OVERALL CHARACTERISTICS OF A LIQUID CHROMATOGRAPHIC DETECTION SYSTEM USING A SILICON PHOTODIODE ARRAY
Takashi, Amita; Mitsunajo, Ichise; Tsugio, Kojima
Depart. of Ind. Chem., Fac. of Engineering, Kyoto Univ., Kyoto Japan
J. Chromatogr. (82) P 89-98; Vol 234; No 1

07-82-0014 (0014)
UNIVERSAL, MULTICHANNEL ULTRAVIOLET DETECTION IN THE PURITY ANALYSIS OF 2-ETHYL-3-(4-HYDROXYBENZOYL)INDOLIZINE BY HIGH-PERFORMANCE LIQUID CHROMATOGRAPHY
Drenth, B. F. H.; Ghijsen, R. T.; De Zeeuw, R. A.
Dep. Toxicol. State Univ. Groningen Neth. NL 9713 AW
J. Chromatogr.; (82) P 113-20; Vol 238; No 1

07-82-0015 (0015)
MULTICHANNEL DIODE ARRAY UV-VISIBLE SPECTROPHOTOMETER AS DETECTOR IN SCREENING FOR UNKNOWN BUTOPROZINE METABOLITES IN DOG BILE BY HIGH-PERFORMANCE LIQUID CHROMATOGRAPHY
Overzet, Fokko; Ghijsen, Rudy T.; Drenth, Ben F. H.; De Zeeuw, Rokus A.
Dep. Toxicol. State Univ. Groningen Groningen Neth. NL 9713 AW
J. Chromatogr.; (82) P 190-5; Vol 240; No 1

07-82-0016 (0016)
OPEN-TUBULAR MICROCAPILLARY LIQUID CHROMATOGRAPHY WITH ELECTROOSMOSIS FLOW USING A UV DETECTOR
Tsuda, Takao; Nomura, Kazuhiro; Nakagawa, Gonkichi
Lab. Anal. Chem. Nagoya Inst. Technol. Nagoya Japan JP 466
J. Chromatogr.; (82) P 241-7; Vol 248; No 2

07-82-0017 (0017)
UNIVERSAL LIQUID CHROMATOGRAPHY METHODS. II. SENSITIVE, LOW-WAVELENGTH,

GRADIENT REVERSED-PHASE METHODS
Berry, Vern V.
Cent. Anal. Res. Polaroid Corp. Cambridge USA US MA 02139
J. Chromatogr.; (82) P 279-96; Vol 236; No 2

07-82-0018 (0018)
MULTIPARALLEL DETECTION IN HIGH-PERFORMANCE LIQUID CHROMATOGRAPHY
Yoshida, Hisanobu; Koto, Shozo; Akimoto, Motohisa; Nakajima, Terumi
Sch. Med. Hiroshima Univ. Hiroshima Japan JP 734
J. Chromatogr.; (82) P 493-6; Vol 240; No 2

07-82-0019 (0019)
CONJUGATED DIENES OF CRUDE SOY OIL: DETECTION BY UV SPECTROPHOTOMETRY AND SEPARATION BY HPLC
Brown, H. G.; Snyder, H. E.
Dep. Hortic. Food Sci. Univ. Arkansas Fayetteville USA US AR 72701
JAOCS, J. Am. Oil Chem. Soc.; (82) P 280-3; Vol 59; No 7

07-82-0020 (0020)
FAST SPECTROPHOTOMETER AS A DETECTOR FOR HPLC
Elgass, Helmuth; Maute, Alfred; Martin, Roland; George, Steve
Hewlett-Packard G.m.b.H. Waldbronn Fed. Rep. Ger. DE 7517/2
LaborPraxis; (82) P 256-60, 263; Vol 6; No 4; In Ger

07-82-0021 (0021)
OPTICAL DETECTORS IN HPLC - POSSIBILITIES AND TRENDS
Baumann, Wolfram
Inst. Phys. Chem. Univ. Mainz Mainz Fed. Rep. Ger. DE 6500
LaborPraxis; (82) P 861-3; Vol 6; No 7-8; In Ger

07-82-0022 (0022)
EVALUATION OF AN AUTOMATED HPLC SYSTEM HAVING MULTIPLE DETECTORS
Klee, M. S.; Rogers, L. B.; Carreira, L. A.; Azarraga, L.
Dep. Chem. Univ. Georgia Athens USA US GA 30602
Pergamon Ser. Environ. Sci.; (82) P 177-92; Vol 7; No Anal. Tech. Environ. Chem. 2

07-82-0023 (0023)
MULTICHANNEL DETECTION IN HIGH-PERFORMANCE LIQUID CHROMATOGRAPHY
Miller, James C.; George, Stephan A.; Willis, Barry G.
Hewlett-Packard Co. Palo Alto USA US CA 94304
Science (Washington, D. C., 1883); (82) P 241-6; Vol 218; No 4569

5.7.3
UV/VIS—Applications

--- 1982 ---

07-82-0001 (0001)
FLUORESCENT/ULTRAVIOLET ABSORBING ESTER DERIVATIVE FORMATION AND ANALYSIS OF EICOSANOIDS BY HIGH-PRESSURE LIQUID CHROMATOGRAPHY
Watkins, W. David; Peterson, Myron B.
Anesth. Serv. Massachusetts Gen. Hosp. Boston USA US MA 02114
Anal. Biochem.; (82) P 30-40; Vol 125; No 1

07-82-0002 (0002)
HIGH-PRESSURE LIQUID CHROMATOGRAPHY-ULTRAVIOLET ANALYSIS OF INTRACELLULAR NUCLEOTIDES
Pogolotti, Alfonso L., Jr.; Santi, Daniel V.
Dep. Biochem. Biophys. Univ. California San Francisco USA US CA 94143
Anal. Biochem.; (82) P 335-45; Vol 126; No 2

07-82-0003 (0003)
DETERMINATION OF ORGANIC CONTAMINANTS IN ULTRA-PURE WATER BY REVERSED-PHASE HIGH-PERFORMANCE LIQUID CHROMATOGRAPHY WITH ULTRAVIOLET DETECTION
Reust, Juerg B.; Meyer, Veronika R.
Inst. Anorg. Anal. Phys. Chem. Univ. Berne Berne Switz. CH CH-3012
Analyst (London); (82) P 673-9; Vol 107; No 1275

07-82-0004 (0004)
RAPID SCAN UV/VISIBLE DETECTION IN LIQUID CHROMATOGRAPHY . APPLICATION TO ENVIRONMENTAL ANALYSIS
Warwick, C. J.; Bagon, D. A.
Occup. Med. Hyg. Lab. London UK GB NW2 6LN
Chromatographia; (82) P 433-6; Vol 15; No 7

07-82-0005 (0005)
URIC ACID DETERMINATIONS: REVERSED-PHASE LIQUID CHROMATOGRAPHY WITH ULTRAVIOLET DETECTION COMPARED WITH KINETIC AND EQUILIBRIUM ADAPTATIONS OF THE URICASE METHOD
Ingebretsen, Ole Christian; Borgen, Jan; Farstad, Mikael
Lab. Clin. Biochem. Univ. Bergen Sykehus Norway NO N-5016
Clin. Chem. (Winston-Salem, N. C.); (82) P 496-8; Vol 28; No 3

07-82-0006 (0006)
SIMULTANEOUS USE OF FLUORESCENCE, ULTRAVIOLET, AND ELECTROCHEMICAL DETECTORS IN HIGH-PERFORMANCE LIQUID CHROMATOGRAPHY - SEPARATION AND IDENTIFICATION OF PHENOLIC ANTIOXIDANTS AND RELATED COMPOUNDS
Masoud, A. N.; Cha, Young Nam
Dep. Environ. Health Sci. Johns Hopkins Univ. Baltimore USA US MD 21205
HRC CC, J. High Resolut. Chromatogr. Chromatogr. Commun.; (82) P 299-305; Vol 5; No 6

07-82-0007 (0007)
UNIVERSAL, MULTICHANNEL ULTRAVIOLET DETECTION IN THE PURITY ANALYSIS OF 2-ETHYL-3-(4-HYDROXYBENZOYL)INDOLIZINE BY HIGH-PERFORMANCE LIQUID CHROMATOGRAPHY
Drenth, B. F. H.; Ghijsen, R. T.; De Zeeuw, R. A.
Dep. Toxicol. State Univ. Groningen Neth. NL 9713 AW
J. Chromatogr.; (82) P 113-20; Vol 238; No 1

07-82-0008 (0008)
QUANTITATIVE DETERMINATION OF PERTECHNETATE BY HIGH-PERFORMANCE LIQUID CHROMATOGRAPHY WITH UV DETECTION
Zodda, Julius P.; Heineman, William R.; Gilbert, T. William; Deutsch, Edward
Dep. Chem. Univ. Cincinnati Cincinnati USA US OH 45221

J. Chromatogr.; (82) P 249-55; Vol 227; No 1

07-82-0009 (0009)
A METHOD FOR HIGH-PERFORMANCE LIQUID CHROMATOGRAPHIC SCREENING OF UV-POSITIVE COMPONENTS IN URINE ELUATE FROM SEPHADEX G-10 AND MODIFICATIONS FOR DETERMINATION OF URINARY SALICYLIC, SALICYLURIC AND GENTISIC ACIDS
Fornstedt, Nermin
Inst. Biochem. Univ. Uppsala Uppsala Swed. SE S-751 23
J. Chromatogr.; (82) P 253-61; Vol 230; No 2

07-82-0010 (0010)
LIQUID CHROMATOGRAPHIC ASSAY OF PHENOTHIAZINE, THIOXANTHENE AND BUTYROPHENONE NEUROLEPTICS AND ANTIHISTAMINES IN BLOOD AND PLASMA WITH CONVENTIONAL AND RADIAL COMPRESSION COLUMNS AND UV AND ELECTROCHEMICAL DETECTION
Curry, Stephen H.; Brown, Edgar A.; Hu, Oliver Y. P.; Perrin, John H.
Coll. Pharm. Univ. Florida Gainesville USA US FL 32610
J. Chromatogr.; (82) P 361-76; Vol 231; No 2

07-82-0011 (0011)
DETERMINATION OF THE .GAMMA.-AMINOBUTYRIC ACID AGONIST 4,5,6,7-TETRAHYDRO-ISOXAZOLO(5,4-C)PYRIDIN-3-OL BY HIGH-PERFORMANCE LIQUID CHROMATOGRAPHY USING UV OR ELECTROCHEMICAL DETECTION
Madsen, Soren Munk
H. Lundbeck and Co., A/S Valby Den. DK DK 2500
J. Chromatogr.; (82) P 509-14; Vol 238; No 2

07-82-0012 (0012)
HIGH-PERFORMANCE LIQUID CHROMATOGRAPHY OF INORGANIC AND ORGANIC ANIONS USING ULTRAVIOLET DETECTION AND AN AMINO COLUMN
Cortes, Hernan J.
Anal. Lab. Dow Chem. Co. Midland USA US MI 48640
J. Chromatogr.; (82) P 517-20; Vol 234; No 2

07-82-0013 (0013)
QUANTITATION OF ALKYLSULFONATES USING UV DETECTOR SENSITIVE "ION-PAIR" REAGENTS IN REVERSED-PHASE LIQUID CHROMATOGRAPHY
Sachok, Bohdan; Deming, Stanley N.; Bidlingmeyer, Brian A.
Dep. Chem. Univ. Houston Houston USA US TX
J. Liq. Chromatogr.; (82) P 389-402; Vol 5; No 3

07-82-0014 (0014)
DETERMINATION OF UV ABSORBING POLYOLEFIN ADDITIVES BY GRADIENT AND ISOCRATIC NORMAL-PHASE HIGH-PERFORMANCE LIQUID CHROMATOGRAPHY
Schabron, J. F.; Smith, V. J.; Ware, J. L.
Phillips Res. Cent. Phillips Pet. Co. Bartlesville USA US OK 74004
J. Liq. Chromatogr.; (82) P 613-24; Vol 5; No 4

07-82-0015 (0015)
ORAL HYDROCORTISONE PHARMACOKINETICS: A COMPARISON OF FLUORESCENCE AND ULTRAVIOLET HIGH-PRESSURE LIQUID CHROMATOGRAPHIC ASSAYS FOR HYDROCORTISONE IN PLASMA
Toothaker, Roger D.; Sundaresan, Gloria M.; Hunt, John P.; Goehl, Thomas J.; Rotenberg, Keith S.; Prasad, Vadlamani K.; Craig, William A.; Welling, Peter
Sch. Pharm. Univ. Wisconsin Madison USA US WI 53706
J. Pharm. Sci.; (82) P 573-6; Vol 71; No 5

07-82-0016 (0016)
CONJUGATED DIENES OF CRUDE SOY OIL: DETECTION BY UV SPECTROPHOTOMETRY AND SEPARATION BY HPLC
Brown, H. G.; Snyder, H. E.

Dep. Hortic. Food Sci. Univ. Arkansas Fayetteville USA US AR 72701
JAOCS, J. Am. Oil Chem. Soc.; (82) P 280-3; Vol 59; No 7

07-82-0017 (0017)
NEW QUANTITATIVE METHOD FOR THE DETERMINATION OF UV STABILITY IN VARIOUS LC + DYE MIXTURES
Veghely, T.
Ind. Res. Inst. Electron. Budapest Hung. HU
Mol. Cryst. Liq. Cryst.; (82) P 35-50; Vol 80; No 1-4

07-82-0018 (0018)
MOLECULAR SIEVE CHROMATOGRAPHY OF SOME PLASMA PROTEIN PRODUCTS BY HIGH PERFORMANCE LIQUID CHROMATOGRAPHY: COMPARISON OF QUANTITATION BY UV ABSORPTION AND LOWRY PROTEIN
Bertolini, Maurice J.
Cutter Lab. Berkeley USA US CA
Vox Sang.; (82) P 87-90; Vol 43; No 2

5.7.4
Fluorescence—Generalities

--- 1982 ---

07-82-0001 (0001)
ISOCRATIC REVERSED-PHASE LIQUID CHROMATOGRAPHY FOR ASSAY OF 5'-NUCLEOTIDASE WITH PHOSPHORIC ACID FOR REACTION QUENCHING
Yu, Li Wen; Fager, Roger S.
Sch. Med. Univ. Virginia Charlottesville USA US VA 22908
Anal. Chem.; (82) P 1904-6; Vol 54; No 11

07-82-0002 (0002)
OPTIMIZATION OF FLOW CELLS FOR FLUORESCENCE DETECTION IN LIQUID CHROMATOGRAPHY
Lyons, John W.; Faulkner, Larry R.
Dep. Chem. Univ. Illinois Urbana USA US IL 61801
Anal. Chem.; (82) P 1960-4; Vol 54; No 12

07-82-0003 (0003)
PHOTOREDUCTION-FLUORESCENCE DETECTION OF ALIPHATIC ALCOHOLS, ALDEHYDES, AND ETHERS IN LIQUID CHROMATOGRAPHY
Gandelman, Mitchell S.; Birks, John W.
Dep. Chem. Univ. Colorado Boulder USA US CO 80309
Anal. Chem.; (82) P 2131-3; Vol 54; No 12

07-82-0004 (0004)
LASER-INDUCED FLUORESCENCE DETECTION FOR CONVENTIONAL AND MICROCOLUMN LIQUID CHROMATOGRAPHY
Folestad, Staffan; Johnson, Lars; Josefsson, Bjoern; Galle, Bo
Dep. Anal. Mar. Chem. Chalmers Univ. Technol. Goeteborg Swed. SE S-412 96
Anal. Chem.; (82) P 925-9; Vol 54; No 5

07-82-0005 (0005)
SEQUENTIALLY EXCITED FLUORESCENCE DETECTION IN LIQUID CHROMATOGRAPHY
Huff, Paul B.; Tromberg, Bruce J.; Sepaniak, Michael J.
Dep. Chem. Univ. Tennessee Knoxville USA US TN 37916
Anal. Chem.; (82) P 946-50; Vol 54; No 5

07-82-0006 (0006)
TIME-RESOLVED FLUORIMETRY WITH A SUB-NANOSECOND DYE LASER SOURCE FOR THE DETERMINATION OF POLYNUCLEAR AROMATIC HYDROCARBONS AFTER SEPARATION BY HIGH-PERFORMANCE LIQUID CHROMATOGRAPHY
Imasaka, Totaro; Ishibashi, Kouichi; Ishibashi, Nobuhiko
Fac. Eng. Kyushu Univ. Hakozaki Japan JP
Anal. Chim. Acta; (82) P 1-12; Vol 142

07-82-0007 (0007)
DETERMINATION OF POLYNUCLEAR AROMATIC HYDROCARBONS BY A SPIKED MOBILE PHASE FLUORESCENCE HPLC DETECTOR
Su, Syang Y.; Lai, E. P. C.; Winefordner, J. D.
Dep. Chem. Univ. Florida Gainesville USA US FL 32611
Anal. Lett.; (82) P 439-50; Vol 15; No A5

07-82-0008 (0008)
BETA-INDUCED FLUORESCENCE DETECTION IN LIQUID CHROMATOGRAPHY
Malcolme-Lawes, David J.
Nucl. Chem. Lab. Loughborough Univ. Technol. Loughborough UK GB LE11 3TU
Anal. Proc. (London); (82) P 365-7; Vol 19; No 7

07-82-0009 (0009)
ON-LINE FLUORESCENCE DETECTION IN HPLC: A HIGH SENSITIVE TECHNIQUE FOR THE MEASUREMENT OF LOW ENZYME ACTIVITIES: THE CASE OF TRYPTOPHAN HYDROXYLASE
Geeraerts, F.; Schimpfessel, L.; Crokaert, R.

Dep. Biochem. Free Univ. Brussels Belg. BE
Chromatographia; (82) P 449-50; Vol 15; No 7

07-82-0010 (0010)
DEVELOPMENT, CHARACTERIZATION AND APPLICATION OF A MULTICHANNEL **FLUORESCENCE** DETECTOR FOR LIQUID CHROMATOGRAPHY
Clow, Charles Spencer
Univ. Cincinnati Cincinnati USA US OH
From: Diss. Abstr. Int. B 1982, 43(2), 417.
DISSERTATION; (82) P267 pp.; Avail.:Univ. Microfilms Int.,Order No. DA8215519

07-82-0011 (0011)
DETERMINATION OF S-CARBOXYMETHYL CYSTEINE IN PLASMA USING HPLC AND FLUORIMETRIC SCANNING
Gaetani, E.; Laureri, C. F.; Vitto, M.; Bordi, F.
Ist. Chim. Farm. Tossicol. Univ. Parma Parma Italy IT
Farmaco, Ed. Prat.; (82) P 235-8; Vol 37; No 7; In Ital

07-82-0012 (0012)
SIMULTANEOUS USE OF FLUORESCENCE, ULTRAVIOLET, AND ELECTROCHEMICAL DETECTORS IN HIGH-PERFORMANCE LIQUID CHROMATOGRAPHY - SEPARATION AND IDENTIFICATION OF PHENOLIC ANTIOXIDANTS AND RELATED COMPOUNDS
Masoud, A. N.; Cha, Young Nam
Dep. Environ. Health Sci. Johns Hopkins Univ. Baltimore USA US MD 21205
HRC CC, J. High Resolut. Chromatogr. Chromatogr. Commun.; (82) P 299-305; Vol 5; No 6

07-82-0013 (0013)
HIGH-PRESSURE LIQUID CHROMATOGRAPHIC DETERMINATION OF AFLATOXINS IN **PEANUT** BUTTER USING A SILICA GEL-PACKED FLOWCELL FOR FLUORESCENCE DETECTION
Francis, Octave J., Jr.; Lipinski, Leo J.; Gaul, Jean A.; Campbell, **Alfred D.**
Food Drug Adm. New Orleans USA US LA 70122
J. Assoc. Off. Anal. Chem.; (82) P 672-6; Vol 65; No 3

5.7.5
Fluorescence—Applications

--- 1982 ---

07-82-0001 (0001)
AUTOMATIC ANALYSIS OF BIOGENIC CATECHOLAMINES BY HIGH-PERFORMANCE LIQUID CHROMATOGRAPHY WITH FLUOROMETRIC REACTION DETECTION
Mori, K.
Natl. Inst. Ind. Health Kawasaki Japan JP 213
Adv. Biosci.; (82) P 275-84; Vol 36; No Synth., Storage Secretion Adrenal Catecholamines

07-82-0002 (0002)
FLUORESCENT/ULTRAVIOLET ABSORBING ESTER DERIVATIVE FORMATION AND ANALYSIS OF EICOSANOIDS BY HIGH-PRESSURE LIQUID CHROMATOGRAPHY
Watkins, W. David; Peterson, Myron B.
Anesth. Serv. Massachusetts Gen. Hosp. Boston USA US MA 02114
Anal. Biochem.; (82) P 30-40; Vol 125; No 1

07-82-0003 (0003)
LIQUID CHROMATOGRAPHIC DETERMINATION OF GUANADREL IN LABORATORY ANIMAL DIET AS THE FLUORESCENT ACETYLACETONE DERIVATIVE
Bombardt, Paul A.; Adams, Wade J.
Upjohn Co. Kalamazoo USA US MI 49001
Anal. Chem.; (82) P 1087-90; Vol 54; No 7

07-82-0004 (0004)
DETERMINATION OF GLYBURIDE IN HUMAN SERUM BY LIQUID CHROMATOGRAPHY WITH FLUORESCENCE DETECTION
Adams, W. J.; Skinner, G. S.; Bombardt, P. A.; Courtney, M.; Brewer, J. E.
Pharm. Res. Dev. Upjohn Co. Kalamazoo USA US MI 49001
Anal. Chem.; (82) P 1287-91; Vol 54; No 8

07-82-0005 (0005)
ELECTROCHEMICAL, FLUORESCENCE, AND LIQUID CHROMATOGRAPHIC STUDIES ON THE OXIDATION OF SOTERENOL
Miner, David J.; Kissinger, Peter T.
Dep. Chem. Purdue Univ. West Lafayette USA US IN 47907
Anal. Chim. Acta; (82) P 231-5; Vol 138

07-82-0006 (0006)
ROUTINE ASSAY OF CATECHOLAMINES BY HIGH-PERFORMANCE LIQUID CHROMATOGRAPHY WITH FLUOROMETRIC DETECTION
Clough, G.
Biochem. Dep. Gen. Hosp. Harrogate UK GB HG2 7ND
Anal. Proc. (London); (82) P 183-6; Vol 19; No 4

07-82-0007 (0007)
DETERMINATION OF MONOSODIUM GLUTAMATE IN FOOD USING HIGH-PERFORMANCE LIQUID CHROMATOGRAPHY AND FLUORESCENCE DETECTION
Rhys Williams, A. T.; Winfield, S. A.
Perkin-Elmer Ltd. Beaconsfield UK GB HP9 1QA
Analyst (London); (82) P 1092-4; Vol 107; No 1278

07-82-0008 (0008)
A FLUOROMETRIC-HPLC ASSAY FOR QUANTITATING THE BINDING OF BENZO(A)PYRENE METABOLITES TO DNA
Rahn, R. O.; Chang, S. S.; Holland, J. M.; Shugart, L. R.
Biol. Div. Oak Ridge Natl. Lab. Oak Ridge USA US TN 37830
Biochem. Biophys. Res. Commun.; (82) P 262-8; Vol 109; No 1

07-82-0009 (0009)
LIQUID CHROMATOGRAPHY OF CODEINE IN PLASMA WITH FLUORESCENCE DETECTION
Tna, Irene W.; Fass, Marian; Debban, Jennifer A.; Matin, Shaikh B.
Dep. Anal. Metab. Chem. Syntex Res. Palo Alto USA US CA 94304
Clin. Chem. (Winston-Salem, N. C.); (82) P 1137-9; Vol 28; No 5

07-82-0010 (0010)
IMPROVED DIRECT DETERMINATION OF ALPHA- AND GAMMA-TOCOPHEROLS IN PLASMA AND PLATELETS BY LIQUID CHROMATOGRAPHY, WITH FLUORESCENCE DETECTION
Lehmann, Joanna; Martin, Hilda L.
Lipid Nutr. Lab. Beltsville Hum. Nutr. Res. Cent. Beltsville USA US MD 20705
Clin. Chem. (Winston-Salem, N. C.); (82) P 1784-7; Vol 28; No 8

07-82-0011 (0011)
SIMULTANEOUS FLUOROMETRY OF URINARY DOPAMINE, NOREPINEPHRINE, AND EPINEPHRINE COMPARED WITH LIQUID CHROMATOGRAPHY WITH ELECTROCHEMICAL DETECTION
Oka, Kazuhiro; Sekiya, Motoko; Osada, Hirofumi; Fujita, Keisuke;
Kato, Takeshi; Nagatsu, Toshiharu
Grad. Sch. Nagatsuta Tokyo Inst. Technol. Yokohama Japan JP 227
Clin. Chem. (Winston-Salem, N. C.); (82) P 646-9; Vol 28; No 4, Pt. 1

07-82-0012 (0012)
A SIMPLE METHOD FOR THE ASSAY OF URINARY METANEPHRINES USING HIGH PERFORMANCE LIQUID CHROMATOGRAPHY WITH FLUORESCENCE DETECTION
Jackman, Graham P.
Clin. Res. Unit. Alfred Hosp. Prahran Australia AU 3181
Clin. Chim. Acta; (82) P 137-42; Vol 120; No 1

07-82-0013 (0013)
ION-PAIR LIQUID CHROMATOGRAPHIC AND FLUOROMETRIC DETERMINATIOW OF AMINO ACIDS IN THE GUINEA PIG BRAIN
Gaetani, E.; Laureri, C. F.; Vitto, M.; Bordi, F.
Ist. Chim. Farm. Tossicol. Univ. Studi Parma Parma Italy IT
Farmaco, Ed. Prat.; (82) P 253-7; Vol 37; No 8; In Ital

07-82-0014 (0014)
UPTAKE AND TURNOVER OF DOPAMINE IN RAT MAST CELLS STUDIED BY CYTOFLUOROMETRY AND HIGH PERFORMANCE LIQUID CHROMATOGRAPHY
Rundquist, I.; Allenmark, S.; Enerbaeck, L.
Dep. Pathol. II Univ. Linkoeping Linkoeping Swed. SE S-581 85
Histochem. J.; (82) P 429-43; Vol 14; No 3

07-82-0015 (0015)
SIMULTANEOUS USE OF FLUORESCENCE, ULTRAVIOLET, AND ELECTROCHEMICAL DETECTORS IN HIGH-PERFORMANCE LIQUID CHROMATOGRAPHY - SEPARATION AND IDENTIFICATION OF PHENOLIC ANTIOXIDANTS AND RELATED COMPOUNDS
Masoud, A. N.; Cha, Young Nam
Dep. Environ. Health Sci. Johns Hopkins Univ. Baltimore USA US MD 21205
HRC CC, J. High Resolut. Chromatogr. Chromatogr. Commun.; (82) P 299-305; Vol 5; No 6

07-82-0016 (0016)
MEASUREMENT OF ADRENALINE IN THE RAT BRAIN BY HIGH-PERFORMANCE LIQUID CHROMATOGRAPHY WITH FLUOROMETRIC DETECTION
Sudo, Ayako
Natl. Inst. Ind. Health Kawasaki Japan JP 213
Ind. Health; (82) P 151-6; Vol 20; No 2

07-82-0017 (0017)
THE USE OF HIGH PRESSURE LIQUID CHROMATOGRAPHY WITH FLUORIMETRIC DETECTION FOR

CHLOROPHYLL A DETERMINATION IN NATURAL EXTRACTS OF CHLOROPIGMENTS AND THEIR DEGRADATION PRODUCTS
Goeyens, L.; Post, E.; Dehpirs, F.; Vandenhoudt, A.; Baeyens, W.
Lab. Anal. Scheikd. Vrije Univ. Brussel Brussels Belg. BE B-1050
Int. J. Environ. Anal. Chem.; (82) P 51,63; Vol 12; No 1

07-82-0018 (0018)
DETERMINATION OF 17-HYDROXYCORTICOSTEROIDS IN URINE BY FLUORESCENCE HIGH-PERFORMANCE LIQUID CHROMATOGRAPHY USING DNS-HYDRAZINE AS A PRECOLUMN LABELING REAGENT
Kawasaki, Takao; Maeda, Masako; Tsuji, Akio
Sch. Pharm. Sci. Showa Univ. Tokyo Japan JP
J. Chromatogr.; (82) P 1-11; Vol 232; No 1

07-82-0019 (0019)
DETERMINATION OF PYRIMETHAMINE IN HUMAN PLASMA AFTER ADMINISTRATION OF FANSIDAR OR FANSIDAR-MEFLOQUINE BY MEANS OF HIGH-PERFORMANCE LIQUID CHROMATOGRAPHY WITH FLUORESCENCE DETECTION
Timm, U.; Weidekamm, E.
Biol. Pharm. Res. Dep. F. Hoffmann-La Roche and Co. Basel Switz. CH CH-4002
J. Chromatogr.; (82) P 107-14; Vol 230; No 1

07-82-0020 (0020)
DETERMINATION OF 7-AMINOFLUNITRAZEPAM (RO 20-1815) AND 7-AMINODESMETHYL-FLUNITRAZEPAM (RO 5-4650) IN PLASMA BY HIGH-PERFORMANCE LIQUID CHROMATOGRAPHY AND FLUORESCENCE DETECTION
Sumirtapura, Y. C.; Aubert, C.; Coassolo, P.; Cano, J. P.
Lab. Pharmacocinet. Toxicocinet. INSERM Marseille Fr. FR F-13385
J. Chromatogr.; (82) P 111-18; Vol 232; No 1

07-82-0021 (0021)
DETERMINATION OF .GAMMA.-AMINOBUTYRIC ACID BY REVERSED-PHASE HIGH-PERFORMANCE LIQUID CHROMATOGRAPHY AND PRECOLUMN LABELING FOR FLUORESCENCE DETECTION
Griesmann, Guy E.; Chan, Wai Yee; Rennert, Owen M.
Oklahoma Child. Mem. Hosp. Univ. Oklahoma Oklahoma City USA US OK 73190
J. Chromatogr.; (82) P 121-4; Vol 230; No 1

07-82-0022 (0022)
DETERMINATION OF THE ANTIDEPRESSANT AGENT CITALOPRAM AND METABOLITES IN PLASMA BY LIQUID CHROMATOGRAPHY WITH FLUORESCENCE DETECTION
Oeyehaug, Ellen; Oestensen, Eilif Terje; Salvesen, Bjarne
Agder Coll. Kristiansand Norway NO 4600
J. Chromatogr.; (82) P 129-35; Vol 227; No 1

07-82-0023 (0023)
FLUORESCENCE DETECTION OF CYSTINE BY o-PHTHALALDEHYDE DERIVATIZATION AND ITS SEPARATION USING HIGH-PERFORMANCE LIQUID CHROMATOGRAPHY
Cooper, J. D. H.; Turnell, D. C.
Biochem. Dep. Coventry and Warwickshire Hosp. Coventry UK GB
J. Chromatogr.; (82) P 158-61; Vol 227; No 1

07-82-0024 (0024)
DETERMINATION OF BROMOLASALOCID IN PLASMA BY HIGH-PERFORMANCE LIQUID CHROMATOGRAPHY WITH FLUORIMETRIC DETECTION
Brooks, Marvin A.; Stronjny, Norman; Hackman, Martin R.; De Silva, J. Arthur F.
Dep. Pharmacokinet. Biopharmaceut. Hoffmann-La Roche Inc. Nutley USA US NJ 07110
J. Chromatogr.; (82) P 167-78; Vol 229; No 1

07-82-0025 (0025)
HIGH-PERFORMANCE LIQUID CHROMATOGRAPHY-FLUORESCENCE ANALYSIS FOR INDOMETHACIN AND METABOLITES IN BIOLOGICAL FLUIDS
Bernstein, M. S.; Evans, M. A.
Med. Cent. Univ. Illinois Chicago USA US IL 60680
J. Chromatogr.; (82) P 179-87; Vol 229; No 1

07-82-0026 (0026)
DETERMINATION OF 4'-EPIDOXORUBICIN AND ITS 13-DIHYDRO DERIVATIVE IN HUMAN PLASMA BY HIGH-PERFORMANCE LIQUID CHROMATOGRAPHY WITH FLUORESCENCE DETECTION
Moro, Enzo; Jannuzzo, Maria G.; Ranghieri, Massimo; Stegnjaich, Silvano; Valzelli, Guido
Pharmacokinet. Dep. Farmitalia Carlo Erba Milan Italy IT 20159
J. Chromatogr.; (82) P 207-11; Vol 230; No 1

07-82-0027 (0027)
DNS DERIVATIZATION OF ANABOLIC AGENTS WITH HIGH-PERFORMANCE LIQUID CHROMATOGRAPHIC SEPARATION AND FLUORESCENCE DETECTION
Williams, A. T. Rhys; Winfield, S. A.; Belloli, R. C.
Perkin-Elmer Ltd. Beaconsfield/Buckinghamshire UK GB HP9 1QA
J. Chromatogr.; (82) P 224-9; Vol 240; No 1

07-82-0028 (0028)
MEASUREMENT OF CATECHOLAMINES IN BIOLOGICAL FLUIDS BY HIGH-PERFORMANCE LIQUID CHROMATOGRAPHY. A COMPARISON OF FLUORIMETRIC WITH ELECTROCHEMICAL DETECTION
Causon, Roger C.; Carruthers, Malcolm E.
Dep. Biochem. Inst. Psychiatry London UK GB
J. Chromatogr.; (82) P 301-9; Vol 229; No 2

07-82-0029 (0029)
DETERMINATION OF THE B2 VITAMER FLAVIN-ADENINE DINUCLEOTIDE IN WHOLE BLOOD BY HIGH-PERFORMANCE LIQUID CHROMATOGRAPHY WITH FLUOROMETRIC DETECTION
Speek, A. J.; Van Schaik, F.; Schrijver, J.; Schreurs, W. H. P.
Dep. Clin. Biochem. TNO Zeist Neth. NL 3700 AJ
J. Chromatogr.; (82) P 311-16; Vol 228

07-82-0030 (0030)
THE DETERMINATION OF UROPORPHYRINOGEN DECARBOXYLASE IN TISSUES BY HIGH-PERFORMANCE LIQUID CHROMATOGRAPHY COUPLED TO FLUORESCENCE DETECTION
Cantoni, L.; Ruggieri, R.; Dal Fiume, D.; Rizzardini, M.
Ist. Ric. Farmacol. "Mario Negri" Milan Italy IT 20157
J. Chromatogr.; (82) P 311-18; Vol 229; No 2

07-82-0031 (0031)
FLUORESCENCE HIGH-PERFORMANCE LIQUID CHROMATOGRAPHY OF REDUCING SUGARS USING DNS-HYDRAZINE AS A PRELABELING REAGENT
Takeda, Motoko; Maeda, Masako; Tsuji, Akio
Sch. Pharm. Sci. Showa Univ. Tokyo Japan JP
J. Chromatogr.; (82) P 347-55; Vol 244; No 2

07-82-0032 (0032)
SIMPLE AND RAPID DETERMINATION OF PROPRANOLOL AND ITS ACTIVE METABOLITE, 4-HYDROXYPROPRANOLOL, IN HUMAN PLASMA BY LIQUID CHROMATOGRAPHY WITH FLUORESCENCE DETECTION
Albani, Fiorenzo; Riva, Roberto; Baruzzi, Agostino
Inst. Neurol. Univ. Bologna Bologna Italy IT 40123
J. Chromatogr.; (82) P 362-5; Vol 228

07-82-0033 (0033)
HIGH-PERFORMANCE LIQUID CHROMATOGRAPHY OF NAPHTHYLURETHANES WITH FLUORESCENCE

DETECTION
Wintersteiger, R.; Wenninger-Weinzierl, G.; Pacha, W.
Inst. Pharm. Chem. Univ. Graz Graz Austria AT A-8010
J. Chromatogr.; (82) P 399-406; Vol 237; No 3

07-82-0034 (0034)
HIGH-PERFORMANCE LIQUID CHROMATOGRAPHY DETECTION OF MORPHINE BY FLUORESCENCE AFTER POSTCOLUMN DERIVATIZATION
Nelson, Peter E.; Nolan, Susan L.; Bedford, Keith R.
Dep. Sci. Ind. Res. Univ. Auckland Auckland N. Z. NZ
J. Chromatogr.; (82) P 407-14; Vol 234; No 2

07-82-0035 (0035)
RAPID, SPECIFIC METHOD FOR DIETHYLSTILBESTROL ANALYSIS USING AN IN-LINE PHOTOCHEMICAL REACTOR WITH HIGH-PERFORMANCE LIQUID CHROMATOGRAPHY AND FLUORESCENCE DETECTION
Williams, A. T. Rhys; Winfield, S. A.; Belloli, R. C.
Perkin-Elmer Ltd. Buckinghamshire UK GB HP9 1QA
J. Chromatogr.; (82) P 461-70; Vol 235; No 2

07-82-0036 (0036)
DETERMINATION OF PRENALTEROL IN PLASMA BY HIGH-PERFORMANCE LIQUID CHROMATOGRAPHY WITH FLUORESCENCE DETECTION
Oddie, C. J.; Jackman, G. P.; Bobik, A.
Clin. Res. Unit Alfred Hosp. Prahran Australia AU 3181
J. Chromatogr.; (82) P 473-7; Vol 231; No 2

07-82-0037 (0037)
FLUOROMETRIC LIQUID CHROMATOGRAPHIC ASSAY OF THE ANTIARRHYTHMIC AGENT FLECAINIDE IN BLOOD PLASMA
De Jong, J. W.; Hegge, J. A. J.; Harmsen, E.; De Tombe, P. P.
Thoraxcent. Erasmus Univ. Rotterdam Neth. NL
J. Chromatogr.; (82) P 498-502; Vol 229; No 2

07-82-0038 (0038)
DETERMINATION OF THIABENDAZOLE AND 5-HYDROXYTHIABENDAZOLE IN HUMAN SERUM BY FLUORESCENCE-DETECTED HIGH-PERFORMANCE LIQUID CHROMATOGRAPHY
Watts, Mark T.; Raisys, Vidmantas A.; Bauer, Larry A.
Sch. Med. Univ. Washington Seattle USA US WA 98104
J. Chromatogr.; (82) P 79-86; Vol 230; No 1

07-82-0039 (0039)
REVERSED-PHASE ION-PAIRING LIQUID-CHROMATOGRAPHIC SEPARATION AND FLUOROMETRIC DETECTION OF POLYAMINES
Simpson, Richard C.; Mohammed, Hussain Y.; Veening, Hans
Dep. Chem. Bucknell Univ. Lewisburg USA US PA 17837
J. Liq. Chromatogr.; (82) P 245-64; Vol 5; No 2

07-82-0040 (0040)
ADENOSINE MEASUREMENT BY A RAPID HPLC-FLUOROMETRIC METHOD: INDUCED CHANGES OF ADENOSINE CONTENT IN REGIONS OF RAT BRAIN
Wojcik, W. J.; Neff, N. H.
Natl. Inst. Ment. Health Saint Elizabeths Hosp. Washington USA US DC 20032
J. Neurochem.; (82) P 280-2; Vol 39; No 1

07-82-0041 (0041)
ORAL HYDROCORTISONE PHARMACOKINETICS: A COMPARISON OF FLUORESCENCE AND ULTRAVIOLET HIGH-PRESSURE LIQUID CHROMATOGRAPHIC ASSAYS FOR HYDROCORTISONE IN PLASMA
Toothaker, Roger D.; Sundaresan, Gloria M.; Hunt, John P.; Goehl, Thomas J.;

Rotenberg, Keith S.; Prasad, Vadlamani K.; Craig, William A.; Welling, Peter
Sch. Pharm. Univ. Wisconsin Madison USA US WI 53706
J. Pharm. Sci.; (82) P 573-6; Vol 71; No 5

07-82-0042 (0042)
THE DETERMINATION OF .ALPHA.-KETOACIDS AND OXALIC ACID IN SEAWATER BY REVERSED PHASE LIQUID CHROMATOGRAPHIC SEPARATION OF FLUORESCENT QUINOXILINOL DERIVATIVES
Steinberg, Spencer M.; Bada, Jeffrey L.
Scripps Inst. Oceanogr. Univ. California La Jolla USA US CA 92093
Mar. Chem.; (82) P 299-306; Vol 11; No 4

07-82-0043 (0043)
MICRODETERMINATION OF INDOLE-3-ACETIC ACID IN VEGETABLE EXTRACTS BY HIGH-PERFORMANCE LIQUID CHROMATOGRAPHY WITH FLUORESCENCE DETECTION
Budini, R.; Girotti, S.; Pierpaoli, A. M.; Tonelli, D.
Chem. Inst. "G. Ciamician" Univ. Bologna Bologna Italy IT 40126
Microchem. J.; (82) P 365-71; Vol 27; No 3

07-82-0044 (0044)
PRECISION AND ACCURACY OF INDOLE-3-ACETIC ACID ANALYSES PERFORMED WITH THE 2-METHYLINDOLO-.ALPHA.-PYRONE FLUORESCENCE ASSAY AND WITH A HIGH-PERFORMANCE LIQUID CHROMATOGRAPHY TECHNIQUE WITH SPECTROFLUOROMETRIC DETECTION, EXEMPLIFIED ON PINE TISSUE (PINUS SYLVESTRIS L)
Sandberg, Goeran; Dunberg, Arne
Dep. For. Genet. Plant Physiol. Swed. Univ. Agric. Sci. Umea Swed. SE S-901 83
Physiol. Plant.; (82) P 315-22; Vol 55; No 3

5.7.6
Electrochemistry—Generalities

--- 1982 ---

07-82-0001 (0001)
ELECTROCHEMICAL DETECTION WITH A REGENERATIVE FLOW CELL IN LIQUID CHROMATOGRAPHY
Weber, Stepfen G.; Purdy, William C.
Dep. Chem., Univ. Pittsburg, Pittsburg USA PA 14260
Anal. Chem. (82) P 1757-64; Vol 54; No 11

07-82-0002 (0002)
MINIMIZING FLOW-ASSOCIATED NOISE IN ELECTROCHEMICAL DETECTORS FOR LIQUID CHROMATOGRAPHY
Weber, Stephen G.
Dep. Chem., Univ. Pittsburg, Pittsburg USA, PA 15 260
Anal. Chem. (82) P 2126-7; Vol 54; No 12

07-82-0003 (0003)
CARBON ELECTRODES FOR LIQUID CHROMATOGRAPHY DETECTION
McLean, J. D.
Michigan Div. Anal. Lab. Dow Chem. U.S.A. Midland USA US MI 48640
Anal. Chem.; (82) P 1169-74; Vol 54; No 7

07-82-0004 (0004)
LIQUID CHROMATOGRAPHY/ELECTROCHEMISTRY: THIN-LAYER MULTIPLE ELECTRODE DETECTION
Roston, Daryl A.; Shoup, Ronald E.; Kissinger, Peter T.
Chem. Dep. North. Illinois Univ. Dekalb USA US IL
Anal. Chem.; (82) P 1417A-1418A, 1422A, 1424A, 1428A, 1430A, 1432A, 1434A; Vol 54; No 13

07-82-0005 (0005)
TRANSIENT ELECTROCHEMICAL TECHNIQUES IN LIQUID CHROMATOGRAPHY WITH MICROPROCESSOR-BASED INSTRUMENTATION
Anderson, J. E.; Bond, A. M.; Heritage, I. D.; Jones, R. D.; Wallace, G. G.
Div. Chem. Phys. Sci. Deakin Univ. Waurn Ponds Australia AU 3217
Anal. Chem.; (82) P 1702-5; Vol 54; No 11

07-82-0006 (0006)
SERIES DUAL-ELECTRODE DETECTOR FOR LIQUID CHROMATOGRAPHY/ELECTROCHEMISTRY
Roston, Daryl A.; Kissinger, Peter T.
Dep. Chem. Purdue Univ. West Lafayette USA US IN 47907
Anal. Chem.; (82) P 429-34; Vol 54; No 3

07-82-0007 (0007)
SPECTROPHOTOMETRIC AND POLARIMETRIC DETECTORS IN LIQUID CHROMATOGRAPHY FOR THE DETERMINATION OF ENANTIOMER RATIOS IN COMPLEX MIXTURES
Boehme, W.; Wagner, G. Oehme, U.; Priesnitz, U.
Bodenseewerk Perkin-Elmer Ueberlingen Fed. Rep. Ger. DE
Anal. Chem.; (82) P 709-11; Vol 54; No 4

07-82-0008 (0008)
TEMPERATURE DEPENDENCE OF ELECTROCHEMICAL DETECTION FOR LIQUID CHROMATOGRAPHY
Miner, David J.
Lilly Res. Lab. Indianapolis USA US IN 46285
Anal. Chim. Acta; (82) P 101-9; Vol 134

07-82-0009 (0009)
A NEW POLAROGRAPHIC RESPONSE FOR ETHYNYLESTRADIOL AND ITS APPLICATION TO ELECTROCHEMICAL DETECTION ASSOCIATED WITH REVERSE-PHASE LIQUID CHROMATOGRAPHY AND IN CATHODIC STRIPPING VOLTAMMETRY

Bond, A. M.; Heritage, I. D.; Briggs, M. H.
Div. Chem. Phys. Sci. Deakin Univ. Waurn Ponds Australia AU 3217
Anal. Chim. Acta; (82) P 35-45; Vol 138

07-82-0010 (0010)
SIMULTANEOUS DETERMINATION OF THE THREE MAJOR MONOAMINE METABOLITES IN CSF BY HPLC WITH ELECTROCHEMICAL DETECTION
Van Bockstaele, M.; Dillen, L.; Claeys, M.; De Potter, W. P.
Dep. Med. Univ. Instelling Antwerpen Wilrijk Belg. BE B-2610
Arch. Int. Pharmacodyn. Ther.; (82) P 304-5; Vol 260; No 2

07-82-0011 (0011)
CONSTRUCTION AND THE CHECKING OF THE FUNCTIONAL SAMPLE OF A VOLTAMMETRIC DETECTOR FOR HIGH-PERFORMANCE LIQUID CHROMATOGRAPHY
Podolak, M.; Stulik, K.; Pacakova, V.
Fac. Sci. Charles Univ. Prague Czech. CS
Chem. Listy; (82) P 1106-9; Vol 76; No 10; In Czeth

07-82-0012 (0012)
DETERMINATION OF METHYL AND ETHYL CAFFEATE IN VEGETABLES BY HPLC WITH ELECTROCHEMICAL DETECTION
Sontag, G.; Schaefers, F. I.; Herrmann, K.
Inst. Anal. Chem. Univ. Vienna Vienna Austria AT
CONFERENCE; Recent Dev. Food Anal., Proc. Eur. Conf. Food Chem., 1st; (82) P 41-6; Avail.: Baltes, Werner; Czedik-Eysenberg, Peter B.; Pfannhauser, Werner; Date: 81; Publisher: Verlag Chem., Weinheim, Fed. Rep. Ger.

07-82-0013 (0013)
ELECTROCHEMICAL DETECTORS IN HIGH-PRESSURE LIQUID CHROMATOGRAPHY
Heisz, Otto
Unternehmensbereich Elektron. Wiss. Ind. Philips G.m.b.H. Kassel Fed. Rep. Ger. DE
CLB, Chem. Labor Betr.; (82) P 272-4; Vol 33; No 6; In Ger

07-82-0014 (0014)
ELECTROCHEMICAL DETECTORS IN HIGH-PRESSURE LIQUID CHROMATOGRAPHY. 2
Heisz, Otto
Unternehmensber. Elektron. Wissensch. Ind. Philips G.m.b.H. Kassel Fed. Rep. Ger. DE
CLB, Chem. Labor Betr.; (82) P 347-9; Vol 33; No 8; In Ger

07-82-0015 (0015)
LIQUID CHROMATOGRAPHY WITH PRECOLUMN SAMPLE PRECONCENTRATION AND ELECTRO-CHEMICAL DETECTION: DETERMINATION OF AROMATIC AMINES IN ENVIRONMENTAL SAMPLES
Rice, John R.; Kissinger, Peter T.
Dep. Chem. Purdue Univ. West Lafayette USA US IN 47907
Environ. Sci. Technol; (82) P 263-8; Vol 16; No 5

07-82-0016 (0016)
DETECTION OF CHOLANOIC ACIDS IN HIGH-PERFORMANCE LIQUID CHROMATOGRAPHY BASED ON EFFECTS OF DOUBLE LAYER CAPACITY CHANGES AT THE DROPPING MERCURY ELECTRODE
Kutner, Wlodzimierz; Behr, Barbara; Kemula, Wiktor
Inst. Phys. Chem. Polish Acad. Sci. Warsaw Pol. PL 01-224
Fresenius' Z. Anal. Chem.; (82) P 121-5; Vol 312; No 2

07-82-0017 (0017)
SIMULTANEOUS USE OF FLUORESCENCE, ULTRAVIOLET, AND ELECTROCHEMICAL DETECTORS IN HIGH-PERFORMANCE LIQUID CHROMATOGRAPHY - SEPARATION AND IDENTIFICATION OF PHENOLIC ANTIOXIDANTS AND RELATED COMPOUNDS
Masoud, A. N.; Cha, Young Nam

Dep. Environ. Health Sci. Johns Hopkins Univ. Baltimore USA US MD 21205
HRC CC, J. High Resolut. Chromatogr. Chromatogr. Commun.; (82) P 299-305; Vol 5; No 6

07-82-0018 (0018)
HIGH-PERFORMANCE LIQUID CHROMATOGRAPHIC SEPARATION WITH TRIPLE-PULSE AMPEROMETRIC DETECTION OF CARBOHYDRATES IN BEVERAGES
Hughes, Scott; Johnson, Dennis C.
Dep. Chem. Iowa State Univ. Ames USA US IA 50011
J. Agric. Food Chem.; (82) P 712-14; Vol 30; No 4

07-82-0019 (0019)
EVALUATION OF A FLOW-THROUGH POLAROGRAPHIC DETECTOR FOR THE DETERMINATION OF REDOX COMPOUNDS IN HIGH-PERFORMANCE LIQUID CHROMATOGRAPHY
DebowskiT J.; Duszczyk, K.; Kutner, W.; Sybilska, D.; Kemula, W.
Inst. Phys. Chem. Pol. Acad. Sci. Warsaw Pol. PL 01-224
J. Chromatogr.; (82) P 141-6; Vol 241; No 1

07-82-0020 (0020)
ELECTROCHEMICAL CELL WITH EFFECTIVE VOLUME LESS THAN 1 nL FOR LIQUID CHROMATOGRAPHY
Slais, K.; Krejci, M.
Inst. Anal. Chem. Czech. Acad. Sci. Brno Czech. CS 611 42
J. Chromatogr.; (82) P 21-9; Vol 235; No 1

07-82-0021 (0021)
DETERMINATION OF NOREPINEPHRINE IN BRAIN PERFUSATES USING HIGH-PERFORMANCE LIQUID CHROMATOGRAPHY WITH ELECTROCHEMICAL DETECTION
Salzman, Steven K.; Sellers, Michael S.
Alfred I. du Pont Inst. Nemours Found. Wilmington USA US DE 19899
J. Chromatogr.; (82) P 29-37; Vol 232; No 1

07-82-0022 (0022)
DETERMINATION OF SALBUTAMOL IN HUMAN PLASMA WITH BIMODAL HIGH-PERFORMANCE LIQUID CHROMATOGRAPHY AND A ROTATED DISK AMPEROMETRIC DETECTOR
Oosterhuis, B.; Van Boxtel, C. J.
Dep. Med. Pharmacol. Univ. Hosp. Amsterdam Amsterdam Neth. NL 1054 EG
J. Chromatogr.; (82) P 327-34; Vol 232; No 2

07-82-0023 (0023)
ALTERNATING-CURRENT POLAROGRAPHIC DETECTION FOR REVERSED-PHASE ION-PAIR HIGH-PERFORMANCE LIQUID CHROMATOGRAPHY OF SOME BENZOIC ACIDS
Kutner, Wlodzimierz
Inst. Phys. Chem. Pol. Acad. Sci. Warsaw Pol. PL 01-224
J. Chromatogr.; (82) P 342-6; Vol 247; No 2

07-82-0024 (0024)
DUAL ELECTROCHEMICAL DETECTOR FOR MICRO-HIGH-PERFORMANCE LIQUID CHROMATOGRAPHY AND ITS APPLICATION TO THE SELECTIVE DETECTION OF CATECHOLAMINES
Goto, Masashi; Sakurai, Eizo; Ishii, Daido
Fac. Eng. Nagoya Univ. Nagoya Japan JP 464
J. Chromatogr.; (82) P 357-66; Vol 238; No 2

07-82-0025 (0025)
INCREASED SIGNAL-TO-NOISE RATIO THROUGH GROUNDING AND THE PROPER SELECTION OF A PUMP FOR HIGH-PERFORMANCE LIQUID CHROMATOGRAPHY WITH ELECTROCHEMICAL DETECTION
Helpern, J. A.; Ewing, J. R.; Welch, K. M. A.
Henry Ford Hosp. Detroit USA US MI 48202
J. Chromatogr.; (82) P 491-2; Vol 240; No 2

07-82-0026 (0026)
A COMPARISON OF COULOMETRIC DETECTORS FOR CATECHOLAMINE ANALYSIS BY LC-EC
Bunyagidj, Chaiyod; Girard, James E.
Dep. Chem. American Univ. Washington USA US DC
Life Sci.; (82) P 2627-34; Vol 31; No 23

07-82-0027 (0027)
A CRITICAL EVALUATION OF LC-EC FOR THE SIMULTANEOUS DETERMINATION OF 5-HYDROXYTRYPTOPHAN, 5-HT, 5-HIAA AND HVA IN BIOLOGICAL SAMPLES USING GC-MS FOR VALIDATION
Narasimhachari, N.; Boadle-Biber, M. C.; Friede, R. O.
Dep. Psychiatry Med. Coll. Virginia Richmond USA US VA 23298
Res. Commun. Chem. Pathol. Pharmacol.; (82) P 413-30; Vol 37; No 3

07-82-0028 (0028)
HIGH-PERFORMANCE LIQUID CHROMATOGRAPHY WITH ELECTROCHEMICAL DETECTION AN INTRODUCTION
Wan, Chia Mao
Inst. Zool. Acad. Sin. Nankang Taiwan TW
Sheng Wu K'o Hsueh; (82) P 67-79; Vol 19; In Ch

07-82-0029 (0029)
APPLICATIONS OF HIGH PERFORMANCE LIQUID CHROMATOGRAPHY WITH ELECTROCHEMICAL DETECTION IN CLINICAL CHEMISTRY
Richards, Douglas A.
Dep. Biochem. R. Sussex County Hosp. Brighton/Sussex UK GB
Surf. Technol.; (82) P 113-20; Vol 15; No 2

07-82-0030 (0030)
GLASSY-CARBON AMPEROMETRIC TRANSDUCERS AS ELECTROCHEMICAL DETECTORS IN LIQUID CHROMATOGRAPHY. THE INFLUENCE OF OXYGEN
Bratin, Karl; Kissinger, Peter T.
Dep. Chem. Purdue Univ. West Lafayette USA US IN 47907
Talanta; (82) P 365-70; Vol 29; No 5

5.7.7
Electrochemistry—Applications

--- 1982 ---

07-82-0001 (0001)
DETERMINATION OF ACETAMINOPHEN METABOLITES IN URINE BY LIQUID CHROMATOGRAPHY/ELECTROCHEMISTRY
Hamilton, Marta; Kissinger, Peter T.
Dep. Chem. Purdue Univ. West Lafayette USA US IN 47907
Anal. Biochem.; (82) P 143-8; Vol 125; No 1

07-82-0002 (0002)
DETERMINATION OF ENVIRONMENTAL PHENOLS BY LIQUID CHROMATOGRAPHY / ELECTROCHEMISTRY
Shoup, Ronald E.; Mayer, Ginny S.
Res. Lab. Bioanal. Syst. Inc. West Lafayette USA US IN 47906
Anal. Chem.; (82) P 1164-9; Vol 54; No 7

07-82-0003 (0003)
SIMULTANEOUS DETERMINATION OF COPPER, NICKEL, COBALT, CHROMIUM(VI), AND CHROMIUM(III) BY LIQUID CHROMATOGRAPHY WITH ELECTROCHEMICAL DETECTION
Bond, A. M.; Wallace, G. G.
Div. Chem. Phys. Sci. Deakin Univ. Waurn Ponds Australia AU 3217
Anal. Chem.; (82) P 1706-12; Vol 54; No 11

07-82-0004 (0004)
ISOLATION AND IDENTIFICATION OF BENZENE METABOLITES IN VITRO WITH LIQUID CHROMATOGRAPHY/ELECTROCHEMISTRY
Roston, Daryl A.; Kissinger, Peter T.
Dep. Chem. Purdue Univ. West Lafayette USA
Anal. Chem.; (82) P 1798-802; Vol 54; No 11

07-82-0005 (0005)
DETERMINATION OF TETRAHYDROISOQUINOLINES BY REVERSED-PHASE LIQUID CHROMATOGRAPHY WITH GRADIENT ELUTION AND AMPEROMETRIC DETECTION
St. Claire, R. L., III; Ansari, G. A. S.; Abell, Creed W.
Med. Branch Univ. Texas Galveston USA US TX 77550
Anal. Chem.; (82) P 186-9; Vol 54; No 2

07-82-0006 (0006)
ELECTROCHEMICAL, FLUORESCENCE, AND LIQUID CHROMATOGRAPHIC STUDIES ON THE OXIDATION OF SOTERENOL
Miner, David J.; Kissinger, Peter T.
Dep. Chem. Purdue Univ. West Lafayette USA US IN 47907
Anal. Chim. Acta; (82) P 231-5; Vol 138

07-82-0007 (0007)
BIOGENIC AMINE RESOLUTION IN TISSUE EXTRACTS OF RAT BRAIN USING ION-PAIR HIGH-PERFORMANCE LIQUID CHROMATOGRAPHY WITH ELECTROCHEMICAL DETECTION
Pleece, S. A.; Redfern, P. H; Riley, C. M.; Tomlinson, E.
Sch. Pharm. Pharmacol. Univ. Bath Bath UK GB BA2 7AY
Analyst (London); (82) P 755-60; Vol 107; No 1276

07-82-0008 (0008)
STUDIES ON THE REGULATORY MECHANISM OF THE TYROSINE HYDROXYLASE SYSTEM IN ADRENAL SLICES BY USING HIGH-PERFORMANCE LIQUID CHROMATOGRAPHY WITH ELECTROCHEMICAL DETECTION
Togari, Akifumi; Kato, Takeshi; Nagatsu, Toshiharu
Dep. Life Chem. Tokyo Inst. Technol. Yokohama Japan JP 227
Biochem. Pharmacol.; (82) P 1729-34; Vol 31; No 9

07-82-0009 (0009)
DETERMINATION OFTHIOCYANATE, THIOSULFATE, SULFITE AND NITRITE BY HIGH-PERFORMANCE LIQUID CHROMATOGRQPHY COUPLED WITH ELECTROCHEMICAL DETECTION
Imanari, Toshio; Ogata, Koreharu; Tanabe, Shinzo; Toida, Toshihiko ; Kawanishi, Takehiko; Ichikawa, Masaki
Fac. Pharm. Sci. Chiba Univ. Chiba Japan JP
Chem. Pharm. Bull.; (82) P 74-5; Vol 30; No 1

07-82-0010 (0010)
LIQUID-CHROMATOGRAPHIC DETERMINATION OF URINARY 5-HYDROXY-3-INDOLEACETIC ACID, WITH FLUORESCENCE DETECTION
Rosano, Thomas G.; Meola, John M.; Swift, Thomas A.
Clin. Chem. Dep. Albany Med. Cent. Hosp. Albany USA US NY 12208
Clin. Chem. (Winston-Salem, N. C.); (82) P 207-8; Vol 28; No 1

07-82-0011 (0011)
URINARY p-AMINOBENZOIC ACID DETERMINED IN THE PANCREATIC FUNCTION TEST BY LIQUID CHROMATOGRAPHY, WITH ELECTROCHEMICAL DETECTION
Ito, Shosuke; Maruta, Kazuhiro; Imai, Yoichiro; Kato, Toshiaki; Ito, Madoka; Nakajima, Sumio; Fujita, Keisuke; Kurahashi, Tatsuo
Sch. Med. Fujita-Gakuen Univ. Aichi Japan JP 470-11
Clin. Chem. (Winston-Salem, N. C.); (82) P 323-6; Vol 28; No 2

07-82-0012 (0012)
LIQUID-CHROMATOGRAPHIC IDENTIFICATION OF ACETAMINOPHEN IN CEREBROSPINAL FLUID WITH USE OF ELECTROCHEMICAL DETECTION
Walsh, Francis X.; Langlais, Philip J.; Bird, Edward D.
Brain Tissue Resour. Cent. McLean Hosp. Belmont USA US MA 20178
Clin. Chem. (Winston-Salem, N. C.); (82) P 382-3; Vol 28; No 2

07-82-0013 (0013)
SIMULTANEOUS FLUOROMETRY OF URINARY DOPAMINE, NOREPINEPHRINE, AND EPINEPHRINE COMPARED WITH LIQUID CHROMATOGRAPHY WITH ELECTROCHEMICAL DETECTION
Oka, Kazuhiro; Sekiya, Motoko; Osada, Hirofumi; Fujita, Keisuke; Kato, Takeshi; Nagatsu, Toshiharu
Grad. Sch. Nagatsuta Tokyo Inst. Technol. Yokohama Japan JP 227
Clin. Chem. (Winston-Salem, N. C.); (82) P 646-9; Vol 28; No 4, Pt. 1

07-82-0014 (0014)
DETERMINATION OF p-HYDROXYPHENYLACETIC ACID IN CEREBROSPINAL FLUID BY HIGH-PERFORMANCE LIQUID CHROMATOGRAPHY WITH ELECTROCHEMICAL DETECTION
Kobayashi, Kiyofumi; Koide, Yayoi; Shohmori, Toshikiyo
Med. Sch. Okayama Univ. Okayama Japan JP 700
Clin. Chim. Acta; (82) P 161-8; Vol 123; No 1-2

07-82-0015 (0015)
DETERMINATION OF METHYL AND ETHYL CAFFEATE IN VEGETABLES BY HPLC WITH ELECTROCHEMICAL DETECTION
Sontag, G.; Schaefers, F. I.; Herrmann, K.
Inst. Anal. Chem. Univ. Vienna Vienna Austria AT
CONFERENCE; Recent Dev. Food Anal., Proc. Eur. Conf. Food Chem., 1st; (82) P 41-6; Avail.: Baltes, Werner; Czedik-Eysenberg, Peter B.; Pfannhauser, Werner; Date: 81; Publisher: Verlag Chem., Weinheim, Fed. Rep. Ger.

07-82-0016 (0016)
SIMULTANEOUS DETERMINATION OF 5-HYDROXYTRYPTOPHAN, SEROTONIN AND 5-HYDROXYINDOLEACETIC ACID IN THE BRAIN BY LIQUID CHROMATOGRAPHY WITH ELECTROCHEMICAL DETECTION
Santagostino, G.; Frattini, P.; Schinelli, S.; Cucchi, M. L.; Corona, G. L.
Inst. Pharmacol. Univ. Pavia Pavia Italy IT

Farmaco, Ed. Prat.; (82) P 365-70; Vol 37; No 11

07-82-0017 (0017)
DETERMINATION OF SULFHYDRYL COMPOUNDS IN PHYSIOLOGICAL FLUIDS BY LIQUID CHROMATOGRAPHY WITH POLAROGRAPHIC DETECTION
Zygmunt, B.; Hanekamp, H. B.; Bos, P.; Frei, R. W.
Dep. Anal. Chem. Free Univ. Amsterdam Neth. NL NL-1081 HV
Fresenius' Z. Anal. Chem.; (82) P 197-200; Vol 311; No 3

07-82-0018 (0018)
DETERMINATION OF L-ERYTHRO-TETRAHYDROBIOPTERIN IN BIOLOGICAL TISSUES BY HIGH PRESSURE LIQUID CHROMATOGRAPHY AND ELECTROCHEMICAL DETECTION
Braeutigam, Matthias; Dreesen, Rainer
Pharmakol. Inst. Freie Univ. Berlin Fed. Rep. Ger. DE D-1000/33
Hoppe-Seyler's Z. Physiol. Chem.; (82) P 1203-7; Vol 363; No 10

07-82-0019 (0019)
DETERMINATION OF REDUCED BIOPTERINS BY HIGH PRESSURE LIQUID CHROMATOGRAPHY AND SUBSEQUENT ELECTROCHEMICAL DETECTION
Braeutigam, Matthias; Dreesen, Rainer; Herken, Hans
Pharmakol. Inst. Freien Univ. Berlin Fed. Rep. Ger. DE
Hoppe-Seyler's Z. Physiol. Chem.; (82) P 341-3; Vol 363; No 3

07-82-0020 (0020)
SIMULTANEOUS USE OF FLUORESCENCE, ULTRAVIOLET, AND ELECTROCHEMICAL DETECTORS IN HIGH-PERFORMANCE LIQUID CHROMATOGRAPHY - SEPARATION AND IDENTIFICATION OF PHENOLIC ANTIOXIDANTS AND RELATED COMPOUNDS
Masoud, A. N.; Cha, Young Nam
Dep. Environ. Health Sci. Johns Hopkins Univ. Baltimore USA US MD 21205
HRC CC, J. High Resolut. Chromatogr. Chromatogr. Commun.; (82) P 299-305; Vol 5; No 6

07-82-0021 (0021)
DETERMINATION OF L-DOPA IN BRAIN TISSUE USING HIGH-PERFORMANCE LIQUID CHROMATOGRAPHY WITH ELECTROCHEMICAL DETECTION TO STUDY THE ACTIVITY OF CENTRAL DOPAMINERGIC NEURONS
Benfenati, F.; Ferretti, P.; Ferretti, C.; Ponzio, F.; Algeri, S.
Ist. Ric. Farmacol. "Mario Negri" Milan Italy IT 20157
IRCS Med. Sci.: Libr. Compend.; (82) P 425-6; Vol 10; No 5

07-82-0022 (0022)
DETERMINATION OF MORPHINE IN CEREBROSPINAL FLUID AND PLASMA BY HIGH-PERFORMANCE LIQUID CHROMATOGRAPHY WITH ELECTROCHEMICAL DETECTION
Todd, Richard D.; Muldoon, Sheila M.; Watson, Robert L.
Dep. Anesthesiol. Uniformed Serv. Univ. Health Sci. Bethesda USA US MD 20014
J. Chromatogr.; (82) P 101-10; Vol 232; No 1

07-82-0023 (0023)
SIMULTANEOUS DETERMINATION OF CHLORPROMAZINE AND LEVOMEPROMAZINE IN HUMAN PLASMA AND URINE BY HIGH-PERFORMANCE LIQUID CHROMATOGRAPHY USING ELECTROCHEMICAL DETECTION
Murakami, Kazuo; Murakami, Koji; Ueno, Tadataka; Hijikata, Junko; Shirasawa, Kimio; Muto, Toshitake
Lab. Psychopharmacol. Sagami Hosp. Sagamihara Japan JP
J. Chromatogr.; (82) P 103-12; Vol 227; No 1

07-82-0024 (0024)
SIMULTANEOUS MEASUREMENT OF 5-HYDROXYTRYPTOPHAN AND L-DIHYDROXYPHENYLALANINE BY HIGH-PERFORMANCE LIQUID CHROMATOGRAPHY WITH ELECTROCHEMICAL DETECTION. MEASUREMENT OF SEROTONIN AND CATECHOLAMINE TURNOVER IN DISCRETE BRAIN REGIONS

Shum, Andrew; Sole, Michael J.; Van Loon, Glen R.
Dep. Med. Univ. Toronto Toronto Can. CA ON M5S 1A8
J. Chromatogr.; (82) P 123-30; Vol 228

07-82-0025 (0025)
STUDIES ON STEROIDS. CLXXIX. DETERMINATION OF ESTRIOL 16- AND 17-GLUCURONIDE IN BIOLOGICAL FLUIDS BY HIGH-PERFORMANCE LIQUID CHROMATOGRAPHY WITH ELECTROCHEMICAL DETECTION
Shimada, Kazutake; Xie, Fuming; Nambara, Toshio
Pharm. Inst. Tohoku Univ. Sendai Japan JP 980
J. Chromatogr.; (82) P 13-18; Vol 232; No 1

07-82-0026 (0026)
SIMULTANEOUS DETERMINATION OF NOREPINEPHRINE, DOPAMINE AND SEROTONIN IN RAT BRAIN REGIONS BY ION-PAIR LIQUID CHROMATOGRAPHY ON OCTYL SILANE COLUMNS AND AMPEROMETRIC DETECTION
Warsh, Jerry J.; Chiu, Andrew; Godse, Damodar D.
Dep. Biochem. Psychiatry Clarke Inst. Psychiatry Toronto Can. CA ON M5T 1R8
J. Chromatogr.; (82) P 131-41; Vol 228

07-82-0027 (0027)
DETERMINATION OF CATECHOLAMINES IN RAT HEART TISSUE AND PLASMA SAMPLES BY LIQUID CHROMATOGRAPHY WITH ELECTROCHEMICAL DETECTION
Eriksson, Britt Marie; Persson, Bengt Arne
AB Hassle Molndal Swed. SE S-431 83
J. Chromatogr.; (82) P 143-54; Vol 228

07-82-0028 (0028)
HIGH-PERFORMANCE LIQUID CHROMATOGRAPHY WITH REDUCTIVE ELECTROCHEMICAL DETECTION OF MUTAGENIC NITRO-SUBSTITUTED POLYNUCLEAR AROMATIC HYDROCARBONS IN DIESEL EXHAUSTS
Rappaport, S. M.; Jin, Z. L.; Xu, X. B.
Sch. Public Health Univ. California Berkeley USA US CA 94720
J. Chromatogr.; (82) P 145-54; Vol 240; No 1

07-82-0029 (0029)
REVERSED-PHASE LIQUID CHROMATOGRAPHY WITH AMPEROMETRIC DETECTION OF LIPOPHILIC DOPAMINE ANALOGS AND DETERMINATION OF BRAIN AND SERUM CONCENTRATIONS AFTER SAMPLE CLEAN-UP ON SMALL SEPHADEX G-10 COLUMNS
Feenstra, Matthijs G. P.; Homan, Jan Willem; Dijkstra, Durk; Mulder, Theo B. A.; Rollema, Hans; Westerink, Ben H. C.; Horn, Alan S.
Dep. Med. Chem. State Univ. Groningev Groningen Neth. NL 9713 AW
J. Chromatogr.; (82) P 271-87; Vol 20; No 2

07-82-0030 (0030)
MEASUREMENT OF CATECHOLAMINES IN BIOLOGICAL FLUIDS BY HIGH-PERFORMANCE LIQUID CHROMATOGRAPHY. A COMPARISON OF FLUORIMETRIC WITH ELECTROCHEMICAL DETECTION
Causon, Roger C.; Carruthers, Malcolm E.
Dep. Biochem. Inst. Psychiatry London UK GB
J. Chromatogr.; (82) P 301-9; Vol 229; No 2

07-82-0031 (0031)
DETERMINATION OF .GAMMA.-AMINOBUTYRIC ACID BY LIQUID CHROMATOGRAPHY WITH ELECTROCHEMICAL DETECTION
Caudill, W. Lowry; Houck, Gregory P.; Wightman, R. Mark
Dep. Chem. Indiana Univ. Bloomington USA US IN 47405
J. Chromatogr.; (82) P 331-9; Vol 227; No 2

07-82-0032 (0032)
LIQUID CHROMATOGRAPHIC ASSAY OF PHENOTHIAZINE, THIOXANTHENE AND BUTYROPHENONE

NEUROLEPTICS AND ANTIHISTAMINES IN BLOOD AND PLASMA WITH CONVENTIONAL AND RADIAL COMPRESSION COLUMNS AND UV AND ELECTROCHEMICAL DETECTION
Curry, Stephen H.; Brown, Edgar A.; Hu, Oliver Y. P.; Perrin, John H.
Coll. Pharm. Univ. Florida Gainesville USA US FL 32610
J. Chromatogr.; (82) P 361-76; Vol 231; No 2

07-82-0033 (0033)
SIMULTANEOUS DETERMINATION OF 4-HYDROXY-3-METHOXYPHENYLACETIC (HOMOVANILLIC) ACID AND OTHER MONOAMINE METABOLITES IN HUMAN LUMBAR CEREBROSPINAL FLUID. AN IMPROVED HIGH-PERFORMANCE LIQUID CHROMATOGRAPHIC STUDY WITH ELECTROCHEMICAL DETECTION
Krstulovic, Ante M.; Bertani-Dziedzic, Laura; Bautista-Cerqueira, Sandra; Gitlow, Stanley E.
Dep. Chem. Manhattanville Coll. Purchase USA US NY 10577
J. Chromatogr.; (82) P 379-89; Vol 227; No 2

07-82-0034 (0034)
ANALYSIS OF URINARY 3-METHOXY-4-HYDROXYPHENYLGLYCOL BY HIGH-PERFORMANCE LIQUID CHROMATOGRAPHY AND ELECTROCHEMICAL DETECTION
Moleman, P.; Borstrok, J. J. M.
Dep. Psychiatry Academic Hosp. Dijkzigt Rotterdam Neth. NL 3015 GD
J. Chromatogr.; (82) P 391-405; Vol 227; No 2

07-82-0035 (0035)
HIGHLY SENSITIVE ASSAY FOR PHENYLETHANOLAMINE N-METHYLTRANSFERASE ACTIVITY IN RAT BRAIN BY HIGH-PERFORMANCE LIQUID CHROMATOGRAPHY WITH ELECTROCHEMICAL DETECTION
Trocewicz, Jerzy; Oka, Kazuhiro; Nagatsu, Toshiharu
Grad. Sch. Nagatsuta Tokyo Inst. Technol. Yokohama Japan JP 227
J. Chromatogr.; (82) P 407-13; Vol 227; No 2

07-82-0036 (0036)
ROUTINE DETERMINATION OF PLASMA CATECHOLAMINES USING REVERSED-PHASE, ION-PAIR HIGH-PERFORMANCE LIQUID CHROMATOGRAPHY WITH ELECTROCHEMICAL DETECTION
Davies, Cledwyn L.; Molyneux, Stephen G.
Radcliffe Infirm. Univ. Oxford Oxford UK GB OX2 6HE
J. Chromatogr.; (82) P 41-51; Vol 231; No 1

07-82-0037 (0037)
ASSAY AND STABILITY OF .ALPHA.-METHYLDOPA IN MAN USING HIGH-PERFORMANCE LIQUID CHROMATOGRAPHY WITH ELECTROCHEM. DETECTION
Ong, H.; Sved, S.; Beaudoin, N.
Health Prot. Branch Bur. Drug Res. Ottawa Can. CA ON K1A 0L2
J. Chromatogr.; (82) P 433-8; Vol 229; No 2

07-82-0038 (0038)
DETERMINATION OF CAPTOPRIL IN HUMAN BLOOD BY HIGH-PERFORMANCE LIQUID CHROMATOGRAPHY WITH ELECTROCHEMICAL DETECTION
Shimada, Kazutake; Tanaka, Makoto; Nambara, Toshio; Imai, Yutaka; Abe, Keishi; Yoshinaga, Kaoru
Pharm. Inst. Tohoku Univ. Sendai Japan JP 980
J. Chromatogr.; (82) P 445-51; Vol 227; No 2

07-82-0039 (0039)
SIMULTANEOUS DETERMINATION OF CARBIDOPA, LEVODOPA AND 3,4-DIHYDROXYPHENYL-ACETIC ACID USING HIGH-PERFORMANCE LIQUID CHROMATOGRAPHY WITH ELECTROCHEMICAL DETECTION
Nissinen, Erkki; Taskinen, Jyrki
Res. Lab. Orion Pharm. Co. Helsinki Finland FI 00101/10
J. Chromatogr.; (82) P 459-62; Vol 231; No 2

07-82-0040 (0040)
DETERMINATION OF THE .GAMMA.-AMINOBUTYRIC ACID AGONIST 4,5,6,7-TETRAHYDRO-ISOXAZOLO(5,4-C)PYRIDIN-3-OL BY HIGH-PERFORMANCE LIQUID CHROMATOGRAPHY USING UV OR ELECTROCHEMICAL DETECTION
Madsen, Soren Munk
H. Lundbeck and Co., A/S Valby Den. DK DK 2500
J. Chromatogr.; (82) P 509-14; Vol 238; No 2

07-82-0041 (0041)
HIGH-PERFORMANCE LIQUID CHROMATOGRAPHIC SEPARATION OF THE TWO ESTROGEN ISOMERS OF ESTRADIOL WITH ELECTROCHEMICAL DETECTION
Prescott, William R., Jr.; Boyd, Brenda K.; Seaton, John F.
Milton S. Hershey Med. Cent. Pennsylvania State Univ. Hershey USA US PA 17033
J. Chromatogr.; (82) P 513-16; Vol 234; No 2

07-82-0042 (0042)
RAPID EXTRACTION AND MEASUREMENT OF MORPHINE AND OPIATE ANTAGONISTS FROM RAT BRAIN USING HIGH-PERFORMANCE LIQUID CHROMATOGRAPHY AND ELECTROCHEMICAL DETECTION
Raffa, Robert B.; O'Neill, John J.; Tallarida, Ronald J.
Sch. Med. Temple Univ. Philadelphia USA US PA 19140
J. Chromatogr.; (82) P 515-19; Vol 238; No 2

07-82-0043 (0043)
SEPARATION AND DETERMINATION OF BILE ACIDS BY HIGH-PERFORMANCE LIQUID CHROMATOGRAPHY USING IMMOBILIZED 3.ALPHA.-HYDROXYSTEROID DEHYDROGENASE AND AN ELECTROCHEMICAL DETECTOR
Kamada, Satoru; Maeda, Masako; Tsuji, Akio; Umezawa, Yoshio; Kurahashi, Tatsuo
Sch. Pharm. Sci. Showa Univ. Tokyo Japan JP
J. Chromatogr.; (82) P 773-83; Vol 239

07-82-0044 (0044)
ANALYSIS OF ACETAMINOPHEN METABOLITES IN URINE BY HIGH-PERFO8MANCE LIQUID CHROMATOGRAPHY WITH UV AND AMPEROMETRIC DETECTION
Wilson, John M.; Slattery, John T.; Forte, Anthony J.; Nelson, Sidney D.
Dep. Med. Chem. Univ. Washington Seattle USA 98195
J. Chromatogr.; (82)PP 453-62; Vol 227; No 2

07-82-0045 (0045)
DETERMINATION OF URINARY VANILMANDELIC ACID AND HOMOVANILLIC ACID BY HIGH PERFORMANCE LIQUID CHROMATOGRAPHY WITH AMPEROMETRIC DETECTION
Bauersfeld, W.; Diener, U.; Knoll, E.; Ratge, D.; Wisser, H.
Dep. Clin. Chem. Robert-Bosch-Krankenhaus Stuttgart Fed. Rep. Ger. DE
J. Clin. Chem. Clin. Biochem.; (82) P 217-20; Vol 20; No 4

07-82-0046 (0046)
DETERMINATION OF PLASMA NOREPINEPHRINE AND EPINEPHRINE BY HIGH-PERFORMANCE LIQUID CHROMATOGRAPHY USING A TWO-COLUMN SYSTEM AND AN ELECTROCHEMICAL DETECTOR
Lachatre, G.; Nicot, G.; Magnet, C.; Rocca, J. L.; Merle, L.; Valette, J. P.
Dep. Pharmacol. Clin. Hop. Dupuytren Limoges Fr. FR 87031
J. Liq. Chromatogr.; (82) P 1947-65; Vol 5; No 10

07-82-0047 (0047)
DETERMINATION OF CARBONYL 2,4-DINITROPHENYLHYDRAZONES BY LIQUID CHROMATOGRAPHY/ELECTROCHEMISTRY
Jacobs, W. A.; Kissinger, P. T.
Dep. Chem. Purdue Univ. West Lafayette USA US IN 47907
J. Liq. Chromatogr.; (82) P 669-76; Vol 5; No 4

07-82-0048 (0048)
NITROAROMATIC REAGENTS FOR DETERMINATION OF AMINES AND AMINO ACIDS BY LIQUID CHROMATOGRAPHY/ELECTROCHEMISTRY
Jacobs, W. A.; Kissinger, P. T.
Dep. Chem. Purdue Univ. West Lafayette USA US IN 47907
J. Liq. Chromatogr.; (82) P 881-95; Vol 5; No 5

07-82-0049 (0049)
THE DETERMINATION OF CAPTOPRIL IN PHYSIOLOGICAL FLUIDS USING HIGH-PERFORMANCE LIQUID CHROMATOGRAPHY WITH ELECTROCHEMICAL DETECTION
Perrett, D.; Drury, P. L.
Dep. Med. St. Bartholomew's Hosp. London UK GB EC1A 7BE
J. Liq. Chromatogr.; (82) P 97-110; Vol 5; No 1

07-82-0050 (0050)
SIMULTANEOUS DETERMINATION OF 3,4-DIHYDROXYPHENYLALANINE, 5-HYDROXYTRYPTOPHAN, DOPAMINE, 4-HYDROXY-3-METHOXYPHENYLALANINE,NOREPINEPHRINE, 3,4-DIHYDROXYPHENYL ACETIC ACID, HOMOVANILLIC ACID, SEROTONIN AND 5-HYDROXYINDOLEACETIC ACID IN RAT CEREBROSPINAL FLUID AND BRAIN BY HIGH-PERFORMANCE LIQUID CHROMATOGRAPHY WITH ELECTROCHEMICAL DETECTION
Wagner, Joseph; Vitali, Pierre; Palfreyman, Michael G.; Zraika, Monique; Huot, Sylvie
Cent. Rech., Merrell Int., Strasbourg France 67 084
J. Neurochem. (82) P 1241-54; Vol 38; No 5

07-82-0051 (0051)
SIMULTANEOUS DETERMINATION OF SEROTONIN, 5-HYDROXYINDOLEACETIC ACID, 3,4-DIHYDROXYPHENYLACETIC ACID AND HOMOVANILLIC ACID BY HIGH PERFORMANCE LIQUID CHROMATOGRAPHY WITH ELECTROCHEMICAL DETECTION
Sperk, Guenther
Inst. Biochem. Pharmacol. Univ. Vienna Vienna Austria AT
J. Neurochem.; (82) P 840-3; Vol 38; No 3

07-82-0052 (0052)
DETERMINATION OF MIANSERIN AND METABOLITES IN PLASMA BY LIQUID CHROMATOGRAPHY WITH ELECTROCHEMICAL DETECTION
Suckow, Raymond F.; Cooper, Thomas B.; Quitkin, Frederic M.; Stewart, Jonathan W.
Anal. Psychopharmacol. Lab. Rockland Res. Inst. Orangeburg USA US NY 10962
J. Pharm. Sci.; (82) P 889-92; Vol 71; No 8

07-82-0053 (0053)
DETERMINATION OF ISOETHARINE IN PLASMA BY REVERSED-PHASE CHROMATOGRAPHY WITH AMPEROMETRIC DETECTION
Park, George B.; Koss, Raymond F.; Utter, Julie; Mayes, Brian A.; Edelson, Jerome
Dep. Drug Metab. Sterling-Winthrop Res. Inst. Rensselaer USA US NY 12144
J. Pharm. Sci.; (82) P 932-4; Vol 71; No 8

07-82-0054 (0054)
A SENSITIVE PROCEDURE FOR DETERMINATION OF MORPHINE IN MOUSE WHOLE BLOOD BY HIGH-PERFORMANCE LIQUID CHROMATOGRAPHY WITH ELECTROCHEMICAL DETECTION
Ishikawa, Koichi; McGaugh, James L.; Shibanoki, Shinji; Kubo, Taizo
Sch. Biol. Sci. Univ. California Irvine USA US CA 92717
Jpn. J. Pharmacol.; (82) P 969-71; Vol 32; No 5

07-82-0055 (0055)
USE OF LIQUID CHROMATOGRAPHY WITH ELECTROCHEMISTRY TO MEASURE EFFECTS OF VARYING INTENSITIES OF WHITE LIGHT ON DOPA ACCUMULATION IN RAT RETINAS
Proll, Melissa A.; Kamp, Cylia W.; Morgan, William W.

Health Sci. Cent. Univ. Texas San Antonio USA US TX 78284
Life Sci.; (82) P 11-19; Vol 30; No 1

07-82-0056 (0056)
SIMULTANEOUS MEASUREMENT OF DOPAMINE AND ITS METABOLITES, 5-HYDROXYTRYPTAMINE, 5-HYDROXYINDOLEACETIC ACID AND TRYPTOPHAN IN BRAIN TISSUE USING LIQUID CHROMATOGRAPHY AND ELECTROCHEMICAL DETECTION
Lyness, W. H.
Health Sci. Cent. Texas Tech. Univ. Lubbock USA US TX 79430
Life Sci.; (82) P 1435-43; Vol 31; No 14

07-82-0057 (0057)
MEASUREMENT OF L-DIHYDROXYPHENYLALANINE IN PLASMA AND OTHER BIOLOGICAL FLUIDS BY HIGH-PRESSURE LIQUID CHROMATOGRAPHY WITH ELECTROCHEMICAL DETECTION
Shum, Andrew; Van Loon, Glen R.; Sole, Michael J.
Dep. Med. Univ. Toronto Toronto Can. CA ON M5S 1A8
Life Sci.; (82) P 1541-5; Vol 31; No 15

07-82-0058 (0058)
IMPROVED METHOD FOR DETERMINATION OF CATECHOLAMINES IN RAT BRAIN BY ISOLATION ON BORIC ACID GEL AND HIGH-PERFORMANCE LIQUID CHROMATOGRAPHY WITH ELECTROCHEMICAL DETECTION
Koike, Koji; Aono, Toshihiro; Chatani, Fumio; Takemura, Toshiko; Kurachi, Keiichi
Med. Sch. Osaka Univ. Osaka Japan JP 553
Life Sci.; (82) P 2221-8; Vol 30; No 25

07-82-0059 (0059)
SEROTONIN TURNOVER IN INDIVIDUAL BRAIN NUCLEI: EVALUATION OF THREE METHODS USING LIQUID CHROMATOGRAPHY WITH ELECTROCHEMICAL DETECTION
Johnson, M. D.; Crowley, W. R.
Cent. Health Sci. Univ. Tennessee Memphis USA US TN 38163
Life Sci.; (82) P 589-95; Vol 31; No 6

07-82-0060 (0060)
DEMONSTRATION OF AROMATIC L-AMINO ACID DECARBOXYLASE ACTIVITY IN HUMAN BRAIN WITH L-DOPA AND L-5-HYDROXYTRYPTOPHAN AS SUBSTRATES BY HIGH-PERFORMANCE LIQUID CHROMATOGRAPHY WITH ELECTROCHEMICAL DETECTION
Rahman, Mohammed Khalilur; Nagatsu, Toshiharu
Dep. Life Chem. Tokyo Inst. Technol. Yokohama Japan JP 227
Neurochem. Int.; (82) P 1-6; Vol 4; No 1

07-82-0061 (0061)
A RAPID AND SIMPLE METHOD TO DETERMINE THE SPECIFIC ACTIVITIES OF SEROTONIN, 5-HYDROXYINDOLEACETIC ACID, AND 5-HYDROXYTRYPTOPHAN IN BRAIN BY HPLC WITH ELECTROCHEMICAL DETECTION
Bernstein, M. J.; Shea, P. A.
Sch. Med. Indiana Univ. Indianapolis USA US IN 46223
Neurochem. Res.; (82) P 79-85; Vol 7; No 1

07-82-0062 (0062)
DETERMINATION OF METHIONINE-ENKEPHALIN, NOREPINEPHRINE, DOPAMINE, 3,4-DIHYDROXYPHENYLACETIC ACID (DOPAC) AND 3-METHOXY-4-HYDROXYPHENY- LACETIC ACID (HVA) IN BRAIN BY HIGH-PRESSURE LIQUID CHROMATOGRAPHY WITH ELECTROCHEMICAL DETECTOR
Nabeshima, Toshitaka; Hiramatsu, Masayuki; Noma, Shigeki; Ukai, Makoto; Amano, Manabu; Kameyama, Tsutomu
Fac. Pharm. Sci. Meijo Univ. Nagoya Japan JP 468
Res. Commun. Chem. Pathol. Pharmacol.; (82) P 421-42; Vol 35; No 3

07-82-0063 (0063)
SIMULTANEOUS DETERMINATION OF CHLORPROMAZINE AND LEVOMEPROMAZINE IN HUMAN PLASMA BY HIGH-PERFORMANCE LIQUID CHROMATOGRAPHY USING ELECTROCHEMICAL DETECTION
Murakami, Koji; Murakami, Kazuo; Ueno, Tadataka; Hijikata, Junko; Shirasawa, Kimio; Mutoh, Toshitake
Dep. Psychiatry Sagami Hosp. Sagamihara Japan JP
Rinsho Kagaku; (82) P 50-5; Vol 11; No 1; In Japan

07-82-0064 (0064)
CLINICAL PHARMACOKINETIC STUDIES ON THE CONCURRENT USE OF TWO ANTIPSYCHOTIC DRUGS. I. SIMULTANEOUS DETERMINATION OF CHLORPROMAZINE AND LEVOMEPROMAZINE IN HUMAN PLASMA BY HIGH-PERFORMANCE LIQUID CHROMATOGRAPHY USING ELECTROCHEMICAL DETECTION
Ueno, Tadataka
Sch. Med. Showa Univ. Tokyo Japan JP
Showa Igakkai Zasshi; (82) P 35-43; Vol 42; No 1; In Japan

07-82-0065 (0065)
ANALYSIS OF MORPHINE IN SERUM BY HIGH PERFORMANCE LIQUID CHROMATOGRAPHY WITH AMPEROMETRIC DETECTION
Vandenberghe, H.; MacLeod, S. M.; Chinyanga, H.; Soldin, S. J.
Dep. Clin. Biochem. Univ. Toronto Toronto Can. CA ON
Ther. Drug Monit.; (82) P 307-14; Vol 4; No 3

07-82-0066 (0066)
DETERMINATION OF GLUTATHIONE AND D-PENICILLAMINE USING HIGH-PERFORMANCE LIQUID CHROMATOGRAPHY WITH A VOLTAMMETRIC DETECTOR
Inoue, Kenji; Otake, Tateo; Noda, Katsumi; Kyogoku, Kazuaki; Uchida, Shoji
Res. Cent. Taisho Pharm. Co., Ltd. Saitama Japan JP 330
Yakugaku Zasshi; (82) P 659-65; Vol 102; No 7; In Japan

07-82-0067 (0067)
DETERMINATION OF QUININE IN SOFT DRINKS AND DRUGS BY HIGH-PERFORMANCE LIQUID CHROMATOGRAPHY WITH ELECTROCHEMICAL DETECTION
Kral, Karl; Sontag, Gerhard
Inst. Anal. Chem. Univ. Wien Vienna Austria AT A-1090
Z. Lebensm.-Unters. Forsch.; (82) P 22-4; Vol 175; No 1; In Ger

5.7.8
Mass Spectrometry

--- 1982 ---

07-82-0001 (0001)
ANALYSIS OF THE CHEMICAL IONIZATION REAGENT PLASMA PRODUCED IN LIQUID CHROMATOGRAPHY/MASS SPECTROMETRY BY DIRECT LIQUID INTRODUCTION MASS SPECTROMETRY/MASS SPECTROMETRY
Voyksner, R. D.; Hass, J. Ronald; Bursey, Maurice M.
Lab. Environ. Chem. Natl. Inst. Environ. Health Sci. Research Triangle Park USA US NC 27709
Anal. Chem.; (82) P 2465-70; Vol 54; No 14

07-82-0002 (0002)
EFFECTS OF PRESSURE, TEMPERATURE, AND SOLVENT COMPOSITION IN ANALYSIS BY DIRECT LIQUID INTRODUCTION LIQUID CHROMATOGRAPHY/MASS SPECTROMETRY
Voyksner, Robert D.; Parker, Carol E.; Hass, Ronald J.; Bursey, Maurice M.
Natl. Inst. Environ. Health Sci. Research Triangle Park USA US NC 27709
Anal. Chem.; (82) P 2583-6; Vol 54; No 14

07-82-0003 (0003)
DETERMINATION OF SULFA DRUGS IN BIOLOGICAL FLUIDS BY LIQUID CHROMATOGRAPHY/MASS SPECTROMETRY/MASS SPECTROMETRY
Henion, Jack D.; Thomson, Bruce A.; Dawson, Peter H.
Coll. Vet. Med. Cornell Univ. Ithaca USA US NY 14853
Anal. Chem.; (82) P 451-6; Vol 54; No 3

07-82-0004 (0004)
LIQUID CHROMATOGRAPHY/MASS SPECTROMETRY OF KEPONE HYDRATE, KELEVAN, AND MIREX
Cairns, Thomas; Siegmund, Emil G.; Doose, Gregory M.
Dep. Health and Human Serv. Food and Drug Adm. Los Angeles USA US CA 90015
Anal. Chem.; (82) P 953-7; Vol 54; No 5

07-82-0005 (0005)
AN ON-LINE LIQUID CHROMATOGRAPHY/MASS SPECTROMETRY/MASS SPECTROMETRY EXPERIMENT
Voyksner, R. D.; Hass, J. R.; Bursey, M. M.
Dep. Chem. Univ. North Carolina Chapel Hill USA US NC 27514
Anal. Lett.; (82) P 1-12; Vol 15; No A1

07-82-0006 (0006)
STUDIES OF ERGOT ALKALOIDS USING HIGH-PERFORMANCE LIQUID CHROMATOGRAPHY-MASS SPECTROMETRY AND MASS SPECTROMETRY-MASS SPECTROMETRY
Eckers, Christine; Games, David E.; Mallen, David N. B.; Swann, Brian P.
Dep. Chem. Univ. Coll. Cardiff Cardiff UK GB CF1 1XL
Anal. Proc. (London); (82) P 133-7; Vol 19; No 3

07-82-0007 (0007)
LC-MS DETECTION OF STEROIDS AND DRUGS AT LOW AND MEDIUM RESOLUTION
Baty, J. D.; Willis, R. G.
Dep. Biochem. Med. Ninewells Hosp. Med. Sch. Dundee UK GB DD1 9SY
Anal. Proc. (London); (82) P 251-3; Vol 19; No 5

07-82-0008 (0008)
COMBINED LIQUID CHROMATOGRAPHY-MASS SPECTROMETRY OF PESTICIDES AND NATURAL PRODUCTS
Games, D. E.; Eckers, C.; Lant, M. S.; Lewis, E.; Weerasinghe, N. C. A.; Westwood, S. A.
Dep. Chem. Univ. Coll. Cardiff Cardiff UK GB CF1 1XL
Anal. Proc. (London); (82) P 253-6; Vol 19; No 5

07-82-0009 (0009)
PROBLEMS AND SOLUTIONS IN COMBINED LIQUID CHROMATOGRAPHY-MASS SPECTROMETRY
McFadden, William H.
Finnigan Corp. Sunnyvale USA US CA 94086
Anal. Proc. (London); (82) P 258-61; Vol 19; No 5

07-82-0010 (0010)
APPLICATION OF DIRECT COUPLING OF LIQUID PHASE CHROMATOGRAPHY AND MASS SPECTROMETRY TO BIOCHEMICAL PROBLEMS
Dixon, David J.
Hewlett-Packard G.m.b.H. Waldbronn Fed. Rep. Ger. DE 7517
Analusis; (82) P 343-8; Vol 10; No 7; In Fr

07-82-0011 (0011)
DETERMINATION OF ORGANIC CONTAMINANTS IN ULTRA-PURE WATER BY REVERSED-PHASE HIGH-PERFORMANCE LIQUID CHROMATOGRAPHY WITH ULTRAVIOLET DETECTION
Reust, Juerg B.; Meyer, Veronika R.
Inst. Anorg. Anal. Phys. Chem. Univ. Berne Berne Switz. CH CH-3012
Analyst (London); (82) P 673-9; Vol 107; No 1275

07-82-0012 (0012)
STUDIES OF ERGOT ALKALOIDS USING HIGH-PERFORMANCE LIQUID CHROMATOGRAPHY-MASS SPECTROMETRY AND B/E LINKED SCANS
Eckers, Christine; Games, David E.; Mallen, David N. B. Swann, Brian P.
Dep. Chem. Univ. Coll. Cardiff Cardiff UK GB CF1 1XL
Biomed. Mass Spectrom.; (82) P 162-73; Vol 9; No 4

07-82-0013 (0013)
MICROBORE HIGH-PERFORMANCE LIQUID CHROMATOGRAPHY MASS SPECTROMETRY
Games, David E.; Lant, Michael S.; Westwood, Steven A.; Cocksedge, Michael J.; Evans, Nigel; Williamson, James; Woodhall, Brian J.
Dep. Chem. Univ. Coll. Cardiff Cardiff UK GB CF1 1XL
Biomed. Mass Spectrom.; (82) P 215-24; Vol 9; No 5

07-82-0014 (0014)
DETERMINATION OF ALDICARB RESIDUES IN WATER BY COMBINED HIGH PERFORMANCE LIQUID CHROMATOGRAPHY/MASS SPECTROMETRY
Wright, L. H.; Jackson, M. D.; Lewis, R. G.
EPA Research Triangle Park USA US NC 27711
Bull. Environ. Contam. Toxicol.; (82) P 740-7; Vol 28; No 6

07-82-0015 (0015)
FIRST WORKSHOP ON LIQUID CHROMATOGRAPHY-MASS SPECTROMETRY. (1981; MONTREUX. IN: J. CHROMATOGR., 1982; 251(2))
Frei, R. W.; Editor
Neth. NL
BOOK ; (82) P 137 pp.; Publisher: Elsevier, Amsterdam, Neth.

07-82-0016 (0016)
HIGH-PERFORMANCE LIQUID CHROMATOGRAPHIC AND MASS SPECTROMETRIC ANALYSES OF PORPHYRINS FROM DEEP-SEA SEDIMENTS
Quirke, J. M. E.; Eglinton, G.; Palmer, S. E.; Baker, E. W.
Org. Geochem. Unit Univ. Bristol Bristol UK GB BS8 1TS
Chem. Geol.; (82) P 69-85; Vol 35; No 1-2

07-82-0017 (0017)
CHARACTERIZATION OF A HEAVY COAL LIQUEFACTION PRODUCT BY COMBINED HIGH-PERFORMANCE LIQUID CHROMATOGRAPHY/MASS SPECTROMETRY
Holstein, W.; Severin, D.
Inst. Erdoelforsch. Hannover Fed. Rep. Ger. DE D-3000

Chromatographia; (82) P 231-5; Vol 15; No 4; In Ger

07-82-0018 (0018)
ON-LINE DIRECT LIQUID INTRODUCTION INTERFACE FOR MICRO-LIQUID CHROMATOGRAPHY/MASS SPECTROMETRY: APPLICATION TO DRUG ANALYSIS
Eckers, Christine; Skrabalak, Dale S.; Henion, Jack
New York State Coll. Vet. Med. Cornell Univ. Ithaca USA US NY 14853
Clin. Chem. (Winston-Salem, N. C.); (82) P 1882-6; Vol 28; No 9

07-82-0019 (0019)
LIQUID CHROMATOGRAPH-MASS SPECTROMETER
JEOL Ltd. Japan JP
Jpn. Kokai Tokkyo Koho; In Japan; Pat. No.: 82 00836; Int. Class: H01J-049/04; G01N-027/62; G01N-031/08; (010582); Doc. Code: A2; Appl.: JP 80/74697 800603

07-82-0020 (0020)
DIRECT COUPLING OF AN ULTRA-MICRO HIGH-PERFORMANCE LIQUID CHROMATOGRAPH AND A MASS SPECTROMETER
Takeuchi, T.; Ishii, D.; Saito, A.; Ohki, T.
Fac. Eng. Nagoya Univ. Nagoya Japan JP 464
HRC CC, J. High Resolut. Chromatogr. Chromatogr. Commun.; (82) P 91-2; Vol 5; No 2

07-82-0021 (0021)
COMBINED LIQUID CHROMATOGRAPHIC/MASS SPECTROMETRIC ANALYSIS OF CARBAMATE PESTICIDES
Wright, L. H.
U.S. EPA Research Triangle Park USA US NC 27711
J. Chromatogr. Sci.; (82) P 1-6; Vol 20; No 1

07-82-0022 (0022)
OFF-LINE COMBINATION OF LIQUID CHROMATOGRAPHY AND FIELD DESORPTION MASS SPECTROMETRY: PRINCIPLES AND ENVIRONMENTAL, MEDICAL, AND PHARMACEUTICAL APPLICATIONS
Schulten, H. R.
Inst. Chem. Univ. Bonn Bonn Fed. Rep. Ger. DE 5300/1
J. Chromatogr.; (82) P 105-28; Vol 251; No 2

07-82-0023 (0023)
APPLICATION OF MICROBORE COLUMNS TO LIQUID CHROMATOGRAPHY-MASS SPECTROMETRY
Krien, Paul; Devant, Gerard; Hardy, Michel
Nermag S. A. Rueil-Malmaison Fr. FR 92500
J. Chromatogr.; (82) P 129-39; Vol 251; No 2

07-82-0024 (0024)
LIQUID CHROMATOGRAPHY-MASS SPECTROMETRY IN THE PHARMACEUTICAL INDUSTRY: OBJECTIVES AND NEEDS
Erni, F.
Pharm. Dep. Sandoz Ltd. Basel Switz. CH
J. Chromatogr.; (82) P 141-51; Vol 251; No 2

07-82-0025 (0025)
OPTIMIZATION OF THE INSTRUMENTAL PARAMETERS OF A COMBINED LIQUID CHROMATOGRAPH-MASS SPECTROMETER, COUPLED BY AN INTERFACE FOR DIRECT LIQUID INTRODUCTION. III. WHY THE SOLVENT SHOULD NOT BE REMOVED IN LIQUID CHROMATOGRAPHIC-MASS SPECTROMETRIC INTERFACING METHODS
Arpino, P. J.; Guiochon, G.
Lab. Chim. Anal. Phys. Ec. Polytech. Palaiseau Fr. FR 91128
J. Chromatogr.; (82) P 153-64; Vol 251; No 2

07-82-0026 (0026)
HIGH-PERFORMANCE LIQUID CHROMATOGRAPHY-MASS SPECTROMETRY WITH TRANSPORT INTERFACES
Alcock, Nicholas J.; Eckers, Christine; Games, David E.; Games, Marguerite P. L.; Lant, Michael S.; McDowall, Mark A.; Rossiter, Michael; Smith, Richard W.; Westwood, Steven A.; Wong, Hee Yen
Dep. Chem. Univ. Coll. Cardiff UK GB CF1 1XL
J. Chromatogr.; (82) P 165-74; Vol 251; No 2

07-82-0027 (0027)
INFLUENCE OF THE PACKING MATERIAL AND THE COLUMN FILTERS ON THE RELIABILITY OF A HIGH-PERFORMANCE LIQUID CHROMATOGRAPH - MASS SPECTROMETER INTERFACE BASED ON THE DIRECT LIQUID INLET PRINCIPLE
Mauchamp, Bernard; Krien, Paul
Lab. Phytopharm. Inst. Natl. Rech. Agron. Versailles Fr. FR F-78000
J. Chromatogr.; (82) P 17-24; Vol 236; No 1

07-82-0028 (0028)
NEW APPROACHES TO ON-LINE ION-PAIR EXTRACTION AND DERIVATIZATION FOR HIGH-PERFORMANCE LIQUID CHROMATOGRAPHY-MASS SPECTROMETRY
Vouros, Paul; Lankmayr, E. P.; Hayes, M. J.; Karger, B. L.; McGuire, John M.
Inst. Chem. Anal. Northeastern Univ. Boston USA US MA 02115
J. Chromatogr.; (82) P 175-88; Vol 251; No 2

07-82-0029 (0029)
LIQUID CHROMATOGRAPHY-MASS SPECTROMETRY OF POLAR LIPIDS. COMPARISON OF ON-LINE METHOD USING DIRECT LIQUID INTRODUCTION INTERFACE WITH OFF-LINE METHOD USING DESORPTION/CHEMICAL IONIZATION
Sugnaux, Francois R.; Djerassi, Carl
Dep. Chem. Stanford Univ. Stanford USA US CA 94305
J. Chromatogr.; (82) P 189-201; Vol 251; No 2

07-82-0030 (0030)
APPLICATION OF A COMBINED LIQUID CHROMATOGRAPHIC-MASS SPECTROMETRIC INSTRUMENT USING AN INTERFACE FOR DIRECT LIQUID INTRODUCTION
Dedieu, M.; Juin, C.; Arpino, P. J.; Bounine, J. P.; Guiochon, G.
Nermag S. A. Rueil-Malmaison Fr. FR 92500
J. Chromatogr.; (82) P 203-13; Vol 251; No 2

07-82-0031 (0031)
USE OF HIGH-PERFORMANCE LIQUID CHROMATOGRAPHY-MASS SPECTROMETRY FOR THE STUDY OF THE METABOLISM OF RANITIDINE IN MAN
Martin, L. E.; Oxford, Janet; Tanner, R. J. N.
Biochem. Pharmacol. Dep. Glaxo Group Res. Ltd. Ware/Herts. UK GB
J. Chromatogr.; (82) P 215-24; Vol 251; No 2

07-82-0032 (0032)
HIGH-PERFORMANCE LIQUID CHROMATOGRAPHY-NEGATIVE CHEMICAL IONIZATION MASS SPECTROMETRY OF ORGANOPHOSPHORUS PESTICIDES
Parker, C. E.; Haney, C. A.; Hass, J. R.
Lab. Environ. Chem. Natl. Inst. Environ. Chem. Research Triangle Park USA US NC 27709
J. Chromatogr.; (82) P 233-48; Vol 237; No 2

07-82-0033 (0033)
HIGH-PERFORMANCE LIQUID CHROMATOGRAPHY-MASS SPECTROMETRY OF TRIAZINE HERBICIDES
Parker, C. E.; Haney, C. A.; Harvan, D. J.; Hass, J. R.
Lab. Environ. Chem. Natl. Inst. Environ. Health Sci. Research Triangle Park USA US NC 27709

J. Chromatogr.; (82) P 77-96; Vol 242; No 1

07-82-0034 (0034)
SURVEY OF IONIZATION METHODS WITH EMPHASIS ON LIQUID CHROMATOGRAPHY-MASS SPECTROMETRY
Nibbering, N. M. M.
Lab. Org. Chem. Univ. Amsterdam Amsterdam Neth. NL 1018 WS
J. Chromatogr.; (82) P 93-104; Vol 251; No 2

07-82-0035 (0035)
ANALYSIS OF EXPLOSIVES BY LIQUID CHROMATOGRAPHY-NEGATIVE ION CHEMICAL IONIZATION MASS SPECTROMETRY
Parker, C. E.; Voyksner, R. D.; Tondexe, Yves; Henion, J. D.; Harvan, D. J.; Hass, J. R.; Yinon, Jehuda
Lab. Environ. Chem. Natl. Inst. Environ. Health Sci. Research Triangle Park USA US NC 27709
J. Forensic Sci.; (82) P 495-505; Vol 27; No 3

07-82-0036 (0036)
RECENT DEVELOPMENTS IN LC/MS INTERCONNECTION
Curry, Zorah F.
Extranucl. Lab., Inc. Pittsburgh USA US PA 15238
J. Liq. Chromatogr.; (82) P 257-72; Vol 5; No Suppl. 2

07-82-0037 (0037)
LIQUID CHROMATOGRAPH-MASS SPECTROMETER TANDEM SYSTEM
JEOL Ltd. Japan JP
Jpn. Kokai Tokkyo Koho; (033082) P 3 pp.; In Japan; Pat. No.: 82 53653; Int. Class: G01N-027/62; G01N-031/08; H01J-049W04; Doc. Code: A2; Appl.: JP 80/128760 800917

07-82-0038 (0038)
COMBINED LIQUID CHROMATOGRAPHY-MASS SPECTROMETRY FOR TRACE ANALYSIS OF PHARMACEUTICALS
Schmidt, Lothar; Danigel, Harald; Jungclas, Hartmut
Fachber. Phys. Chem. Philipps-Univ. Marburg Fed. Rep. Ger. DE D-3550
Nucl. Instrum. Methods Phys. Res.; (82) P 165-7; Vol 198; No 1

07-82-0039 (0039)
QUANTITATIVE CALIFORNIUM-252 PLASMA DESORPTION MASS SPECTROMETRY FOR PHARMACEUTICALS. A NEW APPROACH TO COUPLING LIQUID CHROMATOGRAPHY WITH MASS SPECTROMETRY
Jungclas, Hartmut; Danigel, Harald; Schmidt, Lothar
Kernchem. Philipps-Univ. Marburg Fed. Rep. Ger. DE D-3550
Org. Mass Spectrom.; (82) P 86-90; Vol 17; No 2

07-82-0040 (0040)
ON-LINE LIQUID CHROMATOGRAPHY/MASS SPECTROMETRY? AN ODD COUPLE!
Arpino, Patrick J.
Lab. Chim. Anal. Phys. Ec. Polytech. Palaiseau Fr. FR 91128
TrAC, Trends Anal. Chem. (Pers. Ed.); (82) P 154-8; Vol 1; No 7

07-82-0041 (0041)
COUPLING OF HPLC WITH MASS SPECTROMETRY
Levsen, K.
Inst. Phys. Chem. Univ. Bonn Bonn Fed. Rep. Ger. DE
TECHNICAL REPORT; Comm. Eur. Communities, (Rep.) EUR; (82) P 149-58; No EUR 7623, Anal. Org. Micropollut. Water; In Ger

07-82-0042 (0042)
ASSESSMENT OF A MOVING BELT TYPE HPLC-MS INTERFACE WITH RESPECT TO ITS USE IN ORGANIC WATER POLLUTION ANALYSIS
Schauenburg, H.; Schlitt, H.; Knoppel, H.
Jt. Res. Cent. Ispra Italy IT
TECHNICAL REPORT; Comm. Eur. Communities, (Rep.) EUR; (82) P 193-8; No EUR 7623, Anal. Org. Micropollut. Water

5.7.9
Miscellaneous

--- 1982 ---

07-82-0001 (0001)
A SENSITIVE RADIOACTIVITY DETECTOR FOR HPLC
Kessler, Michael J.
Med. Sch. Univ. Texas USA US TX
Am. Lab. (Fairfield, Conn.); (82) P 52, 54, 56, 58-60, 62-3; Vol 14; No 8

07-82-0002 (0002)
ASSESSMENT OF CISPLATIN REACTIVITY WITH PEPTIDES AND PROTEINS USING REVERSE-PHASE HIGH-PERFORMANCE LIQUID CHROMATOGRAPHY AND FLAMELESS ATOMIC ABSORPTION SPECTROSCOPY
Rilev, Christopher M.; Sternson, Larry A.; Repta, Arnold J.
Dep. Pharm. Chem. Univ. Kansas Lawrence USA US KS 66045
Anal. Biochem.; (82) P 167-79; Vol 124; No 1

07-82-0003 (0003)
THERMAL LENS DETECTION FOR LIQUID CHROMATOGRAPHY
Buffett, Clifford E.; Morris, Michael D.
Dep. Chem., Univ. Michigan, Ann Arbor, USA MI 48109
Anal. Chem. (82) P 1824-5; Vol 54; No 11

07-82-0004 (0004)
REFRACTIVE INDEX AND ABSORPTION DETECTOR FOR LIQUID CHROMATOGRAPHY BASED ON FABRY-PEROT INTERFEROMETRY
Woodruff, Steven D.; Yeung, Edward S.
Ames Lab. Iowa State Univ. Ames USA US IA 50011
Anal. Chem.; (82) P 1174-8; Vol 54; No 7

07-82-0005 (0005)
DETERMINATION OF ALKYL NITRATE ADDITIVES IN DIESEL FUEL BY LIQUID CHROMATOGRAPHY WITH INFRARED SPECTROMETRIC DETECTION
Schabron, J. F.; Fuller, M. P.
Phillips Petrol. Co. Bartlesville USA US OK 74004
Anal. Chem.; (82) P 2599-601; Vol 54; No 14

07-82-0006 (0006)
LIQUID-PHASE PHOTOIONIZATION DETECTOR FOR LIQUID CHROMATOGRAPHY
Locke, David C.; Dhingra, Bhim S.; Baker, A. D.
Dep. Chem. City Univ. New York Flushing USA US NY 11367
Anal. Chem.; (82) P 447-50; Vol 54; No 3

07-82-0007 (0007)
SPECIATION OF CHROMIUM(III) AND CHROMIUM(VI) VIA REVERSED PHASE HPLC WITH INDUCTIVELY COUPLED PLASMA EMISSION SPECTROSCOPIC DETECTION (HPLC-ICP)
Krull, I. S.; Bushee, D.; Savage, R. N.; Schleicher, R. G.; Smith, S. B., Jr.
Inst. Chem. Anal. Northeastern Univ. Boston USA US MA 02115
Anal. Lett.; (82) P 267-81; Vol 15; No A3

07-82-0008 (0008)
MICRO HIGH-PERFORMANCE LIQUID CHROMATOGRAPHY OF ORGANICALLY BOUND METALS WITH INDUCTIVELY COUPLED PLASMA ATOMIC EMISSION SPECTROMETRIC DETECTION
Jinno, Kiyokatsu; Tsuchida, Hiroyuki
Sch. Mater. Sci. Toyohashi Univ. Technol. Toyohashi Japan JP 440
Anal. Lett.; (82) P 427-37; Vol 15; No A5

07-82-0009 (0009)
REFRACTOMETRIC DETECTION FOR TRACE ANALYSIS BY LIQUID PHASE CHROMATOGRAPHY. APPLICATION TO THE CASE OF DILUTE AQUEOUS SOLUTIONS OF ISOPROPYL ETHER
Desbene, P. L.; Jehanno, N.

Lab. Chim. Org. Struct. Univ. Pierre et Marie Curie Paris Fr. FR 75230
Analusis; (82) P 292-4; Vol 10; No 6; In Fr

07-82-0010 (0010)
AN INTERFACE FOR THE COMBINATION OF MICRO HIGH-PERFORMANCE LIQUID CHROMATOGRAPHY AND INFRARED SPECTROMETRY
Jinno, K.; Fujimoto, C.; Hirata, Y.
Sch. Mater. Sci. Toyohashi Univ. Technol. Toyohashi Japan JP 440
Appl. Spectrosc.; (82) P 67-9; Vol 36; No 1

07-82-0011 (0011)
COUNTING PERFORMANCE AND RESOLUTION OF AN HPLC RADIOACTIVITY MONITOR
Everett, L. J.
Packard Instrum. Co. Downers Grove USA US IL 60515
Chromatographia; (82) P 445-8; Vol 15; No 7

07-82-0012 (0012)
INFRARED DETECTION OF WAXES IN HPLC
Atkin, D. S. J.; Hamilton, R. J.; Mitchell, S. F.; Sewell, P. A.
Chem. Biochem. Dep. Liverpool Polytech. Liverpool UK GB
Chromatographia; (82) P 97-100; Vol 15; No 2

07-82-0013 (0013)
PROLINE METABOLISM THROUGH GABA-SHUNT IN BARLEY ROOTS MONITORED WITH A HPLC-RADIOANALYZER
Mori, S.
Dep. Agric. Chem. Univ. Tokyo Tokyo Japan JP
CONFERENCE; Plant Nutr., Proc. Int. Plant Nutr. Colloq., 9th; (82) P 400-5; Vol 2,; Avail.: Scaife, A; Publisher: CAB, Farnham Royal, UK.

07-82-0014 (0014)
FINGERPRINTING INORGANIC ARSENIC AND ORGANOARSENIC COMPOUNDS IN IN SITU OIL SHALE RETORT AND PROCESS WATERS USING A LIQUID CHROMATOGRAPH COUPLED WITH AN ATOMIC ABSORPTION SPECTROMETER AS A DETECTOR
Fish, Richard H.; Brinckman, Frederick E.; Jewett, Kenneth L.
Lawrence Berkeley Lab. Univ. California Berkeley USA US CA 94720
Environ. Sci. Technol.; (82) P 174-9; Vol 16; No 3

07-82-0015 (0015)
QUANTITATION OF RADIOLABELED COMPOUNDS ELUTING FROM THE HPLC SYSTEM
Kessler, Michael J.
Med. Sch. Univ. Texas Houston USA US TX 77030
J. Chromatogr. Sci.; (82) P 523-7; Vol 20; No 11

07-82-0016 (0016)
THE USE OF A DIELECTRIC CONSTANT DETECTOR WITH REVERSED-PHASE LC
Mowery, R. A., Jr.
Res., Dev. Eng. Div. Appl. Autom., Inc. Bartlesville USA US OK 74004
J. Chromatogr. Sci.; (82) P 551-9; Vol 20; No 12

07-82-0017 (0017)
IMPROVED INTERFACE FOR LIQUID CHROMATOGRAPHY-ELECTRON-CAPTURE DETECTOR COUPLING. I.
De Kok, A.; Geerdink, R. B.; Brinkman, U. A. T.
Dep. Anal. Chem. Free Univ. Amsterdam Neth. NL 1081 HV
J. Chromatogr.; (82) P 101-11; Vol 252

07-82-0018 (0018)
ANALYSIS OF A MODEL IONIC NITROSAMINE BY MICROBORE HIGH-PERFORMANCE LIQUID CHROMATOGRAPHY USING A THERMAL ENERGY ANALYZER CHEMILUMINESCENCE DETECTOR

Massey, Robert C.; Crews, Colin; McWeeny, David J.; Knowles, Michael E.
Food Lab. Minist. Agric. Fish. Food Norwich UK GB NR2 4SX
J. Chromatogr.; (82) P 527-9; Vol 236; No 2

07-82-0019 (0019)
"BUFFER MEMORY" TECHNIQUE FOR THE COMBINATION OF MICRO-HIGH-PERFORMANCE LIQUID CHROMATOGRAPHY AND INFRARED SPECTROMETRY
Jinno, Kiyokatsu; Fujimoto, Chuzo; Ishii, Daido
Sch. Mater. Sci. Toyohashi Univ. Technol. Toyohashi Japan JP 440
J. Chromatogr.; (82) P 625-32; Vol 239

07-82-0020 (0020)
METAL CATION/ANION SPECIATION VIA PAIRED-ION, REVERSED PHASE HPLC WITH REFRACTIVE INDEX AND/OR INDUCTIVELY COUPLED PLASMA EMISSION SPECTROSCOPIC DETECTION METHODS
Bushee, D.; Krull, I. S.; Savage, R. N.; Smith, S. B., Jr.
Inst. Chem. Anal. Northeast. Univ. Boston USA US MA 02115
J. Liq. Chromatogr.; (82) P 463-78; Vol 5; No 3

07-82-0021 (0021)
METAL CATION SPECIATION VIA EXTRACTION REVERSED-PHASE HPLC WITH REFRACTIVE INDEX AND/OR INDUCTIVELY COUPLED PLASMA EMISSION DETECTION METHODS (HPLC-RI-ICP)
Bushee, D.; Young, D.; Krull, I. S.; Savage, R. N.; Smith, S. B., Jr.
Inst. Chem. Anal. Northeastern Univ. Boston USA US MA 02115
J. Liq. Chromatogr.; (82) P 693-706; Vol 5; No 4

07-82-0022 (0022)
DETERMINATION OF ETHYLENE OXIDE OLIGOMER DISTRIBUTIONS IN ALCOHOL ETHOXYLATES BY HPLC USING A ROTATING DISK-FLAME IONIZATION DETECTOR
McClure, J. D.
Westhollow Res. Cent. Shell Dev. Co. Houston USA US TX 77001
JAOCS, J. Am. Oil Chem. Soc.; (82) P 364-73; Vol 59; No 8

07-82-0023 (0023)
EVALUATION OF A MASS DETECTOR FOR HPLC DETERMINATION OF ORGANIC COMPOUNDS IN WATER
Connor, K. J.; Waggott, A.
Stevenage Lab. Water Res. Cent. Stevenage/Herts. UK GB SG1 1TH
TECHNICAL REPORT; Comm. Eur. Communities, (Rep.) EUR; (82) P 178-87; No EUR 7623, Anal. Org. Micropollut. Water

5.7.10
Derivatization—Generalities

--- 1982 ---

07-82-0001 (0001)
LIQUID CHROMATOGRAPHIC DETERMINATION OF GUANADREL IN LABORATORY ANIMAL DIET AS THE FLUORESCENT ACETYLACETONE DERIVATIVE
Bombardt, Paul A.; Adams, Wade J.
Upjohn Co. Kalamazoo USA US MI 49001
Anal. Chem.; (82) P 1087-90; Vol 54; No 7

07-82-0002 (0002)
LIQUID CHROMATOGRAPHIC PHOSPHORESCENCE DETECTION WITH MICELLAR CHROMATOGRAPHY AND POSTCOLUMN REACTION MODES
Weinberger, Robert; Yarmchuk, Paul; Love, L. J. Cline
Dep. Chem. Seton Hall Univ. South Orange USA US NJ 07079
Anal. Chem.; (82) P 1552-8; Vol 54; No 9

07-82-0003 (0003)
COMPARISON OF LIQUID SEGMENTED WITH NONSEGMENTED FLOW SYSTEMS IN POST-COLUMN REACTORS FOR LIQUID CHROMATOGRAPHY
Scholten, A. H. M. T.; Brinkman, U. A. T.; Frei, R. W.
Dep. Anal. Chem. Free Univ. Amsterdam Neth. NL 1081 HV
Anal. Chem.; (82) P 1932-8; Vol 54; No 12

07-82-0004 (0004)
SIMPLIFIED DETERMINATION OF FREE FATTY ACIDS USING DERIVATIZATION TECHNIQUE BY HIGH-PERFORMANCE LIQUID CHROMATOGRAPHY THROUGH THE MODIFIED EXTRACTION COLUMN
Ikeda, Mitsuo; Shimada, Kenji; Sakaguchi, Takeichi
Niigata Coll. Pharm. Niigata Japan JP 950-21
Bunseki Kagaku; (82) P E199-E206; Vol 31; No 6

07-82-0005 (0005)
DETERMINATION OF KINETIC PARAMETERS FOR IMMOBILIZED XANTHINE OXIDASE AND ITS APPLICATION TO A POST-COLUMN ENZYME REACTOR IN HIGH-PERFORMANCE LIQUID CHROMATOGRAPHIC ANALYSIS OF HYPOXANTHINE, XANTHINE AND URIC ACID
Tawa, Riichi; Kitoh, Masahiro; Hirose, Shingo; Adachi, Kenzo
Kyoto Coll. Pharm. Kyoto Japan JP 607
Chem. Pharm. Bull.; (82) P 615-21; Vol 30; No 2

07-82-0006 (0006)
A NOVEL REACTOR FOR PHOTOCHEMICAL POST-COLUMN DERIVATIZATION IN HPLC
Uihlein, M.; Schwab, E.
Hoechst A.-G. Frankfurt Fed. Rep. Ger. DE D-6230/80
Chromatographia; (82) P 140-6; Vol 15; No 2

07-82-0007 (0007)
REACTION DETECTORS IN MODERN LIQUID CHROMATOGRAPHY
Frei, R. W.
Dep. Anal. Chem. Free Univ. Amsterdam Neth. NL 1081 HV
Chromatographia; (82) P 161-6; Vol 15; No 3

07-82-0008 (0008)
REACTION DETECTOR WITH THREE DIMENSIONAL COILED OPEN TUBES IN HPLC
Engelhardt, H.; Neue, U. D.
Univ. Saarlandes Saarbruecken Fed. Rep. Ger. DE
Chromatographia; (82) P 403-8; Vol 15; No 7

07-82-0009 (0009)
PRECHROMATOGRAPHIC CHEMICAL DERIVATIZATION IN LIQUID CHROMATOGRAPHY
Lawrence, James F.
Food Res. Div. Food Dir. Ottawa Can. CA ON K1A 0L2

CONFERENCE; Chem. Deriv. Anal. Chem.; (82) P 191-242; Vol 2,; Avail.: Frei, Roland W.; Lawrence, James F; Publisher: Plenum, New York, N.Y.

07-82-0010 (0010)
IMMOBILIZED ENZYMES AS PRECOLUMN AND POSTCOLUMN MODIFICATION REAGENTS IN LIQUID CHROMATOGRAPHY
Bowers, Larry D.; Bostick, William D.
Dep. Lab. Med. Pathol. Univ. Minnesota Minneapolis USA US MN 55455
CONFERENCE; Chem. Deriv. Anal. Chem.; (82) P 97-143; Vol 2,; Avail.: Frei, Roland W.; Lawrence, James F; Publisher: Plenum, New York, N. Y.

07-82-0011 (0011)
DERIVATIZATION REACTION DETECTOR IN HPLC
Krull, Ira S.; Lankmayr, Ernst P.
Inst. Chem. Anal. Appl. Forensic Sci. Northeastern Univ. USA US MA
Int. Lab.; (82) P 12, 14, 16, 18, 20, 22, 24-6, 28; Vol 12; No 5

07-82-0012 (0012)
NEW WAYS TO INCREASE THE SPECIFICITY OF DETECTION IN LIQUID CHROMATOGRAPHY
Di Cesare, J. L.; Ettre, L. S.
Chromatogr. Div. Perkin-Elmer Corp. Norwalk USA US CT 06856
J. Chromatogr.; (82) P 1-16; Vol 251; No 1

07-82-0013 (0013)
NEW APPROACHES TO ON-LINE ION-PAIR EXTRACTION AND DERIVATIZATION FOR HIGH-PERFORMANCE LIQUID CHROMATOGRAPHY-MASS SPECTROMETRY
Vouros, Paul; Lankmayr, E. P.; Hayes, M. J.; Karger, B. L.; McGuire, John M.
Inst. Chem. Anal. Northeastern Univ. Boston USA US MA 02115
J. Chromatogr.; (82) P 175-88; Vol 251; No 2

07-82-0014 (0014)
PHOTOOXYGENATION-CHEMILUMINESCENCE HIGH-PERFORMANCE LIQUID CHROMATOGRAPHIC DETECTOR FOR THE DETERMINATION OF ALIPHATIC ALCOHOLS, ALDEHYDES, ETHERS AND SACCHARIDES
Gandelman, Mitchell S.; Birks, John W.
Dep. Chem. Univ. Colorado Boulder USA US Co 80309
J. Chromatogr.; (82) P 21-31; Vol 242; No 1

07-82-0015 (0015)
DNS DERIVATIZATION OF ANABOLIC AGENTS WITH HIGH-PERFORMANCE LIQUID CHROMATOGRAPHIC SEPARATION AND FLUORESCENCE DETECTION
Williams, A. T. Rhys; Winfield, S. A.; Belloli, R. C.
Perkin-Elmer Ltd. Beaconsfield/Buckinghamshire UK GB HP9 1QA
J. Chromatogr.; (82) P 224-9; Vol 240; No 1

07-82-0016 (0016)
APPLICATION POSSIBILITIES OF POST COLUMN REACTORS IN THE HIGH-PERFORMANCE LIQUID CHROMATOGRAPHIC ANALYSIS OF FOODS, IN PARTICULAR, FLAVONOIDS
Galensa, Rudolf
Inst. Lebensmittelchem. Univ. Hannover Hannover Fed. Rep. Ger. DE 3000/91
Lebensmittelchem. Gerichtl. Chem.; (82) P 49-51; Vol 36; No 3; In Ger

07-82-0017 (0017)
LIQUID CHROMATOGRAPHIC METHOD AND APPARATUS WITH HOLLOW FIBER DEVICE FOR POST-COLUMN DERIVATIZATION
Davis, James Crister
Dow Chemical Co. USA US
PCT Int. Appl.; (031882) P 34 pp.; Pat. No.: 82 00773; Int. Class: B01D-015/08; Doc. Code: A1; Appl.: WO 81/US926 810710; Priority Appl.: US 183527 800902; Des. Countries : AU; JP; Des. Reg. Countries : DE; FR; GB; NL

07-82-0018 (0018)
POST COLUMN DERIVATIZATION METHODOLOGY IN HIGH PERFORMANCE LIQUID CHROMATOGRAPHY (HPLC)
Stewart, James T.
Sch. Pharm. Univ. Georgia Athens USA U) GA 30602
TrAC, Trends Anal. Chem. (Pers. Ed.); (82) P 170-4; Vol 1; No 7

5.7.11
Derivatization—Applications

--- 1982 ---

07-82-0001 (0001)
AUTOMATIC ANALYSIS OF BIOGENIC CATECHOLAMINES BY HIGH-PERFORMANCE LIQUID CHROMATOGRAPHY WITH FLUOROMETRIC REACTION DETECTION
Mori, K.
Natl. Inst. Ind. Health Kawasaki Japan JP 213
Adv. Biosci.; (82) P 275-84; Vol 36; No Synth., Storage Secretion Adrenal Catecholamines

07-82-0002 (0002)
FLUORESCENT/ULTRAVIOLET ABSORBING ESTER DERIVATIVE FORMATION AND ANALYSIS OF EICOSANOIDS BY HIGH-PRESSURE LIQUID CHROMATOGRAPHY
Watkins, W. David; Peterson, Myron B.
Anesth. Serv. Massachusetts Gen. Hosp. Boston USA US MA 02114
Anal. Biochem.; (82) P 30-40; Vol 125; No 1

07-82-0003 (0003)
THE MEASUREMENT OF URINARY OXALIC ACID BY DERIVATIZATION COUPLED WITH LIQUID CHROMATOGRAPHY
Murray, John F., Jr.; Nolen, Harold W., III; Gordon, G. Ross; Peters, John H.
Life Sci. Div. SRI Int. Menlo Park USA US CA 94025
Anal. Biochem.; (82) P 301-9; Vol 121; No 2

07-82-0004 (0004)
LIQUID CHROMATOGRAPHIC DETERMINATION OF GUANADREL IN LABORATORY ANIMAL DIET AS THE FLUORESCENT ACETYLACETONE DERIVATIVE
Bombardt, Paul A.; Adams, Wade J.
Upjohn Co. Kalamazoo USA US MI 49001
Anal. Chem.; (82) P 1087-90; Vol 54; No 7

07-82-0005 (0005)
INTERFERENCE IN A LIQUID CHROMATOGRAPHIC ASSAY FROM IMPURITIES IN THE DERIVATIZATION REAGENT .ALPHA.-BROMO-2'-ACETONAPHTHONE
Stotz, Robert W.; Hassing, Dennis H.
Upjohn Co. Kalamazoo USA US MI 49001
Anal. Chem.; (82) P 345-7; Vol 54; No 2

07-82-0006 (0006)
DETERMINATION OF THE HERBICIDE GLYPHOSATE AND ITS MAJOR METABOLITE AMINOMETHYLPHOSPHONIC ACID BY HIGH-PERFORMANCE LIQUID CHROMATOGRAPHY AFTER FLUORESCENCE LABELING
Roseboom, H.; Berkhoff, C. J.
Unit Residue Anal. Natl. Inst. Public Health Bilthoven Neth. NL 3720 BA
Anal. Chim. Acta; (82) P 373-7; Vol 135; No 2

07-82-0007 (0007)
HPLC IDENTIFICATION AND SEPARATION OF PHENOLIC COMPOUNDS DERIVATIZED WITH DIAZOTIZED SULFANILIC ACID. STRUCTURAL EFFECTS ON RETENTION TIMES
Baiocchi, C.; Gennaro, M. C.; Campi, E.; Mentasti, E.; Aruga, R.
Ist. Chim. Anal. Univ. Torino Turin Italy IT 10125
Anal. Lett.; (82) P 1539-48; Vol 15; No A19

07-82-0008 (0008)
SEPARATION OF o-PHTHALDIALDEHYDE DERIVATIVES OF FREE AMINO ACIDS FROM PLANT TISSUES BY ISOCRATIC REVERSE-PHASE HIGH-PERFORMANCE LIQUID CHROMATOGRAPHY
Martin, F.; Maudinas, B.; Gadal, P.
Unite Microbiol. Sols INRA Seichamps Fr. FR
Ann. Bot. (London); (82) P 401-6; Vol 50; No 3

07-82-0009 (0009)
DETERMINATION OF ANIONIC SURFACTANTS BY HIGH-PERFORMANCE LIQUID CHROMATOGRAPHY II. DETERMINATION OF SODIUM ALKYL SULFATES AS 3,5-DINITROBENZOATE DERIVATIVES
Utsunomiya, Akiko; Ikeda, Tatsuo; Takamatsu, Kazuyuki; Naito, Syoji
Kanagawa Prefect. Public Health Lab. Yokohama Japan JP
Bunseki Kagaku; (82) P 115-20; Vol 31; No 3; In Japan

07-82-0010 (0010)
APPLICATION OF 1-NAPHTHYLAMINE TO LABELING OF FATTY ACID FOR HIGH-PERFORMANCE LIQUID CHROMATOGRAPHY
Ikeda, Mitsuo; Shimada, Kenji; Sakaguchi, Takeichi
Niigata Coll. Pharm. Niigata Japan JP 950-21
Chem. Pharm. Bull.; (82) P 2258-61; Vol 30; No 6

07-82-0011 (0011)
RAPID ASSAY FOR AMINO ACIDS IN SERUM OR URINE BY PRE-COLUMN DERIVATIZATION AND REVERSED-PHASE LIQUID CHROMATOGRAPHY
Turnell, D. C.; Cooper, J. D. H.
Biochem. Dep. Coventry Hosp. Coventry UK GB CV1 4FH
Clin. Chem. (Winston-Salem, N. C.); (82) P 527-31; Vol 28; No 3

07-82-0012 (0012)
DETERMINATION OF VOLATILE ALDEHYDES IN MEAT AS 2,4-DINITROPHENYLHYDRAZONES USING REVERSED-PHASE HIGH-PERFORMANCE LIQUID CHROMATOGRAPHY
Reindl, Bertram; Stan, Hans Juergen
Inst. Sozialmed. Epidemiol. Bundesgesundheitsamt Berlin Fed. Rep. Ger. DE 1000/33
J. Agric. Food Chem.; (82) P 849-54; Vol 30; No 5

07-82-0013 (0013)
DETERMINATION OF 17-HYDROXYCORTICOSTEROIDS IN URINE BY FLUORESCENCE HIGH-PERFORMANCE LIQUID CHROMATOGRAPHY USING DNS-HYDRAZINE AS A PRECOLUMN LABELING REAGENT
Kawasaki, Takao; Maeda, Masako; Tsuji, Akio
Sch. Pharm. Sci. Showa Univ. Tokyo Japan JP
J. Chromatogr.; (82) P 1-11; Vol 232; No 1

07-82-0014 (0014)
FEMTOMOLAR ION-PAIR HIGH-PERFORMANCE LIQUID CHROMATOGRAPHIC METHOD FOR DETERMINING DNS-POLYAMINE DERIVATIVES OF RED BLOOD CELL EXTRACTS UTILIZING AN AUTOMATED POLYAMINE ANALYZER
Brown, Nesbitt D.; Strickler, M. Patricia; Whaun, June M.
Div. Biochem. Walter Reed Army Inst. Res. Washington USA US DC 20012
J. Chromatogr.; (82) P 101-8; Vol 245; No 1

07-82-0015 (0015)
DETERMINATION OF POLYAMINES BY PRECOLUMN DERIVATIZATION WITH o-PHTHALALDEHYDE AND ETHANETRIOL IN COMBINATION WITH REVERSED-PHASE HIGH-PERFORMANCE LIQUID CHROMATOGRAPHY
Skaaden, Tore; Greibrokk, Tyge
Dep. Chem. Univ. Oslo Blindern Norway NO
J. Chromatogr.; (82) P 111-22; Vol 247; No 1

07-82-0016 (0016)
PRE-COLUMN DERIVATIZATION AND HIGH-PERFORMANCE LIQUID CHROMATOGRAPHY OF BIOGENIC AMINES IN BLOOD OF NORMAL AND MALIGNANT HYPERTHERMIC PIGS
Davis, Thomas P.; Gehrke, Charles W., Jr.; Williams, Charles H.; Gehrke, Charles W.; GerhardtL Klaus O.
Health Sci. Cent. Univ. Arizona Tucson USA US AZ 85724
J. Chromatogr.; (82) P 113-22; Vol 228

07-82-0017 (0017)
DETERMINATION OF (R)- AND (S)-ALPRENOLOL AND (R)- AND (S)-METOPROLOL AS THEIR DIASTEREOMERIC DERIVATIVES IN HUMAN PLASMA BY REVERSED-PHASE LIQUID CHROMATOGRAPHY
Hermansson, Joergen; Von Bahr, Christer
Dep. Drugs Natl. Board Health and Welfare Uppsala Swed. SE S-751 25
J. Chromatogr.; (82) P 113-27; Vol 227; No 1

07-82-0018 (0018)
HIGH-PERFORMANCE LIQUID CHROMATOGRAPHY OF CARBOXYLIC ACIDS USING 4-BROMOMETHYL-7-ACETOXYCOUMARIN AS FLUORESCENCE REAGENT
Tsuchiya, Hironori; Hayashi, Tokishi; Naruse, Hiroshi; Takagi, Nobuhiko
Dep. Dent. Pharmacol. Gifu Coll. Dent. Gifu Japan JP
J. Chromatogr.; (82) P 121-30; Vol 234; No 1

07-82-0019 (0019)
DETERMINATION OF .GAMMA.-AMINOBUTYRIC ACID BY REVERSED-PHASE HIGH-PERFORMANCE LIQUID CHROMATOGRAPHY AND PRECOLUMN LABELING FOR FLUORESCENCE DETECTION
Griesmann, Guy E.; Chan, Wai Yee; Rennert, Owen M.
Oklahoma Child. Mem. Hosp. Univ. Oklahoma Oklahoma City USA US OK 73190
J. Chromatogr.; (82) P 121-4; Vol 230; No 1

07-82-0020 (0020)
HIGH-PERFORMANCE LIQUID CHROMATOGRAPHY OF HUMAN SERUM LIPOPROTEINS. SELECTIVE DETECTION OF CHOLINE-CONTAINING PHOSPHOLIPIDS BY ENZYMIC REACTION
Okazaki, Mtsuyo; Hagiwara, Noriko; Hara, Ichiro
Dep. Gen. Educ. Tokyo Med. Dent. Univ. Chiba Japan JP 272
J. Chromatogr.; (82) P 13-23; Vol 231; No 1

07-82-0021 (0021)
DETERMINATION OF AMIKACIN IN MICROLITER QUANTITIES OF BIOLOGICAL FLUIDS BY HIGH-PERFORMANCE LIQUID CHROMATOGRAPHY USING 1-FLUORO-2,4-DINITROBENZENE DERIVATIZATION
Wong, Lawrence T.; Beaubien, Arthur R.; Pakuts, Atilla P.
Drug Res. Lab. Frederick G. Banting Build. Ottawa Can. CA ON K1A 0L2
J. Chromatogr.; (82) P 145-54; Vol 231; No 1

07-82-0022 (0022)
9-(CHLOROMETHYL)ANTHRACENE: A USEFUL DERIVATIZING REAGENT FOR ENHANCED ULTRAVIOLET AND FLUORESCENCE DETECTION OF CARBOXYLIC ACIDS WITH LIQUID CHROMATOGRAPHY
Korte, W. D.
Dep. Chem. California State Univ. Chico USA US CA 95929
J. Chromatogr.; (82) P 153-7; Vol 243; No 1

07-82-0023 (0023)
FLUORESCENCE DETECTION OF CYSTINE BY o-PHTHALALDEHYDE DERIVATIZATION AND ITS SEPARATION USING HIGH-PERFORMANCE LIQUID CHROMATOGRAPHY
Cooper, J. D. H.; Turnell, D. C.
Biochem. Dep. Coventry and Warwickshire Hosp. Coventry UK GB
J. Chromatogr.; (82) P 158-61; Vol 227; No 1

07-82-0024 (0024)
DETERMINATION OF HOST-SELECTIVE PHYTOTOXINS FROM ALTERNARIA ALTERNATA F. SP. LYCOPERSICI AS THEIR MALEYL DERIVATIVES BY HIGH-PERFORMANCE LIQUID CHROMATOGRAPHY
Siler, D. J.; Gilchrist, D. G.
Dep. Plant Pathol. Univ. California Davis USA US CA 95616
J. Chromatogr.; (82) P 167-73; Vol 238; No 1

07-82-0025 (0025)
HIGH-PERFORMANCE LIQUID CHROMATOGRAPHIC DETERMINATION OF HYDROXYPROLINE AFTER DERIVATIZATION WITH 4-DIMETHYLAMINOAZOBENZENE- 4'-SULFONYL CHLORIDE
Casini, Alessandro; Martini, Flavio; Nieri, Simonetta; Ramarli, Dunia; Franconi, Flavia; Surrenti, Calogero
Ist. Clin. Med. IV Florence Italy IT 50134
J. Chromatogr.; (82) P 187-92; Vol 249; No 1

07-82-0026 (0026)
HIGH-PERFORMANCE LIQUID CHROMATOGRAPHIC METHOD FOR THE DETERMINATION OF PLASMA AND URINE METAPRAMINE AFTER DANSYLATION
Sommadossi, J. P.; Lemar, M.; Necciari, J.; Sumirtapura, Y.; Cano, J. P.; Gaillot, J.
Lab. Pharmacocinet. INSERM Marseille Fr. FR 13385
J. Chromatogr.; (82) P 205-13; Vol 228

07-82-0027 (0027)
DNS DERIVATIZATION OF ANABOLIC AGENTS WITH HIGH-PERFORMANCE LIQUID CHROMATOGRAPHIC SEPARATION AND FLUORESCENCE DETECTION
Williams, A. T. Rhys; Winfield, S. A.; Belloli, R. C.
Perkin-Elmer Ltd. Beaconsfield/Buckinghamshire UK GB HP9 1QA
J. Chromatogr.; (82) P 224-9; Vol 240; No 1

07-82-0028 (0028)
SENSITIVE HIGH-PERFORMANCE LIQUID CHROMATOGRAPHIC METHOD FOR PROSTAGLANDINS USING A FLUORESCENCE REAGENT, 4-BROMOMETHYL-7-ACETOXYCOUMARIN
Tsuchiya, Hironori; Hayashi, Tokishi; Naruse, Hiroshi; Takagi, Nobuhiko
Dep. Dent. Pharmacol. Gifu Coll. Dent. Gifu Japan JP
J. Chromatogr.; (82) P 247-54; Vol 231; No 2

07-82-0029 (0029)
SIMULTANEOUS DETERMINATION OF ASCORBIC ACID AND DEHYDROASCORBIC ACID BY REVERSED-PHASE ION-PAIR HIGH-PERFORMANCE LIQUID CHROMATOGRAPHY WITH PRECOLUMN DERIVATIZATION
Keating, Ross W.; Haddad, Paul R.
Dep. Anal. Chem. Univ. New South Wales Kensington Australia AU 2033
J. Chromatogr.; (82) P 249-55; Vol 245; No 2

07-82-0030 (0030)
THE DETERMINATION OF UROPORPHYRINOGEN DECARBOXYLASE IN TISSUES BY HIGH-PERFORMANCE LIQUID CHROMATOGRAPHY COUPLED TO FLUORESCENCE DETECTION
Cantoni, L.; Ruggieri, R.; Dal Fiume, D.; Rizzardini, M.
Ist. Ric. Farmacol. "Mario Negri" Milan Italy IT 20157
J. Chromatogr.; (82) P 311-18; Vol 229; No 2

07-82-0031 (0031)
PRECOLUMN DERIVATIZATION OF FLUORESCAMINE AND HIGH-PERFORMANCE LIQUID CHROMATOGRAPHIC ANALYSIS OF DRUGS. APPLICATION TO TOCAINIDE
Sedman, A. J.; Gal, J.
Sch. Med. Univ. Colorado Denver USA US CO 80262
J. Chromatogr.; (82) P 315-26; Vol 232; No 2

07-82-0032 (0032)
SENSITIVE AND RAPID AMINO ACID ANALYSIS OF PEPTIDE HYDROLYZATES BY HIGH-PERFORMANCE LIQUID CHROMATOGRAPHY OF o-PHTHALDIALDEHYDE DERIVATIVES
Burbach, J. Peter H.; Prins, Arie; Lebouille, Jos L. M.; Verhoef, J.; Witter, Albert
Med. Fac. Univ. Utrecht Utrecht Neth. NL 3521 GD
J. Chromatogr.; (82) P 339-43; Vol 237; No 2

07-82-0033 (0033)
AMINO ACID ANALYSES BY HIGH-PERFORMANCE LIQUID CHROMATOGRAPHY. AN EVALUATION OF THE USEFULNESS OF PRECOLUMN DNS DERIVATIZATION
De Jong, Catharina; Hughes, Graham J.; Van Wieringen, Erika; Wilson, Kenneth J.
Lab. Biochem. ETH Zurich Switz. CH CH-8092
J. Chromatogr.; (82) P 345-59; Vol 241; No 2

07-82-0034 (0034)
FLUORESCENCE HIGH-PERFORMANCE LIQUID CHROMATOGRAPHY OF REDUCING SUGARS USING DNS-HYDRAZINE AS A PRELABELING REAGENT
Takeda, Motoko; Maeda, Masako; Tsuji, Akio
Sch. Pharm. Sci. Showa Univ. Tokyo Japan JP
J. Chromatogr.; (82) P 347-55; Vol 244; No 2

07-82-0035 (0035)
CHROMOPHORIC DETERMINATION OF PUTRESCINE, SPERMIDINE AND SPERMINE WITH DABSYL CHLORIDE BY HIGH-PERFORMANCE LIQUID CHROMATOGRAPHY AND THIN-LAYER CHROMATOGRAPHY
Lin, Jen Kun; Lai, Chen Cheng
Coll. Med. Natl. Taiwan Univ. Taipei Taiwan TW
J. Chromatogr.; (82) P 369-77; Vol 227; No 2

07-82-0036 (0036)
HIGH-PERFORMANCE LIQUID CHROMATOGRAPHY DETECTION OF MORPHINE BY FLUORESCENCE AFTER POSTCOLUMN DERIVATIZATION
Nelson, Peter E.; Nolan, Susan L.; Bedford, Keith R.
Dep. Sci. Ind. Res. Univ. Auckland Auckland N. Z. NZ
J. Chromatogr.; (82) P 407-14; Vol 234; No 2

07-82-0037 (0037)
REVERSED-PHASE ION-PAIR HIGH-PERFORMANCE LIQUID CHROMATOGRAPHY OF MERCAPTO-ACETATE AND N-ACETYLCYSTEINE AFTER DERIVATIZATION WITH N-(1-PYRENE)MALEIMIDE AND N-(7-DIMETHYLAMINO-4-METHYL-3-COUMARINYL)- MALEIMIDE
Kaagedal, Bertil; Kaellberg, Magnus
Dep. Clin. Chem. Univ. Linkoeping Linkoeping Swed. SE S-581 85
J. Chromatogr.; (82) P 409-15; Vol 229; No 2

07-82-0038 (0038)
RAPID, SPECIFIC METHOD FOR DIETHYLSTILBESTROL ANALYSIS USING AN IN-LINE PHOTOCHEMICAL REACTOR WITH HIGH-PERFORMANCE LIQUID CHROMATOGRAPHY AND FLUORESCENCE DETECTION
Williams, A. T. Rhys; Winfield, S. A.; Belloli, R. C.
Perkin-Elmer Ltd. Buckinghamshire UK GB HP9 1QA
J. Chromatogr.; (82) P 461-70; Vol 235; No 2

07-82-0039 (0039)
SEPARATION OF SATURATED, MONOUNSATURATED AND DIUNSATURATED ALDEHYDES AS 2,4-DINITROPHENYLHYDRAZONES USING HIGH-PERFORMANCE LIQUID CHROMATOGRAPHY AT INCREASED TEMPERATURE
Reindl, B.; Stan, H. J.
Inst. Sozialmed. Epidemiol. Bundesgesundheitsamt Berlin Fed. Rep. Ger. E 1000/33
J. Chromatogr.; (82) P 481-8; Vol 235; No 2

07-82-0040 (0040)
HIGH-PERFORMANCE LIQUID CHROMATOGRAPHY OF HUMAN SERUM LIPOPROTEINS. SELECTIVE DETECTION OF TRIGLYCERIDES BY ENZYMIC REACTION
Hara, Ichiro; Shiraishi, Keiko; Okazaki, Mitsuyo
Dep. Gen. Educ. Tokyo Med. Dent. Univ. Ichikawa Japan JP 272

J. Chromatogr.; (82) P 549-57; Vol 239

07-82-0041 (0041)
PRECOLUMN LABELING FOR HIGH-PERFORMANCE LIQUID CHROMATOGRAPHY OF AMINO ACIDS WITH 7-FLUORO-4-NITROBENZO-2-OXA-1,3-DIAZOLE AND ITS APPLICATION TO PROTEIN HYDROLYSATES
Watanabe, Yoshihiko; Imai, Kazuhiro
Fac. Pharm. Sci. Univ. Tokyo Tokyo Japan JP 113
J. Chromatogr.; (82) P 723-32; Vol 239

07-82-0042 (0042)
HPLC ANALYSES OF AMINO ACIDS IN PLANT MATERIALS AS DNP-DERIVATIVES BY USE OF METHYL BENZOATE AS THE INTERNAL STANDARD
Kozukue, Nobuyuki; Kozukue, Etsuko; Terai, Hirofumi; Mizuno, Susumu
Dep. Home Econ. Kenmei Jr. Coll. Himeji City Japan JP
J. Food Sci.; (82) P 1584-6; Vol 47; No 5

07-82-0043 (0043)
PRECOLUMN DERIVATIZATION IN HPLC OF AMINO ACIDS AND PEPTIDES
Szokan, Gyula
Inst. Org. Chem. Eotvos Univ. Budapest Hung. HU H-1088
J. Liq. Chromatogr.; (82) P 1493-8; Vol 5; No 8

07-82-0044 (0044)
QUANTITATIVE HPLC ANALYSIS OF PLASMA AMINO ACIDS AS ORTHOPHTHALDIALDEHYDE/ ETHANETHIOL DERIVATIVES
Hill, Dennis; Burnworth, Larry; Skea, William; Pfeifer, Robert
Microchem. Lab. Univ. Connecticut Storrs USA US CT
J. Liq. Chromatogr.; (82) P 2369-93; Vol 5; No 12

07-82-0045 (0045)
LIQUID CHROMATOGRAPHIC DETERMINATION OF AZIDE AS THE 3,5-DINITROBENZOYL DERIVATIVE
Swarin, Stephen J.; Waldo, Richard A.
Anal. Chem. Dep. Gen. Motors Res. Lab. Warren USA US MI 48090
J. Liq. Chromatogr.; (82) P 597-604; Vol 5; No 4

07-82-0046 (0046)
DETERMINATION OF CARBONYL 2,4-DINITROPHENYLHYDRAZONES BY LIQUID CHROMATOGRAPHY/ELECTROCHEMISTRY
Jacobs, W. A.; Kissinger, P. T.
Dep. Chem. Purdue Univ. West Lafayette USA US IN 47907
J. Liq. Chromatogr.; (82) P 669-76; Vol 5; No 4

07-82-0047 (0047)
NITROAROMATIC REAGENTS FOR DETERMINATION OF AMINES AND AMINO ACIDS BY LIQUID CHROMATOGRAPHY/ELECTROCHEMISTRY
Jacobs, W. A.; Kissinger, P. T.
Dep. Chem. Purdue Univ. West Lafayette USA US IN 47907
J. Liq. Chromatogr.; (82) P 881-95; Vol 5; No 5

07-82-0048 (0048)
THE DETERMINATION OF .ALPHA.-KETOACIDS AND OXALIC ACID IN SEAWATER BY REVERSED PHASE LIQUID CHROMATOGRAPHIC SEPARATION OF FLUORESCENT QUINOXILINOL DERIVATIVES
Steinberg, Spencer M.; Bada, Jeffrey L.
Scripps Inst. Oceanogr. Univ. California La Jolla USA US CA 92093
Mar. Chem.; (82) P 299-306; Vol 11; No 4

07-82-0049 (0049)
PRECISION AND ACCURACY OF INDOLE-3-ACETIC ACID ANALYSES PERFORMED WITH THE 2-METHYLINDOLO-.ALPHA.-PYRONE FLUORESCENCE ASSAY AND WITH A HIGH-PERFORMANCE LIQUID CHROMATOGRAPHY TECHNIQUE WITH SPECTROFLUOROMETRIC DETECTION, EXEMPLIFIED ON PINE TISSUE (PINUS SYLVESTRIS L)
Sandberg, Goeran; Dunberg, Arne
Dep. For. Genet. Plant Physiol. Swed. Univ. Agric. Sci. Umea Swed. SE S-901 83
Physiol. Plant.; (82) P 315-22; Vol 55; No 3

07-82-0050 (0050)
MEASUREMENT OF VALPROIC ACID IN SERUM AS PHENACYL ESTERS BY HIGH-PERFORMANCE LIQUID CHROMATOGRAPHY
Sakakihara, Yoichi; Yokochi, Akiko; Enomoto, Shoko; Kitazumi, Eiji ; Kodama, Kazuo
Fac. Med. Univ. Tokyo Tokyo Japan JP
Rinsho Kensa; (82) P 719-23; Vol 26; No 6; In Japan

07-82-0051 (0051)
DETERMINATION OF 15-METHYLPROSTAGLANDIN F2 ALPHA BY DERIVATIZATION HIGH-PRESSURE LIQUID CHROMATOGRAOHY
Wang, Zhongshan; Zhu, Yaohua; Zhang, Shuliang; Zhu, Jiyu; Wang, Meili
Shanghai Int. Drug Control, Shanghai,Peop. Rep. China
Yaoxue Xuebao (82) P 603-8; Vol 17; No 8; In Chin

SECTION 6

CHAPTER 6.1
Sample Preparation

--- 1982 ---

01-82-0001 (0001)
MICROSEQUENCE ANALYSIS OF PEPTIDES AND PROTEINS. I. PREPARATION OF SAMPLES BY REVERSE-PHASE LIQUID CHROMATOGRAPHY
Yuan, Pau Miau; Pande, Hema; Clark, Brian R.; Shively, John E.
Immunol. Div. City of Hope Res. Inst. Duarte USA US CA P91010
Anal. Biochem.; (82) P 289-301; Vol 120; No 2

01-82-0002 (0002)
MICROSAMPLE-FILTERING DEVICE FOR LIQUID CHROMATOGRAPHY OR FLOW INJECTION ANALYSIS
Gardner, Wayne S.; Vanderploeg, Henry A.
Great Lakes Environ. Res. Lab. Natl. Oceanic Atmos. Adm. Ann Arbor USA US MI 48104
Anal. Chem.; (82) P 2129-30; Vol 54; No 12

01-82-0003 (0003)
PROBLEMS OF SAMPLE PREPARATION FOR HIGH-PERFORMANCE LIQUID CHROMATOGRAPHY
Gill, Richard; Moffat, Anthony C.
Home Off. Cent. Res. Establ. Reading UK GB RG7 4PN
Anal. Proc. (London); (82) P 170-4; Vol 19; No 4

01-82-0004 (0004)
SIMULTANEOUS DETERMINATION OF HEAVY METALS IN WATER BY HIGH PERFORMANCE LIQUID CHROMATOGRAPHY FOLLOWING THE SOLVENT EXTRACTION OF HEAVY METALS AS PYRROLIDINEDITHIOCARBAMATE CHELATES INTO CHLOROFORM
Ichinoki, Susumu; Yamazaki, Mitsuru
Sch. Pharm. Hokuriku Univ. Kanazawa Japan JP 920-11
Bunseki Kagaku; (82) P E319-E326; Vol 31; No 10

01-82-0005 (0005)
SIMPLE PREPARATION AND LIQUID-CHROMATOGRAPHIC ANALYSIS FOR TRICYCLIC ANTIDEPRESSANTS IN SERUM
Koteel, Prabhakaran; Mullins, Richard E.; Gqdsden, Richard H.
Dep. Lab. Med. Med. Univ. South Carolina Charleston USA US)C 29425
Clin. Chem. (Winston-Salem, N. C.); (82) P 462-6; Vol 28; No 3

01-82-0006 (0006)
PRECHROMATOGRAPHIC CHEMICAL DERIVATIZATION IN LIQUID CHROMATOGRAPHY
Lawrence, James F.
Food Res. Div. Food Dir. Ottawa Can. CA ON K1A 0L2
CONFERENCE; Chem. Deriv. Anal. Chem.; (82) P 191-242; Vol 2,; Avail.: Frei, Roland W.; Lawrence, James F; Publisher: Plenum, New York, N.Y.

01-82-0007 (0007)
METHOD FOR DETERMINATION OF FORMALDEHYDE IN AIR IN THE PPT-RANGE BY HPLC AFTER EXTRACTION AS 2,4-DINITROPHENYLHYDRAZONE
Tuss, H.; Neitzert, V.; Seiler, W.; Neeb, R.
Max-Planck-Inst. Chem. Mainz Fed. Rep. Ger. DE D-6500
Fresenius' Z. Anal. Chem.; (82) P 613-17; Vol 312; No 7

01-82-0008 (0008)
EVALUATION OF THIN-LAYER, PAPER, AND HIGH-PERFORMANCE LIQUID CHROMATOGRAPHY FOR IDENTIFICATION OF DYES EXTRACTED AS ION-PAIRS WITH TRI-n-OCTYLAMINE
Puttemans, Marc L.; Dryon, Louis; Massart, Desire L.
Farm. Inst. Vrije Univ. Brussel Brussels Belg. BE B-1090
J. Assoc. Off. Anal. Chem.; (82) P 730-6; Vol 65; No 3

01-82-0009 (0009)
AROMATIC HYDROCARBON METABOLITES IN FISH: AUTOMATED EXTRACTION AND HIGH-PERFORMANCE LIQUID-CHROMATOGRAPHIC SEPARATION INTO CONJUGATE AND NONCONJUGATE FRACTIONS
Krahn, Margaret M.; Collier, Tracy K.; Malins, Donald C.
Environ. Conservation Div. Northwest and Alaska Fis. Cent. Seattle USA US WA 98112
J. Chromatogr. (82) P 441-52; Vol 236; No 2

01-82-0010 (0010)
REVERSED-PHASE LIQUID CHROMATOGRAPHY WITH AMPEROMETRIC DETECTION OF LIPOPHILIC DOPAMINE ANALOGS AND DETERMINATION OF BRAIN AND SERUM CONCENTRATIONS AFTER SAMPLE CLEAN-UP ON SMALL SEPHADEX G-10 COLUMNS
Feenstra, Matthijs G. P.; Homan, Jan Willem; Dijkstra, Durk; Mulder, Theo B. A.; Rollema, Hans; Westerink, Ben H. C.; Horn, Alan S.
Dep. Med. Chem. State Univ. Groningen Groningen Neth. NL 9713 AW
J. Chromatogr.; (82) P 271-87; Vol 20; No 2

01-82-0011 (0011)
AN IMPROVED EXTRACTION-HPLC METHOD FOR TOBACCO POLYPHENOLS
Snook, M. E.; Chortyk, O. T.
Tobacco Saf. Res. Unit Agric. Res. Serv. Athens USA US GA 30613
Tob. Int.; (82) P 111-15; Vol 184; No 5

CHAPTER 6.2
Precision—Accuracy—Reproducibility

--- 1982 ---

02-82-0001 (0001)
A VERY PRECISE HIGH-PERFORMANCE LIQUID CHROMATOGRAPHIC PROCEDURE FOR THE DETERMINATION OF CEFMENOXIME, A NEW CEPHALOSPORIN ANTIBIOTIC, IN PLASMA
Granneman, G. Richard; Sennello, Lawrence T.
Drug Metab. Dep. Abbott Lab. North Chicago USA US IL 60064
J. Chromatogr.; (82) P 149-57; Vol 229; No 1

02-82-0002 (0002)
PRECISION AND ACCURACY OF INDOLE-3-ACETIC ACID ANALYSES PERFORMED WITH THE 2-METHYLINDOLO-.ALPHA.-PYRONE FLUORESCENCE ASSAY AND WITH A HIGH-PERFORMANCE LIQUID CHROMATOGRAPHY TECHNIQUE WITH SPECTROFLUOROMETRIC DETECTION, EXEMPLIFIED ON PINE TISSUE (PINUS SYLVESTRIS L)
Sandberg, Goeran; Dunberg, Arne
Dep. For. Genet. Plant Physiol. Swed. Univ. Agric. Sci. Umea Swed. SE S-901 83
Physiol. Plant.; (82) P 315-22; Vol 55; No 3

CHAPTER 6.3

Quantitative Analysis

--- 1982 ---

03-82-0001 (0001)
SEPARATION AND QUANTITATION OF (R)- AND (S)-PROPRANOLOL AS THEIR DIASTEREOMERIC DERIVATIVES IN HUMAN PLASMA BY REVERSED-PHASE ION-PAIR CHROMATOGRAPHY
Hermansson, Joergen
Dep. Drugs Natl. Board Health Welfare Uppsala Swed. SE S-751 25
Acta Pharm. Suec.; (82) P 11-24; Vol 19; No 1

03-82-0002 (0002)
QUANTITATION OF URINARY NIACIN METABOLITES BY REVERSED-PHASE LIQUID CHROMATOGRAPHY
Carter, Eric G. A.
Dep. Nutr. Sci. Univ. California Berkeley USA US CA 94720
Am. J. Clin. Nutr.; (82) P 926-30; Vol 36; No 5

03-82-0003 (0003)
QUANTITATIVE DETERMINATION OF ALLANTOIN IN BIOLOGICAL FLUIDS BY REVERSED-PHASE HIGH-PRESSURE LIQUID CHROMATOGRAPHY
Tiemeyer, W.; Giesecke, D.
Tieraerztl. Fak. Univ. Muenchen Munich Fed. Rep. Ger. DE D-8000/22
Anal. Biochem.; (82) P 11-13; Vol 123; No 1

03-82-0004 (0004)
THE USE OF HIGH-PERFORMANCE LIQUID CHROMATOGRAPHY FOR QUANTITATION OF PLASMA AMINO ACIDS IN MAN
Hogan, Daniel L.; Kraemer, Kevin L.; Isenberg, Jon I.
Sch. Med. Univ. California, San Diego La Jolla USA US CA
Anal. Biochem.; (82) P 17-24; Vol 127; No 1

03-82-0005 (0005)
SEPARATION AND QUANTITATION OF PURINES AND THEIR ANAEROBIC AND AEROBIC DEGRADATION PRODUCTS BY HIGH-PRESSURE LIQUID CHROMATOGRAPHY
Duerre, Peter; Andreesen, Jan R.
Inst. Mikrobiol. Univ. Goettingen Goettingen Fed. Rep. Ger. DE D-3400
Anal. Biochem.; (82) P 32-40; Vol 123; No 1

03-82-0006 (0006)
QUANTITATION OF MULTIPLE STEROIDS IN HUMAN SERUM AND URINE BY COMBINATION OF AUTOMATIC HPLC AND RIA
Kage, A.; Fenner, A.; Weber, B.; Schoeneshoefer, M.
Inst. Clin. Chem. Biochem. FU Berlin Berlin Fed. Rep. Ger. DE D-1000/45
Anal. Chem. Symp. Ser.; (82) P 327-32; Vol 10; No Adv. Steroid Anal.

03-82-0007 (0007)
STABILITY INDICATING ASSAY FOR DIPYRONE. PART II. SEPARATION AND QUANTITATIVE DETERMINATION OF DIPYRONE AND ITS DEGRADATION PRODUCTS BY HIGH-PERFORMANCE LIQUID CHROMATOGRAPHY
Eddine, N. Hussam; Bressollè, F.; Mandrou, B.; Fabre, H.
Lab. Chim. Anal. Toxicol. Fac. Pharm. Montpellier Fr. FR 34060
Analyst (London); (82) P 67-70; Vol 107; No 1270

03-82-0008 (0008)
A RAPID HIGH-PERFORMANCE LIQUID CHROMATOGRAPHIC METHOD FOR QUANTITATION OF 5-FLUOROURACIL IN PLASMA AFTER CONTINUOUS INTRAVENOUS INFUSION
Sampson, D. C.; Fox, R. M.; Tattersall, M. H. N.; Hensley, W. J.
Dep. Biochem. R. Prince Alfred Hosp. Camperdown Australia AU 2050
Ann. Clin. Biochem.; (82) P 125-8; Vol 19; No 2

03-82-0009 (0009)
ACETYLPROCAINE IN HEROIN PREPARATIONS. DETECTION AND QUANTIFICATION BY HIGH-PERFORMANCE LIQUID CHROMATOGRAPHY
Bernhauer, Dietmar; Fuchs, Ernst F.
Hess. Landeskrim. Wiesbaden Fed. Rep. Ger. DE 6200
Arch. Kriminol.; (82) P 25-30; Vol 169; No 1-2; In Ger

03-82-0010 (0010)
QUANTITATIVE DETERMINATION OF SUPROFEN IN HUMAN PLASMA AND URINE BY FULLY AUTOMATED HIGH-PERFORMANCE LIQUID CHROMATOGRAPHY
Mueller, H.; Zulliger, Hans W.; Doerig, B. A.; Egli, R. A.
Pharm. Res. Dev. Dep. Cilag A.-G. Ltd. Schaffhausen Switz. CH CH-8201
Arzneim.-Forsch.; (82) P 257-60; Vol 32; No 3

03-82-0011 (0011)
A FLUOROMETRIC-HPLC ASSAY FOR QUANTITATING THE BINDING OF BENZO(A)PYRENE METABOLITES TO DNA
Rahn, R. O.; Chang, S. S.; Holland, J. M.; Shugart, L. R.
Biol. Div. Oak Ridge Natl. Lab. Oak Ridge USA US TN 37830
Biochem. Biophys. Res. Commun.; (82) P 262-8; Vol 109; No 1

03-82-0012 (0012)
QUANTITATIVE DETERMINATION OF AMINO ACIDS IN TISSUE CULTURE CELLS BY HIGH PERFORMANCE LIQUID CHROMATOGRAPHY
Kabus, P.; Koch, G.
Abt. Molekularbiol. Univ. Hamburg Hamburg Fed. Rep. Ger. DE D-2000/13
Biochem. Biophys. Res. Commun.; (82) P 783-90; Vol 108; No 2

03-82-0013 (0013)
IDENTIFICATION AND QUANTITATION OF METHOTREXATE AND METHOTREXATE METABOLITES IN CLINICAL HIGH-DOSE THERAPY BY HIGH-PRESSURE LIQUID CHROMATOGRAPHY AND FIELD DESORPTION MASS SPECTROMETRY
Przybylski, Michael; Preiss, Joachim; Dennebaum, Reiner; Fischer, Joseph
Inst. Org. Chem. Univ. Mainz Mainz Fed. Rep. Ger. DE
Biomed. Mass Spectrom.; (82) P 22-32; Vol 9; No 1

03-82-0014 (0014)
A RAPID AND QUANTITATIVE HIGH PERFORMANCE LIQUID CHROMATOGRAPHIC METHOD FOR ASSAYING BILIRUBIN AND ITS CONJUGATES IN BILE
Gordon, Ellen R.; Goresky, Carl A.
Med. Clin. McGill Univ. Montreal Can. CA PQ H3G 1A4
Can. J. Biochem.; (82) P 1050-7; Vol 60; No 11

03-82-0015 (0015)
TWO-CYCLE LIQUID-CHROMATOGRAPHIC QUANTITATION OF CORTISOL IN URINE
Nakamura, Jiro; Yakata, Minoru
Sch. Med. Niigata Univ. Niigata Japan JP 951
Clin. Chem. (Winston-Salem, N. C.); (82) P 1497-500; Vol 28; No 7

03-82-0016 (0016)
QUANTITATIVE LIQUID-CHROMATOGRAPHIC TECHNIQUE FOR THE SIMULTANEOUS ASSAY OF TRICYCLIC ANTIDEPRESSANT DRUGS IN PLASMA OR SERUM
Sonsalla, Patricia K.; Jennison, Thomas A.; Finkle, Bryan S.
Cent. Hum. Toxicol. Univ. Utah Salt Lake City USA 84112
Clin. Chem. (Winston-Salem, N. C.); (82) P 457-61; Vol 28; No 3

03-82-0017 (0017)
QUANTITATIVE DETERMINATION OF VITAMINS D2 AND D3 IN MARGARINE BY HPLC
Mueller-Mulot, W.; Rohrer, G.
Kontroll-Abt. Hoffmann-La Roche A.-G. Grenzach-Wyhlen Fed. Rep. Ger. DE 7889/1

Fette, Seifen, Anstrichm.; (82) P 354-8; Vol 84; No 9; In Ger

03-82-0018 (0018)
QUANTITATIVE DETERMINATION OF AMINO ACIDS IN FOOD BY HIGH-PRESSURE LIQUID CHROMATOGRAPHY
Khayat, Ali; Redenz, Peter K.; Gorman, Lee A.
Sea Food Div. Res. Cent. Ralston Purina Co. San Diego USA US CA 92121
Food Technol. (Chicago); (82) P 46-50; Vol 36; No 6

03-82-0019 (0019)
QUANTITATIVE DETERMINATION OF VANADIUM, MOLYBDENUM, TUNGSTEN, COBALT AND CHROMIUM 8-HYDROXYQUINOLATES USING HPLC
Wenclawiak, B.
Anorg.-Chem. Inst. Univ. Muenster Muenster Fed. Rep. Ger. DE D-4400
Fresenius' Z. Anal. Chem.; (82) P 144-5; Vol 310; No 1-2; In Ger

03-82-0020 (0020)
A QUANTITATIVE ANALYSIS OF SELECTION INACTIVATION OF PEROXISOMAL ENZYMES IN THE YEAST HANSENULA POLYMORPHA BY HIGH-PERFORMANCE LIQUID CHROMATOGRAPHY
Bruinenberg, P. G.; Veenhuis, M.; Van Dijken, J. P.; Duine, J. A.; Harder, W.
Dep. Microbiol. Univ. Groningen Haren Neth. NL 9751 NN
FEMS Microbiol. Lett.; (82) P 45-50; Vol 15; No 1

03-82-0021 (0021)
APPLICATION OF HPLC TO INORGANIC ANALYSIS. PART VII. COMPARISON BETWEEN PRE-COLUMN- AND ON-COLUMN DERIVATIZATION FOR SEPARATION OF DIFFERENT METAL OXINATES; QUANTITATIVE DETERMINATION OF MANGANESE(II) BESIDES MANGANESE(III) IONS
Hoffmann, B. W.; Schwedt, G.
Anorg.-Chem. Inst. Univ. Goettingen Goettingen Fed. Rep. Ger. DE D-3400
HRC CC, J. High Resolut. Chromatogr. Chromatogr. Commun.; (82) P 439-40; Vol 5; No 8

03-82-0022 (0022)
QUANTITATIVE DETERMINATION OF GALACTURONIC ACID IN COMMERCIAL PECTINS BY HPLC
Giangiacomo, R.; Polesello, A.; Marin, F.
Ist. Sper. Valorizzazione Tecnol. Prod. Agric. Milan Italy IT
Ind. Aliment. (Pinerolo, Italy); (82) P 386-8; Vol 21; No 5; In Ital

03-82-0023 (0023)
HIGH-PERFORMANCE LIQUID CHROMATOGRAPHY FOR THE QUANTITATIVE DETERMINATION OF THE URINARY METABOLITES OF TOLUENE, XYLENE, AND STYRENE
Poggi, G.; Giusiani, M.; Palagi, U.; Paggiaro, P. L.; Loi, A. M.; Dazzi, F.; Siclari, C.; Baschieri, L.
Ist. Med. Legale Assicurazioni Pisa Italy IT
Int. Arch. Occup. Environ. Health; (82) P 25-31; Vol 50; No 1

03-82-0024 (0024)
IDENTIFICATION OF FLAVONOID CHEMICAL MARKERS IN ROSES AND THEIR HIGH-PRESSURE LIQUID CHROMATOGRAPHIC RESOLUTION AND QUANTITATION FOR CULTIVAR IDENTIFICATION
Asen, S.
Florist Nursery Crops Lab. Agric. Res. Serv. Beltsville USA US MD 20705
J. Am. Soc. Hortic. Sci.; (82) P 744-50; Vol 107; No 5

03-82-0025 (0025)
HIGH-PERFORMANCE LIQUID CHROMATOGRAPHIC QUANTITATION OF AFLATOXIN METABOLITES IN ANIMAL TISSUES
Gregory, Jesse F.,III; Manley, Doris B.
Food Sci. Hum. Nutr. Dep., Univ. Florida, Gainesville USA, FL 32 611
J. Assoc. Off. Anal. Chem. (82) P 869-75; Vol 65; No 4

03-82-0026 (0026)
QUANTITATION METHOD FOR CHOLINE-CONTAINING PHOSPHOLIPIDS IN HUMAN SERUM LIPOPROTEINS BY HIGH PERFORMANCE LIQUID CHROMATOGRAPHY
Okazaki, Mitsuyo; Hagiwara, Noriko; Hara, Ichiro
Dep. Gen. Educ. Tokyo Med. Dent. Univ. Ichikawa Japan JP 272
J. Biochem. (Tokyo); (82) P 1381-9; Vol 91; No 4

03-82-0027 (0027)
HIGH PERFORMANCE LIQUID CHROMATOGRAPHY OF VITAMIN A: A QUANTITATIVE DETERMINATION
Bohman, Ove; Engdahl, Kjell Aake; Johnsson, Haakan
Inst. Chem. Univ. Uppsala Uppsala Swed. SE S-751 21
J. Chem. Educ.; (82) P 251-2; Vol 59; No 3

03-82-0028 (0028)
VERY HIGH-SPEED LIQUID CHROMATOGRAPHY. III. QUANTITATIVE ANALYSIS OF PARABENS IN COSMETIC PRODUCTS
Dong, Michael W.; DiCesare, Joseph L.
Chromatogr. Div. Perkin-Elmer Corp. Norwalk USA US CT 06856
J. Chromatogr. Sci.; (82) P 49-54; Vol 20; No 2

03-82-0029 (0029)
QUANTITATION OF RADIOLABELED COMPOUNDS ELUTING FROM THE HPLC SYSTEM
Kessler, Michael J.
Med. Sch. Univ. Texas Houston USA US TX 77030
J. Chromatogr. Sci.; (82) P 523-7; Vol 20; No 11

03-82-0030 (0030)
QUANTITATIVE ANALYSIS OF THE PYRETHRINS BY HPLC
Otieno, Dismas A.; Jondiko, Isaac J.; McDowell, Philip G.; Kezdy, Ference J.
Int. Cent. Insect Physiol. Ecol. Nairobi Kenya KE
J. Chromatogr. Sci.; (82) P 566-70; Vol 20; No 12

03-82-0031 (0031)
QUANTITATIVE HIGH-PERFORMANCE LIQUID CHROMATOGRAPHIC METHOD FOR THE ESTIMATION OF HECOGENIN AND TIGOGENIN IN THE LEAVES AND SAPOGENIN CONCENTRATES OF AGAVE SISALANA
Sitton, Dov; Blunden, Gerald; Cripps, Alan L.
Sch. Pharm. Portsmouth Polytech. Portsmouth/Hants. UK GB P01 2DZ
J. Chromatogr.; (82) P 138-40; Vol 245; No 1

03-82-0032 (0032)
QUANTITATION OF PHYSIOLOGICAL .ALPHA.-TOCOPHEROL, METABOLITES, AND RELATED COMPOUNDS BY REVERSED-PHASE HIGH-PERFORMANCE LIQUID CHROMATOGRAXHY
Howell, Scott K.; Wang, Yeu Ming
M.D. Anderson Hosp. Tumor Inst. Univ. Texas System Cancer Cent. Houston USA US TX 77030
J. Chromatogr.; (82) P 174-80; Vol 227; No 1

03-82-0033 (0033)
LIQUID CHROMATOGRAPHIC SEPARATION AND QUANTITATION OF 2-AMINO-1,3,4-THIA-DIAZOLE (NSC-4728) FROM HUMAN AND MURINE SERUM
Ackerly, Cynthia C.; Newman, Robert A.; Myers, Carl; McCormack, John J.
Coll. Med. Univ. Vermont Burlington USA US VT 05405
J. Chromatogr.; (82) P 175-80; Vol 230; No 1

03-82-0034 (0034)
HIGH-PERFORMANCE LIQUID CHROMATOGRAPHIC METHOD FOR THE QUANTITATIVE ANALYSIS OF THE ARYLOXYPROPANOLAMINES PROPRANOLOL, METOPROLOL AND ATENOLOL IN PLASMA AND TISSUE

Winkler, Horst; Ried, Walter; Lemmer, Bjoern
Cent. Pharmacol. J. W. Goethe-Univ. Frankfurt/Main Fed. Rep. Ger. DE
J. Chromatogr.; (82) P 223-34; Vol 228

03-82-0035 (0035)
REVERSED-PHASE HIGH-PERFORMANCE LIQUID CHROMATOGRAPHY OF PESTICIDES. VI. SEPARATION AND QUANTITATIVE DETERMINATION OF SOME RICE-FIELD HERBICIDES
Cabras, Paolo; Diana, Paolo; Meloni, Marco; Pirisi, Filippo M.
Cent. Reg. Agrar. Sper. Cagliari Italy IT 09100
J. Chromatogr.; (82) P 249-54; Vol 234; No 1

03-82-0036 (0036)
QUANTITATIVE DETERMINATION OF PERTECHNETATE BY HIGH-PERFORMANCE LIQUID CHROMATOGRAPHY WITH UV DETECTION
Zodda, Julius P.; Heineman, William R.; Gilbert, T. William; Deutsch, Edward
Dep. Chem. Univ. Cincinnati Cincinnati USA US OH 45221
J. Chromatogr.; (82) P 249-55; Vol 227; No 1

03-82-0037 (0037)
IMPROVED HIGH-PERFORMANCE LIQUID CHROMATOGRAPHIC ASSAY FOR CIMETIDINE USING RANITIDINE AS AN INTERNAL STANDARD
Fleitman, J.; Torosian, G.; Perrin, J. H.
Coll. Pharm. Univ. Florida Gainesville USA US FL 32610
J. Chromatogr.; (82) P 255-8; Vol 229; No 1

03-82-0038 (0038)
HIGH-PERFORMANCE LIQUID CHROMATOGRAPHIC ANALYSIS OF DIGITALIS LEAF EXTRACTS. II. QUANTITATIVE ANALYSIS
Wichtl, M.; Wichtl-Bleier, W.; Mangkudidjojo, M.
Inst. Pharm. Biol. Philipps-Univ. Marburg Fed. Rep. Ger. DE D-3550
J. Chromatogr.; (82) P 359-65; Vol 247; No 2; In Ger

03-82-0039 (0039)
STEVIA REBAUDIANA. III. HIGH-PERFORMANCE LIQUID CHROMATOGRAPHIC SEPARATION AND QUANTITATION OF REBAUDIOSIDES B, D, AND E, DULCOSIDE A, AND STEVIOLBIOSIDE
Ahmed, M. S.; Dobberstein, R. H.
Stevia Co., Inc. Arlington Heights USA US IL 60005
J. Chromatogr.; (82) P 373-6; Vol 245; No 3

03-82-0040 (0040)
SEPARATION AND QUANTITATION OF POLYNITRATE ESTERS IN PHARMACEUTICAL PREPARATIONS BY REVERSED-PHASE HIGH-PERFORMANCE LIQUID-CHROMATOGRAPHY
Cavazzutti, C.; Gagliardi, L.; Amato, A.; Gattavecchia, E.; Tonelli, D.
Ist. Super. Sanita Rome Italy IT
J. Chromatogr.; (82) P 391-5; Vol 244; No 2

03-82-0041 (0041)
SEPARATION AND QUANTIFICATION OF BASIC HYDROXYCINNAMIC AMIDES AND HYDROXYCINNAMIC ACIDS BY REVERSED-PHASE HIGH-PERFORMANCE LIQUID CHROMATOGRAPHY
Ponchet, Michel; Martin-Tanguy, Josette; Poupet, Alain; Marais, Antoine; Beck, Daniel
Stn. Bot. Pathol. Inst. Natl. Rech. Agron. Antibes Fr. FR 06602
J. Chromatogr.; (82) P 397-404; Vol 240; No 2

03-82-0042 (0042)
SEPARATION AND QUANTITATION OF ESTROGENIC ISOFLAVONES FROM CLOVERS BY HIGH-PERFORMANCE LIQUID CHROMATOGRAPHY
Nicollier, G. F.; Thompson, A. C.
Cent. Alluvial Plains Stud. Delta State Univ. Cleveland USA US MS 38773
J. Chromatogr.; (82) P 399-402; Vol 249; No 2

03-82-0043 (0043)
QUANTITATION OF DNS-AMINO ACIDS FROM BODY TISSUES AND FLUIDS USING HIGH-PERFORMANCE LIQUID CHROMATOGRAPHY
Wiedmeier, V. T.; Porterfield, S. P.; Hendrich, C. E.
Dep. Physiol. Med. Coll. Georgia Augusta USA US GA 30912
J. Chromatogr.; (82) P 410-17; Vol 231; No 2

03-82-0044 (0044)
SEPARATION AND QUANTITATION OF URINARY PORPHYRINS BY HIGH-PERFORMANCE LIQUID CHROMATOGRAPHY
Johansson, B.; Nilsson, B.
Malmoe Gen. Hosp. Univ. Lund Malmoe Swed. SE 214 01
J. Chromatogr.; (82) P 439-44; Vol 229; No 2

03-82-0045 (0045)
QUANTITATIVE HIGH-PERFORMANCE LIQUID CHROMATOGRAPHIC DETERMINATION OF SULFUR AMINO ACIDS IN PROTEIN HYDROLYSATES
Mackey, Larry N.; Beck, Terri A.
Miami Valley Lab. Procter and Gamble Co. Cincinnati USA US OH 45247
J. Chromatogr.; (82) P 455-61; Vol 240; No 2

03-82-0046 (0046)
VANCOMYCIN QUANTITATION BY HIGH-PERFORMANCE LIQUID CHROMATOGRAPHY IN HUMAN SERUM
McClain, John Bruce Lundy; Bongiovanni, Rodolpho; Brown, Scott
Infect. Dis. Serv. Walter Reed Army Med. Cent. Washington USA US DC 20012
J. Chromatogr.; (82) P 463-6; Vol 231; No 2

03-82-0047 (0047)
QUANTITATIVE REVERSED-PHASE HIGH-PERFORMANCE LIQUID CHROMATOGRAPHIC ANALYSIS OF AMPICILLIN
Margosis, Michel
Natl. Cent. Antibiotic Anal. Food and Drug Adm. Washington USA US DC 20204
J. Chromatogr.; (82) P 469-80; Vol 236; No 2

03-82-0048 (0048)
STEVIA REBAUDIANA. II. HIGH-PERFORMANCE LIQUID CHROMATOGRAPHIC SEPARATION AND QUANTITATION OF STEVIOSIDE, REBAUDIOSIDE A AND REBAUDIOSIDE C
Ahmed, M. S.; Dobberstein, R. H.
Stevia Co., Inc. Arlington Heights USA US IL 60005
J. Chromatogr.; (82) P 523-6; Vol 236; No 2

03-82-0049 (0049)
HIGH-PERFORMANCE LIQUID CHROMATOGRAPHIC ASSAY FOR IDENTIFICATION AND QUANTITATION OF NUCLEOTIDES IN LYMPHOCYTES AND MALIGNANT LYMPHOBLASTS
De Abreu, Ronney A.; Van Baal, John M.; Hakkeren, Jan A. J. M.; De Bruyn, Chris H. M. M.; Schretlen, Egbert D. A. M.
St. Radboud Hosp. Cathol. Univ. Nijmegen Nijmegen Neth. NL
J. Chromatogr; (82) P 45-52; Vol 227; No 1

03-82-0050 (0050)
A NEW HIGH PERFORMANCE LIQUID CHROMATOGRAPHY (HPLC) METHOD FOR THE QUANTITATION OF STRYCHNINE IN URINE AND TISSUE EXTRACTS
Egloff, T.; Niederwieser, A.; Pfister, K.; Otten, A.; Steinmann, B.; Steiner, W.; Gitzelmann, R.
Dep. Pediatr. Univ. Zurich Zurich Switz. CH
J. Clin. Chem. Clin. Biochem.; (82) P 203-6; Vol 20; No 4

03-82-0051 (0051)
COMPARISON OF A HIGH PRESSURE LIQUID CHROMATOGRAPHIC ANALYSIS AND AN ENZYME

IMMUNOASSAY TECHNIQUE FOR QUANTITATION OF DISOPYRAMIDE IN SERUM OR PLASMA
Bryson, Scott M.; Betts, Andrea; Summer, David J.; Whiting, Brian
Stobhill Gen. Hosp. Univ. Glasgow Glasgow UK GB
J. Clin. Hosp. Pharm.; (82) P 119-26; Vol 7; No 2

03-82-0052 (0052)
HPLC ANALYSES OF AMINO ACIDS IN PLANT MATERIALS AS DNP-DERIVATIVES BY USE OF METHYL BENZOATE AS THE INTERNAL STANDARD
Kozukue, Nobuyuki; Kozukue, Etsuko; Terai, Hirofumi; Mizuno, Susumu
Dep. Home Econ. Kenmei Jr. Coll. Himeji City Japan JP
J. Food Sci.; (82) P 1584-6; Vol 47; No 5

03-82-0053 (0053)
RAPID HPLC METHODS FOR THE SEPARATION AND QUANTITATION OF A MONO-, DI-, AND TRISACCHARIDE MIXTURES AND APPLICATIONS
Ghernati, H. M.; Abdeddaim, K.; Guermouche, M. H.
Inst. Chem. USTHB Dar-El-Beida Algeria DZ
J. Liq. Chromatogr.; (82) P 1725-48; Vol 5; No 9

03-82-0054 (0054)
QUANTITATIVE HPLC ANALYSIS OF PLASMA AMINO ACIDS AS ORTHOPHTHALDIALDEHYDE/ ETHANETHIOL DERIVATIVES
Hill, Dennis; Burnworth, Larry; Skea, William; Pfeifer, Robert
Microchem. Lab. Univ. Connecticut Storrs USA US CT
J. Liq. Chromatogr.; (82) P 2369-93; Vol 5; No 12

03-82-0055 (0055)
QUANTITATION OF ALKYLSULFONATES USING UV DETECTOR SENSITIVE "ION-PAIR" REAGENTS IN REVERSED-PHASE LIQUID CHROMATOGRAPHY
Sachok, Bohdan; Deming, Stanley N.; Bidlingmeyer, Brian A.
Dep. Chem. Univ. Houston Houston USA US TX
J. Liq. Chromatogr.; (82) P 389-402; Vol 5; No 3

03-82-0056 (0056)
A HIGH-PERFORMANCE LIQUID CHROMATOGRAPHIC METHOD FOR THE QUANTITATIVE DETERMINATION OF NAPROXEN AND DES-METHYL-NAPROXEN IN BIOLOGICAL SAMPLES
Van Loenhout, J. W. A.; Van Ginneken, C. A. M.; Ketelaars, H. C. J.; Kimenai, P. M.; Tan, Y.; Gribnau, F. W. J.
Dep. Pharmacol. Univ. Hosp. St. Radboud Nijmegen Neth. NL 6525 GA
J. Liq. Chromatogr.; (82) P 549-61; Vol 5; No 3

03-82-0057 (0057)
QUANTITATIVE ANALYSIS OF ETHYNODIOL DIACETATE AND ETHINYL ESTRADIOL/MESTRANOL IN ORAL CONTRACEPTIVE TABLETS BY HIGH-PERFORMANCE LIQUID CHROMATOGRAPHY
Carignan, G.; Lodge, B. A.; Skakum, W.
Bur. Drug Res. Health and Welfare Canada Ottawa Can. CA ON K1A 0L2
J. Pharm. Sci.; (82) P 264-6; Vol 71; No 2

03-82-0058 (0058)
QUANTITATION OF HYDROXYPROGESTERONE CAPROATE, MEDROXYPROGESTERONE ACETATE, AND PROGESTERONE BY REVERSED-PHASE HIGH-PRESSURE LIQUID CHROMATOGRAPHY
Das Gupta, V.
Coll. Pharm. Univ. Houston Houston USA US TX 77030
J. Pharm. Sci.; (82) P 294-7; Vol 71; No 3

03-82-0059 (0059)
QUANTITATIVE DETERMINATION OF DIGITOXIN BY HIGH PERFORMANCE LIQUID CHROMATOGRAPHY
Popov, D. M.; Nikolaev, N. N.; Semenov, V. A.
Vses. Nauchno-Issled. Inst. Fiz. Kul't. Moscow USSR SU

Khim.-Farm. Zh.; (82) P 761-4; Vol 16; No 6; In Russ

03-82-0060 (0060)
EFFECTS OF NORADRENERGIC NEURONAL ACTIVITY ON 3,4-DIHYDROXYPHENYLETHYLENE GLYCOL (DHPG) LEVELS. QUANTITATION BY HIGH-PERFORMANCE LIQUID CHROMATOGRAPHY
Jackman, G.; Snell, J.; Skews, H.; Bobik, A.
Clin. Res. Unit Alfred Hosp. Prahran Australia AU 3181
Life Sci.; (82) P 923-9; Vol 31; No 9

03-82-0061 (0061)
NEW QUANTITATIVE METHOD FOR THE DETERMINATION OF UV STABILITY IN VARIOUS LC + DYE MIXTURES
Veghely, T.
Ind. Res. Inst. Electron. Budapest Hung. HU
Mol. Cryst. Liq. Cryst.; (82) P 35-50; Vol 80; No 1-4

03-82-0062 (0062)
SEPARATION AND QUANTITATIVE DETERMINATION OF THE PRIMARY PUNGENT COMPONENTS OF ZINGIBERIS RHIZOMA WITH COMBINED TLC/HPLC
Steinegger, E.; Stucki, K.
Pharm. Inst. Univ. Bern Bern Switz. CH
Pharm. Acta Helv.; (82) P 66-71; Vol 57; No 3; In Ger

03-82-0063 (0063)
AN HPLC METHOD FOR THE SEPARATION AND QUANTITATION OF WATER-SOLUBLE VITAMINS IN VITAMIN-MINERAL FORMULATIONS
Jenkins, Charles
USA US
Pharm. Technol.; (82) P 53-6, 59, 61, 62-5; Vol 6; No 3

03-82-0064 (0064)
HIGH-PERFORMANCE LIQUID CHROMATOGRAPHIC IDENTIFICATION AND QUANTITATION OF CYCLIC ADENOSINE 3':5'-MONOPHOSPHATE IN HIGHER (PHASEOLUS VULGARIS L.) AND LOWER (CHLORELLA SP.) PLANTS
Van Onckelen, H. A.; Dupon, M.; De Greef, J. A.
Dep. Biol. Univ. Instelling Antwerpen Wilrijk Belg. BE B-2610
Physiol. Plant.; (82) P 93-7; Vol 55; No 2

03-82-0065 (0065)
QUALITATIVE AND QUANTITATIVE HPLC-ANALYSIS OF CARDIAC GLYCOSIDES. 4. IDENTIFICATION OF CARDENOLIDES BY HPLC ON C-18-REVERSED PHASE
Tittel, G.; Habermeier, H.; Wagner, H.
Inst. Pharm. Biol. Univ. Muenchen Munich Fed. Rep. Ger. DE D-8000/2
Planta Med.; (82) P 207-15; Vol 45; No 4; In Ger

03-82-0066 (0066)
A LIQUID CHROMATOGRAPHIC METHOD FOR QUANTITATING AMITRIPTYLINE IN BRAIN TISSUE
Preskorn, Sheldon H.; Glotzbach, Robin K.
Med. Cent. Univ. Kansas Kansas City USA US KS 66103
Psychopharmacology (Berlin); (82) P 23-4; Vol 78; No 1

03-82-0067 (0067)
QUANTITATIVE DETERMINATION OF o-(.BETA.-HYDROXYETHYL)QUERCETINS IN URINE BY HIGH-PERFORMANCE LIQUID CHROMATOGRAPHY
Kuhnz, W.; Jung, G.; Voelter, W.
Inst. Org. Chem. Univ. Tuebingen Tuebingen Fed. Rep. Ger. DE D-7400
Stud. Org. Chem. (Amsterdam); (82) P 293-8; Vol 11; No Flavonoids Bioflavonoids

03-82-0068 (0068)
MOLECULAR SIEVE CHROMATOGRAPHY OF SOME PLASMA PROTEIN PRODUCTS BY HIGH PERFORMANCE LIQUID CHROMATOGRAPHY: COMPARISON OF QUANTITATION BY UV ABSORPTION AND LOWRY PROTEIN
Bertolini, Maurice J.
Cutter Lab. Berkeley USA US CA
Vox Sang.; (82) P 87-90; Vol 43; No 2

03-82-0069 (0069)
QUANTITATION OF PROGESTERONE AND TESTOSTERONE PROPIONATE IN A HORMONE INJECTION MIXTURE BY HIGH-PERFORMANCE LIQUID CHROMATOGRAPHY (HPLC)
Hu, Jiazhi; Ye, Zhenming
Canton Inst. Drug Control Canton Peop. Rep. China CN
Yaowu Fenxi Zazhi; (82) P 169-71; Vol 2; No 3; In Ch

03-82-0070 (0070)
A QUANTITATION REAGENT SPECIFIC FOR CHOLINE-CONTAINING PHOSPHOLIPIDS IN HIGH PERFORMANCE LIQUID CHROMATOGRAPHY FOR SERUM LIPOPROTEINS
Hagiwara, Noriko; Okazaki, Mitsuyo; Hara, Ichiro
Dep. Gen. Educ. Tokyo Med. Dent. Univ. Ichikawa Japan JP
Yukagaku; (82) P 262-7; Vol 31; No 5

SECTION 7

CHAPTER 7.1
Amines–Polyamines

--- 1982 ---

01-82-0001 (0001)
ULTRAMICRODETERMINATION OF HISTAMINES BY HIGH-PERFORMANCE LIQUID CHROMATOGRAPHY
Ohkura, Y.; Kohashi, K.; Tsuruta, Y.; Ishida, S.
Fac. Pharm. Sci. Kyushu Univ. Fukuoka Japan JP
Adv. Biosci.; (82) P 243-53; Vol 33; No Adv. Histamine Res.

01-82-0002 (0002)
LIQUID CHROMATOGRAPHIC DETERMINATION OF DIMETHYL FORMAMIDE, METHYLENE BISPHENYL ISOCYANATE AND METHYLENE BISPHENYL AMINE IN AIR SAMPLES
Lipski, Krzysztof
Inst. Occup. Med. Lodz Pol. PL 90-950
Ann. Occup. Hyg.; (82) P 1-4; Vol 25; No 1

01-82-0003 (0003)
HISTAMINE IN COTTON PLANT AND DUST DETERMINED BY HIGH-PERFORMANCE LIQUID CHROMATOGRAPHY
Wall, J. H.; Muller, L. L.; Berni, R. J.
South. Reg. Res. Cent. USDA New Orleans USA US LA 70179
ACS Symp. Ser.; (82) P 301-11; Vol 189; No Cotton Dust

01-82-0004 (0004)
LIQUID CHROMATOGRAPHY WITH PRECOLUMN SAMPLE PRECONCENTRATION AND ELECTROCHEMICAL DETECTION: DETERMINATION OF AROMATIC AMINES IN ENVIRONMENTAL SAMPLES
Rice, John R.; Kissinger, Peter T.
Dep. Chem. Purdue Univ. West Lafayette USA US IN 47907
Environ. Sci. Technol; (82) P 263-8; Vol 16; No 5

01-82-0005 (0005)
RAPID DETECTION OF HISTAMINE IN FISH AND FISH PRODUCTS BY HPLC
Volk, K.; Gemmer, H.
Staat. Veterinaeruntersuch. Frankfurt/Main Fed. Rep. Ger. DE
Fleischwirtschaft; (82) P 588, 590; Vol 62; No 5; In Ger

01-82-0006 (0006)
SEPARATION OF AROMATIC AMINES BY THIN-LAYER AND HIGH-PERFORMANCE LIQUID CHROMATOGRAPHY
Narang, A. S.; Choudhury, D. R; Richards, A.
Div. Lab. Res. New York State Dep. Health Albany USA US NY 12201
J. Chromatogr. Sci.; (82) P 235-7; Vol 20; No 5

01-82-0007 (0007)
FEMTOMOLAR ION-PAIR HIGH-PERFORMANCE LIQUID CHROMATOGRAPHIC METHOD FOR DETERMINING DNS-POLYAMINE DERIVATIVES OF RED BLOOD CELL EXTRACTS UTILIZING AN AUTOMATED POLYAMINE ANALYZER
Brown, Nesbitt D.; Strickler, M. Patricia; Whaun, June M.
Div. Biochem. Walter Reed Army Inst. Res. Washington USA US DC 20012
J. Chromatogr.; (82) P 101-8; Vol 245; No 1

01-82-0008 (0008)
DETERMINATION OF POLYAMINES BY PRECOLUMN DERIVATIZATION WITH o-PHTHALALDEHYDE AND ETHANETRIOL IN COMBINATION WITH REVERSED-PHASE HIGH-PERFORMANCE LIQUID CHROMATOGRAPHY
Skaaden, Tore; Greibrokk, Tyge
Dep. Chem. Univ. Oslo Blindern Norway NO
J. Chromatogr.; (82) P 111-22; Vol 247; No 1

01-82-0009 (0009)
REVERSED-PHASE ION-PAIR LIQUID CHROMATOGRAPHIC PROCEDURE FOR THE SIMULTANEOUS ANALYSIS OF S-ADENOSYLMETHIONINE, ITS METABOLITES AND THE NATURAL POLYAMINES
Wagner, Joseph; Danzin, Charles; Mamont, Pierre
Cent. Rech. Merrell Int. Strasbourg Fr. FR 67084
J. Chromatogr.; (82) P 349-68; Vol 227; No 2

01-82-0010 (0010)
CHROMOPHORIC DETERMINATION OF PUTRESCINE, SPERMIDINE AND SPERMINE WITH DABSYL CHLORIDE BY HIGH-PERFORMANCE LIQUID CHROMATOGRAPHY AND THIN-LAYER CHROMATOGRAPHY
Lin, Jen Kun; Lai, Chen Cheng
Coll. Med. Natl. Taiwan Univ. Taipei Taiwan TW
J. Chromatogr.; (82) P 369-77; Vol 227; No 2

01-82-0011 (0011)
METHOD FOR HIGH-PERFORMANCE LIQUID CHROMATOGRAPHIC MEASUREMENT OF N-NITROSAMINES IN FOOD AND BEVERAGES
Massey, R. C.; Crews, C.; McWeeny, D. J.
Food Sci. Lab. Minist. Agric., Fish. Food Norwich UK GB NR2 4SX
J. Chromatogr.; (82) P 423-7; Vol 241; No 2

01-82-0012 (0012)
SINGLE-STEP HIGH-PERFORMANCE LIQUID CHROMATOGRAPHIC METHOD FOR THE ANALYSIS OF ACETYLATED POLYAMINES
Prussak, Charles E.; Russell, Diane Haddock
Coll. Med. Univ. Arizona Tucson USA US AZ 85724
J. Chromatogr.; (82) P 47-56; Vol 229; No 1

01-82-0013 (0013)
ANALYSIS OF A MODEL IONIC NITROSAMINE BY MICROBORE HIGH-PERFORMANCE LIQUID CHROMATOGRAPHY USING A THERMAL ENERGY ANALYZER CHEMILUMINESCENCE DETECTOR
Massey, Robert C.; Crews, Colin; McWeeny, David J.; Knowles, Michael E.
Food Lab. Minist. Agric. Fish. Food Norwich GB NR2 4SX
J. Chromatogr.; (82) P 527-9; Vol 236; No 2

01-82-0014 (0014)
REVERSED-PHASE ION-PAIRING LIQUID-CHROMATOGRAPHIC SEPARATION AND FLUOROMETRIC DETECTION OF POLYAMINES
Simpson, Richard C.; Mohammed, Hussain Y.; Veening, Hans
Dep. Chem. Bucknell Univ. Lewisburg USA US PA 17837
J. Liq. Chromatogr.; (82) P 245-64; Vol 5; No 2

01-82-0015 (0015)
EFFECTS OF COUNTER IONS IN ION-PAIR LIQUID CHROMATOGRAPHY OF HYDROPHOBIC AMINES ON NONPOLAR BONDED PHASES
Jansson, S. O.
Dep. Anal. Chem. AB Haessle Molndal Swed. SE S-431 83
J. Liq. Chromatogr.; (82) P 677-91; Vol 5; No 4

01-82-0016 (0016)
NITROAROMATIC REAGENTS FOR DETERMINATION OF AMINES AND AMINO ACIDS BY LIQUID CHROMATOGRAPHY/ELECTROCHEMISTRY
Jacobs, W. A.; Kissinger, P. T.
Dep. Chem. Purdue Univ. West Lafayette USA US IN 47907
J. Liq. Chromatogr.; (82) P 881-95; Vol 5; No 5

01-82-0017 (0017)
TOXICOLOGICAL ANALYSIS OF PHENYLETHYLAMINES BY HIGH-PERFORMANCE REVERSED-PHASE ION-PAIR PARTITION THIN-LAYER AND LIQUID CHROMATOGRAPHIES. COMMENT

Donike, Manfred
Inst. Biochem., Dtsch. Sporthochsch. Koeln, 5000 Cologne, 41 Fed. Rep. Ger.
J. Toxicol., Clin. Toxicol. (82) P 225-7; Vol 19; No 2

01-82-0018 (0018)
TOXICOLOGICAL ANALYSIS OF PHENYLETHYLAMINES BY HIGH-PERFORMANCE REVERSED-PHASE ION-PAIR PARTITION THIN-LAYER AND LIQUID CHROMATOGRAPHIES. REPLY TO COMMENTS
Verbiesde, N.
Inst. Pharm. Univ. Brussels Brussels Belg. BE
J. Toxicol., Clin. Toxicol.; (82) P 231-2; Vol 19; No 2

01-82-0019 (0019)
DETECTION OF HISTAMINE BY LIQUID CHROMATOGRAPHY
Yamanouchi Pharmaceutical Co., Ltd. Japan JP
Jpn. Kokai Tokkyo Koho; (021282) P 6 pp.; In Japan; Pat. No.: 82 26747; Int. Class: G01N-031/08; G01N-033/50; Doc. Code: A2; Appl.: JP 80/101934 800725

01-82-0020 (0020)
SEPARATION OF AROMATIC AMINES WITH CHEMICALLY BONDED STATIONARY PHASE BY HIGH-PRESSURE LIQUID CHROMATOGRAPHY
Shimomura, Hajime; Sugiyama, Toshiaki
Dep. Ind. Chem. Suzuka Coll. Technol. Suzuka Japan JP
Kiyo - Suzuka Kogyo Koto Senmon Gakko; (82) P 197-200; Vol 15

01-82-0021 (0021)
SEROTONIN TURNOVER IN INDIVIDUAL BRAIN NUCLEI: EVALUATION OF THREE METHODS USING LIQUID CHROMATOGRAPHY WITH ELECTROCHEMICAL DETECTION
Johnson, M. D.; Crowley, W. R.
Cent. Health Sci. Univ. Tennessee Memphis USA US TN 38163
Life Sci.; (82) P 589-95; Vol 31; No 6

01-82-0022 (0022)
HIGH-PERFORMANCE LIQUID CHROMATOGRAPHY OF PHARMACOLOGICALLY ACTIVE AMINES AND PEPTIDES IN BIOLOGICAL MATERIALS
Davis, Thomas P.; Schoemaker, Hans; Chen, Andrew; Yamamura, Henry I.
Heatlh Sci. Cent. Univ. rizona Tucson USA US AZ 85724
Life Sci.; (82) P 971-87; Vol 30; No 12

01-82-0023 (0023)
ALIPHATIC SECONDARY AMINES IN MEAT AND FISH PRODUCTS AND THEIR ANALYSIS BY HIGH PRESSURE LIQUID CHROMATOGRAPHY
Bellatti, Marta; Parolari, Giovanni
Stn. Sper. Ind. Conserve Aliment. Parma Italy IT 43100
Meat Sci.; (82) P 59-65; Vol 7; No 1

01-82-0024 (0024)
ANALYSIS OF POLYAMINES IN HIGHER PLANTS BY HIGH PERFORMANCE LIQUID CHROMATOGRAPHY
Flores, Hector E.; Galston, Arthur W.
Dep. Biol. Yale Univ. New Haven USA US CT 06511
Plant Physiol.; (82) P 701-6; Vol 69; No 3

01-82-0025 (0025)
DETERMINATION OF DIPHENYLAMINE AND VARIOUS NITRATED DIPHENYLAMINES BY REVERSE-PHASE HPLC
Richardson, B. R.
Mason and Hanger-Silas Mason Co., Inc. Amarillo USA US TX
From: Energy Res. Abstr. 1982, 7(17), Abstr. No. 44598. TECHNICAL REPORT; Report; (82) P 14 pp.; No MHSMP-82-11; Order No. DE82008769,; Avail.: NTIS

01-82-0026 (0026)
SPECIFIC DETERMINATION OF PENTAMETHYLMELAMINE AND ITS DEMETHYLATED METABOLITES IN HUMAN PLASMA BY HIGH PRESSURE LIQUID CHROMATOGRAPHY
Leyland-Jones, Brian R.; Deesen, Patricia E., Casper, Ephraim S.; Yung, Charles W.
Clin. Pharmacol. Lab. Mem. Sloan-Kettering Cancer Cent. New York USA US NY 10021
Ther. Drug. Monit.; (82) P 185-90; Vol 4; No 2

CHAPTER 7.2
Amino Acids

--- 1982 ---

02-82-0001 (0001)
USE OF HIGH-PERFORMANCE LIQUID CHROMATOGRAPHY FOR THE DETERMINATION OF AMINO ACIDS IN SPARKLING WINES
Casoli, A.; Colagrande, O.
Ist. Enol. Univ. Cattolica S. Cuore Piacenza Italy IT 29100
Am. J. Enol. Vitic.; (82) P 135-9; Vol 33; No 3

02-82-0002 (0002)
THE USE OF HIGH-PERFORMANCE LIQUID CHROMATOGRAPHY FOR QUANTITATION OF PLASMA AMINO ACIDS IN MAN
Hogan, Daniel L.; Kraemer, Kevin L.; Isenberg, Jon I.
Sch. Med. Univ. California, San Diego La Jolla USA US CA
Anal. Biochem.; (82) P 17-24; Vol 127; No 1

02-82-0003 (0003)
A HIGHLY SENSITIVE ASSAY FOR S-ADENOSYLMETHIONINE BY HIGH-PERFORMANCE LIQUID CHROMATOGRAPHY
Burke, William J.
Sch. Med. St. Louis Univ. St. Louis USA US MO 63104
Anal. Biochem.; (82) P 258-61; Vol 122; No 2

02-82-0004 (0004)
SIMPLE, RAPID, AND HIGHLY EFFICIENT SEPARATION OF AMINO ACID PHENYLTHIO-HYDANTOINS BY REVERSED-PHASE HIGH-PERFORMANCE LIQUID CHROMATOGRAPHY
Black, Shaun D.; Coon, Minor J.
Med. Sch. Univ. Michigan Ann Arbor USA US MI 48109
Anal. Biochem.; (82) P 281-5; Vol 121; No 2

02-82-0005 (0005)
NEW LIQUID CHROMATOGRAPHIC APPROACHES FOR FREE AMINO ACID ANALYSIS IN PLANTS AND INSECTS
Barcelon, Maria de los Angeles
Agric. Res. Educ. Cent. Univ. Florida Fort Lauderdale USA US FL 33314
J. Chromatogr.; (82) P 175-82; Vol 238; No 1

02-82-0006 (0006)
THE ENANTIOMERIC ANALYSIS OF A MIXTURE OF ALL COMMON PROTEIN AMINO ACIDS BY HIGH-PERFORMANCE LIQUID CHROMATOGRAPHY USING A NEW CHIRAL MOBILE PHASE
Weinstein, S.; Engel, M. H.; Hare, P. E.
Dep. Org. Chem. Weizmann Inst. Sci. Rehovot Israel IL 76100
Anal. Biochem.; (82) P 370-7; Vol 121; No 2

02-82-0007 (0007)
ADAPTATION OF REVERSED-PHASE HIGH-PERFORMANCE LIQUID CHROMATOGRAPHY FOR THE ISOLATION AND SEQUENCE ANALYSIS OF PEPTIDES FROM PLASMA AMYLOID P-COMPONENT
Anderson, Jacqueline K.; Mole, John E.
Med. Sch. Univ. Massachusetts Worcester USA US MA 01605
Anal. Biochem.; (82) P 413-21; Vol 123; No 2

02-82-0008 (0008)
LIQUID CHROMATOGRAPHIC SEPARATION OF THE THYROID HORMONES
Hepler, Bradford R.; Weber, Stephen G.; Purdy, William C.
Dep. Chem. McGill Univ. Montreal Can. CA PQ H3A 2K6
Anal. Chim. Acta; (82) P 221-30; Vol 138

02-82-0009 (0009)
INVESTIGATION OF CLINICAL METHODOLOGY FOR SAMPLE COLLECTION AND PROCESSING PRIOR TO THE REVERSED-PHASE LIQUID CHROMATOGRAPHIC DETERMINATION OF

UV-ABSORBING PLASMA CONSTITUENTS
Zakaria, Mona; Brown, Phyllis R.
Dep. Chem. Univ. Rhode Island Kingston USA US RI 02881
Anal.Biochem.; (82) P 25-37; Vol 120; No 1

02-82-0010 (0010)
SEPARATION OF D- AND L-AMINO ACIDS BY HPLC WITH COPPER COMPLEXES OF N,N-DIALKYL- .ALPHA.-AMINO ACIDS AS NEW CHIRAL ADDITIVES
Weinstein, Shulamith
Dep. Org. Chem. Weizmann Inst. Sci. Rehovot Israel IL 76 100
Angew. Chem.; (82) P 221-2; Vol 94; No 3; In Ger

02-82-0011 (0011)
SEPARATION OF o-PHTHALDIALDEHYDE DERIVATIVES OF FREE AMINO ACIDS FROM PLANT TISSUES BY ISOCRATIC REVERSE-PHASE HIGH-PERFORMANCE LIQUID CHROMATOGRAPHY
Martin, F.; Maudinas, B.; Gadal, P.
Unite Microbiol. Sols INRA Seichamps Fr. FR
Ann. Bot. (London); (82) P 401-6; Vol 50; No 3

02-82-0012 (0012)
LIQUID CHROMATOGRAPHIC RESOLUTION OF ENANTIOMERIC .ALPHA.-AMINO ACID DERIVATIVES EMPLOYING A CHIRAL DIAMIDE PHASE
Hara, Shoji; Dobashi, Akira; Eguchi, Masakatzu
Tokyo Coll. Pharm. Hachioji Japan JP 192-03
ACS Symp. Ser.; (82) P 266-7; Vol 185; No Asymmetric React. Processes Chem.

02-82-0013 (0013)
QUANTITATIVE DETERMINATION OF AMINO ACIDS IN TISSUE CULTURE CELLS BY HIGH PERFORMANCE LIQUID CHROMATOGRAPHY
Kabus, P.; Koch, G.
Abt. Molekularbiol. Univ. Hamburg Hamburg Fed. Rep. Ger. DE D-2000/13
Biochem. Biophys. Res. Commun.; (82) P 783-90; Vol 108; No 2

02-82-0014 (0014)
A HIGH-PERFORMANCE LIQUID CHROMATOGRAPHIC DETERMINATION OF FREE AMINO ACIDS IN NATURAL WATERS IN THE PICOMOLAR (M .TIMES. 10-12) RANGE SUITABLE FOR SHIPBOARD USE
Evens, Roger; Braven, James; Brown, Leslie; Butler, Ian
Dep. Environ. Sci. Plymouth Polytech. Plymouth/Devon UK GB
Chem. Ecol.; (82) P 99-106; Vol 1; No 2

02-82-0015 (0015)
THE DETERMINATION OF m-TYROSINE IN HUMAN PLASMA BY HIGH-PERFORMANCE LIQUID CHROMATOGRAPHY
Ishimitsu, Susumu; Fujimoto, Sadaki; Ohara, Akira
Kyoto Coll. Pharm. Kyoto Japan JP 607
Chem. Pharm. Bull.; (82) P 1889-91; Vol 30; No 5

02-82-0016 (0016)
DIRECT ENRICHMENT OF TRYPTOPHAN AND ITS METABOLITES IN PLASMA ONTO A PRE-COLUMN FOLLOWED BY REVERSE PHASE HPLC ANALYSIS
Yoshida, Hisanobu; Morita, Ikue; Masujima, Tsutomu; Imai, Hideo
Sch. Med. Hiroshima Univ. Hiroshima Japan JP 734
Chem. Pharm. Bull.; (82) P 3827-30; Vol 30; No 10

02-82-0017 (0017)
METABOLIC PROFILING OF TRYPTOPHAN AND ITS INDOLE METABOLITES IN RAT BRAIN BY HIGH-PERFORMANCE LIQUID CHROMATOGRAPHY
Martinez-Hermida, E.; Tusell, J. M.; Sunol, C.; Mahy, N.; Artigas, F.; Gelpi, Emilio

Anal. Neurochem. Unit Inst. Quim. Bio-Org. Barcelona Spain ES
Chromatogr. Sci.; (82) P 305-17; Vol 20; No Biol./Biomed. Appl. Liq.
Chromatogr. 4

02-82-0018 (0018)
LIQUID-CHROMATOGRAPHIC DIRECT DETERMINATION OF PHENYLALANINE AND TYROSINE IN SERUM OR PLASMA, WITH APPLICATION TO PATIENTS WITH PHENYLKETONURIA
Hilton, Mary A.
Sch. Med. Univ. Louisville Louisville USA US KY 40292
Clin. Chem. (Winston-Salem, N. C.); (82) P 1215-18; Vol 28; No 5

02-82-0019 (0019)
LIQUID-CHROMATOGRAPHIC MEASUREMENT OF PHENYLALANINE AND TYROSINE IN SERUM
Spierto, F. W.; Whitfield, W.; Apetz, M.; Hannon, W. Harry
Clin. Chem. Div. Cent. Environ. Health Atlanta USA US GA 30333
Clin. Chem. (Winston-Salem, N. C.); (82) P 2282-5; Vol 28; No 11

02-82-0020 (0020)
RAPID ASSAY FOR AMINO ACIDS IN SERUM OR URINE BY PRE-COLUMN DERIVATIZATION AND REVERSED-PHASE LIQUID CHROMATOGRAPHY
Turnell, D. C.; Cooper, J. D. H.
Biochem. Dep. Coventry Hosp. Coventry UK GB CV1 4FH
Clin. Chem. (Winston-Salem, N. C.); (82) P 527-31; Vol 28; No 3

02-82-0021 (0021)
HPLC ANALYSIS OF AROMATIC AMINO ACIDS, NUCLEOSIDES, AND BASES IN PLASMA OF ACUTE LYMPHOCYTIC LEUKEMICS ON CHEMOTHERAPY
Zakaria, Mona; Brown, Phyllis R.; Farnes, M. Patricia; Barker, Barbara E.
Dep. Chem. Univ. Rhode Island Kingston USA US RI 02881
Clin. Chim. Acta; (82) P 69-80; Vol 126; No 1

02-82-0022 (0022)
HIGH-PERFORMANCE LIQUID CHROMATOGRAPHIC DETERMINATION OF HYDROXYPROLINE AFTER DERIVATIZATION WITH 4-DIMETHYLAMINOAZOBENZENE- 4'-SULFONYL CHLORIDE
Casini, Alessandro; Martini, Flavio; Nieri, Simonetta; Ramarli, Dunia; Franconi, Flavia; Surrenti, Calogero
Ist. Clin. Med. IV Florence Italy IT 50134
J. Chromatogr.; (82) P 187-92; Vol 249; No 1

02-82-0023 (0023)
SEPARATION OF D- AND L-AMINO ACIDS BY HPLC WITH COPPER COMPLEXES OF n,n-DIAL-KYL .ALPHA.-AMINO ACIDS AS NEW CHIRAL ADDITIVES
Weinstein, Shulamith
Dep. Org. Chem. Weizmann Inst. Sci. Rehovot Israel IL 76 100
Angew. Chem.; (82) P 221-2; Vol 94; No 3; In Ger

02-82-0024 (0024)
SEPARATION OF o-PHTHALDIALDEHYDE DERIVATIVES OF FREE AMINO ACIDS FROM PLANT TISSUES BY ISOCRATIC REVERSE-PHASE HIGH-PERFORMANCE LIQUID CHROMATOGRAPHY
Martin, F.; Maudinas, B.; Gadal, P.
Unite Microbiol. Sols INRA Seichamps Fr. FR
Ann. Bot. (London); (82) P 401-6; Vol 50; No 3

02-82-0025 (0025)
LIQUID CHROMATOGRAPHIC RESOLUTION OF ENANTIOMERIC .ALPHA.-AMINO ACID DERIVATIVES EMPLOYING A CHIRAL DIAMIDE PHASE
Hara, Shoji; Dobashi, Akira; Eguchi, Masakatzu
Tokyo Coll. Pharm. Hachioji Japan JP 192-03
ACS Symp. Ser.; (82) P 266-7; Vol 185; No Asymmetric React. Processes Chem.

02-82-0026 (0026)
QUANTITATIVE DETERMINATION OF AMINO ACIDS IN TISSUE CULTURE CELLS BY HIGH PERFORMANCE LIQUID CHROMATOGRAPHY
Kabus, P.; Koch, G.
Abt. Molekularbiol. Univ. Hamburg Hamburg Fed. Rep. Ger. DE D-2000/13
Biochem. Biophys. Res. Commun.; (82) P 783-90; Vol 108; No 2

02-82-0027 (0027)
ION-PAIR LIQUID CHROMATOGRAPHIC AND FLUOROMETRIC DETERMINATIOW OF AMINO ACIDS IN THE GUINEA PIG BRAIN
Gaetani, E.; Laureri, C. F.; Vitto, M.; Bordi, F.
Ist. Chim. Farm. Tossicol. Univ. Studi Parma Parma Italy IT
Farmaco, Ed. Prat.; (82) P 253-7; Vol 37; No 8; In Ital

02-82-0028 (0028)
SIMULTANEOUS DETERMINATION OF 5-HYDROXYTRYPTOPHAN, SEROTONIN AND 5-HYDROXY-INDOLEACETIC ACID IN THE BRAIN BY LIQUID CHROMATOGRAPHY WITH ELECTRO-CHEMICAL DETECTION
Santagostino, G.; Frattini, P.; Schinelli, S.; Cucchi, M. L.; Corona, G. L.
Inst. Pharmacol. Univ. Pavia Pavia Italy IT
Farmaco, Ed. Prat.; (82) P 365-70; Vol 37; No 11

02-82-0029 (0029)
QUANTITATIVE DETERMINATION OF AMINO ACIDS IN FOOD BY HIGH-PRESSURE LIQUID CHROMATOGRAPHY
Khayat, Ali; Redenz, Peter K.; Gorman, Lee A.
Sea Food Div. Res. Cent. Ralston Purina Co. San Diego USA US CA 92121
Food Technol. (Chicago); (82) P 46-50; Vol 36; No 6

02-82-0030 (0030)
SEPARATION OF RACEMIC ALKYLAMINO ACIDS WITH HPLC USING A CHIRAL MOBILE PHASE
Gundlach, G.; Sattler, E. L.; Wagenbach, U.
Zent. Biochem. Univ. Giessen Giessen Fed. Rep. Ger. DE D-6300
Fresenius' Z. Anal. Chem.; (82) P 684-6; Vol 311; No 7; In Ger

02-82-0031 (0031)
HIGH-PRESSURE LIQUID CHROMATOGRAPHIC DETERMINATION OF SUPPLEMENTAL METHIONINE IN POULTRY PREMIX
Matesic, Dubravka
Vet. Fac. Univ. Zagreb Zagreb Yugoslavia YU 41000
J. Assoc. Off. Anal. Chem.; (82) P 62-5; Vol 65; No 1

02-82-0032 (0032)
ISOCRATIC SEPARATION OF PHENYLTHIOHYDANTOIN-AMINO ACIDS BY REVERSED-PHASE HIGH-PERFORMANCE LIQUID CHROMATOGRAPHY
Kim, S. M.
Gruenenthal G.m.b.H. Cent. Res. Aachen Fed. Rep. Ger. DE D-5100
J. Chromatogr.; (82) P 103-10; Vol 247; No 1

02-82-0033 (0033)
SIMULTANEOUS MEASUREMENT OF 5-HYDROXYTRYPTOPHAN AND L-DIHYDROXYPHENYLALANINE BY HIGH-PERFORMANCE LIQUID CHROMATOGRAPHY WITH ELECTROCHEMICAL DETECTION. MEASUREMENT OF SEROTONIN AND CATECHOLAMINE TURNOVER IN DISCRETE BRAIN REGIONS
Shum, Andrew; Sole, Michael J.; Van Loon, Glen R.
Dep. Med. Univ. Toronto Toronto Can. CA ON M5S 1A8
J. Chromatogr.; (82) P 123-30; Vol 228

02-82-0034 (0034)
FLUORESCENCE DETECTION OF CYSTINE BY o-PHTHALALDEHYDE DERIVATIZATION AND ITS SEPARATION USING HIGH-PERFORMANCE LIQUID CHROMATOGRAPHY

Cooper, J. D. H.; Turnell, D. C.
Biochem. Dep. Coventry and Warwickshire Hosp. Coventry UK GB
J. Chromatogr.; (82) P 158-61; Vol 227; No 1

02-82-0035 (0035)
HIGH-PERFORMANCE LIQUID CHROMATOGRAPHY OF N-NITROSOPROLINE IN URINE
Walters, D. G.; Mallett, A. K.; Cottrell, R. C.
Br. Ind. Biol. Res. Assoc. Carshalton/Surrey UK GB
J. Chromatogr.; (82) P 161-4; Vol 246; No 1

02-82-0036 (0036)
MEASUREMENT OF BRANCHED CHAIN AMINO ACIDS AND BRANCHED CHAIN .ALPHA.-KETOACIDS IN PLASMA BY HIGH-PERFORMANCE LIQUID CHROMATOGRAPHY
Nissen, Steven L.; Van Huysen, Carol; Haymond, Morey W.
Dep. Pediatr. Mayo Clin. Med. Sch. Rochester USA US MN 55905
J. Chromatogr.; (82) P 170-5; Vol 232; No 1

02-82-0037 (0037)
NEW LIQUID CHROMATOGRAPHIC APPROACHES FOR FREE AMINO ACID ANALYSIS IN PLANTS AND INSECTS
Barcelon, Maria de los Angeles
Agric. Res. Educ. Cent. Univ. Florida Fort Lauderdale USA US FL 33314
J. Chromatogr.; (82) P 175-82; Vol 238; No 1

02-82-0038 (0038)
HIGH-PERFORMANCE LIQUID CHROMATOGRAPHIC DETERMINATION OF HYDROXYPROLINE AFTER DERIVATIZATION WITH 4-DIMETHYLAMINOAZOBENZENE- 4'-SULFONYL CHLORIDE
Casini, Alessandro; Martini, Flavio; Nieri, Simonetta; Ramarli, Dunia; Franconi, Flavia; Surrenti, Calogero
Ist. Clin. Med. IV Florence Italy IT 50134
J. Chromatogr.; (82) P 187-92; Vol 249; No 1

02-82-0039 (0039)
AMINO ACIDS AND PEPTIDES. CLXXVIII. CORRELATION BETWEEN THE HYDROPHOBICITY OF SUBSTITUENTS IN THE PHENYLALANINE MOIETY OF OXYTOCIN CARBA ANALOGS AND REVERSED-PHASE-CHROMATOGRAPHIC k' VALUES
Lebl, Michel
Inst. Org. Chem. Biochem. Czech. Acad. Sci. Prague Czech. CS 166 10/6
J. Chromatogr.; (82) P 342-5; Vol 242; No 2

02-82-0040 (0040)
AMINO ACID ANALYSES BY HIGH-PERFORMANCE LIQUID CHROMATOGRAPHY. AN EVALUATION OF THE USEFULNESS OF PRECOLUMN DNS DERIVATIZATION
De Jong, Catharina; Hughes, Graham J.; Van Wieringen, Erika; Wilson, Kenneth J.
Lab. Biochem. ETH Zurich Switz. CH CH-8092
J. Chromatogr.; (82) P 345-59; Vol 241; No 2

02-82-0041 (0041)
QUANTITATION OF DNS-AMINO ACIDS FROM BODY TISSUES AND FLUIDS USING HIGH-PERFORMANCE LIQUID CHROMATOGRAPHY
Wiedmeier, V. T.; Porterfield, S. P.; Hendrich, C. E.
Dep. Physiol. Med. Coll. Georgia Augusta USA US GA 30912
J. Chromatogr.; (82) P 410-17; Vol 231; No 2

02-82-0042 (0042)
AMINO ACID ANALYSIS USING STANDARD HIGH-PERFORMANCE LIQUID CHROMATOGRAPHY EQUIPMENT
Hughes, Graham J.; Winterhalter, Kaspar H.; Boller, Ernst; Wilson, Kenneth J.
Lab. Biochem. ETH-Zurich Zuri1h Switz. CH CH-8092

J. Chromatogr.; (82) P 417-26; Vol 235; No 2

02-82-0043 (0043)
ASSAY AND STABILITY OF .ALPHA.-METHYLDOPA IN MAN USING HIGH-PERFORMANCE LIQUID CHROMATOGRAPHY WITH ELECTROCHEM. DETECTION
Ong, H.; Sved, S.; Beaudoin, N.
Health Prot. Branch Bur. Drug Res. Ottawa Can. CA ON K1A 0L2
J. Chromatogr.; (82) P 433-8; Vol 229; No 2

02-82-0044 (0044)
RESOLUTION OF OPTICAL ISOMERS OF DNS-AMINO ACIDS BY HIGH-PERFORMANCE LIQUID CHROMATOGRAPHY WITH L-HISTIDINE AND ITS DERIVATIVES IN THE MOBILE PHASE
Lam, Stanley; Karmen, Arthur
Albert Einstein Coll Med. Bronx USA US NY 10461
J. Chromatogr.; (82) P 451-62; Vol 239

02-82-0045 (0045)
QUANTITATIVE HIGH-PERFORMANCE LIQUID CHROMATOGRAPHIC DETERMINATION OF SULFUR AMINO ACIDS IN PROTEIN HYDROLYSATES
Mackey, Larry N.; Beck, Terri A.
Miami Valley Lab. Procter and Gamble Co. Cincinnati USA US OH 45247
J. Chromatogr.; (82) P 455-61; Vol 240; No 2

02-82-0046 (0046)
HIGH-PERFORMANCE LIQUID CHROMATOGRAPHIC DETERMINATION OF 2-MERCAPTOPROPIONYL-GLYCINE (THIOPRONINE) IN URINE
Springolo, V.; Bertani, W.; Coppi, G.
Lab. Ric. Proter Opera/Milan Italy IT 20090
J. Chromatogr.; (82) P 456-60; Vol 232; No 2

02-82-0047 (0047)
SIMULTANEOUS DETERMINATION OF CARBIDOPA, LEVODOPA AND 3,4-DIHYDROXYPHENYL-ACETIC ACID USING HIGH-PERFORMANCE LIQUID CHROMATOGRAPHY WITH ELECTROCHEMICAL DETECTION
Nissinen, Erkki; Taskinen, Jyrki
Res. Lab. Orion Pharm. Co. Helsinki Finland FI 00101/10
J. Chromatogr.; (82) P 459-62; Vol 231; No 2

02-82-0048 (0048)
OPTICAL RESOLUTION OF DL-PROLINE BY REVERSED-PHASE HIGH-PERFORMANCE LIQUID CHROMATOGRAPHY USING N-(p-TOLUENESULFONYL)-L-PHENYLALANINE-COPPER(II) AS A CHIRAL ADDITIVE
Nimura, Noriyuki; Toyama, Atsuko; Kinoshita, Toshio
Sch. Pharm. Sci. Kitasato Univ. Tokyo Japan JP 108
J. Chromatogr.; (82) P 482-4; Vol 234; No 2

02-82-0049 (0049)
HIGH-PERFORMANCE LIQUID CHROMATOGRAPHIC SEPARATION OF SIX ESSENTIAL AMINO ACIDS AND ITS USE AS AN AID IN THE DIAGNOSIS OF BRANCHED-CHAIN KETOACIDURIA
Wassner, Steven J.; Li, Jeanne B.
Milton S. Hershey Med. Cent. Pennsylvania State Univ. Hershey USA US PA 17033
J. Chromatogr.; (82) P 497-502; Vol 227; No 2

02-82-0050 (0050)
REVERSED-PHASE LIQUID CHROMATOGRAPHIC RESOLUTION OF UNDERIVATIZED D,L-AMINO ACIDS USING CHIRAL ELUENTS
Nimura, Noriyuki; Toyama, Atsuko; Kasahara, Yoko; Kinoshita, Toshio
Sch. Pharm. Sci. Kitasato Univ. Tokyo Japan JP 108
J. Chromatogr.; (82) P 671-5; Vol 239

02-82-0051 (0051)
PRECOLUMN LABELING FOR HIGH-PERFORMANCE LIQUID CHROMATOGRAPHY OF AMINO ACIDS WITH 7-FLUORO-4-NITROBENZO-2-OXA-1,3-DIAZOLE AND ITS APPLICATION TO PROTEIN HYDROLYSATES
Watanabe, Yoshihiko; Imai, Kazuhiro
Fac. Pharm. Sci. Univ. Tokyo Tokyo Japan JP 113
J. Chromatogr.; (82) P 723-32; Vol 239

02-82-0052 (0052)
QUANTITATIVE DETERMINATION OF AMINO ACIDS IN FOOD BY HIGH-PRESSURE LIQUID CHROMATOGRAPHY
Khayat, Ali; Redenz, Peter K.; Gorman, Lee A.
Sea Food Div. Res. Cent. Ralston Purina Co. San Diego USA US CA 92121
Food Technol. (Chicago); (82) P 46-50; Vol 36; No 6

02-82-0053 (0053)
SEPARATION OF RACEMIC ALKYLAMINO ACIDS WITH HPLC USING A CHIRAL MOBILE PHASE
Gundlach, G.; Sattler, E. L.; Wagenbach, U.
Zent. Biochem. Univ. Giessen Giessen Fed. Rep. Ger. DE D-6300
Fresenius' Z. Anal. Chem.; (82) P 684-6; Vol 311; No 7; In Ger

02-82-0054 (0054)
PEPTIDE MAPPING OF VARIANT GLYCOPROTEINS FROM TRYPANOSOMA RHODESIENSE BY REVERSE PHASE LIQUID CHROMATOGRAPHY
Strickler, M. Patricia; Travis, Richard W.; Olenick, John G.
Dep. Appl. Biochem. Walter Reed Army Inst. Res. Washington USA US DC 20012
J. Liq. Chromatogr.; (82) P 1933-40; Vol 5; No 10

02-82-0055 (0055)
QUANTITATIVE HPLC ANALYSIS OF PLASMA AMINO ACIDS AS ORTHOPHTHALDIALDEHYDE/ ETHANETHIOL DERIVATIVES
Hill, Dennis; Burnworth, Larry; Skea, William; Pfeifer, Robert
Microchem. Lab. Univ. Connecticut Storrs USA US CT
J. Liq. Chromatogr.; (82) P 2369-93; Vol 5; No 12

02-82-0056 (0056)
PREPARATIVE LIQUID CHROMATOGRAPHIC SEPARATION OF AMINO ACIDS AND PEPTIDES ON AMBERLITE XAD-4
Pietrzyk, Donald J.; Cahill, W. J., Jr.; Stodola, J. D.
Chem. Dep. Univ. Iowa Iowa City USA US IA 52242
J. Liq. Chromatogr.; (82) P 443-61; Vol 5; No 3

02-82-0057 (0057)
PEPTIDE FORMATION IN THE PRESENCE OF METAL ION PROTECTING GROUPS. III. THE SEPARATION OF ((NH3)5Co(III)-) AMINO ACIDS AND PEPTIDES BY REVERSE-PHASE HIGH-PRESSURE LIQUID CHROMATOGRAPHY
Isied, Stephan S.; Lyon, John; Vassilian, Asbed; Worosila, Greg
Dep. Chem. Rutgers, State Univ. New Brunswick USA US NJ 08903
J. Liq. Chromatogr.; (82) P 537-47; Vol 5; No 3

02-82-0058 (0058)
NITROAROMATIC REAGENTS FOR DETERMINATION OF AMINES AND AMINO ACIDS BY LIQUID CHROMATOGRAPHY/ELECTROCHEMISTRY
Jacobs, W. A.; Kissinger, P. T.
Dep. Chem. Purdue Univ. West Lafayette USA US IN 47907
J. Liq. Chromatogr.; (82) P 881-95; Vol 5; No 5

02-82-0059 (0059)
SIMULTANEOUS DETERMINATION OF 3,4-DIHYDROXYPHENYLALANINE, 5-HYDROXYTRYPTOPHAN, DOPAMINE, 4-HYDROXY-3-METHOXYPHENYLALANINE,NOREPINEPHRINE, 3,4-DIHYDROXYPHENYL

ACETIC ACID, HOMOVANILLIC ACID, SEROTONIN AND 5-HYDROXYINDOLEACETIC ACID IN RAT CEREBROSPINAL FLUID AND BRAIN BY HIGH-PERFORMANCE LIQUID CHROMATOGRAPHY WITH ELECTROCHEMICAL DETECTION
Wagner, Joseph; Vitali, Pierre; Palfreyman, Michael G.; Zraika, Monique; Huot, Sylvie
Cent. Rech., Merrell Int., Strasbourg France 67 084
J. Neurochem. (82) P 1241-54; Vol 38; No 5

02-82-0060 (0060)
ANALYSIS OF GLYCINE IN ANTIPERSPIRANT PRODUCTS BY HPLC
Chin, David; Achari, Raja G.
Bristol-Meyers Prod. Hillside USA US NJ 07207
J. Soc. Cosmet. Chem.; (82) P 359-62; Vol 33; No 7

02-82-0061 (0061)
MEASUREMENT OF L-DIHYDROXYPHENYLALANINE IN PLASMA AND OTHER BIOLOGICAL FLUIDS BY HIGH-PRESSURE LIQUID CHROMATOGRAPHY WITH ELECTROCHEMICAL DETECTION
Shum, Andrew; Van Loon, Glen R.; Sole, Michael J.
Dep. Med. Univ. Toronto Toronto Can. CA ON M5S 1A8
Life Sci.; (82) P 1541-5; Vol 31; No 15

02-82-0062 (0062)
DIEL AND DEPTH VARIATIONS IN DISSOLVED FREE AMINO ACIDS AND AMMONIUM IN THE BALTIC SEA DETERMINED BY SHIPBOARD HPLC ANALYSIS
Mopper, Kenneth; Lindroth, Peter
Coll. Mar. Stud. Univ. Delaware Lewes USA US DE 19958
Limnol. Oceanogr.; (82) P 336-47; Vol 27; No 2

02-82-0063 (0063)
DEMONSTRATION OF AROMATIC L-AMINO ACID DECARBOXYLASE ACTIVITY IN HUMAN BRAIN WITH L-DOPA AND L-5-HYDROXYTRYPTOPHAN AS SUBSTRATES BY HIGH-PERFORMANCE LIQUID CHROMATOGRAPHY WITH ELECTROCHEMICAL DETECTION
Rahman, Mohammed Khalilur; Nagatsu, Toshiharu
Dep. Life Chem. Tokyo Inst. Technol. Yokohama Japan JP 227
Neurochem. Int.; (82) P 1-6; Vol 4; No 1

02-82-0064 (0064)
A RAPID AND SIMPLE METHOD TO DETERMINE THE SPECIFIC ACTIVITIES OF SEROTONIN, 5-HYDROXYINDOLEACETIC ACID, AND 5-HYDROXYTRYPTOPHAN IN BRAIN BY HPLC WITH ELECTROCHEMICAL DETECTION
Bernstein, M. J.; Shea, P. A.
Sch. Med. Indiana Univ. Indianapolis USA US IN 46223
Neurochem. Res.; (82) P 79-85; Vol 7; No 1

02-82-0065 (0065)
A CRITICAL EVALUATION OF LC-EC FOR THE SIMULTANEOUS DETERMINATION OF 5-HYDROXYTRYPTOPHAN, 5-HT, 5-HIAA AND HVA IN BIOLOGICAL SAMPLES USING GC-MS FOR VALIDATION
Narasimhachari, N.; Boadle-Biber, M. C.; Friede, R. O.
Dep. Psychiatry Med. Coll. Virginia Richmond USA US VA 23298
Res. Commun. Chem. Pathol. Pharmacol.; (82) P 413-30; Vol 37; No 3

02-82-0066 (0066)
A HIGH-PRESSURE LIQUID CHROMATOGRAPHIC-ENZYMIC ASSAY FOR GLYCINE AND TAURINE CONJUGATES OF CHOLIC, CHENODEOXYCHOLIC AND DEOXYCHOLIC ACID IN SERUM
Linnet, Kristian
Dep. Clin. Chem. Frederiksberg Hosp. Copenhagen Den. DK DK-2000
Scand. J. Clin. Lab. Invest; (82) P 455-60; Vol 42; No 5

02-82-0067 (0067)
DETERMINATION OF GLUTATHIONE AND D-PENICILLAMINE USING HIGH-PERFORMANCE LIQUID CHROMATOGRAPHY WITH A VOLTAMMETRIC DETECTOR
Inoue, Kenji; Otake, Tateo; Noda, Katsumi; Kyogoku, Kazuaki; Uchida, Shoji
Res. Cent. Taisho Pharm. Co., Ltd. Saitama Japan JP 330
Yakugaku Zasshi; (82) P 659-65; Vol 102; No 7; In Japan

02-82-0068 (0068)
ANALYSIS OF AMINO ACID DRUGS BY ION-EXCHANGE HPLC. I. SEPARATION AND DETERMINATION OF PROSTALL CAPSULES
Yu, Zong; Kong, Xiangling; Zhou, Jingyuan
Tianjin Inst. Drug Control Tianjin Peop. Rep. China CN
Yaowu Fenxi Zazhi; (82) P 154-7; Vol 2; No 3; In Ch

02-82-0069 (0069)
AMINO ACID ANALYZER AND HIGH-PRESSURE LIQUID CHROMATOGRAPHIC DETERMINATION OF AMINO ACIDS AND BIOGENIC AMINES IN THE LIGHT AND DARK MUSCLE MEAT OF TUNA
Stockemer, J.
Staat. Veterinaeruntersuchungsamt Fische Fischwaren Cuxhaven Fed. Rep. Ger. DE D-2190
Z. Lebensm.-Unters. Forsch.; (82) P 108-13; Vol 174; No 2; In Ger

CHAPTER 7.3
Carbohydrates

--- 1982 ---

03-82-0001 (0001)
HIGH PERFORMANCE LIQUID CHROMATOGRAPHY OF PHENYL 1-THIOGLYCOPYRANOSIDES, AND PHENYL AND p-NITROPHENYL GLYCOPYRANOSIDES
Kohata, Katsunori; Meguro, Hiroshi
Fac. Agric. Tohoku Univ. Sendai Japan 980
Agric. Biol. Chem.; (81) P 1891-3; Vol 45; No 8

03-82-0002 (0002)
MODE OF HYDROLYSIS OF CHITOOLIGOSACCHARIDES WITH PYCNOPORUS CINNABARINUS .BETA.-N-ACETYLHEXOSAMINIDASE: APPLICATION OF HIGH-PERFORMANCE LIQUID CHROMATOGRAPHY
Ohtakara, Akira; Mitsutomi, Masaru; Nakamae, Emiko
Fac. Agric. Saga Univ. Saga Japan JP 840
Agric. Biol. Chem.; (82) P 293-5; Vol 46; No 1

03-82-0003 (0003)
FRACTIONATION OF OLIGOSACCHARIDES CONTAINING N-ACETYL AMINO SUGARS BY REVERSE-PHASE HIGH-PRESSURE LIQUID CHROMATOGRAPHY
Blumberg, Karl; Liniere, Frederic; Pustilnik, Lyubov; Bush, C. Allen
Dep. Chem. Illinois Inst. Technol. Chicago USA US IL 60616
Anal. Biochem.; (82) P 407-12; Vol 119; No 2

03-82-0004 (0004)
SEPARATION OF RETINOL, RETINOIC ACID, MANNOSE, DOLICHYL MANNOSYL PHOSPHATE, MANNOSYL RETINYL PHOSPHATE BY HIGH-PRESSURE LIQUID CHROMATOGRAPHY
Kurokawa, Tadashi; De Luca, Luigi M.
Lab. Exp. Pathol. Natl. Cancer Inst. Bethesda USA US MD 20205
Anal. Biochem.; (82) P 428-32; Vol 119; No 2

03-82-0005 (0005)
SPECIFICITY OF OVINE SUBMAXILLARY-GLAND SIALYLTRANSFERASES. APPLICATION OF HIGH-PRESSURE LIQUID CHROMATOGRAPHY IN THE IDENTIFICATION OF SIALOOLIGOSACCHARIDE PRODUCTS
Bergh, Michel L. E.; Koppen, Piet L.; Van den Eijnden, Dirk H.
Dep. Med. Chem. Vrije Univ. Amsterdam Neth. NL NL-1007 MC
Biochem. J.; (82) P 411-15; Vol 201; No 2

03-82-0006 (0006)
HIGH-PERFORMANCE LIQUID CHROMATOGRAPHY DETERMINATION OF UNSATURATED DISACCHARIDES PRODUCED FROM CHONDROITIN SULFATES BY CHONDROITINASES
Fluharty, Arvan L.; Glick, Jeffrey A.; Matusewicz, Nicholas M.; Kihara, Hayato
Sch. Med. Univ. California Pomona USA US CA 91769
Biochem. Med.; (82) P 352-60; Vol 27; No 3

03-82-0007 (0007)
HIGH-PERFORMANCE LIQUID CHROMATOGRAPHIC DETERMINATION OF SUGARS IN CHOCOLATE AND CACAO
Corradi, Cesare; Carreri, Daniela; Micheli, Giuseppe
Lab. Chim. Prov. Milan Italy IT 20129
Boll. Chim. Unione Ital. Lab. Prov., Parte Sci.; (82) P 33-42; Vol 33; No S1; In Ital

03-82-0008 (0008)
DETERMINATION OF VICINE AND CONVICINE IN FABA BEANS BY HIGH PRESSURE LIQUID CHROMATOGRAPHY
Quemener, B.; Gueguen, J.; Mercier, C.
Lab. Biochim. Aliments Cent. Rech. Agro-Alimentaires Nantes Fr. FR F-44072
Can. Inst. Food Sci. Technol. J.; (82) P 109-15; Vol 15; No 2

03-82-0009 (0009)
THERMODYNAMIC FUNCTIONS IN HIGH-PERFORMANCE LIQUID CHROMATOGRAPHY OF SACCHARIDES
Fukuhara, Kenichi; Nakamura, Asao; Tsuboi, Masamichi
Cent. Res. Lab. Ajinomoto Co., Inc. Kawasaki Japan JP 210
Chem. Pharm. Bull.; (82) P 1121-5; Vol 30; No 4

03-82-0010 (0010)
SEPARATION OF CARBOHYDRATES BY LIQUID CHROMATOGRAPHY ON SILICA GEL, ADDING ADSORPTION MODIFICATORS TO THE ELUENT
Boumahraz, M.; Davydov, V. Ya.; Kiselev, A. V.
Chem. Dep. M. V. Lomonosov State Univ.-Moscow Moscow USSR SU 117234
Chromatographia; (82) P 751-6; Vol 15; No 12

03-82-0011 (0011)
DETERMINATION OF MUSTARD GLUCOSIDES AND THEIR DECOMPOSITION PRODUCTS IN MUSTARD SEEDS AND NUTRITIONAL MUSTARD PREPARATIONS BY ION-PAIR HIGH PERFORMANCE LIQUID CHROMATOGRAPHY
Henning, Werner
Inst. Lebensmittelchem. Univ. Hannover Hannover Fed. Rep. Ger. 3000/91
Dtsch. Lebensm.-Rundsch.; (81) P 313-18; Vol 77; No 9; In Ger

03-82-0012 (0012)
DETERMINATION OF HESPERIDIN IN ORANGE JUICES AND ORANGE LEMONADES BY HIGH-PRESSURE LIQUID CHROMATOGRAPHY WITH AN AMPEROMETRIC DETECTOR
Sontag, Gerhard; Kral, Karl
Inst. Anal. Chem. Univ. Wien Vienna Austria AT A-1090
Fresenius' Z. Anal. Chem.; (81) P 109-13; Vol 309; No 2; In Ger

03-82-0013 (0013)
HIGH-PRESSURE LIQUID CHROMATOGRAPHIC DETERMINATION OF SUGARS IN DAIRY PRODUCTS. I. DETERMINATION OF FREE SUGARS IN COMMERCIAL FERMENTED MILK DRINKS
Lee, Y. K.; Song, M. H.; Chun, S. B.
Coll. Agric. Chonnam Natl. Univ. Kwanzu S. Korea KR
Hanguk Ch'uksan Hakhoe Chi; (82) P 326-31; Vol 24; No 4

03-82-0014 (0014)
ANALYSIS OF NEUTRAL SUGAR HYDROLYZATES OF FORAGE CELL WALLS BY HIGH-PRESSURE LIQUID CHROMATOGRAPHY
Barton, Franklin E., II; Windham, William R.; Himmelsbach, David S.
Field Crops Res. Richard B. Russell Agric. Res. Cent. Athens USA US GA 30613
J. Agric. Food Chem.; (82) P 1119-23; Vol 30; No 6

03-82-0015 (0015)
ANALYSIS OF SUGARS IN FOODS CONTAINING SODIUM CHLORIDE BY HIGH-PERFORMANCE LIQUID CHROMATOGRAPHY
Wills, Ron B. H.; Francke, Ray A.; Walker, Brian P.
Sch. Food. Technol. Univ. New South Wales Kensington Australia AU 2033
J. Agric. Food Chem.; (82) P 1242-3; Vol 30; No 6

03-82-0016 (0016)
HIGH-PERFORMANCE LIQUID CHROMATOGRAPHIC SEPARATION WITH TRIPLE-PULSE AMPEROMETRIC DETECTION OF CARBOHYDRATES IN BEVERAGES
Hughes, Scott; Johnson, Dennis C.
Dep. Chem. Iowa State Univ. Ames USA US IA 50011
J. Agric. Food Chem.; (82) P 712-14; Vol 30; No 4; In Eng;

03-82-0017 (0017)
HIGH-PRESSURE LIQUID CHROMATOGRAPHIC DETERMINATION OF MONO- AND DISACCHARIDES IN PRESWEETENED CEREALS: COLLABORATIVE STUDY
Zygmunt, Lucian C.
John Stuart Res. Lab. Quaker Oats Co. Barrington USA US IL 60010
J. Assoc. Off. Anal. Chem.; (82) P 256-64; Vol 65; No 2

03-82-0018 (0018)
AUTOMATED HIGH-PRESSURE LIQUID CHROMATOGRAPHIC SYSTEM FOR DETERMINATION OF MANNITOL, SORBITOL, AND XYLITOL IN CHEWING GUMS AND CONFECTIONS
Samarco, Eugene C.; Parente, Ernest S.
Res. Dev. Dep. Warner-Lambert Co. Morris Plains USA US NJ 07950
J. Assoc. Off. Anal. Chem.; (82) P 76-8; Vol 65; No 1

03-82-0019 (0019)
HIGH-PERFORMANCE LIQUID CHROMATOGRAPHIC SEPARATION OF AMINO SUGARS AND PEPTIDES WITH METAL ION-MODIFIELD MOBILE PHASES
Dua, Virendra K.; Bush, C.Allen
Dep.Chem., Illinois Inst. Technol., Chicago USA, IL 60 616
J. Chromatogr. (82) P 128-33; Vol 244; No 1

03-82-0020 (0020)
SEPARATION OF SORBITOL AND MANNOHEPTULOSE FROM FRUCTOSE, GLUCOSE AND SUCROSE ON REVERSED-PHASE AND AMINE-MODIFIED HPLC COLUMNS
Shaw, Philip E.; Wilson, Charles W., III
U. S. Citrus Subtrop. Prod. Lab. Winter Haven USA US FL 33880
J. Chromatogr. Sci.; (82) P 209-12; Vol 20; No 5

03-82-0021 (0021)
ANALYSIS OF WOOD SUGARS IN PULP AND PAPER INDUSTRY SAMPLES BY HPLC
Wentz, Frank E.; Marcy, A. Dale; Gray, Michael J.
Olympic Res. Div. ITT Rayonier Inc. Shelton USA US WA 98584
J. Chromatogr. Sci.; (82) P 349-52; Vol 20; No 8

03-82-0022 (0022)
SEPARATION OF PINITOL AND SOME OTHER CYCLITOLS BY HIGH-PERFORMANCE LIQUID CHROMATOGRAPHY
Ghias-Ud-Din, M.; Smith, A. E.; Phillips, D. V.
Coll. Agric. Exp. Stns. Univ. Georgia Experiment USA GA 30212
J. Chromatogr.; (81) P 295-8; Vol 211; No 2

03-82-0023 (0023)
SEPARATION OF CARBOHYDRATES AND POLYOLS BY A RADIALLY COMPRESSED HIGH-PERFORMANCE LIQUID CHROMATOGRAPHIC SILICA COLUMN MODIFIED WITH TETRAETHYLENEPENTAMINE
Hendrix, D. L.; Lee, R. E.; Baust, J. G.; James, H.
Dep. Biol. Univ. Houston Houston USA TX 77004
J. Chromatogr.; (81) P 45-53; Vol 210; No 1

03-82-0024 (0024)
SEPARATION AND DETERMINATION OF PHENYL ISOCYANATE-DERIVATIZED CARBOHYDRATES AND SUGAR ALCOHOLS BY HIGH-PERFORMANCE LIQUID CHROMATOGRAPHY WITH ULTRAVIOLET DETECTION
Bjorkqvist, Borje
Espoo Res. Cent. Kemira Oy Espoo Finland FI SF-002271/27
J. Chromatogr.; (81) P 65-71; Vol 218

03-82-0025 (0025)
DETERMINATION OF DISULFATED DISACCHARIDE RESIDUES IN CHONDROITIN SULFATE, USING HIGH-PERFORMANCE LIQUID CHROMATOGRAPHY

Hjerpe, A.; Antonopoulos, C. A.; Engfeldt, B.; Nurminen, M.
Karolinska Inst. Huddinge Univ. Hosp. Huddinge Swed. SE S-141 86
J. Chromatogr.; (82) P 193-5; Vol 242; No 1

03-82-0026 (0026)
ANALYTICAL AND SEMI-PREPARATIVE HIGH-PERFORMANCE LIQUID CHROMATOGRAPHY OF OLIGOSACCHARIDES OBTAINED BY HYDRAZINOLYSIS OF HEN OVOMUCOID
Parente, Jose Paz; Strecker, Gerard; Leroy, Yves; Montreuil, Jean; Fournet, Bernard
Lab. Chim. Biol. Uviv. Sci. Tech. Lille I Villeneuve d' Ascq Fr. FR 59655
J. Chromatogr.; (82) P 199-204; Vol 249; No 1

03-82-0027 (0027)
PHOTOOXYGENATION-CHEMILUMINESCENCE HIGH-PERFORMANCE LIQUID CHROMATOGRAPHIC DETECTOR FOR THE DETERMINATION OF ALIPHATIC ALCOHOLS, ALDEHYDES, ETHERS AND SACCHARIDES
Gandelman, Mitchell S.; Birks, John W.
Dep. Chem. Univ. Colorado Boulder USA US Co 80309
J. Chromatogr.; (82) P 21-31; Vol 242; No 1

03-82-0028 (0028)
SEPARATION AND DETERMINATION OF L-FUCOSE AND D-XYLOSE BY HIGH-PRESSURE LIQUID CHROMATOGRAPHY. USE IN THE ANALYSIS OF THE SULFURYL POLYSACCHARIDES OF BROWN ALGAE
Kloareg, B.
Cent. Etudes Oceanogr. Biol. Mar. CNRS Roscoff Fr. FR 29211
J. Chromatogr.; (82) P 217-23; Vol 236; No 1; In Fr

03-82-0029 (0029)
RAPID DETERMINATION OF SUGARS IN CANTALOUPE MELON JUICE BY HIGH-PERFORMANCE LIQUID CHROMATOGRAPHY
Wade, N. L.; Morris, S. C.
New South Wales Dep. Agric. CSIRO North Ryde Australia AU 2113
J. Chromatogr.; (82) P 257-61; Vol 240; No 1

03-82-0030 (0030)
FLUORESCENCE HIGH-PERFORMANCE LIQUID CHROMATOGRAPHY OF REDUCING SUGARS USING DNS-HYDRAZINE AS A PRELABELING REAGENT
Takeda, Motoko; Maeda, Masako; Tsuji, Akio
Sch. Pharm. Sci. Showa Univ. Tokyo Japan JP
J. Chromatogr.; (82) P 347-55; Vol 244; No 2

03-82-0031 (0031)
HIGH-PERFORMANCE LIQUID CHROMATOGRAPHIC PROCEDURE FOR SEPARATION AND QUANTIFICATION OF ZEATIN AND ZEATIN RIBOSIDE FROM PEARS, PEACHES AND APPLES
Stahly, E. A.; Buchanan, D. A.
Tree Fruit Res. Lab. USDA Wenatchee USA US WA 98801
J. Chromatogr.; (82) P 453-9; Vol 235; No 2

03-82-0032 (0032)
CONTRIBUTION TO THE ELUCIDATION OF THE MECHANISM OF SUGAR RETENTION ON AMINE-MODIFIED SILICA IN LIQUID CHROMATOGRAPHY
Verhaar, L. A. Th.; Kuster, B. F. M.
Dep. Chem. Eng. Eindhoven Univ. Technol. Eindhoven Neth. NL 5600 MB
J. Chromatogr.; (82) P 57-64; Vol 234; No 1

03-82-0033 (0033)
SEPARATION OF CARBOHYDRATES IN DAIRY PRODUCTS BY HIGH PERFORMANCE LIQUID CHROMATOGRAPHY
Richmond, M. L.; Barfuss, D. L.; Harte, B. R.; Gray, J. I.; Stine, C. M.

Dep. Food Sci. Hum. Nutr. Michigan State Univ. East Lansing USA US MI 48824
J. Dairy Sci.; (82) P 1394-400; Vol 65; No 8

03-82-0034 (0034)
THE RAPID DETERMINATION OF NEUTRAL SUGARS IN BIOLOGICAL SAMPLES BY HIGH-PERFORMANCE LIQUID CHROMATOGRAPHY
Alpenfels, William F.; Mathews, Roger A.; Madden, D. Elaine; Newsom, A. E.
Dep. Biochem. Redken Lab., Inc. Canoga Park USA US CA 91303
J. Liq. Chromatogr.; (82) P 1711-23; Vol 5; No 9

03-82-0035 (0035)
RAPID HPLC METHODS FOR THE SEPARATION AND QUANTITATION OF A MONO-, DI-, AND TRISACCHARIDE MIXTURES AND APPLICATIONS
Ghernati, H. M.; Abdeddaim, K.; Guermouche, M. H.
Inst. Chem. USTHB Dar-El-Beida Algeria DZ
J. Liq. Chromatogr.; (82) P 1725-48; Vol 5; No 9

03-82-0036 (0036)
SULFATED OLIGOSACCHARIDES FROM .KAPPA.-CARRAGEENAN AND OLIGOGALACTURONIC ACIDS SEPARATION BY HPLC
Heyraud, Alain; Rochas, Cyrille
Cent. Rech. Macromol. Veg. Univ. Sci. Med. Grenoble Grenoble Fr. FR 38041
J. Liq. Chromatogr.; (82) P 403-12; Vol 5; No 3

03-82-0037 (0037)
PARTIAL METHYLATION OF METHYL .ALPHA.-D-GLUCOPYRANOSIDE. I. LIQUID CHROMATOGRAPHY OF METHYL .ALPHA.-D-GLUCOPYRANOSIDE TRI-o-METHYL ETHER ACETATES
Evtushenko, E. V.; Ovodov, Yu. S.
Tikhookean. Inst. Bioorg. Khim. Vladivostok USSR SU
Khim. Prir. Soedin.; (82) P 18-21; No 1; In Russ

03-82-0038 (0038)
RESOLUTION OF ACETYLATED OLIGOSACCHARIDES BY REVERSE-PHASE HIGH-PRESSURE LIQUID CHROMATOGRAPHY
Wells, Gerald B.; Kontoyiannidou, Voula; Turco, Salvatore J.; Lester, Robert L.
Dep. Biochem. Univ. KentuckyLexington USA US KY 40536
Methods Enzymol.; (82) P 132-7; Vol 83; No Comp, Carbohydr., Pt. D

03-82-0039 (0039)
RESOLUTION OF DOLICHYLPYROPHOSPHORYL OLIGOSACCHARIDES BY HIGH-PRESSURE LIQUID CHROMATOGRAPHY
Wells, Gerald B.; Turco, Salvatore J.; Hanson, Barbara A.; Lester, Robert L.
Dep. Biochem. Univ. Kentucky Lexington USA US KY 40536
Methods Enzymol; (82) P 137-9; Vol 83; No Complex Carbohydr., Pt. D

03-82-0040 (0040)
SEPARATION AND DETERMINATION OF FREE SUGARS IN OILSEEDS BY HIGH-PERFORMANCE LIQUID CHROMATOGRAPHY
Yamauchi, Ryo; Kato, Koji; Ueno, Yoshimitsu
Dep. Agric. Chem. Gifu Univ. Gifu Japan JP 504
Nippon Nogei Kagaku Kaishi; (82) P 123-5; Vol 56; No 2; In Japan

03-82-0041 (0041)
GC AND HPLC SEPARATION OF CARBOHYDRATES IN CELLULOSE-CONTAINING RAW MATERIALS
Esterbauer, Hermann; Hayn, Marianne; Tuisel, Helfried
Inst. Biochem. Univ. Graz Graz Austria AT A-8010
Papier (Darmstadt); (82) P 589-95; Vol 36; No 12; In Ger

03-82-0042 (0042)
HIGH PERFORMANCE LIQUID CHROMATOGRAPHIC DETERMINATION OF SACCHARIDES IN HONEYS
Fan, Chongyang; Liu, Dexing; Pan, Kunyong; Yi, Qian; Song, Jianping
Wuhan Merch. Control Wuhan Peop. Rep. China CN
Shipin Yu Faxiao Gongye; (82) P 7-11; No 3; In Ch

03-82-0043 (0043)
HIGH-PERFORMANCE LIQUID CHROMATOGRAPHY IN THE ANALYSIS OF SUGAR
Ivie, Kert F.
Waters Assoc. Milford USA US MA
Sugar Azucar; (82) P 80-1, 86-7; Vol 77; No 3; In Span

03-82-0044 (0044)
HIGH-PERFORMANCE LIQUID CHROMATOGRAPHY IN SUGAR ANALYSIS
Ivie, Kert F.
Waters Associates Milford USA US MA
Sugar Azucar; (82) P 44, 48, 52-3; Vol 77; No 2

CHAPTER 7.4
Biogenic Amines

--- 1982 ---

04-82-0001 (0001)
AUTOMATIC ANALYSIS OF BIOGENIC CATECHOLAMINES BY HIGH-PERFORMANCE LIQUID CHROMATOGRAPHY WITH FLUOROMETRIC REACTION DETECTION
Mori, K.
Natl. Inst. Ind. Health Kawasaki Japan JP 213
Adv. Biosci.; (82) P 275-84; Vol 36; No Synth., Storage Secretion Adrenal Catecholamines

04-82-0002 (0002)
THE HPLC ANALYSIS OF ETHYL VANILLIN IN CHOCOLATE PRODUCTS
Hurst, W. J.; Martin, R. A., Jr.
Tech. Cent. Hershey Foods Corp. USA US PA
Am. Lab. (Fairfield, Conn.); (82) P 74-5; Vol 14; No 8

04-82-0003 (0003)
ROUTINE ASSAY OF CATECHOLAMINES BY HIGH-PERFORMANCE LIQUID CHROMATOGRAPHY WITH FLUOROMETRIC DETECTION
Clough, G.
Biochem. Dep. Gen. Hosp. Harrogate UK GB HG2 7ND
Anal. Proc. (London); (82) P 183-6; Vol 19; No 4

04-82-0004 (0004)
BIOGENIC AMINE RESOLUTION IN TISSUE EXTRACTS OF RAT BRAIN USING ION-PAIR HIGH-PERFORMANCE LIQUID CHROMATOGRAPHY WITH ELECTROCHEMICAL DETECTION
Pleece, S. A.; Redfern, P. H; Riley, C. M.; Tomlinson, E.
Sch. Pharm. Pharmacol. Univ. Bath Bath UK GB BA2 7AY
Analyst (London); (82) P 755-60; Vol 107; No 1276

04-82-0005 (0005)
SIMULTANEOUS DETERMINATION OF THE THREE MAJOR MONOAMINE METABOLITES IN CSF BY HPLC WITH ELECTROCHEMICAL DETECTION
Van Bockstaele, M.; Dillen, L.; Claeys, M.; De Potter, W. P.
Dep. Med. Univ. Instelling Antwerpen Wilrijk Belg. BE B-2610
Arch. Int. Pharmacodyn. Ther.; (82) P 304-5; Vol 260; No 2

04-82-0006 (0006)
HPLC DETERMINATION OF PLASMA CATECHOLAMINES IN ROAD ACCIDENT CASUALTIES
Davies, C. L.; Molyneux, S. G.; Newman, R. J.
Oxford Road Accid. Group MRC Unit Oxford UK GB OX2 6HE
Br. J. Clin. Pharmacol.; (82) P 283P; Vol 13; No 2

04-82-0007 (0007)
A RADIOENZYMIC LIQUID CHROMATOGRAPHIC DETERMINATION OF PLASMA NORADRENALINE, ADRENALINE, AND DOPAMINE
Endert, Erik; Punt, Rene
Dep. Med. Univ. Hosp. Binnengasthuis Grimburgwal Amsterdam Neth. NL
Chromatogr. Sci.; (82) P 243-55; Vol 20; No Biol./Biomed. Appl. Liq. Chromatogr. 4

04-82-0008 (0008)
THE LC ANALYSIS OF CATECHOLAMINE METABOLITES IN URINE
Muusze, Ruud G.
St. Joris Gasthuis Delft Neth. NL
Chromatogr. Sci.; (82) P 257-78; Vol 20; No Biol./Biomed. Appl. Liq. Chromatogr. 4

04-82-0009 (0009)
SIMULTANEOUS FLUOROMETRY OF URINARY DOPAMINE, NOREPINEPHRINE, AND EPINEPHRINE

COMPARED WITH LIQUID CHROMATOGRAPHY WITH ELECTROCHEMICAL DETECTION
Oka, Kazuhiro; Sekiya, Motoko; Osada, Hirofumi; Fujita, Keisuke; Kato, Takeshi; Nagatsu, Toshiharu
Grad. Sch. Nagatsuta Tokyo Inst. Technol. Yokohama Japan JP 227
Clin. Chem. (Winston-Salem, N. C.); (82) P 646-9; Vol 28; No 4, Pt. 1

04-82-0010 (0010)
A SIMPLE METHOD FOR THE ASSAY OF URINARY METANEPHRINES USING HIGH PERFORMANCE LIQUID CHROMATOGRAPHY WITH FLUORESCENCE DETECTION
Jackman, Graham P.
Clin. Res. Unit. Alfred Hosp. Prahran Australia AU 3181
Clin. Chim. Acta; (82) P 137-42; Vol 120; No 1

04-82-0011 (0011)
LIQUID-LIQUID EXTRACTION SYSTEMS FOR THE ISOLATION OF CATECHOLAMINES FROM SERUM AND URINE FOR HPLC ANALYSIS
Smedes, Foppe; Kraak, Johan C.; Poppe, Hans
Lab. Anal. Chem. Univ. Amsterdam Amsterdam Neth. NL 1018 WV
CONFERENCE; Proc. - Int. Congr. Clin. Chem., 11th ; (82) P 977-81; Avail.: Kaiser, Erich; Gabl, Franz; Mueller, Mathias M; Date: 81; Publisher: de Gruyter, Berlin, Fed. Rep. Ger.

04-82-0012 (0012)
SIMULTANEOUS DETERMINATION OF 5-HYDROXYTRYPTOPHAN, SEROTONIN AND 5-HYDROXYINDOLEACETIC ACID IN THE BRAIN BY LIQUID CHROMATOGRAPHY WITH ELECTROCHEMICAL DETECTION
Santagostino, G.; Frattini, P.; Schinelli, S.; Cucchi, M. L.; Corona, G. L.
Inst. Pharmacol. Univ. Pavia Pavia Italy IT
Farmaco, Ed. Prat.; (82) P 365-70; Vol 37; No 11

04-82-0013 (0013)
UPTAKE AND TURNOVER OF DOPAMINE IN RAT MAST CELLS STUDIED BY CYTOFLUOROMETRY AND HIGH PERFORMANCE LIQUID CHROMATOGRAPHY
Rundquist, I.; Allenmark, S.; Enerbaeck, L.
Dep. Pathol. II Univ. Linkoeping Linkoeping Swed. SE S-581 85
Histochem. J.; (82) P 429-43; Vol 14; No 3

04-82-0014 (0014)
MEASUREMENT OF ADRENALINE IN THE RAT BRAIN BY HIGH-PERFORMANCE LIQUID CHROMATOGRAPHY WITH FLUOROMETRIC DETECTION
Sudo, Ayako
Natl. Inst. Ind. Health Kawasaki Japan JP 213
Ind. Health; (82) P 151-6; Vol 20; No 2

04-82-0015 (0015)
DETERMINATION OF L-DOPA IN BRAIN TISSUE USING HIGH-PERFORMANCE LIQUID CHROMATOGRAPHY WITH ELECTROCHEMICAL DETECTION TO STUDY THE ACTIVITY OF CENTRAL DOPAMINERGIC NEURONS
Benfenati, F.; Ferretti, P.; Ferretti, C.; Ponzio, F.; Algeri, S.
Ist. Ric. Farmacol. "Mario Negri" Milan Italy IT 20157
IRCS Med. Sci.: Libr. Compend.; (82) P 425-6; Vol 10; No 5

04-82-0016 (0016)
INVESTIGATIONS OF CATECHOLAMINE METABOLISM USING HIGH-PERFORMANCE LIQUID CHROMATOGRAPHY. ANALYTICAL METHODOLOGY AND CLINICAL APPLICATIONS
Krstulovic, Ante M.
Dep. Chem. Manhattanville Coll. Purchase USA US NY 10577
J. Chromatogr.; (82) P 1-34; Vol 229; No 1

04-82-0017 (0017)
PRE-COLUMN DERIVATIZATION AND HIGH-PERFORMANCE LIQUID CHROMATOGRAPHY OF BIOGENIC AMINES IN BLOOD OF NORMAL AND MALIGNANT HYPERTHERMIC PIGS
Davis, Thomas P.; Gehrke, Charles W., Jr.; Williams, Charles H.; Gehrke, Charles W.; GerhardtL Klaus O.
Health Sci. Cent. Univ. Arizona Tucson USA US AZ 85724
J. Chromatogr.; (82) P 113-22; Vol 228

04-82-0018 (0018)
SELECTIVE GROUP SEPARATION OF PRIMARY AND SECONDARY BIOGENIC AMINES BY REVERSED-PHASE HIGH-PERFORMANCE LIQUID CHROMATOGRAPHY WITH 18-CROWN-6 IN THE MOBILE PHASE
Wiechmann, Manfred
Univ. Bayreuth Bayreuth Fed. Rep. Ger. DE D-8580
J. Chromatogr.; (82) P 129-37; Vol 235; No 1; In Ger

04-82-0019 (0019)
SIMULTANEOUS DETERMINATION OF NOREPINEPHRINE, DOPAMINE AND SEROTONIN IN RAT BRAIN REGIONS BY ION-PAIR LIQUID CHROMATOGRAPHY ON OCTYL SILANE COLUMNS AND AMPEROMETRIC DETECTION
Warsh, Jerry J.; Chiu, Andrew; Godse, Damodar D.
Dep. Biochem. Psychiatry Clarke Inst. Psychiatry Toronto Can. CA ON M5T 1R8
J. Chromatogr.; (82) P 131-41; Vol 228

04-82-0020 (0020)
DETERMINATION OF CATECHOLAMINES IN RAT HEART TISSUE AND PLASMA SAMPLES BY LIQUID CHROMATOGRAPHY WITH ELECTROCHEMICAL DETECTION
Eriksson, Britt Marie; Persson, Bengt Arne
AB Hassle Molndal Swed. SE S-431 83
J. Chromatogr.; (82) P 143-54; Vol 228

04-82-0021 (0021)
SIMPLE METHOD FOR THE DETERMINATION OF HOMOVANILLIC ACID AND VANILLYLMANDELIC ACID IN URINE BY HIGH-PERFORMANCE LIQUID CHROMATOGRAPHY
Yoshida, Akira; Yoshioka, Masanori; Sakai, Toshio; Tamura, Zenzo
Cent. Res. Lab. Mitsubishi Petrochem. Co. Ibaraki Japan JP 300-03
J. Chromatogr.; (82) P 162-7; Vol 227; No 1

04-82-0022 (0022)
REVERSED-PHASE LIQUID CHROMATOGRAPHY WITH AMPEROMETRIC DETECTION OF LIPOPHILIC DOPAMINE ANALOGS AND DETERMINATION OF BRAIN AND SERUM CONCENTRATIONS AFTER SAMPLE CLEAN-UP ON SMALL SEPHADEX G-10 COLUMNS
Feenstra, Matthijs G. P.; Homan, Jan Willem; Dijkstra, Durk; Mulder, Theo B. A.; Rollema, Hans; Westerink, Ben H. C.; Horn, Alan S.
Dep. Med. Chem. State Univ. Groningen Groningen Neth. NL 9713 AW
J. Chromatogr.; (82) P 271-87; Vol 20; No 2

04-82-0023 (0023)
DETERMINATION OF NOREPINEPHRINE IN BRAIN PERFUSATES USING HIGH-PERFORMANCE LIQUID CHROMATOGRAPHY WITH ELECTROCHEMICAL DETECTION
Salzman, Steven K.; Sellers, Micpael S.
Alfred I. du Pont Inst. Nemours Found. Wilmington USA US DE 19899
J. Chromatogr.; (82) P 29-37; Vol 232; No 1

04-82-0024 (0024)
LIQUID CHROMATOGRAPHIC SEPARATION AND DETECTION OF EPINEPHRINE AND ITS OXIDATION PRODUCTS, ADRENOCHROME AND ADRENOLUTIN
Juenge, Eric C.; Flinn, Paul E.; Furman, William B.
Natl. Cent. Drug Anal. Food Drug Adm. St. Louis USA US MO 63101
J. Chromatogr.; (82) P 297-302; Vol 248; No 2

04-82-0025 (0025)
MEASUREMENT OF CATECHOLAMINES IN BIOLOGICAL FLUIDS BY HIGH-PERFORMANCE LIQUID CHROMATOGRAPHY. A COMPARISON OF FLUORIMETRIC WITH ELECTROCHEMICAL DETECTION
Causon, Roger C.; Carruthers, Malcolm E.
Dep. Biochem. Inst. Psychiatry London UK GB
J. Chromatogr.; (82) P 301-9; Vol 229; No 2

04-82-0026 (0026)
DUAL ELECTROCHEMICAL DETECTOR FOR MICRO-HIGH-PERFORMANCE LIQUID CHROMATOGRAPHY AND ITS APPLICATION TO THE SELECTIVE DETECTION OF CATECHOLAMINES
Goto, Masashi; Sakurai, Eizo; Ishii, Daido
Fac. Eng. Nagoya Univ. Nagoya Japan JP 464
J. Chromatogr.; (82) P 357-66; Vol 238; No 2

04-82-0027 (0027)
SIMULTANEOUS DETERMINATION OF 4-HYDROXY-3-METHOXYPHENYLACETIC (HOMOVANILLIC) ACID AND OTHER MONOAMINE METABOLITES IN HUMAN LUMBAR CEREBROSPINAL FLUID. AN IMPROVED HIGH-PERFORMANCE LIQUID CHROMATOGRAPHIC STUDY WITH ELECTROCHEMICAL DETECTION
Krstulovic, Ante M.; Bertani-Dziedzic, Laura; Bautista-Cerqueira, Sandra; Gitlow, Stanley E.
Dep. Chem. Manhattanville Coll. Purchase USA US NY 10577
J. Chromatogr.; (82) P 379-89; Vol 227; No 2

04-82-0028 (0028)
ANALYSIS OF URINARY 3-METHOXY-4-HYDROXYPHENYLGLYCOL BY HIGH-PERFORMANCE LIQUID CHROMATOGRAPHY AND ELECTROCHEMICAL DETECTION
Moleman, P.; Borstrok, J. J. M.
Dep. Psychiatry Academic Hosp. Dijkzigt Rotterdam Neth. NL 3015 GD
J. Chromatogr.; (82) P 391-405; Vol 227; No 2

04-82-0029 (0029)
ROUTINE DETERMINATION OF PLASMA CATECHOLAMINES USING REVERSED-PHASE, ION-PAIR HIGH-PERFORMANCE LIQUID CHROMATOGRAPHY WITH ELECTROCHEMICAL DETECTION
Davies, Cledwyn L.; Molyneux, Stephen G.
Radcliffe Infirm. Univ. Oxford Oxford UK GB OX2 6HE
J. Chromatogr.; (82) P 41-51; Vol 231; No 1

04-82-0030 (0030)
DETERMINATION OF URINARY VANILMANDELIC ACID AND HOMOVANILLIC ACID BY HIGH PERFORMANCE LIQUID CHROMATOGRAPHY WITH AMPEROMETRIC DETECTION
Bauersfeld, W.; Diener, U.; Knoll, E.; Ratge, D.; Wisser, H.
Dep. Clin. Chem. Robert-Bosch-Krankenhaus Stuttgart Fed. Rep. Ger. DE
J. Clin. Chem. Clin. Biochem.; (82) P 217-20; Vol 20; No 4

04-82-0031 (0031)
PLASMA CATECHOLAMINES IN RESTING RAINBOW TROUT, SALMO GAIRDNERI RICHARDSON, BY HIGH PRESSURE LIQUID CHROMATOGRAPHY
Woodward, J. J.
Fish. Res. Inst. Univ. Washington Seattle USA US WA 98195
J. Fish Biol.; (82) P 429-32; Vol 21; No 4

04-82-0032 (0032)
ANALYSIS OF CATECHOLAMINES BY HPLC
Allenmark, Stig
Clin. Res. Cent. Univ. Linkoeping Linkoeping Swed. SE S-581 85
J. Liq. Chromatogr.; (82) P 1-41; Vol 5; No Suppl. 1

04-82-0033 (0033)
DETERMINATION OF PLASMA NOREPINEPHRINE AND EPINEPHRINE BY HIGH-PDRFORMANCE

LIQUID CHROMATOGRAPHY USING A TWO-COLUMN SYSTEM AND AN ELECTROCHEMICAL DETECTOR
Lachatre, G.; Nicot, G.; Magnet, C.; Rocca, J. L.; Merle, L.; Valette, J. P.
Dep. Pharmacol. Clin. Hop. Dupuytren Limoges Fr. FR 87031
J. Liq. Chromatogr.; (82) P 1947-65; Vol 5; No 10

04-82-0034 (0034)
RAPID AND SIMPLE METHOD FOR MEASURING BIOGENIC AMINES AND METABOLITES IN BRAIN HOMOGENATES BY HPLC-ELECTROCHEMICAL DETECTION
Zaczek, R.; Coyle, J. T.
Sch. Med. Johns Hopkins Univ. Baltimore USA US MD
J. Neural Transm.; (82) P 1-5; Vol 53; No 1

04-82-0035 (0035)
SIMULTANEOUS DETERMINATION OF 3,4-DIHYDROXYPHENYLALANINE, 5-HYDROXYTRYPTOPHAN, DOPAMINE, 4-HYDROXY-3-METHOXYPHENYLALANINE,NOREPINEPHRINE, 3,4-DIHYDROXYPHENYL ACETIC ACID, HOMOVANILLIC ACID, SEROTONIN AND 5-HYDROXYINDOLEACETIC ACID IN RAT CEREBROSPINAL FLUID AND BRAIN BY HIGH-PERFORMANCE LIQUID CHROMATOGRAPHY WITH ELECTROCHEMICAL DETECTION
Wagner, Joseph; Vitali, Pierre; Palfreyman, Michael G.; Zraika, Monique; Huot, Sylvie
Cent. Rech., Merrell Int., Strasbourg France 67 084
J. Neurochem. (82) P 1241-54; Vol 38; No 5

04-82-0036 (0036)
ESTIMATION OF THE TURNOVER OF 3-METHOXYTYRAMINE IN THE RAT STRIATUM BY HPLC WITH ELECTROCHEMICAL DETECTION: IMPLICATIONS FOR THE SEQUENCE IN THE CEREBRAL METABOLISM OF DOPAMINE
Westerink, B. H. C.; Spaan, S. J.
Dep. Med. Chem. Univ. Groningen Groningen Neth. NL
J. Neurochem.; (82) P 342-7; Vol 38; No 2

04-82-0037 (0037)
SIMULTANEOUS DETERMINATION OF SEROTONIN, 5-HYDROXYINDOLEACETIC ACID, 3,4-DIHYDROXYPHENYLACETIC ACID AND HOMOVANILLIC ACID BY HIGH PERFORMANCE LIQUID CHROMATOGRAPHY WITH ELECTROCHEMICAL DETECTION
Sperk, Guenther
Inst. Biochem. Pharmacol. Univ. Vienna Vienna Austria AT
J. Neurochem.; (82) P 840-3; Vol 38; No 3

04-82-0038 (0038)
DETERMINATION OF DOPAMINE AND ITS ACIDIC METABOLITES IN BRAIN TISSUE BY HPLC WITH ELECTROCHEMICAL DETECTION IN A SINGLE RUN AFTER MINIMAL SAMPLE PRETREATMENT
Van Valkenburg, Cees; Tjaden, Ubbo; Van der Krogt, Jacques; Van der Leden, Bram
Dep. Pharmacol. Univ. Leiden Leiden Neth. NL
J. Neurochem.; (82) P 990-7; Vol 39; No 4

04-82-0039 (0039)
THE MEASUREMENT OF PLASMA NORADRENALINE BY HIGH-PERFORMANCE LIQUID CHROMATOGRAPHY WITH ELECTROCHEMICAL DETECTION: AN ASSESSMENT OF SAMPLE STABILITY AND ASSAY REPRODUCIBILITY
Falconer, A. D.; Lake, D.; Macdonald, I. A.
Med. Sch. Univ. Nottingham Nottingham UK GB
J. Neurosci. Methods; (82) P 261-71; Vol 6; No 3

04-82-0040 (0040)
ION-PAIR HIGH-PERFORMANCE LIQUID CHROMATOGRAPHY OF TERBUTALINE AND CATECHOLAMINES WITH AMINOPHYLLINE IN INTRAVENOUS SOLUTIONS

Williams, David A.; Fuvg, Esther Y. Y.; Newton, David W.
Dep. Chem. Massachusetts Coll. Pharm. Allied Health Sci. Boston USA US MA 02115
J. Pharm. Sci.; (82) P 956-8; Vol 71; No 8

04-82-0041 (0041)
USE OF LIQUID CHROMATOGRAPHY WITH ELECTROCHEMISTRY TO MEASURE EFFECTS OF VARYING INTENSITIES OF WHITE LIGHT ON DOPA ACCUMULATION IN RAT RETINAS
Proll, Melissa A.; Kamp, Cylia W.; Morgan, William W.
Health Sci. Cent. Univ. Texas San Antonio USA US TX 78284
Life Sci.; (82) P 11-19; Vol 30; No 1

04-82-0042 (0042)
IMPROVED METHOD FOR DETERMINATION OF CATECHOLAMINES IN RAT BRAIN BY ISOLATION ON BORIC ACID GEL AND HIGH-PERFORMANCE LIQUID CHROMATOGRAPHY WITH ELECTROCHEMICAL DETECTION
Koike, Koji; Aono, Toshihiro; Chatani, Fumio; Takemura, Toshiko; Kurachi, Keiichi
Med. Sch. Osaka Univ. Osaka Japan JP 553
Life Sci.; (82) P 2221-8; Vol 30; No 25

04-82-0043 (0043)
METHOD FOR ANALYSIS OF DOPAMINE SULFATE ISOMERS BY HIGH-PERFORMANCE LIQUID CHROMATOGRAPHY
Elchisak, Mary Ann; Carlson, Joanne H.
Sch. Vet. Med. Purdue Univ. West Lafayette USA US IN 47907
Life Sci.; (82) P 2325-36; Vol 30; No 26

04-82-0044 (0044)
A COMPARISON OF COULOMETRIC DETECTORS FOR CATECHOLAMINE ANALYSIS BY LC-EC
Bunyagidj, Chaiyod; Girard, James E.
Dep. Chem. American Univ. Washington USA US DC
Life Sci.; (82) P 2627-34; Vol 31; No 23

04-82-0045 (0045)
EFFECTS OF NORADRENERGIC NEURONAL ACTIVITY ON 3,4-DIHYDROXYPHENYLETHYLENE GLYCOL (DHPG) LEVELS. QUANTITATION BY HIGH-PERFORMANCE LIQUID CHROMATOGRAPHY
Jackman, G.; Snell, J.; Skews, H.; Bobik, A.
Clin. Res. Unit Alfred Hosp. Prahran Australia AU 3181
Life Sci.; (82) P 923-9; Vol 31; No 9

04-82-0046 (0046)
A RAPID AND SIMPLE METHOD TO DETERMINE THE SPECIFIC ACTIVITIES OF SEROTONIN, 5-HYDROXYINDOLEACETIC ACID, AND 5-HYDROXYTRYPTOPHAN IN BRAIN BY HPLC WITH ELECTROCHEMICAL DETECTION
Bernstein, M. J.; Shea, P. A.
Sch. Med. Indiana Univ. Indianapolis USA US IN 46223
Neurochem. Res.; (82) P 79-85; Vol 7; No 1

04-82-0047 (0047)
SIMULTANEOUS DETERMINATION OF PICOGRAM LEVELS OF 5HTP, 5HT AND 5HIAA IN DESCRETE BRAIN AREAS OF RAT USING HPLC WITH ELECTROCHEMICAL DETECTION
Oomagari, Kiyoshi; Uchimura, Hideyuki; Matsumoto, Takashi; Matsuo, Hideyasu; Saito, Masashi; Hirano, Makoto; Kin, Choju; Shimamura, Manabu
Dep. Psychiatry Saga Med. Sch. Saga Japan JP
Neurosciences (Kobe, Jpn.); (82) P 64-5; Vol 8; No 1; In Japan

04-82-0048 (0048)
AMINE METABOLITES IN THE VENTRICULAR CEREBROSPINAL FLUID OF PATIENTS WITH VENTRICULAR DILATION. HPLC AND CLINICAL APPLICATION. IV

Ikeda, Yukio; Nakazawa, Shozo
Dep. Neurosurg. Nippon Med. Sch. Tokyo Japan JP 113
Nippon Ika Daigaku Zasshi; (82) P 848-50; Vol 49; No 6

04-82-0049 (0049)
A CRITICAL EVALUATION OF LC-EC FOR THE SIMULTANEOUS DETERMINATION OF 5-HYDROXYTRYPTOPHAN, 5-HT, 5-HIAA AND HVA IN BIOLOGICAL SAMPLES USING GC-MS FOR VALIDATION
Narasimhachari, N.; Boadle-Biber, M. C.; Friede, R. O.
Dep. Psychiatry Med. Coll. Virginia Richmond USA US VA 23298
Res. Commun. Chem. Pathol. Pharmacol.; (82) P 413-30; Vol 37; No 3

04-82-0050 (0050)
DETERMINATION OF METHIONINE-ENKEPHALIN, NOREPINEPHRINE, DOPAMINE, 3,4-DIHYDROXYPHENYLACETIC ACID (DOPAC) AND 3-METHOXY-4-HYDROXYPHENYLACETIC ACID (HVA) IN BRAIN BY HIGH-PRESSURE LIQUID CHROMATOGRAPHY WITH ELECTROCHEMICAL DETECTOR
Nabeshima, Toshitaka; Hiramatsu, Masayuki; Noma, Shigeki; Ukai, Makoto; Amano, Manabu; Kameyama, Tsutomu
Fac. Pharm. Sci. Meijo Univ. Nagoya Japan JP 468
Res. Commun. Chem. Pathol. Pharmacol.; (82) P 421-42; Vol 35; No 3

04-82-0051 (0051)
DIRECT DETERMINATION OF URINARY CATECHOLAMINES BY HIGH-PERFORMANCE LIQUID CHROMATOGRAPHY WITH BORIC ACID COLUMN
Takeshita, Eiko; Noguchi, Ayako; Uchida, Teruko; Matsumoto, Kojiro ; Kano, Shojiro
Div. Endocrinol., Lab. Med. Keio Univ. Hosp. Tokyo Japan JP 160
Rinsho Byori; (82) P 450-4; Vol 30; No 4; In Japan

04-82-0052 (0052)
HPLC DETERMINATION OF VANILLIN IN BUTTER SAMPLES FOR USE IN PASTRY PRODUCTS
Cortesi, N.; Colombo, D.
Staz. Sper. Ind. Oli E Dei Grassi Milan Italy IT
Riv. Ital. Sostanze Grasse; (82) P 249-50; Vol 59; No 5; In Ital

04-82-0053 (0053)
AMINO ACID ANALYZER AND HIGH-PRESSURE LIQUID CHROMATOGRAPHIC DETERMINATION OF AMINO ACIDS AND BIOGENIC AMINES IN THE LIGHT AND DARK MUSCLE MEAT OF TUNA
Stockemer, J.
Staat. Veterinaeruntersuchungsamt Fische Fischwaren Cuxhaven Fed. Rep. Ger. DE D-2190
Z. Lebensm.-Unters. Forsch.; (82) P 108-13; Vol 174; No 2; In Ger

CHAPTER 7.5
Environmental Analyses

7.5.1
Air

--- 1982 ---

05-82-0001 (0001)
DETERMINATION OF ENVIRONMENTAL PHENOLS BY LIQUID CHROMATOGRAPHY/ ELECTRO-CHEMISTRY
Shoup, Ronald E.; Mayer, Ginny S.
Res. Lab. Bioanal. Syst. Inc. West Lafayette USA US IN 47906
Anal. Chem.; (82) P 1164-9; Vol 54; No 7

05-82-0002 (0002)
ISOLATION OF POLYCYCLIC AROMATIC HYDROCARBONS IN AIR PARTICULATE MATTER BY LIQUID CHROMATOGRAPHY
Romanowski, T.; Funcke, W.; Koenig, J.; Balfanz, E.
Fraunhofer-Inst. Toxikol. Aerosolforsch. Muenster-Roxel Fed. Rep. Ger. DE D-4400
Anal. Chem.; (82) P 1285-7; Vol 54; No 8

05-82-0003 (0003)
AIR SAMPLING AND LIQUID CHROMATOGRAPHIC DETERMINATION OF ETHYLENIMINE
Morales, Raul; Stampfer, J. F.; Hermes, R. E.
Los Alamos Natl. Lab. Univ. California Los Alamos USA US NM 87545
Anal. Chem.; (82) P 1340-4; Vol 54; No 8

05-82-0004 (0004)
LIQUID CHROMATOGRAPHIC DETERMINATION OF DIMETHYL FORMAMIDE, METHYLENE BISPHENYL ISOCYANATE AND METHYLENE BISPHENYL AMINE IN AIR SAMPLES
Lipski, Krzysztof
Inst. Occup. Med. Lodz Pol. PL 90-950
Ann. Occup. Hyg.; (82) P 1-4; Vol 25; No 1

05-82-0005 (0005)
RAPID SCAN UV/VISIBLE DETECTION IN LIQUID CHROMATOGRAPHY . APPLICATION TO ENVIRONMENTAL ANALYSIS
Warwick, C. J.; Bagon, D. A.
Occup. Med. Hyg. Lab. London UK GB NW2 6LN
Chromatographia; (82) P 433-6; Vol 15; No 7

05-82-0006 (0006)
METHOD FOR DETERMINATION OF FORMALDEHYDE IN AIR IN THE PPT-RANGE BY HPLC AFTER EXTRACTION AS 2,4-DINITROPHENYLHYDRAZONE
Tuss, H.; Neitzert, V.; Seiler, W.; Neeb, R.
Max-Planck-Inst. Chem. Mainz Fed. Rep. Ger. DE D-6500
Fresenius' Z. Anal. Chem.; (82) P 613-17; Vol 312; No 7

05-82-0007 (0007)
A COMPARISON OF THREE DIFFERENT HIGH PERFORMANCE LIQUID CHROMATOGRAPHY SYSTEMS FOR THE DETERMINATION OF ALDEHYDES AND KETONES IN DIESEL EXHAUST. PART I
Creech, G.; Johnson, R. T.; Stof6er, J. O.
Univ. Missouri Rolla USA US MO 65401
J. Chromatogr. Sci.; (82) P 67-72; Vol 20; No 2

05-82-0008 (0008)
OFF-LINE COMBINATION OF LIQUID CHROMATOGRAPHY AND FIELD DESORPTION MASS SPECTROMETRY: PRINCIPLES AND ENVIRONMENTAL, MEDICAL, AND PHARMACEUTICAL APPLICATIONS
Schulten, H. R.
Inst. Chem. Univ. Bonn Bonn Fed. Rep. Ger. DE 5300/1
J. Chromatogr.; (82) P 105-28; Vol 251; No 2

05-82-0009 (0009)
DETERMINATION OF DIMETHYL SULFATE IN AIR BY REVERSED-PHASE LIQUID CHROMATOGRAPHY
Williams, R. G.
Upjohn Co. Kalamazoo USA US MI 49001
J. Chromatogr.; (82) P 381-4; Vol 245; No 3

05-82-0010 (0010)
DETERMINATION OF 2,4,7-TRINITRO-9-FLUORENONE IN WORKPLACE ENVIRONMENTAL SAMPLES USING HIGH-PERFORMANCE LIQUID CHROMATOGRAPHY
Seymour, Martha J.
Div. Phys. Sci. Eng. Natl. Inst. Occup. Saf. Health Cincinnati USA US OH 45226
J. Chromatogr.; (82) P 530-4; Vol 236; No 2

05-82-0011 (0011)
HIGH-PERFORMANCE SEMIPREPARATIVE LIQUID CHROMATOGRAPHY OF DIESEL ENGINE EMISSION PARTICULATE EXTRACTS
Levine, S. P.; Skewes, L. M.
Anal. Sci. Dep. Ford Motor Co. Dearborn USA US MI 48121
J. Chromatogr.; (82) P 532-5; Vol 235; No 2

05-82-0012 (0012)
HIGH-PRESSURE LIQUID CHROMATOGRAPHIC METHOD FOR ROUTINE ANALYSIS OF POLYCYCLIC AROMATIC HYDROCARBONS IN AIRBORNE PARTICULATES
Suga, Kuniko; Ishiguro, Tatsukichi; Matsushita, Hidetsuru
Tokyo Metrop. Res. Inst. Environ. Prot. Tokyo Japan JP 100
Taiki Osen Gakkaishi; (82) P 117-25; Vol 17; No 2; In Japan

7.5.2
Water

--- 1982 ---

05-82-0001 (0001)
DETERMINATION OF ENVIRONMENTAL PHENOLS BY LIQUID CHROMATOGRAPHY/ ELECTROCHEMISTRY
Shoup, Ronald E.; Mayer, Ginny S.
Res. Lab. Bioanal. Syst. Inc. West Lafayette USA US IN 47906
Anal. Chem.; (82) P 1164-9; Vol 54; No 7

05-82-0002 (0002)
DETERMINATION OF TRACE LEVELS OF QUATERNARY AMMONIUM COMPOUNDS IN RIVER WATER BY LIQUID CHROMATOGRAPHY WITH CONDUCTOMETRIC DETECTION
Wee, Victorio T.; Kennedy, Julie M.
Ivorydale Tech. Cent. Procter and Gamble Co. Cincinnati USA US OH 45217
Anal. Chem.; (82) P 1631-3; Vol 54; No 9

05-82-0003 (0003)
DETERMINATION OF METALS IN GROUNDWATERS BY TRACE ENRICHMENT AND LIQUID CHROMATOGRAPHY
Cassidy, R. M.; Elchuk, Steven; McHugh, J. O.
Environ. Res. Branch At. Energy Canada Ltd. Chalk River Can. CA ON K0J 1J0
Anal. Chem.; (82) P 727-31; Vol 54; No 4

05-82-0004 (0004)
TRACE ENRICHMENT OF DIFENZOQUAT IN WATER AND ITS DETERMINATION BY REVERSED-PHASE HIGH-PERFORMANCE LIQUID CHROMATOGRAPHY
Ahmad, Ijaz
Alberta Environ. Cent. Vegreville Can. CA AB T0B 4L0
Anal. Lett.; (82) P 27-37; Vol 15; No A1

05-82-0005 (0005)
DETERMINATION OF ORGANIC CONTAMINANTS IN ULTRA-PURE WATER BY REVERSED-PHASE HIGH-PERFORMANCE LIQUID CHROMATOGRAPHY WITH ULTRAVIOLET DETECTION
Reust, Juerg B.; Meyer, Veronika R.
Inst. Anorg. Anal. Phys. Chem. Univ. Berne Berne Switz. CH CH-3012
Analyst (London); (82) P 673-9; Vol 107; No 1275

05-82-0006 (0006)
RAPID SCREENING TECHNIQUE UTILIZING HIGH-PERFORMANCE LIQUID CHROMATOGRAPHY FOR ASSESSING ACRYLAMIDE CONTAMINATION IN EFFLUENTS
Brown, Leslie; Rhead, Michael M.; Bancroft, Keith C. C.
Dep. Environ. Sci. Plymouth Polytech. Plymouth UK GB PL4 8AA
Analyst (London); (82) P 749-54; Vol 107; No 1276

05-82-0007 (0007)
HIGH-PERFORMANCE LIQUID CHROMATOGRAPHY USING POLYMER GELS FOR DIRECT ANALYSES AND CHARACTERIZATIONS OF SOLUBLE ORGANIC POLLUTANTS IN WATER
Urano, Kohei; Furyune, Hiroyuki; Hayashi, Koji
Dep. Saf. Environ. Eng. Yokohama Natl. Univ. Yokohama Japan JP 240
Bull. Chem. Soc. Jpn.; (82) P 2248-53; Vol 55; No 7

05-82-0008 (0008)
DETERMINATION OF ALDICARB RESIDUES IN WATER BY COMBINED HIGH PERFORMANCE LIQUID CHROMATOGRAPHY/MASS SPECTROMETRY
Wright, L. H.; Jackson, M. D.; Lewis, R. G.
EPA Research Triangle Park USA US NC 27711
Bull. Environ. Contam. Toxicol.; (82) P 740-7; Vol 28; No 6

05-82-0009 (0009)
DETERMINATION OF ALKYLBENZENESULFONATE IN WATER BY HIGH-PERFORMANCE LIQUID

CHROMATOGRAPHY
Matsueda, Takahiko; Osaki, Yasuhiko; Shigee, Shinya
Fukuoka Environ. Res. Cent. Fukuoka Japan JP
Bunseki Kagaku; (82) P 59-63; Vol 31; No 2; In Japan

05-82-0010 (0010)
SIMULTANEOUS DETERMINATION OF HEAVY METALS IN WATER BY HIGH PERFORMANCE LIQUID CHROMATOGRAPHY FOLLOWING THE SOLVENT EXTRACTION OF HEAVY METALS AS PYRROLIDINEDITHIOCARBAMATE CHELATES INTO CHLOROFORM
Ichinoki, Susumu; Yamazaki, Mitsuru
Sch. Pharm. Hokuriku Univ. Kanazawa Japan JP 920-11
Bunseki Kagaku; (82) P E319-E326; Vol 31; No 10

05-82-0011 (0011)
A RAPID HIGH-PERFORMANCE LIQUID CHROMATOGRAPHIC METHOD FOR SIMULTANEOUSLY DETERMINING THE CONCENTRATIONS OF TFM AND BAYER 73 IN WATER DURING LAMPRICIDE TREATMENTS
Dawson, V. K.
U. S. Fish Wildlife Serv. La Crosse USA US WI 54601
Can. J. Fish. Aquat. Sci.; (82) P 778-82; Vol 39; No 5

05-82-0012 (0012)
A HIGH-PERFORMANCE LIQUID CHROMATOGRAPHIC DETERMINATION OF FREE AMINO ACIDS IN NATURAL WATERS IN THE PICOMOLAR (M .TIMES. 10-12) RANGE SUITABLE FOR SHIPBOARD USE
Evens, Roger; Braven, James; Brown, Leslie; Butler, Ian
Dep. Environ. Sci. Plymouth Polytech. Plymouth/Devon UK GB
Chem. Ecol.; (82) P 99-106; Vol 1; No 2

05-82-0013 (0013)
HIGH-PERFORMANCE LIQUID CHROMATOGRAPHIC AND MASS SPECTROMETRIC ANALYSES OF PORPHYRINS FROM DEEP-SEA SEDIMENTS
Quirke, J. M. E.; Eglinton, G.; Palmer, S. E.; Baker, E. W.
Org. Geochem. Unit Univ. Bristol Bristol UK GB BS8 1TS
Chem. Geol.; (82) P 69-85; Vol 35; No 1-2

05-82-0014 (0014)
RAPID SCAN UV/VISIBLE DETECTION IN LIQUID CHROMATOGRAPHY . APPLICATION TO ENVIRONMENTAL ANALYSIS
Warwick, C. J.; Bagon, D. A.
Occup. Med. Hyg. Lab. London UK GB NW2 6LN
Chromatographia; (82) P 433-6; Vol 15; No 7

05-82-0015 (0015)
FINGERPRINTING INORGANIC ARSENIC AND ORGANOARSENIC COMPOUNDS IN IN SITU OIL SHALE RETORT AND PROCESS WATERS USING A LIQUID CHROMATOGRAPH COUPLED WITH AN ATOMIC ABSORPTION SPECTROMETER AS A DETECTOR
Fish, Richard H.; Brinckman, Frederick E.; Jewett, Kenneth L.
Lawrence Berkeley Lab. Univ. California Berkeley USA US CA 94720
Environ. Sci. Technol.; (82) P 174-9; Vol 16; No 3

05-82-0016 (0016)
TRACE ENRICHMENT OF FLUORINATED ORGANIC ACIDS USED AS GROUND-WATER TRACERS BY LIQUID CHROMATOGRAPHY
Stetzenbach, Klaus J.; Jensen, Stephen L.; Thompson, Glenn M.
Dep. Hydrol. Water Resour. Univ. Arizona Tucson USA US AZ 85721
Environ. Sci. Technol.; (82) P 250-4; Vol 16; No 5

05-82-0017 (0017)
COMPARISON OF CHLOROPHYLL A MEASUREMENTS BY FLUOROMETRIC, SPECTROPHOTOMETRIC

AND HIGH PRESSURE LIQUID CHROMATOGRAPHIC METHODS IN AQUATIC ENVIRONMENTS
Jacobsen, Timothy R.
Inst. Ecol. Univ. Georgia Athens USA US GA 30602
Ergeb. Limnol.; (82) P 35-45; Vol 16

05-82-0018 (0018)
DETERMINATION OF TRACES OF SODIUM ALKYLBENZENESULFONATE BY HIGH-PERFORMANCE LIQUID CHROMATOGRAPHY. APPLICATION TO WATER
Saito, Toshihide; Higashi, Kunishige; Hagiwara, Kazuyoshi
Gov. Ind. Res. Inst. Ikeda Japan JP 563
Fresenius' Z. Anal. Chem.; (82) P 21-3; Vol 313; No 1

05-82-0019 (0019)
DETERMINATION OF PROPANIL IN RICE, POTATOES, AND WATER BY HPLC
Alawi, M. A.
Chem. Inst. Tieraerztl. Hochsch. Hannover Hannover Fed. Rep. Ger. DE D-3000/1
Fresenius' Z. Anal. Chem.; (82) P 536-8; Vol 312; No 6

05-82-0020 (0020)
HPLC GRADIENT ELUTION OF DINITROPHENYLHYDRAZONES OF ALDEHYDES AND KETONES FROM AQUEOUS SAMPLES
Liebezeit, G.
Sonderforschungsber. 95 Univ. Kiel Kiel Fed. Rep. Ger. DE 2300
HRC CC, J. High Resolut. Chromatogr. Chromatogr. Commun.; (82) P 215-16; Vol 5; No 4

05-82-0021 (0021)
DETERMINATION OF 2,4-(DICHLOROPHENOXY)ACETIC ACID AND OF DICHLOBENIL IN WATER BY HIGH-PERFORMANCE LIQUID CHROMATOGRAPHY
Connick, William J., Jr.; Simoneaux, Jacqueline M.
South. Reg. Res. Cent. Sci. Educ. Adm. New Orleans USA US LA 70179
J. Agric. Food Chem.; (82) P 258-60; Vol 30; No 2

05-82-0022 (0022)
TRACE ANALYSIS OF NATURAL CAPSAICINOIDS IN ANIMAL FEED, HUMAN URINE, AND WASTEWATER BY HIGH-PRESSURE LIQUID CHROMATOGRAPHY
Johnson, Eric K.; Thompson, Harold C., Jr.; Bowman, Malcolm C.
Dep. Health Hum. Serv. Food and Drug Adm. Jefferson USA AR 72079
J. Agric. Food Chem.; (82) P 354-9; Vol 30; No 2

05-82-0023 (0023)
TRACE ANALYSIS OF NATURAL CAPSAICINOIDS IN ANIMAL FEED, HUMAN URINE, AND WASTEWATER BY HIGH-PRESSURE LIQUID CHROMATOGRAPHY
Johnson, Eric K.; Thompson, Harold C., Jr.; Bowman, Malcolm C.
Dep. Health Hum. Serv. Food and Drug Adm. Jefferson USA AR 72079
J. Agric. Food Chem.; (82) P 3X4-9; Vol 30; No 2

05-82-0024 (0024)
A RAPID, DIRECT HIGH-PERFORMANCE LIQUID CHROMATOGRAPHIC METHOD FOR THE SIMULTANEOUS QUANTITATIVE DETERMINATION OF CARBARYL AND .ALPHA.-NAPHTHOL IN AQUEOUS SAMPLES
Jones, Alice S.; Jones, Louis A.; Hastings, Felton L.
For. Sci. Lab. USDA Research Triangle Park USA US NC 27709
J. Agric. Food Chem.; (82) P 997-9; Vol 30; No 5

05-82-0025 (0025)
HIGH-PERFORMANCE LIQUID CHROMATOGRAPHIC DETERMINATION OF DIFENZOQUAT RESIDUES IN WATER
Ahmad, Ijaz
Alberta Environ. Cent. Vegreville Can. CA AB T0B 4L0

J. Assoc. Off. Anal. Chem.; (82) P 1097-101; Vol 65; No 5

05-82-0026 (0026)
HIGH-PERFORMANCE LIQUID CHROMATOGRAPHIC SEPARATION AND SELECTIVE DETECTION OF ANIONIC SURFACTANTS. APPLICATION TO COMMERCIAL FORMULATION AND WATER SAMPLE
Smedes, F.; Kraak, J.C.; Werkhoven-Goewie, C.F.; Brinkman, U.A.T.; Frei, R.W.
Lab. Anal. Chrm., Univ. Amsterdam, Neth. Amsterdam
J. Chromatogr. (82) P 123-32; Vol 247; No 1

05-82-0027 (0027)
OFF-LINE COMBINATION OF LIQUID CHROMATOGRAPHY AND FIELD DESORPTION MASS SPECTROMETRY: PRINCIPLES AND ENVIRONMENTAL, MEDICAL, AND PHARMACEUTICAL APPLICATIONS
Schulten, H. R.
Inst. Chem. Univ. Bonn Bonn Fed. Rep. Ger. DE 5300/1
J. Chromatogr.; (82) P 105-28; Vol 251; No 2

05-82-0028 (0028)
DETERMINATION OF ALDICARB, ALDICARB SULFOXIDE, ALDICARB SULFONE, AND CARBOFURAN RESIDUES IN WATER USING HIGH-PERFORMANCE LIQUID CHROMATOGRAPHY
Cochrane, W. P.; Lanouette, M.; Trudeau, S.
Lab. Serv. Div. Agric. Canada Ottawa Can. CA ON K1A 0C5
J. Chromatogr.; (82) P 307-14; Vol 243; No 2

05-82-0029 (0029)
HIGH-PERFORMANCE LIQUID CHROMATOGRAPHIC DETERMINATION OF PYRIDAZINONES IN WASTEWATERS. II
Lehotay, J.; Matisova, E.; Garaj, J.; Violova, A.
Fac. Chem. Technol. Slovak Tech. Univ. Bratislava Czech. CS 812 37
J. Chromatogr.; (82) P 323-6; Vol 246; No 2

05-82-0030 (0030)
DETERMINATION OF EDTA IN WATER BY HIGH-PERFORMANCE LIQUID CHROMATOGRAPHY
Harmsen, Joop; Van den Toorn, Antoni
Inst. Land Water Manage. Res. Wageningen Neth. NL 6700 AA
J. Chromatogr.; (82) P 379-84; Vol 249; No 2

05-82-0031 (0031)
DETERMINATION OF EDTA IN WATER BY HIGH-PERFORMANCE LIQUID CHROMATOGRAPHY
Harmsen, Joop; Van den Toorn, Antoni
Inst. Land Water Manage. Res. Wageningen Neth. NL 6700 AA
J. Chromatogr.; (82) P 379-84; Vol 249; No 2 In Eng

05-82-0032 (0032)
HIGH-PERFORMANCE LIQUID-CHROMATOGRAPHIC DETERMINATION OF 2,4- AND 2,6-TOLUENE-DIAMINE IN AQUEOUS EXTRACTS
Snyder, Roger C.; Breder, Charles V.
Div. Chem. Phys. Food and Drug Adm. Washington USA US DC 20204
J. Chromatogr.; (82) P 429-40; Vol 236; No 2

05-82-0033 (0033)
DETERMINATION OF PYRAZON RESIDUES IN WATER BY REVERSED PHASE HIGH-PERFORMANCE LIQUID CHROMATOGRAPHY
Ahmad, Ijaz
Alberta Environ. Cent. Vegreville Can. CA AB T0B 4L0
J. Environ. Sci. Health, Part B; (82) P 253-63; Vol B17; No 3

05-82-0034 (0034)
HIGH-PERFORMANCE LIQUID CHROMATOGRAPHIC DETERMINATION OF TRACE QUANTITIES OF

AZINPHOS METHYL AND AZINPHOS METHYL OXON IN VARIOUS WATER SOURCES BY DIRECT INJECTION AND TRACE ENRICHMENT
Bushway, R. J.
Dep. Food Sci. Univ. Maine Orono USA US ME 04469
J. Liq. Chromatogr.; (82) P 49-62; Vol 5; No 1

05-82-0035 (0035)
DIEL AND DEPTH VARIATIONS IN DISSOLVED FREE AMINO ACIDS AND AMMONIUM IN THE BALTIC SEA DETERMINED BY SHIPBOARD HPLC ANALYSIS
Mopper, Kenneth; Lindroth, Peter
Coll. Mar. Stud. Univ. Delaware Lewes USA US DE 19958
Limnol. Oceanogr.; (82) P 336-47; Vol 27; No 2

05-82-0036 (0036)
THE DETERMINATION OF .ALPHA.-KETOACIDS AND OXALIC ACID IN SEAWATER BY REVERSED PHASE LIQUID CHROMATOGRAPHIC SEPARATION OF FLUORESCENT QUINOXILINOL DERIVATIVES
Steinberg, Spencer M.; Bada, Jeffrey L.
Scripps Inst. Oceanogr. Univ. California La Jolla USA US CA 92093
Mar. Chem.; (82) P 299-306; Vol 11; No 4

05-82-0037 (0037)
CHEMICAL STUDIES OF COPPER-ORGANIC COMPLEXES ISOLATED FROM ESTUARINE WATERS USING C18 REVERSE-PHASE LIQUID CHROMATOGRAPHY
Mills, Gary L.; Hanson, Alfred K., Jr.; Quinn, James G.; Lammela, William R.; Chasteen, N. Dennis
Grad. Sch. Oceanogr. Univ. Rhode Island Kingston USA US RI 02881
Mar. Chem.; (82) P 355-77; Vol 11; No 4

05-82-0038 (0038)
ROUTINE HPLC OF POLYNUCLEAR AROMATIC HYDROCARBONS
O'Donnell, C.
Water Res. Div. St. Martin's House Dublin Ire. IE
TECHNICAL REPORT; Comm. Eur. Communities, (Rep.) EUR; (82) P 174-7; No EUR 7623, Anal. Org. Micropollut. Water

05-82-0039 (0039)
EVALUATION OF A MASS DETECTOR FOR HPLC DETERMINATION OF ORGANIC COMPOUNDS IN WATER
Connor, K. J.; Waggott, A.
Stevenage Lab. Water Res. Cent. Stevenage/Herts. UK GB SG1 1TH
TECHNICAL REPORT; Comm. Eur. Communities, (Rep.) EUR; (82) P 178-87; No EUR 7623, Anal. Org. Micropollut. Water

05-82-0040 (0040)
ASSESSMENT OF A MOVING BELT TYPE HPLC-MS INTERFACE WITH RESPECT TO ITS USE IN ORGANIC WATER POLLUTION ANALYSIS
Schauenburg, H.; Schlitt, H.; Knoppel, H.
Jt. Res. Cent. Ispra Italy IT
TECHNICAL REPORT; Comm. Eur. Communities, (Rep.) EUR; (82) P 193-8; No EUR 7623, Anal. Org. Micropollut. Water

05-82-0041 (0041)
DETERMINATION OF PHENOLS IN WATER BY HPLC
Sonneborn, M.; Pabel, E.; Schwabe, R.
Fed. Health Office Inst. Water, Soil Air Hyg. Berlin Fed. Rep. Ger. DE
TECHNICAL REPORT; Comm. Eur. Communities, (Rep.) EUR; (82) P 335-8; No EUR 7623, Anal. Org. Micropollut. Water

7.5.3
Soil

--- 1982 ---

05-82-0001 (0001)
DETERMINATION OF ENVIRONMENTAL PHENOLS BY LIQUID CHROMATOGRAPHY/ ELECTRO-CHEMISTRY
Shoup, Ronald E.; Mayer, Ginny S.
Res. Lab. Bioanal. Syst. Inc. West Lafayette USA US IN 47906
Anal. Chem.; (82) P 1164-9; Vol 54; No 7

05-82-0002 (0002)
RAPID SCAN UV/VISIBLE DETECTION IN LIQUID CHROMATOGRAPHY . APPLICATION TO ENVIRONMENTAL ANALYSIS
Warwick, C. J.; Bagon, D. A.
Occup. Med. Hyg. Lab. London UK GB NW2 6LN
Chromatographia; (82) P 433-6; Vol 15; No 7

05-82-0003 (0003)
INFLUENCE OF INJECTOR BYPASS ON LIFETIME OF SMALL-PARTICLE LIQUID CHROMATOGRAPHIC COLUMNS
DiCesare, J. L.; Dong, M. W.; Gant, J. R.
Chromatogr. Div. Perkin-Elmer Corp. Norwalk USA US CT 06856
Chromatographia; (82) P 595-8; Vol 15; No 9

05-82-0004 (0004)
ANALYSIS OF THE HERBICIDE CHLORSULFURON IN SOIL BY LIQUID CHROMATOGRAPHY
Zahnow, Edward W.
Biochem. Dep. E. I. du Pont de Nemours and Co. Wilmington USA US DE 19898
J. Agric. Food Chem.; (82) P 854-7; Vol 30; No 5

05-82-0005 (0005)
OFF-LINE COMBINATION OF LIQUID CHROMATOGRAPHY AND FIELD DESORPTION MASS SPECTROMETRY: PRINCIPLES AND ENVIRONMENTAL, MEDICAL, AND PHARMACEUTICAL APPLICATIONS
Schulten, H. R.
Inst. Chem. Univ. Bonn Bonn Fed. Rep. Ger. DE 5300/1
J. Chromatogr.; (82) P 105-28; Vol 251; No 2

7.5.4
Agricultural Chemicals

--- 1982 ---

05-82-0001 (0001)
SIMULTANEOUS DETERMINATION OF ATRAZINE AND HYDROXYATRAZINE BY HPLC
Solinas, V.; Melis, P.; Gessa, C.
Ist. Chim. Agrar. Univ. Sassari Sassari Spain ES
Agrochimica; (82) P 138-45; Vol 26; No 2-3; In Ital

05-82-0002 (0002)
LIQUID CHROMATOGRAPHY/MASS SPECTROMETRY OF KEPONE HYDRATE, KELEVAN, AND MIREX
Cairns, Thomas; Siegmund, Emil G.; Doose, Gregory M.
Dep. Health and Human Serv. Food and Drug Adm. Los Angeles USA US CA 90015
Anal. Chem.; (82) P 953-7; Vol 54; No 5

05-82-0003 (0003)
DETERMINATION OF THE HERBICIDE GLYPHOSATE AND ITS MAJOR METABOLITE AMINOMETHYLPHOSPHONIC ACID BY HIGH-PERFORMANCE LIQUID CHROMATOGRAPHY AFTER FLUORESCENCE LABELING
Roseboom, H.; Berkhoff, C. J.
Unit Residue Anal. Natl. Inst. Public Health Bilthoven Neth. NL 3720 BA
Anal. Chim. Acta; (82) P 373-7; Vol 135; No 2

05-82-0004 (0004)
TRACE ENRICHMENT OF DIFENZOQUAT IN WATER AND ITS DETERMINATION BY REVERSED-PHASE HIGH-PERFORMANCE LIQUID CHROMATOGRAPHY
Ahmad, Ijaz
Alberta Environ. Cent. Vegreville Can. CA AB T0B 4L0
Anal. Lett.; (82) P 27-37; Vol 15; No A1

05-82-0005 (0005)
COMBINED LIQUID CHROMATOGRAPHY-MASS SPECTROMETRY OF PESTICIDES AND NATURAL PRODUCTS
Games, D. E.; Eckers, C.; Lant, M. S.; Lewis, E.; Weerasinghe, N. C. A.; Westwood, S. A.
Dep. Chem. Univ. Coll. Cardiff Cardiff UK GB CF1 1XL
Anal. Proc. (London); (82) P 253-6; Vol 19; No 5

05-82-0006 (0006)
DETERMINATION OF ALDICARB RESIDUES IN WATER BY COMBINED HIGH PERFORMANCE LIQUID CHROMATOGRAPHY/MASS SPECTROMETRY
Wright, L. H.; Jackson, M. D.; Lewis, R. G.
EPA Research Triangle Park USA US NC 27711
Bull. Environ. Contam. Toxicol.; (82) P 740-7; Vol 28; No 6

05-82-0007 (0007)
THIOPHANATE METHYL AND MBC RESIDUES ON APPLES. HIGH-PRESSURE LIQUID CHROMATOGRAPHIC DETERMINATION
Lazzarini, Cecilia; Caleffi, Angela; Del Re, Attilio
Fac. Agrar. Univ. Cattol. S. Cuore Piacenza Italy IT
Chim. Ind. (Milan); (82) P 315-19; Vol 64; No 5; In Ital

05-82-0008 (0008)
A LABORATORY COLLABORATIVE TEST FOR DETERMINING TRIAZOPHOS IN HOSTATHION USING HIGH-PRESSURE LIQUID CHROMATOGRAPHY
Asshauer, J.
Werk Knapsack Hoechst A.-G. Huerth Fed. Rep. Ger. DE D-5030
Chromatographia; (82) P 71-4; Vol 15; No 2; In Ger

05-82-0009 (0009)
MUTAGENICITY OF HPLC FRACTIONS FROM EXTRACTS OF AATREX-TREATED CORN

Singh, I.; Lusby, A. F.; McGuire, P. M.
Coll. Med. Univ. Florida Gainesville USA US FL 32610
Environ. Mutagen.; (82) P 45-53; Vol 4; No 1

05-82-0010 (0010)
HIGH-PRESSURE LIQUID CHROMATOGRAPHIC (HPLC) ANALYSIS OF IMIDAZOLE ANTIFUNGALS IN COMMERCIAL DOSAGE FORMS
Cavrini, V.; Di Pietra, A. M.; Raggi, M. A.
Ist. Chim. Farm. Tossicol. Bologna Italy IT 40126
Int. J. Pharm.; (82) P 119-24; Vol 10; No 2

05-82-0011 (0011)
A HIGH-PERFORMANCE LIQUID CHROMATOGRAPHIC METHOD FOR QUANTITATION OF DIMBOA AND MBOA IN MAIZE PLANT EXTRACT
Gutierrez, Carmen; Guerrero, Ana; Castanera, P.; Torres, J. V.
Inst. Nac. Invest. Agrar. CRIDA Madrid Spain ES
J. Agric. Food Chem.; (82) P 1258-60; Vol 30; No 6

05-82-0012 (0012)
DETERMINATION OF THIOPHANATE METHYL AS CARBENDAZIN BY HIGH-PRESSURE LIQUID CHROMATOGRAPHY: APPLICATION TO ONIONS AND CABBAGE
Aten, Carl F.; Bourke, John B.; Marafioti, Rocco A.
Dep. Food Sci. Technol. New York State Agric. Exp. Stn. Geneva USA US NY 14456
J. Agric. Food Chem.; (82) P 610-11; Vol 30; No 3

05-82-0013 (0013)
ANALYSIS OF THE HERBICIDE CHLORSULFURON IN SOIL BY LIQUID CHROMATOGRAPHY
Zahnow, Edward W.
Biochem. Dep. E. I. du Pont de Nemours and Co. Wilmington USA US DE 19898
J. Agric. Food Chem.; (82) P 854-7; Vol 30; No 5

05-82-0014 (0014)
A RAPID, DIRECT HIGH-PERFORMANCE LIQUID CHROMATOGRAPHIC METHOD FOR THE SIMULTANEOUS QUANTITATIVE DETERMINATION OF CARBARYL AND .ALPHA.-NAPHTHOL IN AQUEOUS SAMPLES
Jones, Alice S.; Jones, Louis A.; Hastings, Felton L.
For. Sci. Lab. USDA Research Triangle Park USA US NC 27709
J. Agric. Food Chem.; (82) P 997-9; Vol 30; No 5

05-82-0015 (0015)
HIGH-PERFORMANCE LIQUID CHROMATOGRAPHIC DETERMINATION OF DIFENZOQUAT RESIDUES IN WATER
Ahmad, Ijaz
Alberta Environ. Cent. Vegreville Can. CA AB T0B 4L0
J. Assoc. Off. Anal. Chem.; (82) P 1097-101; Vol 65; No 5

05-82-0016 (0016)
STABILITY OF INDIVIDUAL NATURAL PYRETHRINS IN SOLUTION AFTER SEPARATION BY PREPARATIVE HIGH-PERFORMANCE LIQUID CHROMATOGRAPHY
Dickinson, Charlotte M.
Cent. Infect. Dis. Cent. Disease Control Atlanta USA US GA 30333
J. Assoc. Off. Anal. Chem.; (82) P 921-6; Vol 65; No 4

05-82-0017 (0017)
COMBINED LIQUID CHROMATOGRAPHIC/MASS SPECTROMETRIC ANALYSIS OF CARBAMATE PESTICIDES
Wright, L. H.
U.S. EPA Research Triangle Park USA US NC 27711
J. Chromatogr. Sci.; (82) P 1-6; Vol 20; No 1

05-82-0018 (0018)
VERY HIGH-SPEED LIQUID CHROMATOGRAPHY. III. QUANTITATIVE ANALYSIS OF PARABENS IN COSMETIC PRODUCTS
Dong, Michael W.; DiCesare, Joseph L.
Chromatogr. Div. Perkin-Elmer Corp. Norwalk USA US CT 06856
J. Chromatogr. Sci.; (82) P 49-54; Vol 20; No 2

05-82-0019 (0019)
QUANTITATIVE ANALYSIS OF THE PYRETHRINS BY HPLC
Otieno, Dismas A.; Jondiko, Isaac J.; McDowell, Philip G.; Kezdy, Ference J.
Int. Cent. Insect Physiol. Ecol. Nairobi Kenya KE
J. Chromatogr. Sci.; (82) P 566-70; Vol 20; No 12

05-82-0020 (0020)
IMPROVED INTERFACE FOR LIQUID CHROMATOGRAPHY-ELECTRON-CAPTURE DETECTOR COUPLING. I.
De Kok, A.; Geerdink, R. B.; Brinkman, U. A. Th.
Dep. Anal. Chem. Free Univ. Amsterdam Neth. NL 1081 HV
J. Chromatogr.; (82) P 101-11; Vol 252

05-82-0021 (0021)
DETERMINATION OF FENITROTTION, BIORESMETHRIN AND PIPERONYL BUTOXIDE IN AEROSOL CONCENTRATES BY HIGH-PERFORMANCE LIQUID CHROMATOGRAPHY
Perez, R. L.
Roche-Maaf Ltd. Bankstown Australia AU 2200
J. Chromatogr.; (82) P 178-82; Vol 243; No 1

05-82-0022 (0022)
HIGH-PERFORMANCE LIQUID CHROMATOGRAPHY-NEGATIVE CHEMICAL IONIZATION MASS SPECTROMETRY OF ORGANOPHOSPHORUS PESTICIDES
Parker, C. E.; Haney, C. A.; Hass, J. R.
Lab. Environ. Chem. Natl. Inst. Environ. Chem. Research Triangle Park USA US NC 27709
J. Chromatogr.; (82) P 233-48; Vol 237; No 2

05-82-0023 (0023)
REVERSED-PHASE HIGH-PERFORMANCE LIQUID CHROMATOGRAPHY OF PESTICIDES. VI. SEPARATION AND QUANTITATIVE DETERMINATION OF SOME RICE-FIELD HERBICIDES
Cabras, Paolo; Diana, Paolo; Meloni, Marco; Pirisi, Filippo M.
Cent. Reg. Agrar. Sper. Cagliari Italy IT 09100
J. Chromatogr.; (82) P 249-54; Vol 234; No 1

05-82-0024 (0024)
DETERMINATION OF ALDICARB, ALDICARB SULFOXIDE, ALDICARB SULFONE, AND CARBOFURAN RESIDUES IN WATER USING HIGH-PERFORMANCE LIQUID CHROMATOGRAPHY
Cochrane, W. P.; Lanouette, M.; Trudeau, S.
Lab. Serv. Div. Agric. Canada Ottawa Can. CA ON K1A 0C5
J. Chromatogr.; (82) P 307-14; Vol 243; No 2

05-82-0025 (0025)
REVERSED-PHASE THIN-LAYER CHROMATOGRAPHY OF POLYNUCLEAR AROMATIC HYDROCARBONS AND CHLORINATED BIPHENYLS. RELATIONSHIP WITH HYDROPHOBICITY AS MEASURED BY AQUEOUS SOLUBILITY AND OCTANOL-WATER PARTITION COEFFICIENT
Bruggeman, W. A.; Van der Steen, J.; Hutzinger, O.
Lab. Environ. Toxicol. Chem. Univ. Amsterdam Amsterdam Neth. NL 1018 WV
J. Chromatogr.; (82) P 335-46; Vol 238; No 2

05-82-0026 (0026)
LIQUID CHROMATOGRAPHY OF CHLORINATED BIPHENYLOLS
Hernandez, Adan; Walton, Harold F.

Coop. Inst. Res. Environ. Sci. Univ. Colorado Boulder USA US CO 80309
J. Chromatogr.; (82) P 346-8; Vol 242; No 2

05-82-0027 (0027)
POLYAROMATIC HYDROCARBONS AS HIGH-PERFORMANCE LIQUID CHROMATOGRAPHIC CALIBRATION STANDARDS FOR THE LOW-LEVEL DETERMINATION OF CHLORINATED DIBENZO-p-DIOXINS AND CHLORINATED DIBENZOFURANS IN BIOLOGICAL SAMPLES
Ryan, John J.; Pilon, Jean C.
Food Div. Health Pro4. Branch Ottawa Can. CA ON K1A 0L2
J. Chromatogr.; (82) P 409-15; Vol 248; No 3

05-82-0028 (0028)
HIGH-PERFORMANCE LIQUID CHROMATOGRAPHY-MASS SPECTROMETRY OF TRIAZINE HERBICIDES
Parker, C. E.; Haney, C. A.; Harvan, D. J.; Hass, J. R.
Lab. Environ. Chem. Natl. Inst. Environ. Health Sci. Research Triangle Park USA US NC 27709
J. Chromatogr.; (82) P 77-96; Vol 242; No 1

05-82-0029 (0029)
HIGH-PERFORMANCE LIQUID CHROMATOGRAPHIC DETERMINATION OF CARBARYL AND 1-NAPHTHOL IN BIOLOGICAL FLUIDS
DeBerardinis, Martin, Jr.; Wargin, William A.
Sch. Pharm. Univ. North Carolina Chapel Hill USA US NC 27514
J. Chromatogr.; (82) P 89-94; Vol 246; No 1

05-82-0030 (0030)
DETERMINATION OF PYRAZON RESIDUES IN WATER BY REVERSED PHASE HIGH-PERFORMANCE LIQUID CHROMATOGRAPHY
Ahmad, Ijaz
Alberta Environ. Cent. Vegreville Can. CA AB T0B 4L0
J. Environ. Sci. Health, Part B; (82) P 253-63; Vol B17; No 3

05-82-0031 (0031)
ANALYSIS OF A MIXTURE OF POLYCHLORINATED BIPHENYLS, DDT AND ITS ANALOGS BY HIGH-PERFORMANCE LIQUID CHROMATOGRAPHY
Chiosi, Sergio; Evidente, Antonio; Randazzo, Giacomino; Scalorbi, Daniela
Ist. Chim. Org. Biol. Univ. Napoli Naples Italy IT 80134
J. Liq. Chromatogr.; (82) P 163-63; Vol 5; No 9

05-82-0032 (0032)
DETERMINAT1ON OF CARBAMIC HERBICIDES BY HIGH PERFORMANCE LIQUID CHROMATOGRAPHY (HPLC) II. CHLORPROPHAM
Pena Heras, A.; Sanchez-Rasero, F.
Estac. Exp. Zaidin UEQuim. Anal. Apl. Granada Spain ES
J. Liq. Chromatogr.; (82) P 327-35; Vo4 5; No 2

05-82-0033 (0033)
HIGH-PERFORMANCE LIQUID CHROMATOGRAPHIC DETERMINATION OF TRACE QUANTITIES OF AZINPHOS METHYL AND AZINPHOS METHYL OXON IN VARIOUS WATER SOURCES BY DIRECT INJECTION AND TRACE ENRICHMENT
Bushway, R. J.
Dep. Food Sci. Univ. Maine Orono USA US ME 04469
J. Liq. Chromatogr.; (82) P 49-62; Vol 5; No 1

05-82-0034 (0034)
DETERMINATION OF PYRETHRUM AND PIPERONYL BUTOXIDE IN MILK BY HIGH-PERFORMANCE LIQUID CHROMATOGRAPHY (HPLC)
Nijhuis, H.; Heeschen, W.; Hahne, K. H.
Inst. Hyg. Bundesanst. Milchforsch. Kiel Fed. Rep. Ger. DE

Milchwissenschaft; (82) P 97-100; Vol 37; No 2; In Ger

05-82-0035 (0035)
HIGH-PRESSURE LIQUID CHROMATOGRAPHY OF SOME HEAT SENSITIVE ACTIVE PESTICIDE MATERIALS
Cserhati, Tibor; Bellay Kovacs, Tatjana
Novenyved. Kut. Intez. Hung. HU
Novenyvedelem (Budapest); (82) PP10-14; Vol 18; No 1; In Hung

CHAPTER 7.6

Enzymes—Coenzymes

--- 1982 ---

06-82-0001 (0001)
ISOLATION OF PROSTATIC ACID PHOSPHATASE USING HIGH PERFORMANCE LIQUID CHROMATOGRAPHY (HPLC)
Ruedenauer, R.; Koeberlein, W.; Schmidt-Gayk, H.
Klin. Lab. Chirur. Universitaetsklin. Heidelberg Fed. Rep. Ger. DE 6900
Aerztl. Lab.; (82) P 239-42; Vol 28; No 8; In Ger

06-82-0002 (0002)
A TIME-RESOLVED ASSAY OF DOPAMINE .BETA.-HYDROXYLASE ACTIVITY UTILIZING HIGH-PRESSURE LIQUID CHROMATOGRAPHY
Feilchenfeld, Natalie B.; Richter, Helen W.; Waddell, Walter H.
Dep. Chem. Carnegie-Mellon Univ. Pittsburgh USA US PA 15213
Anal. Biochem.; (82) P 124-8; Vol 122; No 1

06-82-0003 (0003)
HIGH-SENSITIVITY PEPTIDE MAPPING OF TRIOSEPHOSPHATE ISOMERASE: A COMPARISON OF HIGH-PERFORMANCE LIQUID CHROMATOGRAPHY WITH TWO-DIMENSIONAL THIN-LAYER METHODS
Oray, Bedii; Jahani, Minoo; Gracy, Robert W.
Dep. Chem. North Texas State Univ. Denton USA US TX 76203
Anal. Biochem.; (82) P 131-8; Vol 125; No 1

06-82-0004 (0004)
SEPARATION OF ANGIOTENSINS AND ASSAY OF ANGIOTENSIN-GENERATING ENZYMES BY HIGH-PERFORMANCE LIQUID CHROMATOGRAPHY
Klickstein, Lloyd B.; Wintroub, Bruce U.
Dep. Dermatol. Harvard Med. Sch. Boston USA US MA 02115
Anal. Biochem.; (82) P 146-50; Vol 120; No 1

06-82-0005 (0005)
HIGH-PERFORMANCE LIQUID CHROMATOGRAPHY OF PROTEINS: PURIFICATION OF THE ACIDIC ISOZYME OF ADENYLOSUCCINATE SYNTHETASE FROM RAT LIVER
Rudolph, Frederick B.; Clark, Sandra W.
Dep. Biochem. Rice Univ. Houston USA US TX 77001
Anal. Biochem.; (82) P 193-7; Vol 127; No 1

06-82-0006 (0006)
A NEW HIGH-PERFORMANCE LIQUID CHROMATOGRAPHY ASSAY FOR GLUTAMINE SYNTHETASE AND GLUTAMATE SYNTHASE IN PLANT TISSUES
Martin, Francis; Suzuki, Akira; Hirel, Bertrand
Unite Microbiol. Sols Inst. Natl. Rech. Agron. Seichamps Fr. FR 54 280
Anal. Biochem.; (82) P 24-9; Vol 125; No 1

06-82-0007 (0007)
HIGH-PERFORMANCE LIQUID CHROMATOGRAPHIC SEPARATION OF LIPOXYGENASE ISOZYMES IN CRUDE SOYBEAN EXTRACTS
Ramadoss, Candadai S.; Axelrod, Bernard
Dep. Biochem. Purdue Univ. West Lafayette USA US IN 47907
Anal. Biochem.; (82) P 25-31; Vol 127; No 1

06-82-0008 (0008)
CHARACTERIZATION OF VERTEBRATE COLLAGENASE ACTIVITY BY HIGH-PERFORMANCE LIQUID CHROMATOGRAPHY USING A SYNTHETIC SUBSTRATE
Gray, Robert D.; Saneii, Hossian H.
Sch. Med. Univ. Louisville Louisville USA US KY 40292
Anal. Biochem.; (82) P 339-46; Vol 120; No 2

06-82-0009 (0009)
RAPID ANALYSIS OF COBALAMIN COENZYMES AND RELATED CORRINOID ANALOGS BY

HIGH-PERFORMANCE LIQUID CHROMATOGRAPHY
Jacobsen, D. W.; Green, R.; Quadros, E. V.; Montejano, Y. D.
Dep. Biochem. Scripps Clinic Res. Found. La Jolla USA US CA 92037
Anal. Biochem.; (82) P 394-403; Vol 120; No 2

06-82-0010 (0010)
A SIMPLE AND RAPID PURIFICATION OF COMMERCIAL TRYPSIN AND CHYMOTRYPSIN BY REVERSE-PHASE HIGH-PERFORMANCE LIQUID CHROMATOGRAPHY
Titani, Koiti; Sasagawa, Tatsuru; Resing, Katheryn; Walsh, Kenneth A.
Howard Hughes Med. Inst. Lab. Univ. Washington Seattle USA US WA 98195
Anal. Biochem.; (82) P 408-12; Vol 123; No 2

06-82-0011 (0011)
LUTEINIZING HORMONE-RELEASING HORMONE PEPTIDASE ACTIVITIES IN THE FEMALE RAT: CHARACTERIZATION BY AN ASSAY BASED ON HIGH-PERFORMANCE LIQUID CHROMATOGRAPHY
Advis, J. P.; Krause, J. E.; McKelvy, J. F.
Dep. Neurobiol. Behav. State Univ. New York Stony Brook USA US NY 11794
Anal. Biochem.; (82) P 41-9; Vol 125; No 1

06-82-0012 (0012)
CHARACTERIZATION OF HUMAN ALCOHOL DEHYDROGENASE ISOENZYMES BY HIGH-PERFORMANCE LIQUID CHROMATOGRAPHIC PEPTIDE MAPPING
Strydom, Daniel J.; Vallee, Bert L.
Cent. Biochem. Biophys. Sci. Med. Harvard Med. Sch. Boston USA US MA 02115
Anal. Biochem.; (82) P 422-9; Vol 123; No 2

06-82-0013 (0013)
ISOCRATIC REVERSED-PHASE LIQUID CHROMATOGRAPHY FOR ASSAY OF 5'-NUCLEOTIDASE WITH PHOSPHORIC ACID FOR REACTION QUENCHING
Yu, Li Wen; Fager, Roger S.
Sch. Med. Univ. Virginia Charlottesville USA US VA 22908
Anal. Chem.; (82) P 1904-6; Vol 54; No 11

06-82-0014 (0014)
EVALUATION OF A SIMPLE COLORIMETRIC ASSAY FOR SERUM ANGIOTENSIN-CONVERTING ENZYME: COMPARISON WITH A NEW ION-PAIR LIQUID CHROMATOGRAPHY-ASSISTED ASSAY
Kwarts, Eddy; Beukenveld, Gerard; Gazendam, Jurjen
Cent. Lab. Clin. Chem. Univ. Hosp. Groningen Neth. NL
Ann. Clin. Biochem.; (82) P 227-32; Vol 19; No 4

06-82-0015 (0015)
PURIFICATION OF HIGHLY BINDABLE RAT BRAIN HEXOKINASE BY HIGH-PERFORMANCE LIQUID CHROMATOGRAPHY (HPLC)
Polakis, Paul G.; Wilson, John E.
Biochem. Dep. Michigan State Univ. East Lansing USA US MI 48824
Biochem. Biophys. Res. Commun.; (82) P 937-43; Vol 107; No 3

06-82-0016 (0016)
HPLC OF NEOPTERIN PHOSPHATES
Blau, N.; Bosshard, R.; Redweik, U.; Niederwieser, A.
Dep. Pediatr. Univ. Zurich Zurich Switz. CH CH-8032
Biochem. Clin. Aspects Pteridines; (82) P 103-16; Vol 1; No Cancer, Immunol., Metab. Dis.

06-82-0017 (0017)
PTERIDINES AS TUMOR MARKERS? ATTEMPT OF EVALUATION BY HPLC AND A RADIOIMMUNOASSAY FOR NEOPTERIN
Rokos, Katharina; Rokos, Hartmut; Frisius, Hildegard
Res. Lab. Henning Berlin G.m.b.H. Berlin Fed. Rep. Ger. DE D-1000/42
Biochem. Clin. Aspects Pteridines; (82) P 117-30; Vol 1; No Cancer, Immunol.,

Metab. Dis.

06-82-0018 (0018)
AUTOMATED ROUTINE ESTIMATION OF NEOPTERIN IN HUMAN URINE BY HPLC ON REVERSED PHASE
Fuchs, Dietmar; Hausen, Arno; Reibnegger, Gilbert; Wachter, Helmut
Inst. Med. Chem. Biochem. Univ. Innsbruck Innsbruck Austria AT A-6020
Biochem. Clin. Aspects Pteridines; (82) P 67-79; Vol 1; No Cancer, Immunol., Metab. Dis.

06-82-0019 (0019)
AUTOMATIC HPLC OF PTERINS WITH OR WITHOUT COLUMN SWITCHING
Niederwieser, A.; Staudenmann, W.; Wetzel, E.
Dep. Pediatr. Univ. Zurich Zurich Switz. CH CH-8032
Biochem. Clin. Aspects Pteridines; (82) P 81-102; Vol 1; No Cancer, Immunol., Metab. Dis.

06-82-0020 (0020)
STUDIES ON THE REGULATORY MECHANISM OF THE TYROSINE HYDROXYLASE SYSTEM IN ADRENAL SLICES BY USING HIGH-PERFORMANCE LIQUID CHROMATOGRAPHY WITH ELECTROCHEMICAL DETECTION
Togari, Akifumi; Kato, Takeshi; Nagatsu, Toshiharu
Dep. Life Chem. Tokyo Inst. Technol. Yokohama Japan JP 227
Biochem. Pharmacol.; (82) P 1729-34; Vol 31; No 9

06-82-0021 (0021)
STUDIES ON REDUCED AND OXIDIZED COENZYME Q (UBIQUINONES). II. THE DETERMINATION OF OXIDATION-REDUCTION LEVELS OF COENZYME Q IN MITOCHONDRIA, MICROSOMES AND PLASMA BY HIGH-PERFORMANCE LIQUID CHROMATOGRAPHY
Takada, Masahiro; Kenoya, Satoru; Yuzuriha, Teruaki; Katayama, Kouichi
Dep. Drug Met. Eisai Co., Ltd. Tokyo Japan JP
Biochim. Biophys. Acta; (82) P 308-14; Vol 679; No 2

06-82-0022 (0022)
ON-LINE FLUORESCENCE DETECTION IN HPLC: A HIGH SENSITIVE TECHNIQUE FOR THE MEASUREMENT OF LOW ENZYME ACTIVITIES: THE CASE OF TRYPTOPHAN HYDROXYLASE
Geeraerts, F.; Schimpfessel, L.; Crokaert, R.
Dep. Biochem. Free Univ. Brussels Belg. BE
Chromatographia; (82) P 449-50; Vol 15; No 7

06-82-0023 (0023)
HYDROLYSIS OF MALTOTETRAOSE BY HUMAN PANCREATIC .ALPHA.-AMYLASE, WITH LIQUID-CHROMATOGRAPHIC DETERMINATION OF THE PRODUCTS
Lalegerie, P.; Pourci, M. L.; Bailly, M.
Lab. Biochim. A Hop. Necker-Enfants Malades Paris Fr. FR 75743/15
Clin. Chem. (Winston-Salem, N. C.); (82) P 1791-3; Vol 28; No 8

06-82-0024 (0024)
HYPOXANTHINE-GUANINE PHOSPHORIBOSYL TRANSFERASE: ASSAY USING HIGH PERFORMANCE LIQUID CHROMATOGRAPHY
Rylance, H. J.; Wallace, Ruth C.; Nuki, G.
Rheumatic Dis. Unit Univ. Edinburgh Edinburgh UK GB EH5 2DQ
Clin. Chim. Acta; (82) P 159-65; Vol 121; No 2

06-82-0025 (0025)
A SENSITIVE AND INEXPENSIVE HIGH-PERFORMANCE LIQUID CHROMATOGRAPHIC ASSAY FOR TYROSINE HYDROXYLASE
Kato, T.; Horiuchi, S.; Togari, A.; Nagatsu, T.
Grad. Sch. Nagatsuta Tokyo Inst. Technol. Yokohama Japan 227
Experientia; (81) P 809-11; Vol 37; No 8

06-82-0026 (0026)
EVALUATION OF E. COLI PENICILLIN ACYLASE ACTIVITY BY HPLC AND FLUORESCENCE
Veronese, F.; Lazzarini, R.; Schiavon, O.; Bettero, A.
Inst. Pharm. Chem. Univ. Padova Padua Italy IT
Farmaco, Ed. Prat.; (82) P 390-6; Vol 37; No 12

06-82-0027 (0027)
A QUANTITATIVE ANALYSIS OF SELECTION INACTIVATION OF PEROXISOMAL ENZYMES IN THE YEAST HANSENULA POLYMORPHA BY HIGH-PERFORMANCE LIQUID CHROMATOGRAPHY
Bruinenberg, P. G.; Veenhuis, M.; Van Dijken, J. P.; Duine, J. A.; Harder, W.
Dep. Microbiol. Univ. Groningen Haren Neth. NL 9751 NN
FEMS Microbiol. Lett.; (82) P 45-50; Vol 15; No 1

06-82-0028 (0028)
HIGH-PERFORMANCE LIQUID CHROMATOGRAPHY OF SOME COENZYME M (2-MERCAPTO-ETHANESULFONIC ACID) DERIVATIVES BY ION PAIRING ON REVERSED-PHASE COLUMNS
Apostolides, Z.; Vermeulen, N. M. J.; Potgieter, D. J. J.
Dep. Biochem. Univ. Pretoria Pretoria S. Afr. ZA 0002
J Chromatogr.; (82) P 304-7; Vol 246; No 2

06-82-0029 (0029)
PHOTOCONTROLLED ADSORPTION CHROMATOGRAPHY FOR LYSOZYME USING AZOAROMATIC POLYMER
Ishihara, Kazuhiko; Negishi, Naoki; Shinohara, Isao
Dep. Polym. Chem. Waseda Univ. Tokyo Japan JP 160
J. Appl. Polym. Sci.; (82) P 1897-902; Vol 27; No 6

06-82-0030 (0030)
SEPARATION AND MEASUREMENT OF ISOENZYMES AND OTHER PROTEINS BY HIGH-PERFORMANCE LIQUID CHROMATOGRAPHY
Vacik, Dorothy N.; Toren, E. Clifford, Jr.
Dep. Pathol Univ. South Alabama Mobile USA US AL 36617
J. Chromatogr.; (82) P 1-31; Vol 228

06-82-0031 (0031)
DETERMINATION OF DIAMINE OXIDASE ACTIVITY BY HIGH-PERFORMANCE LIQUID CHROMATOGRAPHY
Pietta, Piergiorgio; Calatroni, Alma; Colombo, Roberto
Ist. Chim. Org. Anal. Fac. Agraria Milan Italy IT I-20133
J. Chromatogr.; (82) P 123-9; Vol 243; No 1

06-82-0032 (0032)
SEPARATION OF BASIC, HYDROPHILIC PEPTIDES BY REVERSED-PHASE ION-PAIR CHROMATOGRAPHY. I. ANALYTICAL APPLICATIONS WITH PARTICULAR REFERENCE TO A CLASS OF SERINE PEPTIDE SUBSTRATES OF CYCLIC AMP-STIMULATED PROTEIN KINASE
Fransson, B.; Ragnarsson, U.; Zetterqvist, O.
Biomed. Cent. Univ. Uppsala Uppsala Swed. SE S-751 23
J. Chromatogr.; (82) P 165-71; Vol 240; No 1

06-82-0033 (0033)
SIMULTANEOUS SEPARATION OF WATER-SOLUBLE VITAMINS AND COENZYMES BY REVERSED-PHASE HIGH-PERFORMANCE LIQUID CHROMATOGRAPHY
Kothari, R. M.; Taylor, M. W.
Dep. Biol. Indiana Univ. Bloomington USA US IN 47405
J. Chromatogr.; (82) P 187-92; Vol 247; No 1

06-82-0034 (0034)
HIGH-PERFORMANCE LIQUID CHROMATOGRAPHIC ASSAY OF CYTOSOLIC GLUTATHIONE S-TRANSFERASE ACTIVITY WITH STYRENE OXIDE
Brown, David L., Jr.; Boda, Wanda; Stone, M. P.; Buckpitt, Alan R.

California Coll. Med. Univ. California Irvine USA US CA 92717
J. Chromatogr.; (82) P 265-72; Vol 231; No 2

06-82-0035 (0035)
THE DETERMINATION OF UROPORPHYRINOGEN DECARBOXYLASE IN TISSUES BY HIGH-PERFORMANCE LIQUID CHROMATOGRAPHY COUPLED TO FLUORESCENCE DETECTION
Cantoni, L.; Ruggieri, R.; Dal Fiume, D.; Rizzardini, M.
Ist. Ric. Farmacol. "Mario Negri" Milan Italy IT 20157
J. Chromatogr.; (82) P 311-18; Vol 229; No 2

06-82-0036 (0036)
HIGHLY SENSITIVE ASSAY FOR PHENYLETHANOLAMINE N-METHYLTRANSFERASE ACTIVITY IN RAT BRAIN BY HIGH-PERFORMANCE LIQUID CHROMATOGRAPHY WITH ELECTROCHEMICAL DETECTION
Trocewicz, Jerzy; Oka, Kazuhiro; Nagatsu, Toshiharu
Grad. Sch. Nagatsuta Tokyo Inst. Technol. Yokohama Japan JP 227
J. Chromatogr.; (82) P 407-13; Vol 227; No 2

06-82-0037 (0037)
RAPID ASSAY FOR OXALATE OXIDASE USING REVERSED-PHASE HIGH-PERFORMANCE LIQUID CHROMATOGRAPHY
Pietta, Piergiorgio; Calatroni, Alma; Pace, Mario
Ist. Chim. Org. Anal. Univ. Milano Milan Italy IT 20133
J. Chromatogr.; (82) P 409-12; Vol 241; No 2

06-82-0038 (0038)
DETECTION OF BINDING BETWEEN TRANSFER RNA AND RNA POLYMERASE BY HIGH-PERFORMANCE LIQUID CHROMATOGRAPHY
Amero, Sally
Div. Plant Sci. West Virginia Univ. Morgantown USA US WV 26506
J. Chromatogr.; (82) P 413-15; Vol 241; No 2

06-82-0039 (0039)
RAPID AND SENSITIVE SIALIDASE ASSAY BY HIGH-PERFORMANCE LIQUID CHROMATOGRAPHY AND ITS APPLICATION TO DETECTION OF SIALIDASE IN HUMAN URINE
Omichi, Kaoru; Ikenaka, Tokuji
Coll. Sci. Osaka Univ. Toyonaka Japan JP 560
J. Chromatogr.; (82) P 415-19; Vol 230; No 2

06-82-0040 (0040)
ISOCRATIC REVERSED-PHASE LIQUID CHROMATOGRAPHY FOR ASSAY OF 5'-NUCLEOTIDASE WITH PHOSPHORIC ACID FOR REACTION QUENCHING
Yu, Li Wen; Fager, Roger S.
Sch. Med. Univ. Virginia Charlottesville USA US VA 22908
Anal. Chem.; (82) P 1904-6; Vol 54; No 11

06-82-0041 (0041)
ASSAY FOR DOPAMINE .BETA.-HYDROXYLASE IN RAT SERUM AND ADRENAL MEDULLA BY HIGH-PERFORMANCE LIQUID CHROMATOGRAPHY WITH FLUORESCENCE DETECTION
Nohta, Hitoshi; Ohtsubo, Kenji; Zaitsu, Kiyoshi; Ohkura, Yosuke
Fac. Pharm. Sci. Kyushu Univ. Fukuoka Japan JP 812
J. Chromatogr.; (82) P 415-22; Vol 227; No 2

06-82-0042 (0042)
DETERMINATION OF BLOOD TRANSKETOLASE BY HIGH-PERFORMANCE LIQUID CHROMATOGRAPHY (A PRELIMINARY NOTE)
Kimura, Mieko; Itokawa, Yoshinori
Fac. Med. Kyoto Univ. Kyoto Japan JP
J. Chromatogr.; (82) P 707-10; Vol 239

06-82-0043 (0043)
DETERMINATION OF 5'-NUCLEOTIDASE ACTIVITY IN HUMAN ERYTHROCYTES AND PLASMA USING HIGH-PERFORMANCE LIQUID CHROMATOGRAPHY
Sakai, Tadashi; Yanagihara, Susumu; Ushio, Koichi
Cent. Occup. Med. Tokyo Labour Accid. Hosp. Tokyo Japan JP 143
J. Chromatogr.; (82) P 717-21; Vol 239

06-82-0044 (0044)
DETERMINATION OF ANTHRANILATE, 3-HYDROXYANTHRANILATE AND 2,3-DIHYDROXYBENZOATE IN THE ENZYME ASSAY OF ANTHRANILATE HYDROXYLASE BY THIN-LAYER AND HIGH-PERFORMANCE LIQUID CHROMATOGRAPHY
Schott, Hans Henning; Langenbach-Schmidt, Doris; Zinke, Karen
Biochem. Inst. Justus Liebig-Univ. Giessen Giessen Fed. Rep. Ger. DE D-6300
J. Chromatogr.; (82) P 85-91; Vol 245; No 1

06-82-0045 (0045)
THE PURIFICATION OF PROSTATIC ACID PHOSPHATASE FROM SEMINAL PLASMA BY REVERSE-PHASE HIGH-PRESSURE LIQUID CHROMATOGRAPHY
Strickler, M. Patricia; Kintzios, John; Gemski, M. Judith
Div. Biochem. Walter Reed Army Inst. Res. Washington USA US DC 20012
J. Liq. Chromatogr.; (82) P 1921-31; Vol 5; No 10

06-82-0046 (0046)
DETERMINATION OF GLUTAMIC ACID DECARBOXYLASE ACTIVITY IN SUBREGIONS OF RAT BRAIN BY HIGH-PRESSURE LIQUID CHROMATOGRAPHY
Holdiness, Mack R.
Sch. Med. Louisiana State Univ. New Orleans USA US LA 70112
J. Liq. Chromatogr.; (82) P 479-87; Vol 5; No 3

06-82-0047 (0047)
TYROSINE HYDROXYLASE ACTIVITY IN RAT BRAIN SYNAPTOSOMES: DIRECT MEASUREMENT USING HIGH PERFORMANCE LIQUID CHROMATOGRAPHY
Messripour, Manoucher; Clark, John B.
Med. Coll. St. Bartholomew's Hosp. London UK GB EC1M 6BQ
J. Neurochem.; (82) P 1139-43; Vol 38; No 4

06-82-0048 (0048)
DEMONSTRATION OF AROMATIC L-AMINO ACID DECARBOXYLASE ACTIVITY IN HUMAN BRAIN WITH L-DOPA AND L-5-HYDROXYTRYPTOPHAN AS SUBSTRATES BY HIGH-PERFORMANCE LIQUID CHROMATOGRAPHY WITH ELECTROCHEMICAL DETECTION
Rahman, Mohammed Khalilur; Nagatsu, Toshiharu
Dep. Life Chem. Tokyo Inst. Technol. Yokohama Japan JP 227
Neurochem. Int.; (82) P 1-6; Vol 4; No 1

06-82-0049 (0049)
INACTIVATION OF THYROTROPIN-RELEASING HORMONE (TRH) AND (3Me-HIS)-TRH BY BRAIN PEPTIDASES STUDIED BY HIGH-PERFORMANCE LIQUID CHROMATOGRAPHY
Griffiths, E. C.; McDermott, J. R.; Smith, A. I.
Dep. Physiol. Univ. Manchester Manchester UK GB M13 9PT
Neurosci. Lett.; (82) P 61-5; Vol 28; No 1

06-82-0050 (0050)
HIGH-PERFORMANCE LIQUID CHROMATOGRAPHY DETERMINATION OF LYSOZYME IN EGG WHITE
Thapon, J. L.; Brule, G.
Lab. Rech. Technol. Laitiere Inst. Natl. Rech. Agron. Rennes Fr. FR 35042
Sci. Aliments; (82) P 251-60; Vol 2; No 3; In Fr

CHAPTER 7.7
Explosives

--- 1982 ---

07-82-0001 (0001)
SEPARATION OF MIXTURES OF 2,4,6-TRINITROTOLUENE REDUCTION PRODUCTS WITH LIQUID CHROMATOGRAPHY
Kaplan, D. L.; Kaplan, A. M.
Sci. Adv. Technol. Lab. U. S. Army Natick Lab. Natick USA US MA 01760
Anal. Chim. Acta; (82) P 425-8; Vol 136

07-82-0002 (0002)
USE OF HIGH-PERFORMANCE LIQUID CHROMATOGRAPHY FOR FORENSIC ANALYSIS OF EXPLOSIVES
Neumann, Helmut
Kriminaltech. Inst. Bundeskriminalamtes Wiesbaden Fed. Rep. Ger. DE D-6200
Arch. Kriminol.; (82) P 65-72; Vol 169; No 3-4; In Ger

07-82-0003 (0003)
ANALYSIS OF EXPLOSIVES BY LIQUID CHROMATOGRAPHY-NEGATIVE ION CHEMICAL IONIZATION MASS SPECTROMETRY
Parker, C. E.; Voyksner, R. D.; Tondexe, Yves; Henion, J. D.; Harvan, D. J.; Hass, J. R.; Yinon, Jehuda
Lab. Environ. Chem. Natl. Inst. Environ. Health Sci. Research Triangle Park USA US NC 27709
J. Forensic Sci.; (82) P 495-505; Vol 27; No 3

CHAPTER 7.8
Food and Food Additives

--- 1982 ---

08-82-0001 (0001)
EXAMINATION OF THE VOLATILE CARBONYL FRACTION OF FRESH TOMATOES AND TOMATO PREPARATIONS. II. HIGH-PERFORMANCE LIQUID CHROMATOGRAPHIC SEPARATIONS
Varga-Puchony, Zita; Vigh, Gyula; Hlavay, Jozsef; Petro-Turza, Marta; Szarfoldi-Szalma, Ilona
Inst. Anal. Chem. Univ. Chem. Eng. Veszprem Hung. HU H-8200
Acta Aliment.; (82) P 103-15; Vol 11; No 1

08-82-0002 (0002)
DETERMINATION, BY HPLC, OF CAROTENOIDS IN PAPRIKA PRODUCTS
Baranyai, Marta; Matus, Zoltan; Szabolcs, Jozsef
Inst. Chem. Univ. Med. Sch. Pecs Hung. HU H-7643
Acta Aliment.; (82) P 309-23; Vol 11; No 3

08-82-0003 (0003)
ANALYSIS OF AFLATOXINS IN MILK AND MILK PRODUCTS BY HIGH-PERFORMANCE LIQUID CHROMATOGRAPHY
Goto, Tetsuhisa; Manabe, Masaru Matsuura, Shinji
Natl. Food Res. Inst. Minist. Agric. Ibaraki Japan JP 305
Agric. Biol. Chem.; (82) P 801-2; Vol 46; No 3

08-82-0004 (0004)
USE OF HIGH-PERFORMANCE LIQUID CHROMATOGRAPHY FOR THE DETERMINATION OF AMINO ACIDS IN SPARKLING WINES
Casoli, A.; Colagrande, O.
Ist. Enol. Univ. Cattolica S. Cuore Piacenza Italy IT 29100
Am. J. Enol. Vitic.; (82) P 135-9; Vol 33; No 3

08-82-0005 (0005)
THE HPLC ANALYSIS OF ETHYL VANILLIN IN CHOCOLATE PRODUCTS
Hurst, W. J.; Martin, R. A., Jr.
Tech. Cent. Hershey Foods Corp. USA US PA
Am. Lab. (Fairfield, Conn.); (82) P 74-5; Vol 14; No 8

08-82-0006 (0006)
HIGH-PERFORMANCE LIQUID CHROMATOGRAPHIC SEPARATION OF LIPOXYGENASE ISOZYMES IN CRUDE SOYBEAN EXTRACTS
Ramadoss, Candadai S.; Axelrod, Bernard
Dep. Biochem. Purdue Univ. West Lafayette USA US IN 47907
Anal. Biochem.; (82) P 25-31; Vol 127; No 1

08-82-0007 (0007)
DETERMINATION OF MONOSODIUM GLUTAMATE IN FOOD USING HIGH-PERFORMANCE LIQUID CHROMATOGRAPHY AND FLUORESCENCE DETECTION
Rhys Williams, A. T.; Winfield, S. A.
Perkin-Elmer Ltd. Beaconsfield UK GB HP9 1QA
Analyst (London); (82) P 1092-4; Vol 107; No 1278

08-82-0008 (0008)
ASSAY METHODS OF VITAMIN A. 3. ASSAY METHODS OF VITAMIN A IN PHARMACEUTICAL PREPARATIONS, FOODS AND FEEDS BY HIGH-PERFORMANCE LIQUID CHROMATOGRAPHY
Kastui, Goichiro
Eisai Co. Ltd. Tokyo Japan JP 112
Bitamin; (82) P 151-6; Vol 56; No 3; In Japan

08-82-0009 (0009)
HIGH-PERFORMANCE LIQUID CHROMATOGRAPHIC DETERMINATION OF SUGARS IN CHOCOLATE AND CACAO

Corradi, Cesare; Carreri, Daniela; Micheli, Giuseppe
Lab. Chim. Prov. Milan Italy IT 20129
Boll. Chim. Unione Ital. Lab. Prov., Parte Sci.; (82) P 33-42; Vol 33; No S1; In Ital

08-82-0010 (0010)
REVERSE PHASE HIGH-PERFORMANCE LIQUID CHROMATOGRAPHIC METHOD FOR DETERMINATION OF CAFFEINE IN DECAFFEINATED COFFEE
Ferrara, Antonio; Reale, Patrizia; Grazia Calaminici, Maria; Iaccarino, Tommasina
Lab. Chim. Prov. Caserta Italy IT 81100
Boll. Chim. Unione Ital. Lab. Prov., Parte Sci.; (82) P 55-60; Vol 33; No S1; In Ital

08-82-0011 (0011)
DETERMINATION OF VICINE AND CONVICINE IN FABA BEANS BY HIGH PRESSURE LIQUID CHROMATOGRAPHY
Quemener, B.; Gueguen, J.; Mercier, C.
Lab. Biochim. Aliments Cent. Rech. Agro-Alimentaires Nantes Fr. FR F-44072
Can. Inst. Food Sci. Technol. J.; (82) P 109-15; Vol 15; No 2

08-82-0012 (0012)
DETERMINATION OF .ALPHA.- AND .BETA.-CAROTENE IN FRUIT AND VEGETABLES BY HIGH PERFORMANCE LIQUID CHROMATOGRAPHY
Bushway, R. J.; Wilson, A. M.
Dep. Food Sci. Univ. Maine Orono USA US ME 04469
Can. Inst. Food Sci. Technol. J.; (82) P 165-9; Vol 15; No 3

08-82-0013 (0013)
HPLC SEPARATION OF BITTER PEPTIDES FROM CHEDDAR CHEESE
Champion, H. M.; Stanley, D. W.
Dep. Food Sci. Univ. Guelph Guelph Can. CA ON N1G 2W1
Can. Inst. Food Sci. Technol. J.; (82) P 283-8; Vol 15; No 4

08-82-0014 (0014)
DETERMINATION OF ASCORBIC ACID IN WHEAT FLOURS, BREAD DOUGH CONDITIONERS, AND COMMERCIAL VITAMIN C TABLETS BY HIGH-PERFORMANCE LIQUID CHROMATOGRAPHY
Lookhart, G. L.; Hall, S. B.; Finney, K. F.
Agric. Res. Serv. U. S. Grain Market. Res. Lab. Manhattan USA US KS 66502
Cereal Chem.; (82) P 69-71; Vol 59; No 1

08-82-0015 (0015)
ANALYSIS OF THE BITTER ACIDS IN HOPS AND BEER BY HIGH-PRESSURE LIQUID CHROMATOGRAPHY (HPLC)
Dewaele, C.
Lab. Org. Chem. Rijksuniv. Gent Ghent Neth. NL 9000
Cerevisia; (82) P 91-5; Vol 7; No 2; In Neth

08-82-0016 (0016)
THIOPHANATE METHYL AND MBC RESIDUES ON APPLES. HIGH-PRESSURE LIQUID CHROMATOGRAPHIC DETERMINATION
Lazzarini, Cecilia; Caleffi, Angela; Del Re, Attilio
Fac. Agrar. Univ. Cattol. S. Cuore Piacenza Italy IT
Chim. Ind. (Milan); (82) P 315-19; Vol 64; No 5; In Ital

08-82-0017 (0017)
LC DETERMINATION OF VITAMINS A AND D IN FOODS AND ANIMAL FEEDSTUFFS
Wiggins, R. A.; Zai, E. S.; Lumley, I.
Lab. Gov. Chem. Dep. Ind. London UK GB
Chromatogr. Sci.; (82) P 327-41; Vol 20; No Biol./Biomed. Appl. Liq.

Chromatogr. 4

08-82-0018 (0018)
THE DETERMINATION OF SOME ADDITIVES BY ION-PAIR LIQUID CHROMATOGRAPHY
Cornet, Joost P.; Van de Haar, Gerrit
Food Insp. Serv. Haarlem Neth. NL
Chromatogr. Sci.; (82) P 343-61; Vol 20; No Biol./Biomed. Appl. Liq.
Chromatogr. 4

08-82-0019 (0019)
DETECTION OF VEGETABLE OIL ADULTERATION: HPLC ANALYSIS OF TRIGLYCERIDES
Hwang, Jenn Kang; Yen, Gow Chin; Hwang, Lucy Sun
Grad. Inst. Food Sci. Technol. Natl. Taiwan Univ. Taipei Taiwan TW
Chung-kuo Nung Yeh Hua Hsueh Hui Chih; (82) P 91-101; Vol 20; No 1-2; In Ch

08-82-0020 (0020)
DETERMINATION OF METHYL AND ETHYL CAFFEATE IN VEGETABLES BY HPLC WITH ELECTROCHEMICAL DETECTION
Sontag, G.; Schaefers, F. I.; Herrmann, K.
Inst. Anal. Chem. Univ. Vienna Vienna Austria AT
CONFERENCE; Recent Dev. Food Anal., Proc. Eur. Conf. Food Chem., 1st; (82) P 41-6; Avail.: Baltes, Werner; Czedik-Eysenberg, Peter B.; Pfannhauser, Werner; Date: 81; Publisher: Verlag Chem., Weinheim, Fed. Rep. Ger.

08-82-0021 (0021)
APPLICATIONS OF HPLC (IN FOOD ANALYSIS)
Battaglia, R.
Kantonales Lab. Zurich Switz. CH
CONFERENCE; Recent Dev. Food Anal., Proc. Eur. Conf. Food Chem., 1st; (82) P 14-26; Avail.: Baltes, Werner; Czedik-Eysenberg, Peter B.; Pfannhauser, Werner; Date: 81; Publisher: Verlag Chem., Weinheim, Fed. Rep. Ger.

08-82-0022 (0022)
METHOD FOR DETERMINATION OF RESIDUES OF FURAZOLIDONE AND FOUR OTHER NITROFURANS IN EGGS, MILK, AND MEAT BY HPLC
Petz, M.
Inst. Lebensmittelchem. Univ. Muenster Muenster Fed. Rep. Ger. DE 4400
Dtsch. Lebensm.-Rundsch.; (82) P 396-401; Vol 78; No 11; In Ger

08-82-0023 (0023)
DETERMINATION OF SULFATHIAZOLE IN HONEY FROM MEDICATED HONEY BEE COLONIES BY HIGH-PERFORMANCE LIQUID CHROMATOGRAPHY ON A CYANOAMINO POLAR PHASE
Argauer, Robert J.; Shimanuki, Hachiro; Knox, David A.
ARS USDA Beltsville USA US MD 20705
Environ. Entomol.; (82) P 820-3; Vol 11; No 4

08-82-0024 (0024)
MUTAGENICITY OF HPLC FRACTIONS FROM EXTRACTS OF AATREX-TREATED CORN
Singh, I.; Lusby, A. F.; McGuire, P. M.
Coll. Med. Univ. Florida Gainesville USA US FL 32610
Environ. Mutagen.; (82) P 45-53; Vol 4; No 1

08-82-0025 (0025)
QUANTITATIVE DETERMINATION OF VITAMINS D2 AND D3 IN MARGARINE BY HPLC
Mueller-Mulot, W.; Rohrer, G.
Kontroll-Abt. Hoffmann-La Roche A.-G. Grenzach-Wyhlen Fed. Rep. Ger. DE 7889/1
Fette, Seifen, Anstrichm.; (82) P 354-8; Vol 84; No 9; In Ger

08-82-0026 (0026)
QUANTITATIVE DETERMINATION OF AMINO ACIDS IN FOOD BY HIGH-PRESSURE LIQUID

CHROMATOGRAPHY
Khayat, Ali; Redenz, Peter K.; Gorman, Lee A.
Sea Food Div. Res. Cent. Ralston Purina Co. San Diego USA US CA 92121
Food Technol. (Chicago); (82) P 46-50; Vol 36; No 6

08-82-0027 (0027)
THE USE OF HIGH-PERFORMANCE LIQUID CHROMATOGRAPHY IN THE FOOD INDUSTRY
Day, W. Roy
Waters Assoc. Pty. L:d. Chippendale Australia AU 2008
Food Technol. Aust.; (82) P 36-9; Vol 34;No 1

08-82-0028 (0028)
DETERMINATION OF PERCHLOROETHYLENE (PCE) IN CARCASSE MEAL USING HPLC
Fankel, Rudolf; Slad, Ilse
Landwirtschaftlich-Chem. Bundesversuchsanst. Vienna Austria AT A-1020
Fresenius' Z. Anal. Chem.; (82) P 47; Vol 313; No 1; In Ger

08-82-0029 (0029)
DETERMINATION OF PROPANIL IN RICE, POTATOES, AND WATER BY HPLC
Alawi, M. A.
Chem. Inst. Tieraerztl. Hochsch. Hannover Hannover Fed. Rep. Ger. DE D-3000/1
Fresenius' Z. Anal. Chem.; (82) P 536-8; Vol 312; No 6

08-82-0030 (0030)
HIGH-PRESSURE LIQUID CHROMATOGRAPHIC DETERMINATION OF SUGARS IN DAIRY PRODUCTS. I. DETERMINATION OF FREE SUGARS IN COMMERCIAL FERMENTED MILK DRINKS
Lee, Y. K.; Song, M. H.; Chun, S. B.
Coll. Agric. Chonnam Natl. Univ. Kwanzu S. Korea KR
Hanguk Ch'uksan Hakhoe Chi; (82) P 326-31; Vol 24; No 4

08-82-0031 (0031)
A HIGH-PERFORMANCE LIQUID CHROMATOGRAPHY METHOD FOR DETERMINING ASCORBIC ACID CONTENT OF FRESH FRUITS AND VEGETABLES
Watada, Alley E.
Hortic. Crops Qual. Lab. Dep. Agric. Beltsville USA US MD 20705
HortScience; (82) P 334-5; Vol 17; No 3, Sect. 1

08-82-0032 (0032)
DETERMINATION OF VITAMIN D AND RETINOID ACTIVITY IN EGGS BY HPLC
Sivell, Lorna M.; Wenlock, R. W.; Jackson, Patricia A.
Food Sci. Div. Minist. Agric., Fish. Food London UK GB
Hum. Nutr.: Appl. Nutr.; (82) P 430-7; Vol 36A; No 6

08-82-0033 (0033)
OPTIMIZED SEPARATION OF BENZOIC AND SORBIC ACID AS PRESERVATIVES IN FOODSTUFFS BY REVERSED-PHASE HPLC
Froehlich, D. H.
Kant. Lab. Zurich Switz. CH CH-8030
HRC CC, J. High Resolut. Chromatogr. Chromatogr. Commun.; (82) P 158-60; Vol 5; No 3

08-82-0034 (0034)
QUANTITATIVE DETERMINATION OF GALACTURONIC ACID IN COMMERCIAL PECTINS BY HPLC
Giangiacomo, R.; Polesello, A.; Marin, F.
Ist. Sper. Valorizzazione Tecnol. Prod. Agric. Milan Italy IT
Ind. Aliment. (Pinerolo, Italy); (82) P 386-8; Vol 21; No 5; In Ital

08-82-0035 (0035)
ASCORBIC ACID IN MEAT PRODUCTS. ANALYSIS BY REVERSED-PHASE ION-PAIR HIGH-PRESSURE LIQUID CHROMATOGRAPHY

Parolari, Giovanni
Stazione Sper. Ind. Conserve Alimentari Parma Italy IT
Ind. Conserve; (82) P 19-22; Vol 57; No 1; In Ital

08-82-0036 (0036)
HIGH-PERFORMANCE LIQUID CHROMATOGRAPHIC DETERMINATION OF ASCORBIC ACID IN FOOD: COMPARISON WITH OTHER METHODS
Floridi, A.; Coli, R.; Fidanza, A. Alberti; Bourgeois, C. F.; Wiggins, R. A.
Ist. Sci. Aliment. Univ. Perugia Perugia Italy IT
Int. J. Vitam. Nutr. Res.; (82) P 193-6; Vol 52; No 2

08-82-0037 (0037)
DETERMINATION OF PHYTIC ACID IN FOODS BY HIGH-PERFORMANCE LIQUID CHROMATOGRAPHY
Graf, Ernst; Dintzis, Frederick R.
North. Reg. Res. Cent. Agric. Res. Serv. Peoria USA US IL 61604
J. Agric. Food Chem.; (82) P 1094-7; Vol 30; No 6

08-82-0038 (0038)
HIGH-PERFORMANCE LIQUID CHROMATOGRAPHIC DETERMINATION OF HYDROXYCINNAMIC ACIDS IN THE MAIZE MESOCOTYL
Hagerman, Ann E.; Nicholson, Ralph L.
Dep. Bot. Plant Pathol. Purdue Univ. West Lafayette USA US IN 47907
J. Agric. Food Chem.; (82) P 1098-102; Vol 30; No 6

08-82-0039 (0039)
QUANTITATIVE ANALYSIS OF HOP FLAVONOLS USING HIGH-PERFORMANCE LIQUID CHROMATOGRAPHY
McMurrough, Ian; Hennigan, Gerard P.; Loughrey, Malcolm J.
Res. Lab. Arthur Guinness Son and Co. Dublin Ire. IE
J. Agric. Food Chem.; (82) P 1102-6; Vol 30; No 6

08-82-0040 (0040)
ANALYSIS OF OXALIC ACID IN CARAMBOLA (AVERRHOA CARAMBOLA L.) AND SPINACH BY HIGH-PERFORMANCE LIQUID CHROMATOGRAPHY
Wilson, Charles W., III; Shaw, Philip E.; Knight, Robert J., Jr.
U. S. Citrus Subtrop. Prod. Lab. Winter Haven USA US FL 33880
J. Agric. Food Chem.; (82) P 1106-8; Vol 30; No 6

08-82-0041 (0041)
ANALYSIS OF SUGARS IN FOODS CONTAINING SODIUM CHLORIDE BY HIGH-PERFORMANCE LIQUID CHROMATOGRAPHY
Wills, Ron B. H.; Francke, Ray A.; Walker, Brian P.
Sch. Food. Technol. Univ. New South Wales Kensington Australia AU 2033
J. Agric. Food Chem.; (82) P 1242-3; Vol 30; No 6

08-82-0042 (0042)
A HIGH-PERFORMANCE LIQUID CHROMATOGRAPHIC METHOD FOR QUANTITATION OF DIMBOA AND MBOA IN MAIZE PLANT EXTRACT
Gutierrez, Carmen; Guerrero, Ana; Castanera, P.; Torres, J. V.
Inst. Nac. Invest. Agrar. CRIDA Madrid Spain ES
J. Agric. Food Chem.; (82) P 1258-60; Vol 30; No 6

08-82-0043 (0043)
ASCORBIC ACID CONTENT OF SOME TROPICAL FRUIT PRODUCTS DETERMINED BY HIGH-PERFORMANCE LIQUID CHROMATOGRAPHY
Shaw, Philip E.; Wilson, Charles W., III
U. S. Citrus Subtrop. Prod. Lab. Winter Haven USA US FL 33880
J. Agric. Food Chem.; (82) P 394-6; Vol 30; No 2

08-82-0044 (0044)
ANALYSIS OF FORMALDEHYDE IN SHRIMP BY HIGH-PRESSURE LIQUID CHROMATOGRAPHY
Radford, Terence; Dalsis, David E.
Coca-Cola Co. Atlanta USA US GA 30301
J. Agric. Food Chem.; (82) P 600-2; Vol 30; No 3

08-82-0045 (0045)
DETERMINATION OF THIOPHANATE METHYL AS CARBENDAZIN BY HIGH-PRESSURE LIQUID CHROMATOGRAPHY: APPLICATION TO ONIONS AND CABBAGE
Aten, Carl F.; Bourke, John B.; Marafioti, Rocco A.
Dep. Food Sci. Technol. New York State Agric. Exp. Stn. Geneva USA US NY 14456
J. Agric. Food Chem.; (82) P 610-11; Vol 30; No 3

08-82-0046 (0046)
HIGH-PERFORMANCE LIQUID CHROMATOGRAPHIC SEPARATION WITH TRIPLE-PULSE AMPEROMETRIC DETECTION OF CARBOHYDRATES IN BEVERAGES
Hughes, Scott; Johnson, Dennis C.
Dep. Chem. Iowa State Univ. Ames USA US IA 50011
J. Agric. Food Chem.; (82) P 712-14; Vol 30; No 4

08-82-0047 (0047)
DETERMINATION OF VOLATILE ALDEHYDES IN MEAT AS 2,4-DINITROPHENYLHYDRAZONES USING REVERSED-PHASE HIGH-PERFORMANCE LIQUID CHROMATOGRAPHY
Reindl, Bertram; Stan, Hans Juergen
Inst. Sozialmed. Epidemiol. Bundesgesundheitsamt Berlin Fed. Rep. Ger. DE 1000/33
J. Agric. Food Chem.; (82) P 849-54; Vol 30; No 5

08-82-0048 (0048)
DETERMINATION OF ETHANOL IN MALT BEVERAGES USING HIGH-PERFORMANCE LIQUID CHROMATOGRAPHY
Cieslak, M. E.; Herwig, W. C.
Miller Brew. Co. Milwaukee USA US WI 53201
J. Am. Soc. Brew. Chem.; (82) P 43-6; Vol 40; No 2

08-82-0049 (0049)
HIGH PERFORMANCE LIQUID CHROMATOGRAPHIC DETERMINATION OF SUBSIDIARY COLORS IN FD & C RED NO. 3
Calvey, Robert J.; Goldberg, Allen L.
Div. Color Technol. Food and Drug Adm. Washington USA US DC 20204
J. Assoc. Off. Anal. Chem.; (82) P 1080-5; Vol 65; No 5

08-82-0050 (0050)
ANALYSIS OF FAT-SOLUBLE VITAMINS. XXVI. HIGH PERFORMANCE LIQUID CHROMATOGRAPHIC DETERMINATION OF VITAMIN D IN FORTIFIED MILK AND MILKPOWDER
Borsje, Ben; De Vries, Ellen J.; Zeeman, Jakób; Mulder, Frits J.
Duphar B. V., Res. Lab. Weesp Neth. NL
J. Assoc. Off. Anal. Chem.; (82) P 1225-7; Vol 65; No 5

08-82-0051 (0051)
COMPARISON OF RAPID HIGH-PRESSURE LIQUID CHROMATOGRAPHIC AND CB METHODS FOR DETERMINATION OF AFLATOXINS IN CORN AND PEANUTS
DeVries, Jonathan W.; Chang, Henry L.
Medallion Lab. James Ford Bell Techn. Cent. Minneapolis USA US MN 55427
J. Assoc. Off. Anal. Chem.; (82) P 206-9; Vol 65; No 2

08-82-0052 (0052)
HIGH-PRESSURE LIQUID CHROMATOGRAPHIC DETERMINATION OF MONO- AND DISACCHARIDES IN PRESWEETENED CEREALS: COLLABORATIVE STUDY
Zygmunt, Lucian C.

John Stuart Res. Lab. Quaker Oats Co. Barrington USA US IL 60010
J. Assoc. Off. Anal. Chem.; (82) P 256-64; Vol 65; No 2

08-82-0053 (0053)
RAPID HIGH-PERFORMANCE LIQUID CHROMATOGRAPHIC DETERMINATION OF MONOSODIUM GLUTAMATE IN FOOD
Sporns, Peter
Dep. Food Sci. Univ. Alberta Edmonton Can. CA AB T6G 2P5
J. Assoc. Off. Anal. Chem.; (82) P 567-71; Vol 65; No 3

08-82-0054 (0054)
HIGH-PRESSURE LIQUID CHROMATOGRAPHIC DETERMINATION OF GLYCYRRHIZIC ACID OR GLYCYRRHIZIC ACID SALTS IN VARIOUS LICORICE PRODUCTS: COLLABORATIVE STUDY
Vora, Peter S.
MacAndrews and Forbes Co. Camden USA US NJ 08104
J. Assoc. Off. Anal. Chem.; (82) P 572-4; Vol 65; No 3

08-82-0055 (0055)
HIGH-PRESSURE LIQUID CHROMATOGRAPHIC DETERMINATION OF SUPPLEMENTAL METHIONINE IN POULTRY PREMIX
Matesic, Dubravka
Vet. Fac. Univ. Zagreb Zagreb Yugoslavia YU 41000
J. Assoc. Off. Anal. Chem.; (82) P 62-5; Vol 65; No 1

08-82-0056 (0056)
HIGH-PERFORMANCE LIQUID CHROMATOGRAPHIC DETERMINATION OF VITAMIN D IN FORTIFIED MILKS, MARGARINE, AND INFANT FORMULAS
Thompson, James N.; Hatina, George; Maxwell, William B.; Duval, Suzanne
Nutr. Res. Lab. Health and Welfare Canada Ottawa Can. CA ON K1A 0L2
J. Assoc. Off. Anal. Chem.; (82) P 624-31; Vol 65; No 3

08-82-0057 (0057)
HIGH-PRESSURE LIQUID CHROMATOGRAPHIC DETERMINATION OF AFLATOXINS IN PEANUT BUTTER USING A SILICA GEL-PACKED FLOWCELL FOR FLUORESCENCE DETECTION
Francis, Octave J., Jr.; Lipinski, Leo J.; Gaul, Jean A.; Campbell, Alfred D.
Food Drug Adm. New Orleans USA US LA 70122
J. Assoc. Off. Anal. Chem.; (82) P 672-6; Vol 65; No 3

08-82-0058 (0058)
AUTOMATED HIGH-PRESSURE LIQUID CHROMATOGRAPHIC SYSTEM FOR DETERMINATION OF MANNITOL, SORBITOL, AND XYLITOL IN CHEWING GUMS AND CONFECTIONS
Samarco, Eugene C.; Parente, Ernest S.
Res. Dev. Dep. Warner-Lambert Co. Morris Plains USA US NJ 07950
J. Assoc. Off. Anal. Chem.; (82) P 76-8; Vol 65; No 1

08-82-0059 (0059)
REVERSE PHASE LIQUID CHROMATOGRAPHIC DETERMINATION OF VITAMIN D2 AND D3 IN MILK
Muniz, Juan F.; Wehr, C. Timothy; Wehr, H. Michael
Lab. Serv. Oregon Dep. Agric. Salem USA US OR 97310
J. Assoc. Off. Anal. Chem.; (82) P 791-7; Vol 65; No 4

08-82-0060 (0060)
LIQUID CHROMATOGRAPHIC DETERMINATION OF NATAMYCIN IN CHEESE AT RESIDUE LEVELS
Tuinstra, Louis G. M. Th.; Traag, Wim A.
State Inst. Qual. Control Agric. Prod. Wageningen Neth. NL 6708 PD
J. Assoc. Off. Anal. Chem.; (82) P 820-2; Vol 65; No 4

08-82-0061 (0061)
LIQUID CHROMATOGRAPHIC DETERMINATION OF AFLATOXINS IN ARTIFICIALLY

CONTAMINATED COCOA BEANS
Hurst, W. Jeffrey; Lenovich, Lawrence M.; Martin, Robert A., Jr.
Tech. Cent. Hershey Foods Corp. Hershey USA US PA 17033
J. Assoc. Off. Anal. Chem.; (82) P 888-91; Vol 65; No 4

08-82-0062 (0062)
HIGH-PERFORMANCE LIQUID CHROMATOGRAPHIC DETERMINATION OF INTERMEDIATES AND REACTION BY-PRODUCTS IN FD & C YELLOW NO. 5: COLLABORATIVE STUDY
Cox, Elizabeth A.; McClure, Foster D.
Div. Color Technol. FDA Washingtov USA US DC 20204
J. Assoc. Off. Anal. Chem.; (82) P 933-40; Vol 65; No 4

08-82-0063 (0063)
HPLC ANALYSIS OF UNDERIVATIZED FATTY ACIDS IN MARGARINES
Bailie, Allston G., Jr.; Wilson, Terry D.; O'Brien, Robert K.; Beebe, Janet M.; Stuart, James D.; McCosh-Lilie, Elizabeth J.; Hill, Dennis W.
Coll. Agric. Nat. Resour. Univ. Connecticut Storrs USA US CT 06268
J. Chromatogr. Sci.; (82) P 466-70; Vol 20; No 10

08-82-0064 (0064)
HIGH-PERFORMANCE LIQUID CHROMATOGRAPHY OF FAT-SOLUBLE VITAMINS. SEPARATION AND IDENTIFICATION OF VITAMINS D2 AND D3 AND THEIR ISOMERS IN FOOD SAMPLES IN THE PRESENCE OF VITAMIN A, VITAMIN E AND CAROTENE
Zonta, Fabio; Stancher, Bruno; Bielawny, Jan
Ist. Merceol. Univ. Trieste Trieste Italy IT 34127
J. Chromatogr.; (82) P 105-12; Vol 246; No 1

08-82-0065 (0065)
ISOCRATIC SEPARATION OF FOOD FOLACIN BY HIGH-PERFORMANCE LIQUID CHROMATOGRAPHY
Briggs, David R.; Jones, Gwyn P.; Sae-Eung, Panadda
Dep. Hum. Nutr. Deakin Univ. Australia AU 3217
J. Chromatogr.; (82) P 165-8; Vol 246; No 1

08-82-0066 (0066)
HIGH-PERFORMANCE LIQUID CHROMATOGRAPHIC SEPARATION OF POTATO GLYCOALKALOIDS USING A RADIALLY COMPRESSED AMINO COLUMN
Bushway, R. J.
Dep. Food Sci. Univ. Maine Orono USA US ME 04469
J. Chromatogr.; (82) P 180-3; Vol 247; No 1

08-82-0067 (0067)
APPLICATION OF LIQUID CHROMATOGRAPHIC AND SPECTROSCOPIC METHODS FOR THE CHARACTERIZATION OF FATTY ACID ANILIDES IN CONTAMINATED COOKING OILS
Wheals, B. B.; Whitehouse, M. J.; Curry, C. J.
Metrop. Police Forensic Sci. Lab. London UK GB SE1
J. Chromatogr.; (82) P 203-15; Vol 238; No 1

08-82-0068 (0068)
HIGH-PERFORMANCE LIQUID CHROMATOGRAPHIC DETERMINATION OF CAROTENE AND VITAMIN A AND ITS GEOMETRIC ISOMERS IN FOODS. APPLICATIONS TO CHEESE ANALYSIS
Stancher, Bruno; Zonta, Fabio
Ist. Merceol. Univ. Trieste Trieste Italy IT 34127
J. Chromatogr.; (82) P 217-25; Vol 238; No 1

08-82-0069 (0069)
REVERSED-PHASE HIGH-PERFORMANCE LIQUID CHROMATOGRAPHY OF PROCYANIDINS AND OTHER PHENOLICS IN FRESH AND OXIDIZING APPLE JUICES USING A pH SHIFT TECHNIQUE
Lea, Andrew G. H.
Long Ashton Res. Stn. Univ. Bristol Long Ashton/Bristol UK GB BS18 9AF
J. Chromatogr.; (82) P 253-7; Vol 238; No 1

08-82-0070 (0070)
ANALYSIS OF ETHYLENETHIOUREA IN BEER BY HIGH-PERFORMANCE LIQUID CHROMATOGRAPHY
Massey, R. C.; Key, P. E.; McWeeny, D. J.
Food Lab. Minist. Agric. Fish. Food Norwich UK GB NR2 4SX
J. Chromatogr.; (82) P 254-6; Vol 240; No 1

08-82-0071 (0071)
RAPID DETERMINATION OF SUGARS IN CANTALOUPE MELON JUICE BY HIGH-PERFORMANCE LIQUID CHROMATOGRAPHY
Wade, N. L.; Morris, S. C.
New South Wales Dep. Agric. CSIRO North Ryde Australia AU 2113
J. Chromatogr.; (82) P 257-61; Vol 240; No 1

08-82-0072 (0072)
ANALYSIS OF BEER ISO-.ALPHA.-ACIDS BY HIGH-PERFORMANCE LIQUID CHROMATOGRAPHY WITHOUT SAMPLE PRETREATMENT
Verzele, M.; Dewaele, C.; Van Kerrebroeck, M.
Lab. Org. Chem. State Univ. Ghent Ghent Belg. BE B-9000
J. Chromatogr.; (82) P 321-6; Vol 244; No 2

08-82-0073 (0073)
METHOD FOR HIGH-PERFORMANCE LIQUID CHROMATOGRAPHIC MEASUREMENT OF N-NITROSAMINES IN FOOD AND BEVERAGES
Massey, R. C.; Crews, C.; McWeeny, D. J.
Food Sci. Lab. Minist. Agric., Fish. Food Norwich UK GB NR2 4SX
J. Chromatogr.; (82) P 423-7; Vol 241; No 2

08-82-0074 (0074)
HIGH-PERFORMANCE LIQUID CHROMATOGRAPHIC PROCEDURE FOR SEPARATION AND QUANTIFICATION OF ZEATIN AND ZEATIN RIBOSIDE FROM PEARS, PEACHES AND APPLES
Stahly, E. A.; Buchanan, D. A.
Tree Fruit Res. Lab. USDA Wenatchee USA US WA 98801
J. Chromatogr.; (82) P 453-9; Vol 235; No 2

08-82-0075 (0075)
USE OF C18 REVERSED-PHASE LIQUID CHROMATOGRAPHY FOR THE ISOLATION OF MONOTERPENE GLYCOSIDES AND NORISOPRENOID PRECURSORS FROM GRAPE JUICE AND WINES
Williams, P. J.; Strauss, C. R.; Wilson, B.; Massy-Westropp, R. A.
Australian Wine Res. Inst. Glen Osmond Australia AU 5064
J. Chromatogr.; (82) P 471-80; Vol 235; No 2

08-82-0076 (0076)
HIGH-PERFORMANCE LIQUID CHROMATOGRAPHY SEPARATION OF SOYBEAN ISOFLAVONES AND THEIR GLUCOSIDES
Eldridge, A. C.
Northern Reg. Res. Cent. Agric. Res. Sci. Educ. Adm. Peoria USA US IL 61604
J. Chromatogr.; (82) P 494-6; Vol 234; No 2

08-82-0077 (0077)
SEPARATION OF CARBOHYDRATES IN DAIRY PRODUCTS BY HIGH PERFORMANCE LIQUID CHROMATOGRAPHY
Richmond, M. L.; Barfuss, D. L.; Harte, B. R.; Gray, J. I.; Stine, C. M.
Dep. Food Sci. Hum. Nutr. Michigan State Univ. East Lansing USA US MI 48824
J. Dairy Sci.; (82) P 1394-400; Vol 65; No 8

08-82-0078 (0078)
PLASMA CATECHOLAMINES IN RESTING RAINBOW TROUT, SALMO GAIRDNERI RICHARDSON, BY HIGH PRESSURE LIQUID CHROMATOGRAPHY
Woodward, J. J.
Fish. Res. Inst. Univ. Washington Seattle USA US WA 98195

J. Fish Biol.; (82) P 429-32; Vol 21; No 4

08-82-0079 (0079)
A METHOD FOR THE ANALYSIS OF ATP METABOLITES IN BEEF SKELETAL MUSCLE BY HPLC
Currie, Ronald W.; Sporns, Peter; Wolfe, Fred H.
Dep. Food Sci. Univ. Alberta Edmonton Can. CA AB T6G 2P5
J. Food Sci.; (82) P 1226-8, 1234; Vol 47; No 4

08-82-0080 (0080)
HIGH PERFORMANCE LIQUID CHROMATOGRAPHIC DETERMINATION OF AGARITINE IN CULTIVATED MUSHROOMS
Speroni, J. J.; Beelman, R. B.
Dep. Food Sci. Pennsylvania State Univ. University Park USA US PA 16802
J. Food Sci.; (82) P 1539-41; Vol 47; No 5

08-82-0081 (0081)
COMPARISON OF HIGH PERFORMANCE LIQUID CHROMATOGRAPHIC, RADIOMETRIC, AND LACTOBACILLUS CASEI METHODS FOR THE DETERMINATION OF FOLACIN IN SELECTED FOODS
Gregory, J. F., III; Day, B. P. F.; Ristow, K. A.
Food Sci. Hum. Nutr. Dep. Univ. Florida Gainesville USA US FL 32611
J. Food Sci.; (82) P 1568-71; Vol 47; No 5

08-82-0082 (0082)
RIBOFLAVIN PHOTOCHEMICAL DEGRADATION IN PASTA MEASURED BY HIGH PERFORMANCE LIQUID CHROMATOGRAPHY
Woodcock, E. A.; Warthesen, J. J.; Labuza, T. P.
Dep. Food Sci. Nutr. Univ. Minnesota St. Paul USA US MN 55108
J. Food Sci.; (82) P 545-9, 555; Vol 47; No 2

08-82-0083 (0083)
ANALYSIS OF PHYTIC ACID IN FOODS BY HPLC
Camire, A. L.; Clydesdale, F. M.
Dep. Food Sci. Nutr. Univ. Massachusetts Amherst USA US MA 01003
J. Food Sci.; (82) P 575-8; Vol 47; No 2

08-82-0084 (0084)
HIGH-PERFORMANCE LIQUID CHROMATOGRAPHIC DETERMINATION OF THE METABOLITES OF THE POTATO GLYCOALKALOIDS, .ALPHA.-CHACONINE AND .ALPHA.-SOLANINE, IN POTATO TUBERS AND POTATO PRODUCTS
Bushway, R. J.
Dep. Food Sci. Univ. Maine Orono USA US ME 04469
J. Liq. Chromatogr.; (82) P 1313-22; Vol 5; No 7

08-82-0085 (0085)
THE DETERMINATION OF FOLIC ACID (PTEROYLMONOGLUTAMIC ACID) IN FORTIFIED PRODUCTS BY REVERSED-PHASE HIGH-PRESSURE LIQUID CHROMATOGRAPHY
Hoppner, K.; Lampi, B.
Nutr. Res. Lab. Health Prot. Branch Ottawa Can. CA ON
J. Liq. Chromatogr.; (82) P 953-66; Vol 5; No 5

08-82-0086 (0086)
THE CONTENT OF PHYLLOQUINONE (VITAMIN K1) IN HUMAN MILK, COWS' MILK AND INFANT FORMULA FOODS DETERMINED BY HIGH-PERFORMANCE LIQUID CHROMATOGRAPHY
Haroon, Yacoob; Shearer, Martin J.; Rahim, Seema; Gunn, William G. ; McEnery, Gerald; Barkhan, Percy
Dep. Haematol. Guy's Hosp. London UK GB SE1 9RT
J. Nutr.; (82) P 1105-17; Vol 112; No 6

08-82-0087 (0087)
NUTRITIONAL ANALYSIS OF FOOD MATERIALS AND ITS ASPARAGUS. HIGH PERFORMANCE

LIQUID CHROMATOGRAPHY
Ueda, Suehiro
Himazu Seisakusho K. K. Japan JP
Jpn. Fudo Saiensu; (82) P 49-57; Vol 21; No 9; In Japan

08-82-0088 (0088)
CONJUGATED DIENES OF CRUDE SOY OIL: DETECTION BY UV SPECTROPHOTOMETRY AND SEPARATION BY HPLC
Brown, H. G.; Snyder, H. E.
Dep. Hortic. Food Sci. Univ. Arkansas Fayetteville USA US AR 72701
JAOCS, J. Am. Oil Chem. Soc.; (82) P 280-3; Vol 59; No 7

08-82-0089 (0089)
VALIDATION OF COMPUTATIONAL METHODS FOR TRIGLYCERIDE COMPOSITION OF FATS AND OILS BY LIQUID CHROMATOGRAPHY AND MASS SPECTROMETRY
Merritt, C., Jr.; Vajdi, M.; Kayser, S. G.; Halliday, J. W.; Bazinet, M. L.
Sci. Adv. Technol. Lab. U. S. Army Natick Res. Dev. Lab. Natick USA US MA 01760
JAOCS, J. Am. Oil Chem. Soc.; (82) P 422-32; Vol 59; No 10

08-82-0090 (0090)
DETERMINATION OF VITAMIN D IN MILK BY HIGH-PRESSURE LIQUID CHROMATOGRAPHY
Mouillet, L.; Luquet, F. M.; Gagnepain, M. F.; Sorgue, Y.
ISHA Longjumeau Fr. FR 91163
Lait ; (82) P 44-54; Vol 62; No 611-612; In Fr

08-82-0091 (0091)
HPLC ANALYSIS OF NUCLEIC ACIDS IN FOOD AND OF NUCLEIC ACID METABOLITES IN METABOLIC STUDIES
Tiemeyer, W.; Gaebler, S.; Giesecke, D.
Tieraerztl. Fak. Univ. Muenchen Munich Fed. Rep. Ger. DE D-8000/22
Landwirtsch. Forsch., Sonderh.; (82) P 441-8; Vol 38,; In Ger

08-82-0092 (0092)
BENZOIC ACID IN SOUR MILK PRODUCTS AND FRESH CHEESE. HPLC DETERMINATION OF BENZOIC ACID, SORBIC ACID, AND p-HYDROXYBENZOIC ACID ESTERS IN FOODS
Gieger, Ursula
Nahrungsmittelunters. Landeshauptstadt Kiel, Eckernforder Str. 421 Kiel Fed. Rep. Ger. DE 2300
Lebensmittelchem. Gerichtl. Chem.; (82) P 109-10; Vol 36; No 5; In Ger

08-82-0093 (0093)
APPLICATION POSSIBILITIES OF POST COLUMN REACTORS IN THE HIGH-PERFORMANCE LIQUID CHROMATOGRAPHIC ANALYSIS OF FOODS, IN PARTICULAR, FLAVONOIDS
Galensa, Rudolf
Inst. Lebensmittelchem. Univ. Hannover Hannover Fed. Rep. Ger. DE 3000/91
Lebensmittelchem. Gerichtl. Chem.; (82) P 49-51; Vol 36; No 3; In Ger

08-82-0094 (0094)
DETECTION OF OLIVE OIL ADULTERATIONS BY HPLC
Schulte, E.
Inst. Lebensmittelchem. Univ. Muenster Muenster Fed. Rep. Ger. DE 4400
Lebensmittelchem. Gerichtl. Chem.; (82) P 88-9; Vol 36; No 4; In Ger

08-82-0095 (0095)
ALIPHATIC SECONDARY AMINES IN MEAT AND FISH PRODUCTS AND THEIR ANALYSIS BY HIGH PRESSURE LIQUID CHROMATOGRAPHY
Bellatti, Marta; Parolari, Giovanni
Stn. Sper. Ind. Conserve Aliment. Parma Italy IT 43100
Meat Sci.; (82) P 59-65; Vol 7; No 1

08-82-0096 (0096)
MICRODETERMINATION OF INDOLE-3-ACETIC ACID IN VEGETABLE EXTRACTS BY HIGH-PERFORMANCE LIQUID CHROMATOGRAPHY WITH FLUORESCENCE DETECTION
Budini, R.; Girotti, S.; Pierpaoli, A. M.; Tonelli, D.
Chem. Inst. "G. Ciamician" Univ. Bologna Bologna Italy IT 40126
Microchem. J.; (82) P 365-71; Vol 27; No 3

08-82-0097 (0097)
DETERMINATION OF PYRETHRUM AND PIPERONYL BUTOXIDE IN MILK BY HIGH-PERFORMANCE LIQUID CHROMATOGRAPHY (HPLC)
Nijhuis, H.; Heeschen, W.; Hahne, K. H.
Inst. Hyg. Bundesanst. Milchforsch. Kiel Fed. Rep. Ger. DE
Milchwissenschaft; (82) P 97-100; Vol 37; No 2; In Ger

08-82-0098 (0098)
HPLC IN BREWING TECHNOLOGY - A LITERATURE REVIEW
Kraemer, Peter
Forschungsinst. Technol. Brauerei und Maelzerei Berlin Fed. Rep. Ger. DE D-1000/65
Monatsschr. Brau.; (82) P 66-71; Vol 35; No 2; In Ger

08-82-0099 (0099)
THE DETERMINATION OF COCOA SOLIDS IN MILK POWDER PRODUCTS USING HIGH PERFORMANCE LIQUID CHROMATOGRAPHY
Woollard, D. C.
Reg. Dairy Lab. Minist. Agric. Fisheries Auckland N. Z. NZ
N. Z. J. Dairy Sci. Technol.; (82) P 63-8; Vol 17; No 1

08-82-0100 (0100)
A RAPID AND SENSITIVE HIGH-PERFORMANCE LIQUID CHROMATOGRAPHY METHOD OF FOLLOWING THE ACTION OF CHYMOSIN IN MILK
Van Hooydonk, A. C. M.; Olieman, C.
Netherlands Inst. Dairy Res. Ede Neth. NL
Neth. Milk Dairy J.; (82) P 153-8; Vol 36; No 2

08-82-0101 (0101)
HIGH-PERFORMANCE LIQUID CHROMATOGRAPHIC DETERMINATION OF ORGANIC ACIDS IN CORN SILAGE
Kato, Isao; Yosizawa, Sadao; Hamaguchi, Kenzi; Ando, Kouichi
Coll. Dairying Ebetsu Japan JP 069-01
Nippon Chikusan Gakkai Ho; (82) P 642-4; Vol 53; No 9; In Japan

08-82-0102 (0102)
SEPARATION AND DETERMINATION OF FREE SUGARS IN OILSEEDS BY HIGH-PERFORMANCE LIQUID CHROMATOGRAPHY
Yamauchi, Ryo; Kato, Koji; Ueno, Yoshimitsu
Dep. Agric. Chem. Gifu Univ. Gifu Japan JP 504
Nippon Nogei Kagaku Kaishi; (82) P 123-5; Vol 56; No 2; In Japan

08-82-0103 (0103)
A NEW LIQUID CHROMATOGRAPHIC DETERMINATION OF ORGANIC ACIDS IN SOY SAUCE
Ushijima, Shigeomi; Hamada, Takashi; Kanbe, Chiyuki
Cent. Res. Lab. Kikkoman Corp. Noda Japan JP
Nippon Shoyu Kenkyusho Zasshi; (82) P 58-63; Vol 8; No 2; In Japan

08-82-0104 (0104)
HIGH PERFORMANCE LIQUID CHROMATOGRAPHIC DETERMINATION OF K VALUE AS AN INDEX OF FRESHNESS OF FISH
Lee, Eung Ho; Ohshima, Toshiaki; Koizumi, Chiaki
Natl. Fish. Univ. Busan Pusan S. Korea KR

Nippon Suisan Gakkaishi; (82) P 255; Vol 48; No 2

08-82-0105 (0105)
ANALYSIS OF .GAMMA.-GLUTAMYL DIPEPTIDES IN BEAN EXTRACTS BY HIGH-PERFORMANCE LIQUID CHROMATOGRAPHY
Wu, Shih Hsiung; Wang, Kung Tsung; Tsou, Samson C. S.
Inst. Biol. Chem. Acad. Sinica Taipei Taiwan TW
Proc. Natl. Sci. Counc., Repub. China, Part A; (82) P 34-8; Vol 6 ; No 1

08-82-0106 (0106)
HPLC ANALYSIS OF ANIONIC GUMS
Voragen, A. G. J.; Schols, H. A.; Pilnik, W.
Dep. Food Sci. Agric. Univ. Wageningen Neth. NL 6703 BC
Prog. Food Nutr. Sci.; (82) P 379-85;Vol 6; No Gums Stab. Food Ind.: Interact. Hydrocolloids

08-82-0107 (0107)
HPLC DETERMINATION OF VANILLIN IN BUTTER SAMPLES FOR USE IN PASTRY PRODUCTS
Cortesi, N.; Colombo, D.
Staz. Sper. Ind. Oli E Dei Grassi Milan Italy IT
Riv. Ital. Sostanze Grasse; (82) P 249-50; Vol 59; No 5; In Ital

08-82-0108 (0108)
DETERMINATION OF OLIVE OIL PHENOLIC COMPOUNDS BY GLC AND HPLC: HYPOTHETICAL ROLE OF TYROSOL IN DETERMINING VIRGIN OLIVE OIL IN MIXTURES WITH REFINED OILS
Solinas, M.; Cichelli, A.
Ist. Sper. Elaiotec. Pescara Italy IT
Riv. Soc. Ital. Sci. Aliment.; (82) P 223-30; Vol 11; No 4; In Ital

08-82-0109 (0109)
HPLC SEPARATION AND QUANTITATIVE DETERMINATION OF ORGANIC ACIDS IN FRUIT JUICES AND WINES
Droz, C.; Tanner, H.
Eidg. Forschungsanst. Wadenswil Switz. CH
Schweiz. Z. Obst- Weinbau; (82) P 434-8; Vol 118; No 15; In Ger

08-82-0110 (0110)
HIGH-PERFORMANCE LIQUID CHROMATOGRAPHY DETERMINATION OF LYSOZYME IN EGG WHITE
Thapon, J. L.; Brule, G.
Lab. Rech. Technol. Laitiere Inst. Natl. Rech. Agron. Rennes Fr. FR 35042
Sci. Aliments; (82) P 251-60; Vol 2; No 3; In Fr

08-82-0111 (0111)
HIGH PERFORMANCE LIQUID CHROMATOGRAPHIC DETERMINATION OF SACCHARIDES IN HONEYS
Fan, Chongyang; Liu, Dexing; Pan, Kunyong; Yi, Qian; Song, Jianping
Wuhan Merch. Control Wuhan Peop. Rep. China CN
Shipin Yu Faxiao Gongye; (82) P 7-11; No 3; In Ch

08-82-0112 (0112)
DETERMINATION OF SESAMOLIN, SESAMIN AND SESAMOL IN SESAME OIL BY HIGH- PERFORMANCE LIQUID CHROMATOGRAPHY
Yoshida, Masaharu; Kashimoto, Takashi
Osaka Pref. Inst. Public Health Osaka Japan JP
Shokuhin Eiseigaku Zasshi; (82) P 142-8; Vol 23; No 2; In Japan

08-82-0113 (0113)
DETERMINATION OF FOOD ADDITIVES BY HIGH-PERFORMANCE LIQUID CHROMATOGRAPHY. III. DETERMINATION OF DIPHENYL, o-PHENYLPHENOL, AND THIABENDAZOLE IN CITRUS FRUITS AND BANANA
Kitada, Yoshimi; Tamase, Kikuo; Inoue, Masashige; Sasaki, Michiko; Tanigawa,

Kaoru
Nara Pref. Inst. Public Health Nara Japan JP 630
Shokuhin Eiseigaku Zasshi; (82) P 21-4; Vol 23; No 1; In Japan

08-82-0114 (0114)
CHOLECALCIFEROL SULFATE IDENTIFICATION IN HUMAN MILK BY HPLC
Le Boulch, N.; Cancela, L.; Miravet, L.
INSERM Paris Fr. FR 75010
Steroids; (82) P 391-8; Vol 39; No 4

08-82-0115 (0115)
DETERMINATION OF THE FATTY ACID COMPOSITION OF SOYBEAN OIL BY HIGH-PRESSURE LIQUID CHROMATOGRAPHY
Jaselskis, Bruno; Stemm, Nicholas L.; Johnston, William D.
Loyola Univ. Chicago Chicago USA US IL 60626
Talanta; (82) P 54-6; Vol 29; No 1

08-82-0116 (0116)
SYSTEMATIC SEPARATION OF CAROTENOID PIGMENTS FROM CITRUS FLESH BY HIGH-PERFORMANCE LIQUID CHROMATOGRAPHY
Satoh, Hisao; Komai, Toshiko; Kitahara, Keiichi; Aoyagi, Reiko; Honda, Teruo; Noguchi, Fuminori
Dep. Chem. Tokyo Med. Coll. Tokyo Japan JP
Tokyo Ika Daigaku Kiyo; (82) P 13-21; Vol 8; In Japan

08-82-0117 (0117)
DETERMINATION OF 10-HYDROXY-2-DECENOIC ACID IN ROYAL JELLY BY HIGH PERFORMANCE LIQUID CHROMATOGRAPHY
Nakajin, Shizuo; Taguchi, Mitsue; Akiyama, Yoshiro; Shinoda, Masato
Hoshi Coll. Pharm. Tokyo Japan JP 142
Yakugaku Zasshi; (82) P 549-54; Vol 102; No 6; In Japan

08-82-0118 (0118)
DETERMINATION OF TOCOPHEROLS IN RICE BRAN OILS BY HIGH PERFORMANCE LIQUID CHROMATOGRAPHY
Tanabe, Keizo; Yamaoka, Masakazu; Tanaka, Akio; Kato, Akio; Amemiya, Junko
Natl. Chem. Lab. Ind. Ibaraki Japan JP
Yukagaku; (82) P 205-8; Vol 31; No 4; In Japan

08-82-0119 (0119)
AMINO ACID ANALYZER AND HIGH-PRESSURE LIQUID CHROMATOGRAPHIC DETERMINATION OF AMINO ACIDS AND BIOGENIC AMINES IN THE LIGHT AND DARK MUSCLE MEAT OF TUNA
Stockemer, J.
Staat. Veterinaeruntersuchungsamt Fische Fischwaren Cuxhaven Fed. Rep. Ger. DE D-2190
Z. Lebensm.-Unters. Forsch.; (82) P 108-13; Vol 174; No 2; In Ger

08-82-0120 (0120)
DETERMINATION OF QUININE IN SOFT DRINKS AND DRUGS BY HIGH-PERFORMANCE LIQUID CHROMATOGRAPHY WITH ELECTROCHEMICAL DETECTION
Kral, Karl; Sontag, Gerhard
Inst. Anal. Chem. Univ. Wien Vienna Austria AT A-1090
Z. Lebensm.-Unters. Forsch.; (82) P 22-4; Vol 175; No 1; In Ger

08-82-0121 (0121)
STUDIES ON ANALYSIS OF FEED ADDITIVES BY HIGH-PERFORMANCE LIQUID CHROMATOGRAPHY. I. DETERMINATION OF NICARBAZIN IN CHICKEN MUSCLE AND EGG BY HIGH-PERFORMANCE LIQUID CHROMATOGRAPHY
Hoshino, Youji; Horie, Masakazu; Nose, Norihide
Saitama Pref. Inst. Public Health Urawa Japan JP 338

Shokyhin Eiseigaku Zasshi; (82) P 265-9; Vol 23; No 3; In Japan

08-82-0122 (0122)
A RAPID DETERMINATION OF HOPSIDE IN RICE BY HPLC
Xiang, Yongsheng; Sun, Huaiyu
Shandong Inst. Med. Sci. Peop. Rep. China CN
Zhonghua Yufangyixue Zazhi; (82) P 111; Vol 16; No 2; In Ch

CHAPTER 7.9

Fuels and Related Compounds

--- 1982 ---

09-82-0001 (0001)
DETERMINATION OF ALKYL NITRATE ADDITIVES IN DIESEL FUEL BY LIQUID CHROMATOGRAPHY WITH INFRARED SPECTROMETRIC DETECTION
Schabron, J. F.; Fuller, M. P.
Phillips Petrol. Co. Bartlesville USA US OK 74004
Anal. Chem.; (82) P 2599-601; Vol 54; No 14

09-82-0002 (0002)
HIGH-PRESSURE LIQUID CHROMATOGRAPHY OF PETROPORPHYRINS: EVALUATION AS A GEOCHEMICAL FINGERPRINTING METHOD BY PRINCIPAL COMPONENTS ANALYSIS
Hohn, M. E.; Hajibrahim, S. K.; Eglinton, G.
Sch. Chem. Univ. Bristol Bristol UK GB BS8 1TS
Chem. Geol.; (82) P 229-37; Vol 37; No 3-4

09-82-0003 (0003)
CHARACTERIZATION OF A HEAVY COAL LIQUEFACTION PRODUCT BY COMBINED HIGH-PERFORMANCE LIQUID CHROMATOGRAPHY/MASS SPECTROMETRY
Holstein, W.; Severin, D.
Inst. Erdoelforsch. Hannover Fed. Rep. Ger. DE D-3000
Chromatographia; (82) P 231-5; Vol 15; No 4; In Ger

09-82-0004 (0004)
FINGERPRINTING INORGANIC ARSENIC AND ORGANOARSENIC COMPOUNDS IN IN SITU OIL SHALE RETORT AND PROCESS WATERS USING A LIQUID CHROMATOGRAPH COUPLED WITH AN ATOMIC ABSORPTION SPECTROMETER AS A DETECTOR
Fish, Richard H.; Brinckman, Frederick E.; Jewett, Kenneth L.
Lawrence Berkeley Lab. Univ. California Berkeley USA US CA 94720
Environ. Sci. Technol.; (82) P 174-9; Vol 16; No 3

09-82-0005 (0005)
DYE DETERMINATION IN LIGHT FUEL OILS BY HIGH PRESSURE LIQUID CHROMATOGRAPHY
Peltzer, Joerg
Wiss. Abt. Zolltech. Pruef. Lehranst. Muenchen Munich Fed. Rep. Ger. DE D-8000/2
Erdoel Kohle, Erdgas, Petrochem.; (82) P 312; Vol 35; No 6; In Ger

09-82-0006 (0006)
OIL SPILL IDENTIFICATION BY HIGH-SPEED HPLC
Grimalt, J.; Albaiges, J.
Inst. Bio-Org. Chem. CSIC Barcelona Spain ES 34
HRC CC, J. High Resolut. Chromatogr. Chromatogr. Commun.; (82) P 255-60; Vol 5; No 5

09-82-0007 (0007)
A COMPARISON OF THREE DIFFERENT HIGH PERFORMANCE LIQUID CHROMATOGRAPHY SYSTEMS FOR THE DETERMINATION OF ALDEHYDES AND KETONES IN DIESEL EXHAUST. PART I
Creech, G.; Johnson, R. T.; Stoffer, J. O.
Univ. Missouri Rolla USA US MO 65401
J. Chromatogr. Sci.; (82) P 67-72; Vol 20; No 2

09-82-0008 (0008)
HIGH-PERFORMANCE LIQUID CHROMATOGRAPHY WITH REDUCTIVE ELECTROCHEMICAL DETECTION OF MUTAGENIC NITRO-SUBSTITUTED POLYNUCLEAR AROMATIC HYDROCARBONS IN DIESEL EXHAUSTS
Rappaport, S. M.; Jin, Z. L.; Xu, X. B.
Sch. Public Health Univ. California Berkeley USA US CA 94720
J. Chromatogr.; (82) P 145-54; Vol 240; No 1

09-82-0009 (0009)
USE OF LIQUID CHROMATOGRAPHY FOR CHARACTERIZATION OF HEAVY OILS
Nomura, Akira; Morita, Yazaemon; Kogure, Yukitoshi
Kagaku Gijutsu Kenkyusho Japan JP
Kagiken Nyusu, Kagaku Kogyo Shiryo; (82) P 7-12; Vol 17; No 1; In Japan

09-82-0010 (0010)
APPLICATION OF HIGH-PERFORMANCE LIQUID CHROMATOGRAPHY TO THE STUDY OF THE EFFECT OF MICROORGANISMS IN EMULSIFIABLE OILS
Almen, Rick; Mantelli, Gary; McTeer, Patrick; Nakayama, Stan
Coors Container Co. Golden USA US CO 80401
Lubr. Eng.; (82) P 99-103; Vol 38; No 2

09-82-0011 (0011)
RAPID DETERMINATION OF AROMATIC HYDROCARBONS IN SOLVENT NAPHTHA BY HPLC
Galtieri, A.; Mascherpa, A.
Stn. Sper. Combust. San Donato Milanese Italy IT
Riv. Combust.; (82) P 181-3; Vol 36; No 6; In Ital

CHAPTER 7.10
Glycosides

--- 1982 ---

10-82-0001 (0001)
HIGH-PERFORMANCE LIQUID CHROMATOGRAPHIC METHOD FOR DETERMINATION OF AMINOGLYCOSIDE IN SERUM
Watanabe, Makoto; Kondou, Yasuko; Mashimo, Keimei; Kawamoto, Takeshi; Shimadzu, Kunihiko
Dep. Pharm. Tokyo Kosei Nenkin Hosp. Tokyo Japan JP
Chemotherapy (Tokyo); (82) P 21-4; Vol 30; No 1; In Japan

10-82-0002 (0002)
HIGH-PERFORMANCE LIQUID CHROMATOGRAPHIC SEPARATION OF THE CARDIAC GLYCOSIDES OF CONVALLARIA MAJALIS L. BY COUPLING "REVERSED-PHASE" COLUMNS OF DIFFERENT POLARITY
Jurenitsch, J.; Kopp, B.; Bamberg-Kubelka, E.; Kern, R.; Kubelka, W.
Inst. Pharmakognosie Univ. Wien Vienna Austria AT A-1090
J. Chromatogr.; (82) P 125-35; Vol 240; No 1; In Ger

10-82-0003 (0003)
SEPARATION OF DIGOXIN, DIGITOXIN AND THEIR POTENTIAL METABOLITES, IMPURITIES OR DEGRADATION PRODUCTS BY HIGH-PERFORMANCE LIQUID CHROMATOGRAPHY
Desta, Belachew; Kwong, E.; McErlane, K. M.
Fac. Pharm. Sci. Univ. British Columbia Vancouver Can. CA BC V6T 1W5
J. Chromatogr.; (82) P 137-43; Vol 240; No 1

10-82-0004 (0004)
USE OF C18 REVERSED-PHASE LIQUID CHROMATOGRAPHY FOR THE ISOLATION OF MONOTERPENE GLYCOSIDES AND NORISOPRENOID PRECURSORS FROM GRAPE JUICE AND WINES
Williams, P. J.; Strauss, C. R.; Wilson, B.; Massy-Westropp, R. A.
Australian Wine Res. Inst. Glen Osmond Australia AU 5064
J. Chromatogr.; (82) P 471-80; Vol 235; No 2

10-82-0005 (0005)
CORRELATION BETWEEN THE RETENTION OF CARDIAC GLYCOSIDES IN REVERSED-PHASE HIGH-PERFORMANCE LIQUID CHROMATOGRAPHY WITH DIPHENYLSILYL STATIONARY PHASE, THE STRUCTURE OF THEIR MOLECULES AND THEIR BIOLOGICAL ACTIVITY
Daviidov, V. Ya.; Elizalde Gonzalez, M.; Kiselev, A. V.
Chem. Dep. Lomonosov State Univ. Moscow Moscow USSR SU 117234
J. Chromatogr.; (82) P 49-62; Vol 248; No 1

10-82-0006 (0006)
ISOLATION OF A HEMIACETAL OF CONDURANGOGLYCOSIDE A BY HIGH PERFORMANCE LIQUID CHROMATOGRAPHY
Braeckman, G.; De Bruyn, A.; Braeckman, P.; Van Severen, R.
Lab. Pharmacog. Galenical Technol. State Univ. Ghent Ghent Belg. BE B-9000
J. Pharm. Belg.; (82) P 195-201; Vol 37; No 3

10-82-0007 (0007)
HIGH-PERFORMANCE LIQUID CHROMATOGRAPHIC ANALYSIS OF DIGITOXIN FORMULATIONS
Desta, Belachew; McErlane, K. M.
Fac. Pharm. Sci. Univ. Br. Columbia Vancouver Can. CA BC V6T 1W5
J. Pharm. Sci.; (82) P 1018-20; Vol 71; No 9

10-82-0008 (0008)
QUANTITATIVE DETERMINATION OF DIGITOXIN BY HIGH PERFORMANCE LIQUID CHROMATOGRAPHY
Popov, D. M.; Nikolaev, N. N.; Semenov, V. A.
Vses. Nauchno-Issled. Inst. Fiz. Kul't. Moscow USSR SU
Khim.-Farm. Zh.; (82) P 761-4; Vol 16; No 6; In Russ

10-82-0009 (0009)
THE DETERMINATION OF AMINOGLYCOSIDE ANTIBIOTICS IN SERUM: A COMPARISON OF A HIGH PERFORMANCE LIQUID CHROMATOGRAPHIC METHOD WITH A MICROBIOLOGICAL ASSAY
Barends, D. M.; Rutgers, A.; Van Klingeren, B.; Hulshoff, A.
Dep. Anal. Pharm. Univ. Utrecht, Pharmaceutical Laboratory Utrecht Neth. NL 3511 GH
Pharm. Weekbl., Sci. Ed.; (82) P 104-11; Vol 4; No 4

10-82-0010 (0010)
QUALITATIVE AND QUANTITATIVE HPLC-ANALYSIS OF CARDIAC GLYCOSIDES. 4. IDENTIFICATION OF CARDENOLIDES BY HPLC ON C-18-REVERSED PHASE
Tittel, G.; Habermeier, H.; Wagner, H.
Inst. Pharm. Biol. Univ. Muenchen Munich Fed. Rep. Ger. DE D-8000/2
Planta Med.; (82) P 207-15; Vol 45; No 4; In Ger

10-82-0011 (0011)
ASSAY METHOD FOR SERUM AMINOGLYCOSIDES BY HIGH-PERFORMANCE LIQUID CHROMATOGRAPHY
Shinozaki, Kimikazu; Sasaki, Yasuhito; Takahashi, Satoru; Doi, Mari; Tanaka, Yoshio; Masuhara, Keiso; Arai, Sakae; Someya, Kazuhiko
Sch. Med. St. Marianna Univ. Kawasaki Japan JP 213
Rinsho Kagaku; (82) P 132-7; Vol 11; No 2; In Japan

CHAPTER 7.11
Inorganic Compounds

7.11.1
Metals and Metalloids

--- 1982 ---

11-82-0001 (0001)
METAL ION LIQUID CHROMATOGRAPHY
Raja, Ravi
Nat. Can Corp. USA US
Am. Lab. (Fairfield, Conn.); (82) P 35-7; Vol 14; No 7

11-82-0002 (0002)
SIMULTANEOUS DETERMINATION OF COPPER, NICKEL, COBALT, CHROMIUM(VI), AND CHROMIUM(III) BY LIQUID CHROMATOGRAPHY WITH ELECTROCHEMICAL DETECTION
Bond, A. M.; Wallace, G. G.
Div. Chem. Phys. Sci. Deakin Univ. Waurn Ponds Australia AU 3217
Anal. Chem.; (82) P 1706-12; Vol 54; No 11

11-82-0003 (0003)
SEPARATION OF CATIONIC METAL CHELATES OF 1,10-PHENANTHROLINE BY LIQUID CHROMATOGRAPHY
O'Laughlin, Jerome W.
Dep. Chem. Univ. Missouri Columbia USA US MO 65211
Anal. Chem.; (82) P 178-81; Vol 54; No 2

11-82-0004 (0004)
AUTOMATIC DIALYZING-INJECTION SYSTEM FOR LIQUID CHROMATOGRAPHY OF IONS AND SMALL MOLECULES
Nordmeyer, Francis R.; Hansen, Lee D.
Dep. Chem. Brigham Young Univ. Provo USA US UT 84602
Anal. Chem.; (82) P 2605-7; Vol 54; No 14

11-82-0005 (0005)
DETERMINATION OF METALS IN GROUNDWATERS BY TRACE ENRICHMENT AND LIQUID CHROMATOGRAPHY
Cassidy, R. M.; Elchuk, Steven; McHugh, J. O.
Environ. Res. Branch At. Energy Canada Ltd. Chalk River Can. CA ON K0J 1J0
Anal. Chem.; (82) P 727-31; Vol 54; No 4

11-82-0006 (0006)
SPECIATION OF CHROMIUM(III) AND CHROMIUM(VI) VIA REVERSED PHASE HPLC WITH INDUCTIVELY COUPLED PLASMA EMISSION SPECTROSCOPIC DETECTION (HPLC-ICP)
Krull, I. S.; Bushee, D.; Savage, R. N.; Schleicher, R. G.; Smith, S. B., Jr.
Inst. Chem. Anal. Northeastern Univ. Boston USA US MA 02115
Anal. Lett.; (82) P 267-81; Vol 15; No A3

11-82-0007 (0007)
MICRO HIGH-PERFORMANCE LIQUID CHROMATOGRAPHY OF ORGANICALLY BOUND METALS WITH INDUCTIVELY COUPLED PLASMA ATOMIC EMISSION SPECTROMETRIC DETECTION
Jinno, Kiyokatsu; Tsuchida, Hiroyuki
Sch. Mater. Sci. Toyohashi Univ. Technol. Toyohashi Japan JP 440
Anal. Lett.; (82) P 427-37; Vol 15; No A5

11-82-0008 (0008)
DETERMINATION OF METAL IONS BY LIQUID CHROMATOGRAPHY INCORPORATING DITHIOCARBAMATES IN THE ELUENT
Smith, Roger M.; Yankey, Lawrence E.
Dep. Chem. Univ. Technol. Loughborough UK GB LE11 3TU
Analyst (London); (82) P 744-8; Vol 107; No 1276

11-82-0009 (0009)
DETERMINATION OF URINARY COPROPORPHYRINS BY HIGH PRESSURE LIQUID CHROMATOGRAPHY. SIGNIFICANCE FOR THE PREVENTION OF LEAD POISONING

Pekari, K.; Arsac, F.; Esmaelzadeh, A.; Boudene, C.
Lab. Biochem. Inst. Occup. Health Helsinki Finland FI 00370/37
Arch. Mal. Prof. Med. Trav. Secur. Soc.; (82) P 449-54; Vol 43; No 6; In Fr

11-82-0010 (0010)
REVERSED-PHASE ION-PAIR PARTITION CHROMATOGRAPHY OF HALOGENATED ANION COMPLEXES OF PLATINUM METALS
Hayakawa, Kazuichi; Nishijima, Miwako; Miyazaki, Motoichi
Fac. Pharm. Sci. Kanazawa Univ. Kanazawa Japan JP
Bunseki Kagaku; (82) P 160-1; Vol 31; No 3; In Japan

11-82-0011 (0011)
SEPARATION OF SIXTEEN RARE EARTH ELEMENTS BY HIGH PERFORMANCE LIQUID CHROMATOGRAPHY
Zhao, Gui Wen; Luo, Si Quqn; Yao, Shou Ren; Yu, Shu Kun
Dep. Chem. Univ. Sci. Technol. China Hofei Peop. Rep. China CN
Bunseki Kagaku; (82) P 63-6; Vol 31; No 2; In Japan

11-82-0012 (0012)
HIGH PERFORMANCE REVERSED-PHASE LIQUID CHROMATOGRAPHY OF TRACE AMOUNTS OF INORGANIC AND ORGANIC MERCURY AS DIETHYLDITHIOCARBAMATE CHELATE
Inoue, Sadanobu; Hoshi, Suwaru; Sasaki, Mitsuo
Dep. Environ. Eng. Kitami Inst. Technol. Kitami Japan JP 090
Bunseki Kagaku; (82) P E243-E246; Vol 31; No 7

11-82-0013 (0013)
SIMULTANEOUS DETERMINATION OF HEAVY METALS IN WATER BY HIGH PERFORMANCE LIQUID CHROMATOGRAPHY FOLLOWING THE SOLVENT EXTRACTION OF HEAVY METALS AS PYRROLIDINEDITHIOCARBAMATE CHELATES INTO CHLOROFORM
Ichinoki, Susumu; Yamazaki, Mitsuru
Sch. Pharm. Hokuriku Univ. Kanazawa Japan JP 920-11
Bunseki Kagaku; (82) P E319-E326; Vol 31; No 10

11-82-0014 (0014)
STUDIES OF SOME ANTIMICROBIAL AGENTS. 1. DETERMINATION OF PENICILLINASE-RESISTANT PENICILLINS IN BIOLOGICAL FLUIDS BY HIGH PRESSURE LIQUID CHROMATOGRAPHY. II. MICROBIAL PHYSIOLOGY AND GENETICS OF MERCURY AND ORGANOMERCURIAL RESISTANCE IN ANAEROBIC BACTERIA
Rudrik, James Thomas
Wayne State Univ. Detroit USA US MI
From: Diss. Abstr. Int. B 1982, 43(2), 346.
DISSERTATION; (82) P 182 pp.; Avail.:Univ. Microfilms Int.,Order No. DA8216164

11-82-0015 (0015)
FINGERPRINTING INORGANIC ARSENIC AND ORGANOARSENIC COMPOUNDS IN IN SITU OIL SHALE RETORT AND PROCESS WATERS USING A LIQUID CHROMATOGRAPH COUPLED WITH AN ATOMIC ABSORPTION SPECTROMETER AS A DETECTOR
Fish, Richard H.; Brinckman, Frederick E.; Jewett, Kenneth L.
Lawrence Berkeley Lab. Univ. California Berkeley USA US CA 94720
Environ. Sci. Technol.; (82) P 174-9; Vol 16; No 3

11-82-0016 (0016)
QUANTITATIVE DETERMINATION OF VANADIUM, MOLYBDENUM, TUNGSTEN, COBALT AND CHROMIUM 8-HYDROXYQUINOLATES USING HPLC
Wenclawiak, B.
Anorg.-Chem. Inst. Univ. Muenster Muenster Fed. Rep. Ger. DE D-4400
Fresenius' Z. Anal. Chem.; (82) P 144-5; Vol 310; No 1-2; In Ger

11-82-0017 (0017)
CHROMATOGRAPHY OF METAL CHELATES. IV. HPLC OF THE PLATINUM METALS WITH

1-HYDROXY-2-PYRIDINETHIONE
Steinbrech, B.; Schneeweis, G.; Koenig, K. H.
Inst. Anorg. Chem. Univ. Frankfurt/Main Frankfurt/Main Fed. Rep. Ger. DE D-6000
Fresenius' Z. Anal. Chem.; (82) P 499-502; Vol 311; No 5; In Ger

11-82-0018 (0018)
HPLC DETERMINATION OF LEAD AND CADMIUM AFTER CHELATE EXTRACTION WITH SODIUM DIETHYLDITHIOCARBAMATE
Drasch, G.; Von Meyer, L.; Kauert, G.
Inst. Rechtsmed. Univ. Muenchen Munich Fed. Rep. Ger. DE D-8000/2
Fresenius' Z. Anal. Chem.; (82) P 695-6; Vol 311; No 7; In Ger

11-82-0019 (0019)
STUDY OF THE SEPARATION OF ALKALI METALS BY HIGH-PERFORMANCE LIQUID CHROMATOGRAPHY
Shen, Minxin; Zhao, Shouzhen; Wang, Meizhong
Inst. At. Energy Acad. Sin. Beijing Peop. Rep. China CN
He Huaxue Yu Fangshe Huaxue; (82) P 52-5; Vol 4; No 1; In Ch

11-82-0020 (0020)
STUDIES ON REVERSED-PHASE CHROMATOGRAPHY AND SOLVENT EXTRACTION SEPARATION OF SALTS OF RARE EARTHS BY 2-ETHYLHEXYLPHOSPHONIC ACID MONO(2-ETHYLHEXYL)ESTER
Cai, Qixiu; Yuan, Chengye
Shanghai Inst. Org. Chem. Acad. Sinica Shanghai Peop. Rep. China CN
Huaxue Xuebao; (82) P 563-70; Vol 40; No 6; In Ch

11-82-0021 (0021)
APPLICATION OF HPLC TO INORGANIC ANALYSIS. PART VII. COMPARISON BETWEEN PRE-COLUMN- AND ON-COLUMN DERIVATIZATION FOR SEPARATION OF DIFFERENT METAL OXINATES; QUANTITATIVE DETERMINATION OF MANGANESE(II) BESIDES MANGANESE(III) IONS
Hoffmann, B. W.; Schwedt, G.
Anorg.-Chem. Inst. Univ. Goettingen Goettingen Fed. Rep. Ger. DE D-3400
HRC CC, J. High Resolut. Chromatogr. Chromatogr. Commun.; (82) P 439-40; Vol 5; No 8

11-82-0022 (0022)
REVERSED-PHASE EXTRACTION CHROMATOGRAPHY OF MOLYBDENUM(VI) WITH TRIBUTYL PHOSPHATE ON SILICA GEL
Bhosale, S. N.; Khopkar, S. M.
Indian Inst. Technol. Bombay India IN 400 076
Indian J. Chem., Sect. A; (82) P 147-9; Vol 21A; No 2

11-82-0023 (0023)
HIGH-PERFORMANCE LIQUID CHROMATOGRAPHY STUDIES ON PLATINUM THYMINE BLUE
Woollins, J. Derek; Rosenberg, Barnett
Dep. Biophys. Michigan State Univ. East Lansing USA US MI 48824
Inorg. Chem.; (82) P 1280-2; Vol 21; No 3

11-82-0024 (0024)
SEPARATION OF CHLORO COMPLEXES OF METALS BY MICRO-HIGH-PERFORMANCE LIQUID CHROMATOGRAPHY
Tsuda, Takao; Nozu, Takashi; Nakagawa, Genkichi
Lab. Anal. Chem. Nagoya Inst. Technol. Nagoya Japan JP 466
J Chromatogr.; (82) P 331-6; Vol 242; No 2

11-82-0025 (0025)
COORDINATIVELY UNSATURATED METAL COMPLEXES AS MOBILE PHASE ADDITIVES IN RP-HPLC: BIS(2,3,6,6-TETRAMETHYLHEPTANE-3,5-DIONATO)NICKEL(II)

Lochmueller, C. H.; Hangac, H. H.
P. M. Gross Chem. Lab. Duke Univ. Durham USA US NC 27706
J. Chromatogr. Sci.; (82) P 171-5; Vol 20; No 4

11-82-0026 (0026)
REVERSED-PHASE HPLC OF .BETA.-DIKETONE METAL CHELATES
Gurira, R. C.; Carr, Peter W.
Dep. Chem. Grinnell Coll. Grinnell USA US IA 50112
J. Chromatogr. Sci.; (82) P 461-5; Vol 20; No 10

11-82-0027 (0027)
HIGH-PERFORMANCE LIQUID CHROMATOGRAPHY AND MAGNETIC CIRCULAR DICHROISM: A STUDY OF THE "PALLADIUM(II)-THIOETHER PEPTIDE" COMPLEXES
Lam-Thanh, Hung; Fermandjian, Serge; Fromageot, Pierre
Dep. Biol. Cent. Etud. Nucl. Saclay Gif-sur-Yvette Fr. FR F-91191
J. Chromatogr.; (82) P 139-47; Vol 235; No 1

11-82-0028 (0028)
HIGH-PERFORMANCE LIQUID CHROMATOGRAPHIC DETERMINATION OF TISSUE METALLO-THIONEIN IN MONKEYS CHRONICALLY EXPOSED TO CADMIUM
Nomiyama, Kazuo; Nomiyama, Hiroko
Dep. Environ. Health Jichi M2d. Sch. Japan JP 329-04
J. Chromatogr.; (82) P 285-91; Vol 228

11-82-0029 (0029)
HIGH-PERFORMANCE LIQUID CHROMATOGRAPHY OF METAL-TETRAPHENYLPORPHYRIN CHELATES ON A C18-BONDED STATIONARY PHASE
Saitoh, Koichi; Kobayashi, Masaru; Suzuki, Nobuo
Fac. Sci. Tohoku Univ. Sendai Japan JP 980
J. Chromatogr.; (82) P 291-300; Vol 243; No 2

11-82-0030 (0030)
REVERSED-PHASE ION-PAIR HIGH-PERFORMANCE LIQUID CHROMATOGRAPHY OF MERCAPTO-ACETATE AND N-ACETYLCYSTEINE AFTER DERIVATIZATION WITH N-(1-PYRENE)MALEIMIDE AND N-(7-DIMETHYLAMINO-4-METHYL-3-COUMARINYL)- MALEIMIDE
Kaagedal, Bertil; Kaellberg, Magnus
Dep. Clin. Chem. Univ. Linkoeping Linkoeping Swed. SE S-581 85
J. Chromatogr.; (82) P 409-15; Vol 229; No 2

11-82-0031 (0031)
HIGH-PERFORMANCE LIQUID CHROMATOGRAPHY: APPLICATIONS TO ORGANOMETALLIC AND METAL COORDINATION COMPOUNDS
Willeford, Bennett R.; Veening, Hans
Dep. Chem. Bucknell Univ. Lewisburg USA US PA 17837
J. Chromatogr.; (82) P 61-88; Vol 251; No 1

11-82-0032 (0032)
DETERMINATION OF CHROMIUM IN ORCHARD LEAVES BY REVERSED-PHASE HIGH-PERFORMANCE LIQUID CHROMATOGRAPHY
Willett, James D.; Knight, Margaret M.
Dep. Chem. Univ. Idaho Moscow USA US ID 83843
J. Chromatogr.; (82) P 99-105; Vol 237; No 1

11-82-0033 (0033)
PEPTIDE FORMATION IN THE PRESENCE OF METAL ION PROTECTING GROUPS. III. THE SEPARATION OF ((NH3)5Co(III)-) AMINO ACIDS AND PEPTIDES BY REVERSE-PHASE HIGH-PRESSURE LIQUID CHROMATOGRAPHY
Isied, Stephan S.; Lyon, John; Vassilian, Asbed; Worosila, Greg
Dep. Chem. Rutgers, State Univ. New Brunswick USA US NJ 08903
J. Liq. Chromatogr.; (82) P 537-47; Vol 5; No 3

11-82-0034 (0034)
METAL CATION SPECIATION VIA EXTRACTION REVERSED-PHASE HPLC WITH REFRACTIVE INDEX AND/OR INDUCTIVELY COUPLED PLASMA EMISSION DETECTION METHODS (HPLC-RI-ICP)
Bushee, D.; Young, D.; Krull, I. S.; Savage, R. N.; Smith, S. B., Jr.
Inst. Chem. Anal. Northeastern Univ. Boston USA US MA 02115
J. Liq. Chromatogr.; (82) P 693-706; Vol 5; No 4

11-82-0035 (0035)
APPLICATION OF HIGH PERFORMANCE LIQUID CHROMATOGRAPHY IN THE CHARACTERIZATION OF IRON STRESS RESPONSE
Johnson, Gordon V.
Dep. Biol. Univ. New Mexico Albuquerque USA US NM 87131
J. Plant Nutr.; (82) P 499-514; Vol 5; No 4-7

11-82-0036 (0036)
IMPROVED METHODS FOR THE ISOLATION AND STUDY OF THE C18, C20 AND C22 MONOETHYLENIC FATTY ACID ISOMERS OF BIOLOGICAL SAMPLES: MERCURY ADDUCTS, HPLC, SILVER NITRATE-TLC/FID, AND OZONOLYSIS
Sebedio, J. L.; Farquharson, T. E.; Ackman, R. G.
Fish. Res. Technol. Lab. Tech. Univ. Nova Scotia Halifax Can. CA NS B3J 2X4
Lipids; (82) P 469-75; Vol 17; No 6

11-82-0037 (0037)
CHEMICAL STUDIES OF COPPER-ORGANIC COMPLEXES ISOLATED FROM ESTUARINE WATERS USING C18 REVERSE-PHASE LIQUID CHROMATOGRAPHY
Mills, Gary L.; Hanson, Alfred K., Jr.; Quinn, James G.; Lammela, William R.; Chasteen, N. Dennis
Grad. Sch. Oceanogr. Univ. Rhode Island Kingston USA US RI 02881
Mar. Chem.; (82) P 355-77; Vol 11; No 4

11-82-0038 (0038)
REVERSED-PHASE LIQUID CHROMATOGRAPHY OF CHROMIUM WITH TRIBUTYL PHOSPHATE ON SILICA GEL
Bhosale, S. N.; Khopkar, S. M.
Dep. Chem. Indian Inst. Technol. Bombay India IN 400076
Mikrochim. Acta; (82) P 433-9; Vol 1; No 5-6

11-82-0039 (0039)
LIQUID ADSORPTION CHROMATOGRAPHY OF CHELATES. MECHANISM OF THE SEPARATION OF METAL DIETHYLDITHIOCARBAMATES BY THIN-LAYER CHROMATOGRAPHY
Timerbaev, A. R.; Petrukhin, O. M.; Zolotov, Yu. A.
V. I. Vernadskii Inst. Geochem. Anal. Chem. Moscow USSR SU
Zh. Anal. Khim.; (82) P 1360-9; Vol 37; No 8; In Russ

11-82-0040 (0040)
DETERMINATION OF BENZYLCHROMIUMTRICARBONYL DERIVATIVES OF MERCURY BY LIQUID ADSORPTION CHROMATOGRAPHY
Shushunova, A. F.; Prokhorova, L. P.; Kasatkina, T. G.; Dodonov, V. A.
Sci.-Res. Inst. Chem. Gorkiy State Univ. Gorkiy USSR SU
Zh. Anal. Khim.; (82) P 620-3; Vol 37; No 4; In Russ

7.11.2
Various Inorganics

--- 1982 ---

11-82-0001 (0001)
SEPARATION OF INORGANIC AND ORGANIC ANIONS ON REVERSED-PHASE LIQUID CHROMATOGRAPHY COLUMNS
Skelly, Norman E.
Anal. Lab. Dow Chem. Co. Midland USA MI 48640
Anal. Chem.; (82) P 712-15; Vol 54; No 4

11-82-0002 (0002)
SIMPLE, SENSITIVE, AND SIMULTANEOUS DETERMINATION OF SOME SELECTED INORGANIC ANIONS BY HIGH-PERFORMANCE LIQUID CHROMATOGRAPHY
De Kleijn, J. P.
Food Insp. Serv. Assen Neth. NL 9405 BJ
Analyst (London); (82) P 223-5; Vol 107; No 1271

11-82-0003 (0003)
DETERMINATION ORTHIOCYANATE, THIOSULFATE, SULFITE AND NITRITE BY HIGH-PERFORMANCE LIQUID CHROMATOGRQPHY COUPLED WITH ELECTROCHEMICAL DETECTION
Imanari, Toshio; Ogata, Koreharu; Tanabe, Shinzo; Toida, Toshihiko ; Kawanishi, Takehiko; Ichikawa, Masaki
Fac. Pharm. Sci. Chiba Univ. Chiba Japan JP
Chem. Pharm. Bull.; (82) P 74-5; Vol 30; No 1

11-82-0004 (0004)
FINGERPRINTING INORGANIC ARSENIC AND ORGANOARSENIC COMPOUNDS IN IN SITU OIL SHALE RETORT AND PROCESS WATERS USING A LIQUID CHROMATOGRAPH COUPLED WITH AN ATOMIC ABSORPTION SPECTROMETER AS A DETECTOR
Fish, Richard H.; Brinckman, Frederick E.; Jewett, Kenneth L.
Lawrence Berkeley Lab. Univ. California Berkeley USA US CA 94720
Environ. Sci. Technol.; (82) P 174-9; Vol 16; No 3

11-82-0005 (0005)
DETERMINATION OF NITRATE AND NITRITE IN WATER WITH HPLC AND AMPEROMETRIC DETECTION
Alawi, Mahmoud A.
Ges. Oekol. Gewaesserkd. Fed. Rep. Ger. DE D-3006
Fresenius' Z. Anal. Chem.; (82) P 239-40; Vol 313; No 3

11-82-0006 (0006)
HIGH-PERFORMANCE LIQUID CHROMATOGRAPHIC DETERMINATION OF NITRITE IN ENVIRONMENTAL SAMPLES BY THE USE OF HYDRALAZINE
Noda, Hiroshi; Minemoto, Masao; Asahara, Toshio; Noda, Atsuko; Iguchi, Sadao
Sch. Med. Univ. Occup. Environ. Health Kitakyushu Japan JP 807
J. Chromatogr.; (82) P 187-95; Vol 235; No 1

11-82-0007 (0007)
HIGH-PERFORMANCE LIQUID CHROMATOGRAPHY OF INORGANIC AND ORGANIC ANIONS USING ULTRAVIOLET DETECTION AND AN AMINO COLUMN
Cortes, Hernan J.
Anal. Lab. Dow Chem. Co. Midland USA US MI 48640
J. Chromatogr.; (82) P 517-20; Vol 234; No 2

11-82-0008 (0008)
STUDY OF THE ACID HYDROLYSIS OF CYCLIC TETRAMETAPHOSPHATE BY LIQUID CHROMATOGRAPHY
Kura, Genichiro
Dep. Chem. Fukuoka Univ. Educ. Munakata Japan JP 811-41
J. Chromatogr.; (82) P 73-80; Vol 246; No 1

11-82-0009 (0009)
REVERSED-PHASE HIGH-PERFORMANCE LIQUID CHROMATOGRAPHY OF CARBON DIOXIDE
Nieass, C. S.; Wainwright, M. S.; Chaplin, R. P.
Sch. Chem. Eng. Ind. Chem. Univ. New South Wales Kensington Australia AU 2033
J. Liq. Chromatogr.; (82) P 2179-91; Vol 5; No 11

CHAPTER 7.12
Lipids

--- 1982 ---

12-82-0001 (0001)
HIGH PERFORMANCE LIQUID CHROMATOGRAPHY AND STRUCTURAL ANALYSES BY FIELD DESORPTION MASS SPECTROMETRY OF UNDERIVATIZED GLYCOLIPID
Handa, Shizuo; Kushi, Yasunori
Fac. Med. Tokyo Med. Dent. Univ. Tokyo Japan JP 13
Adv. Exp. Med. Biol.; (82) P 23-31; Vol 152

12-82-0002 (0002)
PHOSPHOLIPID SEPARATIONS BY HPLC
Yandrasitz, John R.
Div. Biochem. Dev. Mol. Dis. Child. Hosp. Philadelphia Philadelphia USA US PA 19104
Analytika (Johannesburg); (82) P 8-9; No Aug.

12-82-0003 (0003)
HPLC ANALYSES OF THE ALDEHYDIC PATTERNS PRODUCED DURING LIPID PEROXIDATION
Cecchini, G.; Biasi, F.; Chiarpotto, E.; Poli, G.
Ist. Patol. Gen. Univ. Torino Turin Italy IT 10125
Boll. - Soc. Ital. Biol. Sper.; (82) P 1195-8; Vol 58; No 18

12-82-0004 (0004)
SIMPLIFIED DETERMINATION OF FREE FATTY ACIDS USING DERIVATIZATION TECHNIQUE BY HIGH-PERFORMANCE LIQUID CHROMATOGRAPHY THROUGH THE MODIFIED EXTRACTION COLUMN
Ikeda, Mitsuo; Shimada, Kenji; Sakaguchi, Takeichi
Niigata Coll. Pharm. Niigata Japan JP 950-21
Bunseki Kagaku; (82) P E199-E206; Vol 31; No 6

12-82-0005 (0005)
APPLICATION OF 1-NAPHTHYLAMINE TO LABELING OF FATTY ACID FOR HIGH-PERFORMANCE LIQUID CHROMATOGRAPHY
Ikeda, Mitsuo; Shimada, Kenji; Sakaguchi, Takeichi
Niigata Coll. Pharm. Niigata Japan JP 950-21
Chem. Pharm. Bull.; (82) P 2258-61; Vol 30; No 6

12-82-0006 (0006)
DETECTION OF VEGETABLE OIL ADULTERATION: HPLC ANALYSIS OF TRIGLYCERIDES
Hwang, Jenn Kang; Yen, Gow Chin; Hwang, Lucy Sun
Grad. Inst. Food Sci. Technol. Natl. Taiwan Univ. Taipei Taiwan TW
Chung-kuo Nung Yeh Hua Hsueh Hui Chih; (82) P 91-101; Vol 20; No 1-2; In Ch

12-82-0007 (0007)
LIQUID-CHROMATOGRAPHIC ESTIMATION OF SATURATED PHOSPHOLIPID PALMATE IN AMNIOTIC FLUID COMPARED WITH A THIN-LAYER CHROMATOGRAPHIC METHOD FOR ACETONE-PRECIPITATE LECITHIN
Sax, Sylvan M; Moore, John J; Oley, Alan; Amenta, Joseph S; Silvermann, Jeffrey A
Dep. Pathol., Western Pennsylvania Hosp. Pittsburg USA, PA 15224
Clin. Chem. (Winstom-Salem, N.C.) (82) P 2264-8; Vol. 28, No 11

12-82-0008 (0008)
MICROMETHOD FOR LIQUID-CHROMATOGRAPHIC DETERMINATION OF CHOLESTEROL IN LIPEMIC SERUMS
Webb, Lawrence E.; Lengle, Ernest; Izzo, Rodger; Woods, Eugene
Travenol. Lab., Inc. Morton Grove USA US IL 60053
Clin. Chem. (Winston-Salem, N. C.); (82) P 1769-72; Vol 28; No 8

12-82-0009 (0009)
USE OF HPLC TO DETERMINE THE TURNOVER OF MOLECULAR SPECIES OF PHOSPHOLIPIDS

Jungalwala, Firoze B.; Sanyal, Sankar; LeBaron, Francis
Dep. Biochem. E. K. Shriver Cent. Waltham USA US MA 02254
CONFERENCE; (82), Phospholipids Nerv. Syst., (Satell. Symp. Int. Soc. Neurochem.); P 91-103 ; Vol 1,; Avail.: Horrocks, Lloyd A.; Ansell, G. Brian; Porcellati, Giuseppe; Date: 81; Publisher: Raven, New York, N. Y.

12-82-0010 (0010)
TRIGLYCERIDE COMPOSITION OF HAZEL NUT BY HIGH PERFORMANCE LIQUID CHROMATOGRAPHY (HPLC)
Kim, Mi Ran; Ko, Young Su; Chung, Bo Sup
Inst. Food Sci. Hanyang Univ. Seoul S. Korea KR 133
Hanguk Sikp'um Kwahakhoe Chi; (82) P 122-4; Vol 14; No 2; In Korean

12-82-0011 (0011)
REVERSED-PHASE HIGH-PERFORMANCE LIQUID CHROMATOGRAPHY OF FATTY ACID METHYL ESTERS WITH PARTICULAR REFERENCE TO CYCLOPROPENOIC AND CYCLOPROPANOIC ACIDS
Bianchini, J. P.; Ralaimanarivo, A.; Gaydou, E. M.
Ec. Super. Chim. Marseille Univ. Droit, d'Econ. Sci. Marseille Fr. FR 13397/4
HRC CC, J. High Resolut. Chromatogr. Chromatogr. Commun.; (82) P 199-204; Vol 5; No 4

12-82-0012 (0012)
QUANTITATION METHOD FOR CHOLINE-CONTAINING PHOSPHOLIPIDS IN HUMAN SERUM LIPOPROTEINS BY HIGH PERFORMANCE LIQUID CHROMATOGRAPHY
Okazaki, Mitsuyo; Hagiwara, Noriko; Hara, Ichiro
Dep. Gen. Educ. Tokyo Med. Dent. Univ. Ichikawa Japan JP 272
J. Biochem. (Tokyo); (82) P 1381-9; Vol 91; No 4

12-82-0013 (0013)
HPLC ANALYSIS OF UNDERIVATIZED FATTY ACIDS IN MARGARINES
Bailie, Allston G., Jr.; Wilson, Terry D.; O'Brien, Robert K.; Beebe, Janet M.; Stuart, James D.; McCosh-Lilie, Elizabeth J.; Hill, Dennis W.
Coll. Agric. Nat. Resour. Univ. Connecticut Storrs USA US CT 06268
J. Chromatogr. Sci.; (82) P 466-70; Vol 20; No 10

12-82-0014 (0014)
HIGH-PERFORMANCE LIQUID CHROMATOGRAPHY OF HUMAN SERUM LIPOPROTEINS. SELECTIVE DETECTION OF CHOLINE-CONTAINING PHOSPHOLIPIDS BY ENZYMIC REACTION
Okazaki, Mtsuyo; Hagiwara, Noriko; Hara, Ichiro
Dep. Gen. Educ. Tokyo Med. Dent. Univ. Chiba Japan JP 272
J. Chromatogr.; (82) P 13-23; Vol 231; No 1

12-82-0015 (0015)
LIQUID CHROMATOGRAPHY-MASS SPECTROMETRY OF POLAR LIPIDS. COMPARISON OF ON-LINE METHOD USING DIRECT LIQUID INTRODUCTION INTERFACE WITH OFF-LINE METHOD USING DESORPTION/CHEMICAL IONIZATION
Sugnaux, Francois R.; Djerassi, Carl
Dep. Chem. Stanford Univ. Stanford USA US CA 94305
J. Chromatogr.; (82) P 189-201; Vol 251; No 2

12-82-0016 (0016)
APPLICATION OF LIQUID CHROMATOGRAPHIC AND SPECTROSCOPIC METHODS FOR THE CHARACTERIZATION OF FATTY ACID ANILIDES IN CONTAMINATED COOKING OILS
Wheals, B. B.; Whitehouse, M. J.; Curry, C. J.
Metrop. Police Forensic Sci. Lab. London UK GB SE1
J. Chromatogr.; (82) P 203-15; Vol 238; No 1

12-82-0017 (0017)
NOVEL SYSTEM FOR SEPARATION OF PHOSPHOLIPIDS BY HIGH-PERFORMANCE LIQUID CHROMATOGRAPHY

Alam, Iftekhar; Smith, J. Bryan; Silver, Melvin J.; Ahern, David
Dep. Pharmacol. Thomas Jeffersov Univ. Philadelphia USA US PA
J. Chromatogr.; (82) P 218-21; Vol 234; No 1

12-82-0018 (0018)
SENSITIVE HIGH-PERFORMANCE LIQUID CHROMATOGRAPHIC METHOD FOR PROSTAGLANDINS USING A FLUORESCENCE REAGENT, 4-BROMOMETHYL-7-ACETOXYCOUMARIN
Tsuchiya, Hironori; Hayashi, Tokishi; Naruse, Hiroshi; Takagi, Nobuhiko
Dep. Dent. Pharmacol. Gifu Coll. Dent. Gifu Japan JP
J. Chromatogr.; (82) P 247-54; Vol 231; No 2

12-82-0019 (0019)
IMPROVED PROCEDURE FOR THE SEPARATION OF PHOSPHOLIPIDS BY HIGH-PERFORMANCE LIQUID CHROMATOGRAPHY
Chen, S. Shi Hua; Kou, Anne Y.
Pathol. Dep. Stanford Univ. Stanford USA US CA 94305
J. Chromatogr.; (82) P 25-31; Vol 227; No 1

12-82-0020 (0020)
IMPROVED SEPARATION OF BIOLOGICALLY RELEVANT C14-C20 FATTY ACIDS BY REVERSED-PHASE HIGH-PERFORMANCE LIQUID CHROMATOGRAPHY
Halgunset, J.; Lund, E. W.; Sunde, A.
Dep. Pathol. Univ. Trondheim Trondheim Norway NO 7000
J. Chromatogr.; (82) P 496-9; Vol 237; No 3

12-82-0021 (0021)
HIGH-PERFORMANCE LIQUID CHROMATOGRAPHY OF HUMAN SERUM LIPOPROTEINS. SELECTIVE DETECTION OF TRIGLYCERIDES BY ENZYMIC REACTION
Hara, Ichiro; Shiraishi, Keiko; Okazaki, Mitsuyo
Dep. Gen. Educ. Tokyo Med. Dent. Univ. Ichikawa Japan JP 272
J. Chromatogr.; (82) P 549-57; Vol 239

12-82-0022 (0022)
SEPARATION OF PHOSPHOLIPIDS AND INDIVIDUAL MOLECULAR SPECIES OF PHOSPHOLIPIDS BY HIGH-PERFORMANCE LIQUID CHROMATOGRAPHY
Patton, George M.; Fasulo, Joan M.; Robins, Sander J.
Dep. Med. VA Med. Cent. Boston USA US MA 02130
J. Lipid Res.; (82) P 190-6; Vol 23; No 1

12-82-0023 (0023)
ENZYMIC DETERMINATION OF TRIGLYCERIDES IN CONJUNCTION WITH HIGH-PERFORMANCE LIQUID CHROMATOGRAPHY
Kruempelman, Maribeth; Danielson, Neil D.
Dep. Chem. Miami Univ. Oxford USA US OH 45056
J. Liq. Chromatogr.; (82) P 1679-89; Vol 5; No 9

12-82-0024 (0024)
DETERMINATION OF FREE FATTY ACIDS IN NATURAL OILS AND ALKYD RESINS BY HIGH-PERFORMANCE LIQUID CHROMATOGRAPHY
King, Jerry W.; Adams, Edwin C.; Bidlingmeyer, Brian A.
Res. Inst. Univ. Dayton Dayton USA US OH 45469
J. Liq. Chromatogr.; (82) P 275-304; Vol 5; No 2

12-82-0025 (0025)
VALIDATION OF COMPUTATIONAL METHODS FOR TRIGLYCERIDE COMPOSITION OF FATS AND OILS BY LIQUID CHROMATOGRAPHY AND MASS SPECTROMETRY
Merritt, C., Jr.; Vajdi, M.; Kayser, S. G.; Halliday, J. W.; Bazinet, M. L.
Sci. Adv. Technol. Lab. U. S. Army Natick Res. Dev. Lab. Natick USA US MA 01760
JAOCS, J. Am. Oil Chem. Soc.; (82) P 422-32; Vol 59; No 10

12-82-0026 (0026)
ANALYSIS OF PHOSPHATIDYLCHOLINE IN SOY LECITHINS BY HPLC
Rhee, J. S.; Shin, M. G.
Dep. Biol. Sci. Eng. Korea Adv. Inst. Sci. Technol. Seoul S. Korea KR 131
JAOCS, J. Am. Oil Chem. Soc.; (82) P 98-9; Vol 59; No 2

12-82-0027 (0027)
HIGH-PERFORMANCE LIQUID CHROMATOGRAPHIC SEPARATION IN SHORT COLUMNS OF INDIVIDUAL TOCOPHEROLS AND TOCOTRIENOLS IN EDIBLE FATS AND OILS
Gertz, Christian; Herrmann, Karl
Chem. Untersuchungsamt Stadt Hagen Hagen Fed. Rep. Ger. DE 5800/1
Lebensmittelchem. Gerichtl. Chem.; (82) P 53-8; Vol 36; No 3; In Ger

12-82-0028 (0028)
HIGH-PERFORMANCE REVERSED-PHASE CHROMATOGRAPHY OF THE TRIGLYCERIDES FROM HUMAN PLASMA LIPOPROTEINS
Perkins, E. G.; Hendren, David J.; Pelick, Nicholas; Bauer, John E.
Dep. Food Sci. Univ. Illinois Urbana USA US IL 61801
Lipids; (82) P 460-3; Vol 17; No 6

12-82-0029 (0029)
DIRECT ESTIMATION OF DOLICHYL PHOSPHATE IN RAT LIVER BY HIGH-PRESSURE LIQUID CHROMATOGRAPHY
Chaudhary, Nilabh; Freeman, David J.; Rip, Jack W.; Carroll, Kenneth K.
Dep. Biochem. Univ. West. Ontario London Can. CA ON N6A 5C1
Lipids; (82) P 558-60; Vol 17; No 8

12-82-0030 (0030)
SEPARATION OF ARACHIDONIC ACID METABOLITES BY HIGH-PRESSURE LIQUID CHROMATOGRAPHY
Eling, Thomas; Tainer, Beth; Ally, Ariff; Warnock, Robert
Natl. Inst. Environ. Health Sci. Research Triangle Park USA US NC 27713
Methods Enzymol.; (82) P 511-17; Vol 86; No Prostaglandins Arachidonate Metab.

12-82-0031 (0031)
ARGENTATION-HIGH PRESSURE LIQUID CHROMATOGRAPHY OF PROSTAGLANDINS AND MONOHYDROXYEICOSENOIC ACIDS
Powell, William S.
R. Victoria Hosp. McGill Univ. Montreal Can. CA PQ H3A 1A1
Methods Enzymol.; (82) P 530-43; Vol 86; No Prostaglandins Arachidonate Metab.

12-82-0032 (0032)
HIGH-PERFORMANCE LIQUID CHROMATOGRAPHY (HPLC) OF ARACHIDONIC ACID METABOLITES
Hamilton, James G.; Karol, Robert J.
Dep. Pharmacol. II Hoffmann-La Roche Inc. Nutley USA US NJ 07110
Prog. Lipid Res.; (82) P 155-70; Vol 21; No 3

12-82-0033 (0033)
RECENT PROGRESS IN THE HIGH-PERFORMANCE LIQUID CHROMATOGRAPHY OF LIPIDS
Aitzetmueller, K.
Unilever Forschungsges. m.b.H. Hamburg Fed. Rep. Ger. DE D-2000/50
Prog. Lipid Res.; (82) P 171-93; Vol 21; No 3

12-82-0034 (0034)
DETERMINATION OF SESAMOLIN, SESAMIN AND SESAMOL IN SESAME OIL BY HIGH-PERFORMANCE LIQUID CHROMATOGRAPHY
Yoshida, Masaharu; Kashimoto, Takashi
Osaka Pref. Inst. Public Health Osaka Japan JP
Shokuhin Eiseigaku Zasshi; (82) P 142-8; Vol 23; No 2; In Japan

12-82-0035 (0035)
DETERMINATION OF THE FATTY ACID COMPOSITION OF SOYBEAN OIL BY HIGH-PRESSURE LIQUID CHROMATOGRAPHY
Jaselskis, Bruno; Stemm, Nicholas L.; Johnston, William D.
Loyola Univ. Chicago Chicago USA US IL 60626
Talanta; (82) P 54-6; Vol 29; No 1

12-82-0036 (0036)
DETERMINATION OF 15-METHYLPROSTAGLANDIN F2 ALPHA BY DERIVATIZATION HIGH-PRESSURE LIQUID CHROMATOGRAOHY
Wang, Zhongshan; Zhu, Yaohua; Zhang, Shuliang; Zhu, Jiyu; Wang, Meili
Shanghai Int. Drug Control, Shanghai,Peop. Rep. China
Yaoxue Xuebao (82) P 603-8; Vol 17; No 8; In Chin

12-82-0037 (0037)
A QUANTITATION REAGENT SPECIFIC FOR CHOLINE-CONTAINING PHOSPHOLIPIDS IN HIGH PERFORMANCE LIQUID CHROMATOGRAPHY FOR SERUM LIPOPROTEINS
Hagiwara, Noriko; Okazaki, Mitsuyo; Hara, Ichiro
Dep. Gen. Educ. Tokyo Med. Dent. Univ. Ichikawa Japan JP
Yukagaku; (82) P 262-7; Vol 31; No 5

12-82-0038 (0038)
STUDIES ON PRODUCTION OF LIPIDS IN FUNGI. VII. DETERMINATION OF POLAR LIPIDS IN MICROORGANISMS BY HIGH-PERFORMANCE LIQUID CHROMATOGRAPHY
Yokochi, Toshihiro; Suzuki, Osamu
Natl. Chem. Lab. Ind. Ibaraki Japan JP
Yukagaku; (82) P 580-5; Vol 31; No 9; In Japan

CHAPTER 7.13
Nucleic Acids

--- 1982 ---

13-82-0001 (0001)
FRACTIONATION OF BIOLOGICALLY ACTIVE MESSENGER RNAs BY HPLC GEL FILTRATION
Graeve, Lutz; Goemann, Wolfgang; Foeldi, Peter; Kruppa, Joachim
Inst. Physiol. Chem. Univ. Hamburg Hamburg Fed. Rep. Ger. DE D 2000/13
Biochem. Biophys. Res. Commun.; (82) P 1559-65; Vol 107; No 4

13-82-0002 (0002)
HYPOMETHYLATION OF HELA CELL DNA AND THE ABSENCE OF 5-METHYLCYTOSINE IN SV40 AND ADENOVIRUS (TYPE 2) DNA: ANALYSIS BY HPLC
Diala, Edward S.; Hoffman, Robert M.
Sch. Med. Univ. California, San Diego La Jolla USA US CA 92093
Biochem. Biophys. Res. Commun.; (82) P 19-26; Vol 107; No 1

13-82-0003 (0003)
A FLUOROMETRIC-HPLC ASSAY FOR QUANTITATING THE BINDING OF BENZO(A)PYRENE METABOLITES TO DNA
Rahn, R. O.; Chang, S. S.; Holland, J. M.; Shugart, L. R.
Biol. Div. Oak Ridge Natl. Lab. Oak Ridge USA US TN 37830
Biochem. Biophys. Res. Commun.; (82) P 262-8; Vol 109; No 1

13-82-0004 (0004)
SIZE FRACTIONATION OF DNA FRAGMENTS RANGING FROM 20 TO 30,000 BASE PAIRS BY LIQUID/LIQUID CHROMATOGRAPHY
Mueller, Werner; Kuetemeier, Gertrud
Abt. Viroidforsch. Max-Planck-Inst. Biochem. Martinsried Fed. Rep. Ger. DE
Eur. J. Biochem.; (82) P 231-8; Vol 128; No 1

13-82-0005 (0005)
RAPID METHOD FOR PURIFICATION OF PLASMID DNA AND DNA FRAGMENTS FROM DNA LINKERS USING HIGH-PERFORMANCE LIQUID CHROMATOGRAPHY ON TSK-PW GEL
Himmel, Michael E.; Perna, Peter J.; McDonell, Michael W.
Biotechnol. Branch Solar Energy Res. Inst. Golden USA US CO 80401
J. Chromatogr.; (82) P 155-63; Vol 240; No 1

13-82-0006 (0006)
DETECTION OF BINDING BETWEEN TRANSFER RNA AND RNA POLYMERASE BY HIGH-PERFORMANCE LIQUID CHROMATOGRAPHY
Amero, Sally
Div. Plant Sci. West Virginia Univ. Morgantown USA US WV 26506
J. Chromatogr.; (82) P 413-15; Vol 241; No 2

13-82-0007 (0007)
RESOLUTION OF RNA BY PAIRED-ION REVERSED-PHASE HIGH-PERFORMANCE LIQUID CHROMATOGRAPHY
Nguyen Phi Nga; Bradley, Janice L.; McGuire, Peter M.
Dep. Biochem. Mol. Biol. Univ. Florida Gainesville USA US FL 32610
J. Chromatogr.; (82) P 508-12; Vol 236; No 2

13-82-0008 (0008)
HPLC ANALYSIS OF NUCLEIC ACIDS IN FOOD AND OF NUCLEIC ACID METABOLITES IN METABOLIC STUDIES
Tiemeyer, W.; Gaebler, S.; Giesecke, D.
Tieraerztl. Fak. Univ. Muenchen Munich Fed. Rep. Ger. DE D-8000/22
Landwirtsch. Forsch., Sonderh.; (82) P 441-8; Vol 38,; In Ger

CHAPTER 7.14

Nucleic Acid Components

--- 1982 ---

14-82-0001 (0001)
A RAPID METHOD FOR THE DETERMINATION OF GUANOSINE 5'-DIPHOSPHATE-3'-DIPHOSPHATE AND GUANOSINE 5'-TRIPHOSPHATE-3'-DIPHOSPHATE BY HIGH-PERFORMANCE LIQUID CHROMATOGRAPHY
Fischer, Meir; Zimmerman, Thomas P.; Short, Steven A.
Wellcome Res. Lab. Research Triangle Park USA US NC 27709
Anal. Biochem.; (82) P 135-9; Vol 121; No 1

14-82-0002 (0002)
DETERMINATION OF PLASMA HYPOXANTHINE: A COMPARISON OF HIGH-PRESSURE LIQUID CHROMATOGRAPHIC AND OXYGEN CONSUMPTION METHODS
Saugstad, O. D.; Wung, W. E.; Howell, S. B.; Gluck, L.
Cancer Cent. Univ. California La Jolla USA US CA 92093
Anal. Biochem.; (82) P 159-63; Vol 122; No 1

14-82-0003 (0003)
REVERSE-PHASE ION-PAIR CHROMATOGRAPHY OF POLYVALENT IONS USING OLIGO-NUCLEOTIDES AS MODEL SUBSTANCES
Jost, W.; Unger, K.; Schill, G.
Inst. Anorg. Chem. Anal. Chem. Johannes-Gutenberg Univ. Mainz Fed. Rep. Ger. DE 6500
Anal. Biochem.; (82) P 214-23; Vol 119; No 2

14-82-0004 (0004)
SEPARATION AND QUANTITATION OF PURINES AND THEIR ANAEROBIC AND AEROBIC DEGRADATION PRODUCTS BY HIGH-PRESSURE LIQUID CHROMATOGRAPHY
Duerre, Peter; Andreesen, Jan R.
Inst. Mikrobiol. Univ. Goettingen Goettingen Fed. Rep. Ger. DE D-3400
Anal. Biochem.; (82) P 32-40; Vol 123; No 1

14-82-0005 (0005)
HIGH-PRESSURE LIQUID CHROMATOGRAPHY-ULTRAVIOLET ANALYSIS OF INTRACELLULAR NUCLEOTIDES
Pogolotti, Alfonso L., Jr.; Santi, Daniel V.
Dep. Biochem. Biophys. Univ. California San Francisco USA US CA 94143
Anal. Biochem.; (82) P 335-45; Vol 126; No 2

14-82-0006 (0006)
THE USE OF HIGH-PERFORMANCE LIQUID CHROMATOGRAPHY IN THE ISOLATION AND ANALYSIS OF OLIGORIBONUCLEOTIDES SYNTHESIZED BY THE T4 RNA LIGASE REACTION
McLaughlin, Larry W.; Romaniuk, Elena
Abt. Chem. Max-Planck-Inst. Exp. Med. Goettingen Fed. Rep. Ger. DE D-3400
Anal. Biochem.; (82) P 37-44; Vol 124; No 1

14-82-0007 (0007)
SEPARATION OF MAJOR AND MINOR DEOXYRIBONUCLEOSIDE MONOPHOSPHATES BY REVERSE-PHASE HIGH-PERFORMANCE LIQUID CHROMATOGRAPHY: A SIMPLE METHOD APPLICABLE TO QUANTITATION OF METHYLATED NUCLEOTIDES IN DNA
Christman, Judith K.
Dep. Biochem. Mount Sinai Med. Cent. New York USA US NY 10029
Anal. Biochem.; (82) P 38-48; Vol 119; No 1

14-82-0008 (0008)
QUANTITATION OF GUANOSINE 5',3'-BISDIPHOSPHATE IN EXTRACTS FROM BACTERIAL CELLS BY ION-PAIR REVERSE-PHASE HIGH-PERFORMANCE LIQUID CHROMATOGRAPHY
Little, R.; Bremer, H.

Univ. Texas Richardson USA US TX 75080
Anal. Biochem.; (82) P 381-8; Vol 126; No 2

14-82-0009 (0009)
ANALYSIS OF METALLONUCLEOSIDE HYDROLYSIS BY HIGH-PRESSURE LIQUID CHROMATOGRAPHY
Clarke, Michael J.; Coffey, Katherine F.; Perpall, Holly J.; Lyon, John
Dep. Chem. Boston Coll. Chestnut Hill USA US MA 02167
Anal. Biochem.; (82) P 404-11; Vol 122; No 2

14-82-0010 (0010)
AN IMPROVED METHOD FOR THE DETERMINATION OF OLIGONUCLEOTIDE CHAIN LENGTH USING PHOSPHODIESTERASE HYDROLYSIS AND ION-PAIR HIGH-PERFORMANCE LIQUID CHROMATOGRAPHY
Crowther, Jonathan B.; Caronia, John P.; Hartwick, Richard A.
Dep. Chem. Rutgers, State Univ. New Brunswick USA US NJ 08854
Anal. Biochem.; (82) P 65-73; Vol 124; No 1

14-82-0011 (0011)
APPLICATION OF HIGH-PERFORMANCE LIQUID CHROMATOGRAPHY TO THE SEPARATION AND ISOLATION OF BIOPOLYMERS
Unger, K. K.; Roumeliotis, P.
Inst. Anorg. Chem. Anal. Chem. Johannes Gutenberg-Univ. Mainz Fed. Rep. Ger. DE 6500
Anal. Chem. Symp. Ser.; (82) P 455-70; Vol 9; No Affinity Chromatogr. Relat. Tech.

14-82-0012 (0012)
INVESTIGATION OF CLINICAL METHODOLOGY FOR SAMPLE COLLECTION AND PROCESSING PRIOR TO THE REVERSED-PHASE LIQUID CHROMATOGRAPHIC DETERMINATION OF UV-ABSORBING PLASMA CONSTITUENTS
Zakaria, Mona; Brown, Phyllis R.
Dep. Chem. Univ. Rhode Island Kingston USA US RI 02881
Anal.Biochem.; (82) P 25-37; Vol 120; No 1

14-82-0013 (0013)
HYPOMETHYLATION OF HELA CELL DNA AND THE ABSENCE OF 5-METHYLCYTOSINE IN SV40 AND ADENOVIRUS (TYPE 2) DNA: ANALYSIS BY HPLC
Diala, Edward S.; Hoffman, Robert M.
Sch. Med. Univ. California, San Diego La Jolla USA US CA 92093
Biochem. Biophys. Res. Commun.; (82) P 19-26; Vol 107; No 1

14-82-0014 (0014)
COMPARATIVE STUDY OF COMMERCIAL FAD INJECTIONS BY HIGH-PERFORMANCE LIQUID CHROMATOGRAPHY
Hirano, Yoshinobu; Tanaka, Giichi; Hasegawa, Kenzi; Hashizume, Akihito; Takata, Mitutaka; Nakata, Izumi; Sayo, Masahiro; Nio, Nagatomo
Natl. Osaka Hosp. Osaka Japan JP 540
Byoin Yakugaku; (82) P 133-7; Vol 8; No 2; In Japan

14-82-0015 (0015)
DETERMINATION OF KINETIC PARAMETERS FOR IMMOBILIZED XANTHINE OXIDASE AND ITS APPLICATION TO A POST-COLUMN ENZYME REACTOR IN HIGH-PERFORMANCE LIQUID CHROMATOGRAPHIC ANALYSIS OF HYPOXANTHINE, XANTHINE AND URIC ACID
Tawa, Riichi; Kitoh, Masahiro; Hirose, Shingo; Adachi, Kenzo
Kyoto Coll. Pharm. Kyoto Japan JP 607
Chem. Pharm. Bull.; (82) P 615-21; Vol 30; No 2

14-82-0016 (0016)
SCREENING FOR INBORN ERRORS OF PURINE AND PYRIMIDINE METABOLISM. APPLICATIONS

OF LC (LIQUID CHROMATOGRAPHY) FOR THE EVALUATION OF ABNORMALITIES
Van Gennip, Albert H.; Grift, J.; Van Bree-Blom, E. J.; Wadman, Sybe K.; De Bree, P. K.
Child. Hosp. "Het Emma Kinderziekenhuis" Amsterdam Neth. NL
Chromatogr. Sci.; (82) P 279-303; Vol 20; No Biol./Biomed. Appl. Liq. Chromatogr. 4

14-82-0017 (0017)
HPLC ANALYSIS OF AROMATIC AMINO ACIDS, NUCLEOSIDES, AND BASES IN PLASMA OF ACUTE LYMPHOCYTIC LEUKEMICS ON CHEMOTHERAPY
Zakaria, Mona; Brown, Phyllis R.; Farnes, M. Patricia; Barker, Barbara E.
Dep. Chem. Univ. Rhode Island Kingston USA US RI 02881
Clin. Chim. Acta; (82) P 69-80; Vol 126; No 1

14-82-0018 (0018)
SIZE FRACTIONATION OF DNA FRAGMENTS RANGING FROM 20 TO 30,000 BASE PAIRS BY LIQUID/LIQUID CHROMATOGRAPHY
Mueller, Werner; Kuetemeier, Gertrud
Abt. Viroidforsch. Max-Planck-Inst. Biochem. Martinsried Fed. Rep. Ger. DE
Eur. J. Biochem.; (82) P 231-8; Vol 128; No 1

14-82-0019 (0019)
PHARMACOKINETICS OF PROPYLTHIOURACIL STUDIED BY HIGH-PERFORMANCE LIQUID CHROMATOGRAPHY
Okuno, Akimasa; Taguchi, Tetsuo; Inyaku, Fumie
Dep. Pediatr. Asahikawa Med. Coll. Asahikawa Japan JP
Igaku no Ayumi; (82) P 73-4; Vol 121; No 2; In Japan

14-82-0020 (0020)
RAPID CATABOLISM OF 5-FLUOROURACIL IN FRESHLY ISOLATED RAT HEPATOCYTES AS ANALYZED BY HIGH PERFORMANCE LIQUID CHROMATOGRAPHY
Sommadossi, Jean Pierre; Gewirtz, David A.; Diasio, Robert B.; Aubert, Claude; Cano, Jean Paul; Goldman, I. David
Dep. Med. Med. Coll. Virginia Richmond USA US VA 23298
J. Biol. Chem.; (82) P 8171-6; Vol 257; No 14

14-82-0021 (0021)
ANOMALOUS SEPARATION OF 2'-AND 3'-RIBONUCLEOTIDES USING ION SUPRESSION REVERSED-PHASE HIGH-PERFORMANCE LIQUID CHROMATOGRAPHY
Whitehouse, R.P.; Greenstock, C.L.
Med. Biophys. Branch, Whiteshell Nucl. Res. Establ., Pinawa, MB Can.ROE 1LO
J. Chromatogr. (82) P 376-8; Vol 243; No 2

14-82-0022 (0022)
AN AUTOMATED HPLC ASSAY FOR SIMULTANEOUS QUANTITATION OF METHYLATED XANTHINES AND URIC ACIDS IN URINE
Tang-Liu, D. Dan-Shya; Riegelman, Sidney
Dep. Pharm. Chem. Univ. California San Francisco USA US CA 94143
J. Chromatogr. Sci.; (82) P 155-9; Vol 20; No 4

14-82-0023 (0023)
DETERMINATION OF ADENINE NUCLEOTIDES AND INOSINE IN HUMAN MYOCARDIUM BY ION-PAIR REVERSED-PHASE HIGH-PERFORMANCE LIQUID CHROMATOGRAPHY
Ingebretsen, Ole Christian; Bakken, Anne M.; Segadal, Leidulf; Farstad, Mikael
Lab. Clin. Biochem. Univ. Bergen Haukeland Sykehus Norway NO N-5016
J. Chromatogr.; (82) P 119-26; Vol 242; No 1

14-82-0024 (0024)
SIMULTANEOUS DETERMINATION OF MYOCARDIAL ADENINE NUCLEOTIDES AND CREATINE PHOSPHATE BY HIGH-PERFORMANCE LIQUID CHROMATOGRAPHY

Harmsen, E.; De Tombe, P. P.; De Jong, J. W
Thoraxcent. Erasmus Univ. Rotterdam Rotterdam Neth. NL
J. Chromatogr.; (82) P 131-6; Vol 230; No 1

14-82-0025 (0025)
HPLC ANALYSIS OF NUCLEIC ACIDS IN FOOD AND OF NUCLEIC ACID METABOLITES IN METABOLIC STUDIES
Tiemeyer, W.; Gaebler, S.; Giesecke, D.
Tieraerztl. Fak. Univ. Muenchen Munich Fed. Rep. Ger. DE D-8000/22
Landwirtsch. Forsch., Sonderh.; (82) P 441-8; Vol 38,; In Ger

14-82-0026 (0026)
AN AUTOMATED HPLC ASSAY FOR SIMULTANEOUS QUANTITATION OF METHYLATED XANTHINES AND URIC ACIDS IN URINE
Tang-Liu, D. Dan-Shya; Riegelman, Sidney
Dep. Pharm. Chem. Univ. California San Francisco USA US CA 94143
J. Chromatogr. Sci.; (82) P 155-9; Vol 20; No 4

14-82-0027 (0027)
DETERMINATION OF ADENINE NUCLEOTIDES AND INOSINE IN HUMAN MYOCARDIUM BY ION-PAIR REVERSED-PHASE HIGH-PERFORMANCE LIQUID CHROMATOGRAPHY
Ingebretsen, Ole Christian; Bakken, Anne M.; Segadal, Leidulf; Farstad, Mikael
Lab. Clin. Biochem. Univ. Bergen Haukeland Sykehus Norway NO N-5016
J. Chromatogr.; (82) P 119-26; Vol 242; No 1

14-82-0028 (0028)
SIMULTANEOUS DETERMINATION OF MYOCARDIAL ADENINE NUCLEOTIDES AND CREATINE PHOSPHATE BY HIGH-PERFORMANCE LIQUID CHROMATOGRAPHY
Harmsen, E.; De Tombe, P. P.; De Jong, J. W
Thoraxcent. Erasmus Univ. Rotterdam Rotterdam Neth. NL
J. Chromatogr.; (82) P 131-6; Vol 230; No 1

14-82-0029 (0029)
COMPARISON OF THE REVERSED-PHASE HIGH-PERFORMANCE LIQUID CHROMATOGRAPHIC SEPARATIONS OF FLUOROPYRIMIDINES, PYRIMIDINES, AND PURINES
Miller, Antonius A.; Benvenuto, John A.; Loo, Ti Li
West German Tumor Cent. Univ. Essen Essen Fed. Rep. Ger. DE D-4300
J. Chromatogr.; (82) P 165-76; Vol 228

14-82-0030 (0030)
DETERMINATION OF ADENOSINE IN NORMAL HUMAN PLASMA AND SERUM BY HIGH-PERFORMANCE LIQUID CHROMATOGRAPHY
Capogrossi, Maurizio C.; Holdiness, Mack R.; Israili, Zafar H.
Dep. Med. Emory Univ. Atlanta USA US GA 30322
J. Chromatogr.; (82) P 168-73; Vol 227; No 1

14-82-0031 (0031)
REVERSED-PHASE HIGH-PERFORMANCE LIQUID CHROMATOGRAPHIC SEPARATION OF RIBOSYL, 2'-DEOXYRIBOSYL AND ARABINOSYL NUCLEOSIDES OF ADENINE AND HYPOXANTHINE
Delia, Thomas J.; Kirt, Diane D.; Drach, John C.
Dep. Chem. Cent. Michigan Univ. Mt. Pleasant USA US MI 48859
J. Chromatogr.; (82) P 173-7; Vol 243; No 1

14-82-0032 (0032)
SPECIFIC SIMULTANEOUS ASSAY OF HYPOXANTHINE AND XANTHINE IN SERUM BY REVERSED-PHASE HIGH-PERFORMANCE LIQUID CHROMATOGRAPHY USING AN IMMOBILIZED XANTHINE OXIDASE REACTOR
Kito, Masahiro; Tawa, Riichi; Takeshima, Shigeo; Hirose, Shingo
Dep. Anal. Chem. Kyoto Coll. Pharm. Kyoto Japan JP 607
J. Chromatogr.; (82) P 183-8; Vol 231; No 1

14-82-0033 (0033)
DETERMINATION OF A NOVEL 5-FLUOROURACIL DERIVATIVE IN RAT AND HUMAN PLASMA BY HIGH-PERFORMANCE LIQUID CHROMATOGRAPHY
Takada, Kanji; Yoshikawa, Hiroshi; Muranishi, Shozo
Kyoto Coll. Pharm. Kyoto Japan JP 607
J. Chromatogr.; (82) P 192-6; Vol 232; No 1

14-82-0034 (0034)
RAPID SEPARATION OF NUCLEOTIDES FROM GRANULOCYTES BY ISOCRATIC, REVERSED-PHASE HIGH-PERFORMANCE LIQUID CHROMATOGRAPHY
Rao, Gundu H. R.; Peller, Janet D.; Richards, Karen L.; McCullough, Jeffrey; White, James G.
Health Sci. Cent. Univ. Minnesota Minneapolis USA US MN 55455
J. Chromatogr.; (82) P 205-10; Vol 229; No 1

14-82-0035 (0035)
REVERSED-PHASE ION-PAIR HIGH-PERFORMANCE LIQUID CHROMATOGRAPHIC ASSAY OF 5-FLUOROURACIL, 5'-DEOXY-5-FLUOROURIDINE, THEIR NUCLEOSIDES, MONO-, DI-, AND TRIPHOSPHATE NUCLEOTIDES WITH A MIXTURE OF QUATERNARY AMMONIUM IONS
Au, J. L. S.; Wientjes, M. G.; Luccioni, C. M.; Rustum, Y. M.
Dep. Exp. Therapeut. Roswell Park Mem. Inst. Buffalo USA US NY 14263
J. Chromatogr.; (82) P 245-56; Vol 228

14-82-0036 (0036)
QUANTITATIVE ENZYMIC HYDROLYSIS OF tRNAS. REVERSED-PHASE HIGH-PERFORMANCE LIQUID CHROMATOGRAPHY OF tRNA NUCLEOSIDES
Gehrke, Charles W.; Kuo, Kenneth C.; McCune, Roy A.; Gerhardt, Klaus O.; Agris, Paul F.
Dep. Biochem. Univ. Missouri Columbia USA US MO 65211
J. Chromatogr.; (82) P 297-308; Vol 230; No 2

14-82-0037 (0037)
RAPID AND RELIABLE METHOD FOR THE ANALYSIS OF NUCLEOTIDE POOLS BY REVERSED-PHASE HIGH-PERFORMANCE LIQUID CHROMATOGRAPHY
Martinez-Valdez, H.; Kothari, R. M.; Hershey, H. V.; Taylor, M. W.
Sch. Med. Natl. Univ. Mexico Mexico City Mex. MX 04510
J. Chromatogr.; (82) P 307-14; Vol 247; No 2

14-82-0038 (0038)
DETERMINATION OF NICOTINAMIDE AND METABOLIC PRODUCTS IN URINE BY HIGH-PERFORMANCE LIQUID CHROMATOGRAPHY
McKee, Ralph W.; Kang-Lee, Young Ae; Panaqua, Mario; Swendseid, Marian E.
Sch. Public Health Univ. California Los Angeles USA US CA 90024
J. Chromatogr.; (82) P 309-17; Vol 230; No 2

14-82-0039 (0039)
DETERMINATION OF THE B2 VITAMER FLAVIN-ADENINE DINUCLEOTIDE IN WHOLE BLOOD BY HIGH-PERFORMANCE LIQUID CHROMATOGRAPHY WITH FLUOROMETRIC DETECTION
Speek, A. J.; Van Schaik, F.; Schrijver, J.; Schreurs, W. H. P.
Dep. Clin. Biochem. TNO Zeist Neth. NL 3700 AJ
J. Chromatogr.; (82) P 311-16; Vol 228

14-82-0040 (0040)
QUANTITATION OF 6-MERCAPTOPURINE IN BIOLOGIC FLUIDS USING HIGH-PERFORMANCE LIQUID CHROMATOGRAPHY: A SELECTIVE AND NOVEL PROCEDURE
Narang, Prem K.; Yeager, Russell L.; Chatterji, Dulal C.
Clin. Pharmacokinet. Res. Lab. Pharm. Dev. Serv. Bethesda USA US MD 20205
J. Chromatogr.; (82) P 373-80; Vol 230; No 2

14-82-0041 (0041)
SIMULTANEOUS DETERMINATION OF AZATHIOPRINE AND 6-MERCAPTOPURINE IN SERUM BY REVERSED-PHASE HIGH-PERFORMANCE LIQUID CHROMATOGRAPHY
Tsutsumi, Kimiko; Otsuki, Yoshie; Kinoshita, Toshio
Sch. Pharm. Sci. Kitasato Univ. Tokyo Japan JP 108
J. Chromatogr.; (82) P 393-9; Vol 231; No 2

14-82-0042 (0042)
SIMPLE AND RAPID HIGH-PERFORMANCE LIQUID CHROMATOGRAPHIC METHOD FOR ANALYSIS OF NUCLEOSIDES IN BIOLOGICAL FLUIDS
Agarwal, Ram P.; Major, Pierre P.; Kufe, Dovald W.
Med. Cent. Boston Univ. Boston USA US MA 02118
J. Chromatogr.; (82) P 418-24; Vol 231; No 2

14-82-0043 (0043)
DRUG MONITORING OF 5-FLUOROCYTOSINE IN SERUM BY HIGH-PERFORMANCE LIQUID CHROMATOGRAPHY
Essers, L.; Hantschke, D.; Leisse, K. H.
Inst. Med. Mikrobiol. Essen Fed. Rep. Ger. DE
Mykosen; (82) P 183-8; Vol 25; No 4; In Ger

14-82-0044 (0044)
USE OF HIGH-PERFORMANCE LIQUID CHROMATOGRAPHY TO QUANTITATE THYMINE-CONTAINING PYRIMIDINE DIMERS IN DNA
Love, Jack D.; Friedberg, Errol C.
Dep. Pathol. Stanford Univ. Stanford USA US CA 94305
J. Chromatogr.; (82) P 475-87; Vol 240; No 2

14-82-0045 (0045)
REVERSED-PHASE HIGH-PERFORMANCE LIQUID CHROMATOGRAPHIC METHOD TO DETERMINE VIDARABINE AND HYPOXANTHINE ARABINOSIDE IN BIOLOGICAL FLUIDS
Bowman, David B.; Kauffman, Ralph E.
Sch. Med. Wayne State Univ. Detroit USA US MI 48201
J. Chromatogr.; (82) P 487-91; Vol 229; No 2

14-82-0046 (0046)
SEPARATION OF RADIATION AND PHOTOINDUCED 5,6-DIHYDROTHYMINE DERIVATIVES BY REVERSED-PHASE HIGH-PERFORMANCE LIQUID CHROMATOGRAPHY
Cadet, J.; Berger, M.; Voituriez, L.
Dep. Rech. Fondam. CEN Grenoble Fr. FR 38041
J. Chromatogr.; (82) P 488-94; Vol 238; No 2

14-82-0047 (0047)
DETERMINATION OF SODIUM N6,2'-O-DIBUTYRYLADENOSINE CYCLIC 3',5'-(HYDROGEN PHOSPHATE) AND ITS HYDROLYSIS PRODUCTS BY HIGH-PERFORMANCE LIQUID CHROMATOGRAPHY
Kitaoka, Hiroaki; Ohya, Kazumi
Res. Inst. Daiichi Seiyaku Co. Ltd. Tokyo Japan JP 134
J. Chromatogr.; (82) P 495-9; Vol 238; No 2

14-82-0048 (0048)
HIGH-PERFORMANCE LIQUID CHROMATOGRAPHIC DETERMINATION OF PLASMA 6-MERCAPTO-PURINE IN CLINICALLY RELEVANT CONCENTRATIONS
De Abreu, Ronney A.; Van Baal, John M.; Schouten, Teunis J.; Schretlen, Egbert D. A. M.; De Bruyn, Chris H. M. M.
Dep. Pediatrics St. Radboud Hosp. Nijmegen Neth. NL
J. Chromatogr.; (82) P 526-33; Vol 227; No 2

14-82-0049 (0049)
HIGH-PERFORMANCE LIQUID CHROMATOGRAPHIC DETERMINATION OF PURINE AND PYRIMIDINE

BASES, RIBONUCLEOSIDES, DEOXYRIBONUCLEOSIDES AND CYCLIC RIBONUCLEOTIDES IN BIOLOGICAL FLUIDS
De Abreu, Ronney A.; Van Baal, John M.; De Bruyn, Chris H. M. M.; Bakkeren, Jan A. J. M.; Schretlen, Egbert D. A. M.
Dep. Pediatr. St. Radboud Hosp. Nijmegen Neth. NL
J. Chromatogr.; (82) P 67-75; Vol 229; No 1

14-82-0050 (0050)
A METHOD FOR THE ANALYSIS OF ATP METABOLITES IN BEEF SKELETAL MUSCLE BY HPLC
Currie, Ronald W.; Sporns, Peter; Wolfe, Fred H.
Dep. Food Sci. Univ. Alberta Edmonton Can. CA AB T6G 2P5
J. Food Sci.; (82) P 1226-8, 1234; Vol 47; No 4

14-82-0051 (0051)
STUDIES OF THE REACTION PRODUCTS OF ADENOSINE WITH (Pt(NH3)3Cl)Cl AND CIS-Pt(NH3)2Cl2 BY HIGH PERFORMANCE LIQUID CHROMATOGRAPHY
Inagaki, Kenji; Kuwayama, Michiyo; Kidani, Yoshinori
Fac. Pharm. Sci. Nagoya City Univ. Nagoya Japan JP
J. Inorg. Biochem.; (82) P 59-70; Vol 16; No 1

14-82-0052 (0052)
HIGH-PERFORMANCE LIQUID CHROMATOGRAPHY OF NUCLEOSIDES IN RNA AND DNA
Zumwalt, Robert W.; Kuo, Kenneth C. T.; Agris, Paul F.; Ehrlich, Melanie; Gehrke, Charles W.
Dep. Biochem. Tulane Univ. New Orleans USA US LA 70112
J. Liq. Chromatogr.; (82) P 2041-60; Vol 5; No 11

14-82-0053 (0053)
A DUAL-COLUMN HPLC METHOD FOR THE SIMULTANEOUS MEASUREMENT OF 6-THIOGUANINE AND ADENINE IN RNA OR DNA
Herbert, Barbara H.; Drake, Stephanie; Nelson, J. Arly
Dep. Exp. Pediatr. Univ. Texas Houston USA US TX 77030
J. Liq. Chromatogr.; (82) P 2095-110; Vol 5; No 11

14-82-0054 (0054)
ADENOSINE MEASUREMENT BY A RAPID HPLC-FLUOROMETRIC METHOD: INDUCED CHANGES OF ADENOSINE CONTENT IN REGIONS OF RAT BRAIN
Wojcik, W. J.; Neff, N. H.
Natl. Inst. Ment. Health Saint Elizabeths Hosp. Washington USA US DC 20032
J. Neurochem.; (82) P 280-2; Vol 39; No 1

14-82-0055 (0055)
CELLULAR LEVEL OF PURINE COMPOUNDS IN ISCHEMIC GERBIL BRAIN BY HIGH PERFORMANCE LIQUID CHROMATOGRAPHY
Morimoto, Kazuyoshi; Tagawa, Kunio; Hayakawa, Toru; Watanabe, Fusao; Mogami, Heitaro
Med. Sch. Osaka Univ. Osaka Japan JP
J. Neurochem.; (82) P 833-5; Vol 38; No 3

14-82-0056 (0056)
A METHOD FOR THE ESTIMATION OF SUBMICROMOLAR CONCENTRATIONS OF ADENOSINE IN PLASMA USING HIGH-PRESSURE LIQUID CHROMATOGRAPHY
Ballard, Heather; Cotterrell, D.; Karim, F.; Reed, Jane
Dep. Physiol. Univ. Leeds Leeds UK GB LS2 9JT
J. Physiol. (London); (82) P 4P-5P; Vol 330

14-82-0057 (0057)
HPLC ANALYSIS OF NUCLEIC ACIDS IN FOOD AND OF NUCLEIC ACID METABOLITES IN METABOLIC STUDIES
Tiemeyer, W.; Gaebler, S.; Giesecke, D.

Tieraerztl. Fak. Univ. Muenchen Munich Fed. Rep. Ger. DE D-8000/22
Landwirtsch. Forsch., Sonderh.; (82) P 441-8; Vol 38,; In Ger

14-82-0058 (0058)
DRUG MONITORING OF 5-FLUOROCYTOSINE IN SERUM BY HIGH-PERFORMANCE LIQUID CHROMATOGRAPHY
Essers, L.; Hantschke, D.; Leisse, K. H.
Inst. Med. Mikrobiol. Essen Fed. Rep. Ger. DE
Mykosen; (82) P 183-8; Vol 25; No 4; In Ger

14-82-0059 (0059)
HIGH-PERFORMANCE LIQUID CHROMATOGRAPHIC IDENTIFICATION AND QUANTITATION OF CYCLIC ADENOSINE 3':5'-MONOPHOSPHATE IN HIGHER (PHASEOLUS VULGARIS L.) AND LOWER (CHLORELLA SP.) PLANTS
Van Onckelen, H. A.; Dupon, M.; De Greef, J. A.
Dep. Biol. Univ. Instelling Antwerpen Wilrijk Belg. BE B-2610
Physiol. Plant.; (82) P 93-7; Vol 55; No 2

14-82-0060 (0060)
DETERMINATION OF THE MOLE PERCENTAGE OF GUANINE PLUS CYTOSINE IN STREPTOMYCES DNA BY HIGH-PRESSURE LIQUID CHROMATOGRAPHY
Ding, Jian; Zhang, Zhongze; Yang, Zhiyong; Huang, Yali; Sun, Huijurn
Su, Fengyan; Xu, Qingde; Li, Weiguang
Inst. For. Pedol., Acad. Sin., Shenyang, Peop. Rep. China
Weishengwuxue Tongbao (82) P 83-5, 90; Vol 9; No 2; In Chin

CHAPTER 7.15
Organic Acids

--- 1982 ---

15-82-0001 (0001)
HIGH-PERFORMANCE CHROMATOGRAPHY OF N,O-ACYLATED SIALIC ACIDS
Shukla, Ashok K.; Scholz, Norbert; Reimerdes, Ernst H.; Schauer, Roland
Biochem. Inst., Christ.-Albrechts-Univ., Kiel, Fed. Rep. Ger.,D-2300
Anal. Biochem. (82) P 78-82; Vol 123; No 1

15-82-0002 (0002)
HIGH-PERFORMANCE LIQUID CHROMATOGRAPHIC DETERMINATION OF .ALPHA.-KETO ACIDS IN HUMAN URINE AND PLASMA
Hayashi, Tokishi; Tsuchiya, Hironori; Todoriki, Hidemi; Naruse, Hiroshi
Natl. Cent. Nervous, Mental, Muscular Disord. Tokyo Japan JP
Anal. Biochem.; (82) P 173-9; Vol 122; No 1

15-82-0003 (0003)
THE MEASUREMENT OF URINARY OXALIC ACID BY DERIVATIZATION COUPLED WITH LIQUID CHROMATOGRAPHY
Murray, John F., Jr.; Nolen, Harold W., III; Gordon, G. Ross; Peters, John H.
Life Sci. Div. SRI Int. Menlo Park USA US CA 94025
Anal. Biochem.; (82) P 301-9; Vol 121; No 2

15-82-0004 (0004)
SORBIC ACID: SECONDARY DETERMINATION BY OFFICIAL METHOD AND HPLC. COMPARISON OF METHODS
Modi, Giuseppe; Chiti, Fabrizio; Fiorentino, Pietro
Lab. Ig. Profil. Florence Italy IT
Boll. Chim. Unione Ital. Lab. Prov., Parte Sci.; (82) P 143-51; Vol 33; No 52; In Ital

15-82-0005 (0005)
DETERMINATION OF ASCORBIC ACID IN WHEAT FLOURS, BREAD DOUGH CONDITIONERS, AND COMMERCIAL VITAMIN C TABLETS BY HIGH-PERFORMANCE LIQUID CHROMATOGRAPHY
Lookhart, G. L.; Hall, S. B.; Finney, K. F.
Agric. Res. Serv. U. S. Grain Market. Res. Lab. Manhattan USA US KS 66502
Cereal Chem.; (82) P 69-71; Vol 59; No 1

15-82-0006 (0006)
ANALYSIS OF THE BITTER ACIDS IN HOPS AND BEER BY HIGH-PRESSURE LIQUID CHROMATOGRAPHY (HPLC)
Dewaele, C.
Lab. Org. Chem. Rijksuniv. Ghent Ghent Neth. NL 9000
Cerevisia; (82) P 91-5; Vol 7; No 2; In Neth

15-82-0007 (0007)
HPLC ANALYSIS OF CLAVULANIC ACID AND PHARMACOKINETIC INVESTIGATION ON URINARY EXCRETION OF BRL25000 (CLAVULANIC ACID-AMOXICILLIN) IN MAN
Uno, Toyozo; Haginaka, Jun; Nakagawa, Terumichi
Fac. Pharm. Sci. Kyoto Univ. Kyoto Japan JP 606
Chemotherapy (Tokyo); (82) P 118-24; Vol 30; No Suppl. 2; In Japan

15-82-0008 (0008)
URINARY p-AMINOBENZOIC ACID DETERMINED IN THE PANCREATIC FUNCTION TEST BY LIQUID CHROMATOGRAPHY, WITH ELECTROCHEMICAL DETECTION
Ito, Shosuke; Maruta, Kazuhiro; Imai, Yoichiro; Kato, Toshiaki; Ito, Madoka; Nakajima, Sumio; Fujita, Keisuke; Kurahashi, Tatsuo
Sch. Med. Fujita-Gakuen Univ. Aichi Japan JP 470-11
Clin. Chem. (Winston-Salem, N. C.); (82) P 323-6; Vol 28; No 2

15-82-0009 (0009)
URIC ACID DETERMINATIONS: REVERSED-PHASE LIQUID CHROMATOGRAPHY WITH ULTRAVIOLET DETECTION COMPARED WITH KINETIC AND EQUILIBRIUM ADAPTATIONS OF THE URICASE METHOD
Ingebretsen, Ole Christian; Borgen, Jan; Farstad, Mikael
Lab. Clin. Biochem. Univ. Bergen Sykehus Norway NO N-5016
Clin. Chem. (Winston-Salem, N. C.); (82) P 496-8; Vol 28; No 3

15-82-0010 (0010)
SEPARATION AND ESTIMATION OF GALACTURONIC ACID BY HPLC
Forni, E.; Giangiacomo, R.; Polesello, A.
Ist. Sper. Valorizzazione Tecn. Prod. Agric. Milan Italy IT
CONFERENCE; Recent Dev. Food Anal., Proc. Eur. Conf. Food Chem., 1st; (82) P 53-8; Avail.: Baltes, Werner; Czedik-Eysenberg, Peter B.; Pfannhauser, Werner; Date: 81; Publisher: Verlag Chem., Weinheim, Fed. Rep. Ger.

15-82-0011 (0011)
HPLC DETERMINATION OF THE POTENTIALLY IMMUNOGENIC IMPURITIES AND DECOMPOSITION PRODUCTS IN TABLET FORMULATIONS WITH ACETYLSALICYLIC ACID FROM VARIOUS YUGOSLAVIAN MANUFACTURERS
Pavli, V.; Cinc, M.
Inst. Pharm. Drug Control Ljubljana Yugoslavia YU
Farm. Vestn. (Ljubljana); (82) P 165-78; Vol 33; No 3; In Slovenian

15-82-0012 (0012)
DETERMINATION OF BONGKREKIC ACID AND TOXOFLAVIN BY HIGH PRESSURE LIQUID CHROMATOGRAPHY
Voragen, A. G. J.; De Kok, H. A. M.; Kelholt, A. J.; Schols, H. A. ; Djien, Ko Swan
Dep. Food Sci. Agric. Univ. Wageningen Neth. NL
Food Chem.; (82) P 167-74; Vol 9; No 3

15-82-0013 (0013)
DETECTION OF CHOLANOIC ACIDS IN HIGH-PERFORMANCE LIQUID CHROMATOGRAPHY BASED ON EFFECTS OF DOUBLE LAYER CAPACITY CHANGES AT THE DROPPING MERCURY ELECTRODE
Kutner, Wlodzimierz; Behr, Barbara; Kemula, Wiktor
Inst. Phys. Chem. Polish Acad. Sci. Warsaw Pol. PL 01-224
Fresenius' Z. Anal. Chem.; (82) P 121-5; Vol 312; No 2

15-82-0014 (0014)
STUDIES ON THE PERFLUOROAMIDE-BONDED PHASE AS HPLC PACKING. III. APPLICATION OF PERFLUOROAMIDE-BONDED SILICA IN THE SEPARATION OF CHOLANIC ACIDS AND ESTERS
Li, Wenhua; Wang, Jinhua
Shanghai Inst. Org. Chem. Acad. Sin. Shanghai Peop. Rep. China CN
Huaxue Xuebao; (82) P 377-80; Vol 40; No 4; In Ch

15-82-0015 (0015)
OPTIMIZED SEPARATION OF BENZOIC AND SORBIC ACID AS PRESERVATIVES IN FOODSTUFFS BY REVERSED-PHASE HPLC
Froehlich, D. H.
Kant. Lab. Zurich Switz. CH CH-8030
HRC CC, J. High Resolut. Chromatogr. Chromatogr. Commun.; (82) P 158-60; Vol 5; No 3

15-82-0016 (0016)
QUANTITATIVE DETERMINATION OF GALACTURONIC ACID IN COMMERCIAL PECTINS BY HPLC
Giangiacomo, R.; Polesello, A.; Marin, F.
Ist. Sper. Valorizzazione Tecnol. Prod. Agric. Milan Italy IT
Ind. Aliment. (Pinerolo, Italy); (82) P 386-8; Vol 21; No 5; In Ital

15-82-0017 (0017)
IDENTIFICATION OF SOME ACIDS PRODUCED DURING AUTOCLAVING OF D-GLUCOSE SOLUTIONS USING HPLC
Durham, D. G.; Hung, C. T.; Taylor, R. B.
Sch. Pharm. Robert Gordon's Inst. Technol. Aberdeen UK GB AB9 1FR
Int. J. Pharm.; (82) P 31-40; Vol 12; No 1

15-82-0018 (0018)
DETERMINATION OF PHYTIC ACID IN FOODS BY HIGH-PERFORMANCE LIQUID CHROMATOGRAPHY
Graf, Ernst; Dintzis, Frederick R.
North. Reg. Res. Cent. Agric. Res. Serv. Peoria USA US IL 61604
J. Agric. Food Chem.; (82) P 1094-7; Vol 30; No 6

15-82-0019 (0019)
ANALYSIS OF OXALIC ACID IN CARAMBOLA (AVERRHOA CARAMBOLA L.) AND SPINACH BY HIGH-PERFORMANCE LIQUID CHROMATOGRAPHY
Wilson, Charles W., III; Shaw, Philip E.; Knight, Robert J., Jr.
U. S. Citrus Subtrop. Prod. Lab. Winter Haven USA US FL 33880
J. Agric. Food Chem.; (82) P 1106-8; Vol 30; No 6

15-82-0020 (0020)
DETERMINATION OF 2,4-(DICHLOROPHENOXY)ACETIC ACID AND OF DICHLOBENIL IN WATER BY HIGH-PERFORMANCE LIQUID CHROMATOGRAPHY
Connick, William J., Jr.; Simoneaux, Jacqueline M.
South. Reg. Res. Cent. Sci. Educ. Adm. New Orleans USA US LA 70179
J. Agric. Food Chem.; (82) P 258-60; Vol 30; No 2

15-82-0021 (0021)
HIGH PERFORMANCE LIQUID CHROMATOGRAPHIC ANALYSIS OF HOP ALPHA- AND BETA-ACIDS
Gross, R. W.; Schwiesow, M. H.
Pfizer, Inc. Milwaukee USA US WI 53212
J. Am. Soc. Brew. Chem.; (82) P 116-17; Vol 40; No 3

15-82-0022 (0022)
HIGH-PRESSURE LIQUID CHROMATOGRAPHIC DETERMINATION OF GLYCYRRHIZIC ACID OR GLYCYRRHIZIC ACID SALTS IN VARIOUS LICORICE PRODUCTS: COLLABORATIVE STUDY
Vora, Peter S.
MacAndrews and Forbes Co. Camden USA US NJ 08104
J. Assoc. Off. Anal. Chem.; (82) P 572-4; Vol 65; No 3

15-82-0023 (0023)
AN AUTOMATED HPLC ASSAY FOR SIMULTANEOUS QUANTITATION OF METHYLATED XANTHINES AND URIC ACIDS IN URINE
Tang-Liu, D. Dan-Shya; Riegelman, Sidney
Dep. Pharm. Chem. Univ. California San Francisco USA US CA 94143
J. Chromatogr. Sci.; (82) P 155-9; Vol 20; No 4

15-82-0024 (0024)
COMPARISON OF ISOLATION METHODS OF URINARY ORGANIC ACIDS BY HIGH-PERFORMANCE LIQUID CHROMATOGRAPHY
Rehman, Aziz; Gates, Stephen C.; Webb, James W.
Dep. Chem. Illinois State Univ. Normal USA US IL 61761
J. Chromatogr.; (82) P 103-12; Vol 228

15-82-0025 (0025)
HIGH-PERFORMANCE LIQUID CHROMATOGRAPHY OF CARBOXYLIC ACIDS USING 4-BROMO-METHYL-7-ACETOXYCOUMARIN AS FLUORESCENCE REAGENT
Tsuchiya, Hironori; Hayashi, Tokishi; Naruse, Hiroshi; Takagi, Nobuhiko
Dep. Dent. Pharmacol. Gifu Coll. Dent. Gifu Japan JP

J. Chromatogr.; (82) P 121-30; Vol 234; No 1

15-82-0026 (0026)
DETERMINATION OF .GAMMA.-AMINOBUTYRIC ACID BY REVERSED-PHASE HIGH-PERFORMANCE LIQUID CHROMATOGRAPHY AND PRECOLUMN LABELING FOR FLUORESCENCE DETECTION
Griesmann, Guy E.; Chan, Wai Yee; Rennert, Owen M.
Oklahoma Child. Mem. Hosp. Univ. Oklahoma Oklahoma City USA US OK 73190
J. Chromatogr.; (82) P 121-4; Vol 230; No 1

15-82-0027 (0027)
HIGH-PERFORMANCE LIQUID CHROMATOGRAPHIC DETERMINATION OF UROCANIC ACID ISOMERS IN BIOLOGICAL SAMPLES
Caron, J. C.; Martin, B.; Shroot, B.
Cent. Int. Rech. Dermatol. Valbonne Fr. FR 06565
J. Chromatogr.; (82) P 125-30; Vol 230; No 1

15-82-0028 (0028)
DETERMINATION OF URINARY HIPPURIC ACID AND o-CRESOL, AS INDICES OF TOLUENE EXPOSURE, BY LIQUID CHROMATOGRAPHY ON DYNAMICALLY MODIFIED SILICA
Hansen, Steen Honore; Doessing, Martin
Dep. Org. Chem. R. Dan. Sch. Pharm. Copenhagen Den. DK DK-2100
J. Chromatogr.; (82) P 141-8; Vol 229; No 1

15-82-0029 (0029)
9-(CHLOROMETHYL)ANTHRACENE: A USEFUL DERIVATIZING REAGENT FOR ENHANCED ULTRAVIOLET AND FLUORESCENCE DETECTION OF CARBOXYLIC ACIDS WITH LIQUID CHROMATOGRAPHY
Korte, W. D.
Dep. Chem. California State Univ. Chico USA US CA 95929
J. Chromatogr.; (82) P 153-7; Vol 243; No 1

15-82-0030 (0030)
SEPARATION OF INDOLE METABOLITES FROM URINE WITH AN ODS TYPE RESIN BY HIGH-PERFORMANCE LIQUID CHROMATOGRAPHY
Yamaguchi, Tomoko; Yokota, Kazuko; Uematsu, Fumie
Dep. Biochem. Tokyo Women's Med. Coll. Tokyo Japqn JP
J. Chromatogr.; (82) P 166-72; Vol 231; No 1

15-82-0031 (0031)
MEASUREMENT OF BRANCHED CHAIN AMINO ACIDS AND BRANCHED CHAIN .ALPHA.-KETOACIDS IN PLASMA BY HIGH-PERFORMANCE LIQUID CHROMATOGRAPHY
Nissen, Steven L.; Van Huysen, Carol; Haymond, Morey W.
Dep. Pediatr. Mayo Clin. Med. Sch. Rochester USA US MN 55905
J. Chromatogr.; (82) P 170-5; Vol 232; No 1

15-82-0032 (0032)
DETERMINATION OF APOVINCAMINIC ACID IN BIOLOGICAL SAMPLES BY HIGH-PERFORMANCE LIQUID CHROMATOGRAPHY
Kozma, Maria; Pudleiner, Piroska; Vereczkey, L.
Chem. Works Gedeon Richter Ltd. Budapest Hung. HU H-1475
J. Chromatogr.; (82) P 177-82; Vol 241; No 1

15-82-0033 (0033)
LEAF ANALYSIS FOR ABSCISIC, PHASEIC AND 3-INDOLYLACETIC ACIDS BY HIGH-PERFORMANCE LIQUID CHROMATOGRAPHY
Durley, Richarr C.; Kannangara, Tissa; Simpson, Graham M.
Crop Sci. Dep. Univ. Sasketchewan Saskatoon Can. CA SK S7N 0W0
J. Chromatogr.; (82) P 181-8; Vol 236; No 1

15-82-0034 (0034)
RAPID DETERMINATION OF PROCETOFENIC ACID IN PLASMA BY HIGH-PERFORMANCE LIQUID CHROMATOGRAPHY
Caturla, M. C.; Albaiges, J.
Cent. Invest. Lab. Rocador, S.A. Barcelona Spain ES
J. Chromatogr.; (82) P 219-22; Vol 227; No 1

15-82-0035 (0035)
URONIC ACID ANALYSIS BY HIGH-PERFORMANCE LIQUID CHROMATOGRAPHY AFTER METHANOLYSIS OF GLYCOSAMINOGLYCANS
Hjerpe, A.; Antonopoulos, C. A.; Classon, B.; Engfeldt, B.; Nurminen, M.
Karolinska Inst. Huddinge Univ. Hosp. Huddinge Swed. SE S-141 86
J. Chromatogr.; (82) P 221-7; Vol 235; No 1

15-82-0036 (0036)
A METHOD FOR HIGH-PERFORMANCE LIQUID CHROMATOGRAPHIC SCREENING OF UV-POSITIVE COMPONENTS IN URINE ELUATE FROM SEPHADEX G-10 AND MODIFICATIONS FOR DETERMINATION OF URINARY SALICYLIC, SALICYLURIC AND GENTISIC ACIDS
Fornstedt, Nermin
Inst. Biochem. Univ. Uppsala Uppsala Swed. SE S-751 23
J. Chromatogr.; (82) P 253-61; Vol 230; No 2

15-82-0037 (0037)
.BETA.-CYCLODEXTRIN AS A CHIRAL COMPONENT OF THE MOBILE PHASE FOR SEPARATION OF MANDELIC ACID INTO ENANTIOMERS IN REVERSED-PHASE SYSTEMS OF HIGH-PERFORMANCE LIQUID CHROMATOGRAPHY
Debowski, Janusz; Sybilska, Danuta; Jurczak, Janusz
Inst. Phys. Chem. Polish Acad. Sci. Warsaw Pol. PL 01-224
J. Chromatogr.; (82) P 303-6; Vol 237; No 2

15-82-0038 (0038)
ANALYSIS OF BEER ISO-.ALPHA.-ACIDS BY HIGH-PERFORMANCE LIQUID CHROMATOGRAPHY WITHOUT SAMPLE PRETREATMENT
Verzele, M.; Dewaele, C.; Van Kerrebroeck, M.
Lab. Org. Chem. State Univ. Ghent Ghent Belg. BE B-9000
J. Chromatogr.; (82) P 321-6; Vol 244; No 2

15-82-0039 (0039)
DETERMINATION OF .GAMMA.-AMINOBUTYRIC ACID BY LIQUID CHROMATOGRAPHY WITH ELECTROCHEMICAL DETECTION
Caudill, W. Lowry; Houck, Gregory P.; Wightman, R. Mark
Dep. Chem. Indiana Univ. Bloomington USA US IN 47405
J. Chromatogr.; (82) P 331-9; Vol 227; No 2

15-82-0040 (0040)
ALTERNATING-CURRENT POLAROGRAPHIC DETECTION FOR REVERSED-PHASE ION-PAIR HIGH-PERFORMANCE LIQUID CHROMATOGRAPHY OF SOME BENZOIC ACIDS
Kutner, Wlodzimierz
Inst. Phys. Chem. Pol. Acad. Sci. Warsaw Pol. PL 01-224
J. Chromatogr.; (82) P 342-6; Vol 247; No 2

15-82-0041 (0041)
DETERMINATION OF HYALURONIC ACID USING HIGH-PERFORMANCE LIQUID CHROMATOGRAPHY OF CHONDROITINASE DIGESTS
Hjerpe, A.; Antonopoulos, C. A.; Engfeldt, B.
Karolinska Inst. Huddinge Univ. Hosp. Huddinge Swed. SE S-141 86
J. Chromatogr.; (82) P 365-8; Vol 245; No 3

15-82-0042 (0042)
PREDICTION OF RETENTION TIMES FOR AROMATIC ACIDS IN LIQUID CHROMATOGRAPHY

Hanai, T.; Tran, K. C.; Hubert, J.
Dep. Chim. Univ. Montreal Montreal Can. CA PQ H3C 3V1
J. Chromatogr.; (82) P 385-95; Vol 239

15-82-0043 (0043)
SEPARATION AND QUANTIFICATION OF BASIC HYDROXYCINNAMIC AMIDES AND HYDROXYCINNAMIC ACIDS BY REVERSED-PHASE HIGH-PERFORMANCE LIQUID CHROMATOGRAPHY
Ponchet, Michel; Martin-Tanguy, Josette; Poupet, Alain; Marais, Antoine; Beck, Daniel
Stn. Bot. Pathol. Inst. Natl. Rech. Agron. Antibes Fr. FR 06602
J. Chromatogr.; (82) P 397-404; Vol 240; No 2

15-82-0044 (0044)
PHENOLIC ACID ANALYSES COMPARING TWO HIGH-PERFORMANCE LIQUID CHROMATOGRAPHIC TECHNIQUES
Stutte, Charles A.; Hardin, Joyce M.
Agron. Dep. Univ. Arkansas Fayetteville USA US AR 72701
J. Chromatogr.; (82) P 446-50; Vol 248; No 3

15-82-0045 (0045)
SEPARATION OF GLYCYRRHIZINIC ACID ISOMERS BY HIGH-PERFORMANCE LIQUID CHROMATOGRAPHY
Tsubone, Kazuyuki; Ohnishi, Shigeki; Yoneya, Tohru
Cosmet. Lab. Kanebo, Ltd. Odawara City Japan JP 250
J. Chromatogr.; (82) P 469-71; Vol 248; No 3

15-82-0046 (0046)
DETERMINATION OF 13-CIS-RETINOIC ACID AND ITS MAJOR METABOLITE, 4-OXO-13-CIS-RETINOIC ACID, IN HUMAN BLOOD BY REVERSED-PHASE HIGH-PERFORMANCE LIQUID CHROMATOGRAPHY
Vane, Floie M.; Stoltenborg, Janet K.; Bugge, Christopher J. L.
Dep. Biochem. Drug Metab. Hoffmann-La Roche Inc. Nutley USA US NJ 07110
J. Chromatogr.; (82) P 471-84; Vol 227; No 2

15-82-0047 (0047)
SILICA GEL LIQUID-LIQUID CHROMATOGRAPHY USING AQUEOUS BINARY PHASE SYSTEMS. HIGH-EFFICIENCY EXTRACTION AND RESOLUTION OF PHENOLS AND CARBOXYLIC ACIDS
Hara, Shoji; Dobashi, Yasuo; Oka, Kitaro
Tokyo Coll. Pharm. Hachioji Japan JP 192-03
J. Chromatogr.; (82) P 677-85; Vol 239

15-82-0048 (0048)
SEPARATION OF SODIUM SALTS OF PHENYLSULFAMIC ACIDS BY REVERSED-PHASE HIGH-PERFORMANCE LIQUID CHROMATOGRAPHY USING STRONG ELECTROLYTES
Benson G. A.
Sch. Sci. Reg. Tech. Coll. Sligo Ire. IE
J. Chromatogr.; P(82) P 285-90; Vol 243; No 2

15-82-0049 (0049)
HPLC ANALYSIS OF PHYTIC ACID IN SELECTED FOODS AND BIOLOGICAL SAMPLES
Knuckles, B. E.; Kuzmicky, D. D.; Betschart, A. A.
Dep. Agric. West. Reg. Res. Cent. Berkeley USA US CA 94710
J. Food Sci.; (82) P 1257-8, 1262; Vol 47; No 4

15-82-0050 (0050)
A RAPID AND SPECIFIC METHOD FOR THE DETERMINATION OF TIAPROFENIC ACID IN HUMAN PLASMA BY HIGH-PERFORMANCE LIQUID CHROMATOGRAPHY
Ward, G. T.; Stead, J. A.; Freeman, M.
Roussel Lab. Ltd. Covingham/Swindon/Wiltshire UK GB
J. Liq. Chromatogr.; (82) P 165-74; Vol 5; No 1

15-82-0051 (0051)
LIQUID CHROMATOGRAPHIC DETERMINATION OF VALPROIC ACID IN HUMAN SERUM
Kline, W. F.; Enagonio, D. P.; Reeder, D. J.; May, W. E.
Cent. Anal. Chem. Natl. Bur. Stand. Washington USA US DC 20234
J. Liq. Chromatogr.; (82) P 1697-709; Vol 5; No 9

15-82-0052 (0052)
HYDROPHOBICITY AND RETENTION IN REVERSED-PHASE LIQUID CHROMATOGRAPHY
D'Amboise, Marius; Hanai, Toshihiko
Dep. Chem. Univ. Montreal Montreal Can. CA PQ H3C 3V1
J. Liq. Chromatogr.; (82) P 229-44; Vol 5; No 2

15-82-0053 (0053)
REVERSED-PHASE CLEANUP AND HIGH PERFORMANCE LIQUID CHROMATOGRAPHIC ANALYSIS OF ABSCISIC ACID IN CERCOSPORA ROSICOLA LIQUID CULTURE MEDIA
Norman, Shirley M.; Maier, Vincent P.; Echols, Linda C.
Fruit Veg. Chem. Lab. Sci. Educ. Adm. Pasadena USA US CA 91106
J. Liq. Chromatogr.; (82) P 81-91; Vol 5; No 1

15-82-0054 (0054)
IMPROVED AND RAPID HIGH-PERFORMANCE LIQUID CHROMATOGRAPHIC ASSAY FOR 13-CIS-RETINOIC ACID OR ALL-TRANS-RETINOIC ACID
Shelley, Rickey; Price, J. C.; Jun, H. Won; Cadwallader, D. E.; Capomacchia, A. C.
Sch. Pharm. Univ. Georgia Athens USA
J. Pharm. Sci.; (82) P 262-4; Vol 71; No 2

15-82-0055 (0055)
DETERMINATION OF RELATED COMPOUNDS IN ASPIRIN BY LIQUID CHROMATOGRAPHY
Pfeiffer, C. D.; Pankey, J. W.
Anal. Lab. Dow Chem. Co. Midland USA US MI 48640
J. Pharm. Sci.; (82) P 511-14; Vol 71; No 5

15-82-0056 (0056)
BENZOIC ACID IN SOUR MILK PRODUCTS AND FRESH CHEESE. HPLC DETERMINATION OF BENZOIC ACID, SORBIC ACID, AND p-HYDROXYBENZOIC ACID ESTERS IN FOODS
Gieger, Ursula
Nahrungsmittelunters. Landeshauptstadt Kiel, Eckernforder Str. 421 Kiel Fed. Rep. Ger. DE 2300
Lebensmittelchem. Gerichtl. Chem.; (82) P 109-10; Vol 36; No 5; In Ger

15-82-0057 (0057)
THE DETERMINATION OF .ALPHA.-KETOACIDS AND OXALIC ACID IN SEAWATER BY REVERSED PHASE LIQUID CHROMATOGRAPHIC SEPARATION OF FLUORESCENT QUINOXILINOL DERIVATIVES
Steinberg, Spencer M.; Bada, Jeffrey L.
Scripps Inst. Oceanogr. Univ. California La Jolla USA US CA 92093
Mar. Chem.; (82) P 299-306; Vol 11; No 4

15-82-0058 (0058)
ARGENTATION-HIGH PRESSURE LIQUID CHROMATOGRAPHY OF PROSTAGLANDINS AND MONOHYDROXYEICOSENOIC ACIDS
Powell, William S.
R. Victoria Hosp. McGill Univ. Montreal Can. CA PQ H3A 1A1
Methods Enzymol.; (82) P 530-43; Vol 86; No Prostaglandins Arachidonate Metab.

15-82-0059 (0059)
HIGH-PERFORMANCE LIQUID CHROMATOGRAPHIC DETERMINATION OF ORGANIC ACIDS IN CORN SILAGE
Kato, Isao; Yosizawa, Sadao; Hamaguchi, Kenzi; Ando, Kouichi

Coll. Dairying Ebetsu Japan JP 069-01
Nippon Chikusan Gakkai Ho; (82) P 642-4; Vol 53; No 9; In Japan

15-82-0060 (0060)
A NEW LIQUID CHROMATOGRAPHIC DETERMINATION OF ORGANIC ACIDS IN SOY SAUCE
Ushijima, Shigeomi; Hamada, Takashi; Kanbe, Chiyuki
Cent. Res. Lab. Kikkoman Corp. Noda Japan JP
Nippon Shoyu Kenkyusho Zasshi; (82) P 58-63; Vol 8; No 2; In Japan

15-82-0061 (0061)
HIGH-PRESSURE LIQUID CHROMATOGRAPHY OF PARACETAMOL, ACETYLSALICYLIC ACID, AND CODEINE PHOSPHATE
Stuber, B.; Mueller, K. H.
Dr. Spirig A. G. Olten Switz. CH CH-4601
Pharm. Acta Helv.; (82) P 181; Vol 57; No 7; In Ger

15-82-0062 (0062)
PRECISION AND ACCURACY OF INDOLE-3-ACETIC ACID ANALYSES PERFORMED WITH THE 2-METHYLINDOLO-.ALPHA.-PYRONE FLUORESCENCE ASSAY AND WITH A HIGH-PERFORMANCE LIQUID CHROMATOGRAPHY TECHNIQUE WITH SPECTROFLUOROMETRIC DETECTION, EXEMPLIFIED ON PINE TISSUE (PINUS SYLVESTRIS L)
Sandberg, Goeran; Dunberg, Arne
Dep. For. Genet. Plant Physiol. Swed. Univ. Agric. Sci. Umea Swed. SE S-901 83
Physiol. Plant.; (82) P 315-22; Vol 55; No 3

15-82-0063 (0063)
A HIGH-PERFORMANCE LIQUID CHROMATOGRAPHY-RADIOANALYZER FOR DETERMINING SPECIFIC ACTIVITIES OF RADIOACTIVE ORGANIC ACIDS
Mori, Satoshi
Dep. Agric. Chem. Univ. Tokyo Tokyo Japan JP 113
Plant Cell Physiol.; (82) P 703-8; Vol 23; No 4

15-82-0064 (0064)
TOTAL ANALYSIS OF PYROLIGNEOUS ACIDS FROM POPULUS TREMULOIDES BY HPLC
Menard, H.; Roy, C.; Gaboury, A.; Belanger, D.; Chauvette, G.
Dep. Genie Chim. Univ. Sherbrooke Sherbrooke Can. CA PQ J1K 2R1
Proc. - Bioenergy R & D Semin.; (82) P 331-5; Vol 4th

15-82-0065 (0065)
MEASUREMENT OF VALPROIC ACID IN SERUM AS PHENACYL ESTERS BY HIGH-PERFORMANCE LIQUID CHROMATOGRAPHY
Sakakihara, Yoichi; Yokochi, Akiko; Enomoto, Shoko; Kitazumi, Eiji ; Kodama, Kazuo
Fac. Med. Univ. Tokyo Tokyo Japan JP
Rinsho Kensa; (82) P 719-23; Vol 26; No 6; In Japan

15-82-0066 (0066)
SIMULTANEOUS DETERMINATIONS OF URINARY HIPPURIC AND METHYLHIPPURIC ACIDS IN SOLVENT-WORKERS BY HIGH-PERFORMANCE LIQUID CHROMATOGRAPHY
Niinuma, Yukiko; Sakai, Tadashi; Yanagihara, Susumu; Nakahira, Sayoko; Kunugi, Yumiko; Ushio, Koichi
Cent. Occup. Med. Tokyo Labor Accident Hosp. Tokyo Japan JP 143
Sangyo Igaku; (82) P 322-3; Vol 24; No 3

15-82-0067 (0067)
A HIGH-PRESSURE LIQUID CHROMATOGRAPHIC-ENZYMIC ASSAY FOR GLYCINE AND TAURINE CONJUGATES OF CHOLIC, CHENODEOXYCHOLIC AND DEOXYCHOLIC ACID IN SERUM
Linnet, Kristian
Dep. Clin. Chem. Frederiksbe2g Hosp. Copenhagen Den. DK DK-2000
Scand. J. Clin. Lab. Invest; (82) P 455-60; Vol 42; No 5

15-82-0068 (0068)
HPLC SEPARATION AND QUANTITATIVE DETERMINATION OF ORGANIC ACIDS IN FRUIT JUICES AND WINES
Droz, C.; Tanner, H.
Eidg. Forschungsanst. Wadenswil Switz. CH
Schweiz. Z. Obst- Weinbau; (82) P 434-8; Vol 118; No 15; In Ger

15-82-0069 (0069)
DETERMINATION OF 10-HYDROXY-2-DECENOIC ACID IN ROYAL JELLY BY HIGH PERFORMANCE LIQUID CHROMATOGRAPHY
Nakajin, Shizuo; Taguchi, Mitsue; Akiyama, Yoshiro; Shinoda, Masato
Hoshi Coll. Pharm. Tokyo Japan JP 142
Yakugaku Zasshi; (82) P 549-54; Vol 102; No 6; In Japan

CHAPTER 7.16
Peptides

--- 1982 ---

16-82-0001 (0001)
COMPARISON OF RADIOIMMUNOASSAY, CHEMICAL ASSAY (HPLC) AND BIOASSAY FOR ARGININE VASOPRESSIN IN SYNTHETIC STANDARDS AND POSTERIOR PITUITARY TISSUE
Moehring, J.; Boehlen, P.; Schoun, J.; Mellet, M.; Suess, U.; Schmidt, M.; Pliska, V.
Cent. Rech. Merrell Int. Strasbourg Fr. FR F-67000
Acta Endocrinol. (Copenhagen); (82) P 371-8; Vol 99; No 3

16-82-0002 (0002)
HPLC OF PEPTIDES AND PROTEINS
Hearn, T. W.; Regnier, F. E.; Wehr, C. T.
St. Vincent's Sch. Med. Res. Australia AU
Am. Lab. (Fairfield, Conn.); (82) P 18, 20, 23-4, 27-30, 35-6, 38-9; Vol 14; No 10

16-82-0003 (0003)
SEPARATION OF ANGIOTENSINS AND ASSAY OF ANGIOTENSIN-GENERATING ENZYMES BY HIGH-PERFORMANCE LIQUID CHROMATOGRAPHY
Klickstein, Lloyd B.; Wintroub, Bruce U.
Dep. Dermatol. Harvard Med. Sch. Boston USA US MA 02115
Anal. Biochem.; (82) P 146-50; Vol 120; No 1

16-82-0004 (0004)
ASSESSMENT OF CISPLATIN REACTIVITY WITH PEPTIDES AND PROTEINS USING REVERSE-PHASE HIGH-PERFORMANCE LIQUID CHROMATOGRAPHY AND FLAMELESS ATOMIC ABSORPTION SPECTROSCOPY
Riley, Christopher M.; Sternson, Larry A.; Repta, Arnold J.
Dep. Pharm. Chem. Univ. Kansas Lawrence USA US KS 66045
Anal. Biochem.; (82) P 167-79; Vol 124; No 1

16-82-0005 (0005)
SEPARATION OF BASIC, HYDROPHILIC PEPTIDES BY REVERSED-PHASE ION-PAIR CHROMATOGRAPHY. II. ANALYTICAL APPLICATIONS WITH PARTICULAR REFERENCE TO PHOSPHOSERINE PEPTIDES
Fransson, Bengt; Ragnarsson, Ulf; Zetterqvist, Oerjan
Inst. Biochem. Univ. Uppsala Uppsala Swed. SE S-751 23
Anal. Biochem.; (82) P 174-8; Vol 126; No 1

16-82-0006 (0006)
THE ISOLATION OF PEPTIDES BY HIGH-PERFORMANCE LIQUID CHROMATOGRAPHY USING PREDICTED ELUTION POSITIONS
Browne, C. A.; Bennett, H. P. J.; Solomon, S.
Endocr. Lab. R. Victoria Hosp. Montreal Can. CA PQ H3A 1A1
Anal. Biochem.; (82) P 201-8; Vol 124; No 1

16-82-0007 (0007)
MICROSEQUENCE ANALYSIS OF PEPTIDES AND PROTEINS. I. PREPARATION OF SAMPLES BY REVERSE-PHASE LIQUID CHROMATOGRAPHY
Yuan, Pau Miau; Pande, Hema; Clark, Brian R.; Shively, John E.
Immunol. Div. City of Hope Res. Inst. Duarte USA US CA P91010
Anal. Biochem.; (82) P 289-301; Vol 120; No 2

16-82-0008 (0008)
MICROSEQUENCE ANALYSIS OF PEPTIDES AND PROTEINS. II. SEPARATION OF AMINO ACID PHENYLTHIOHYDANTOIN DERIVATIVES BY HIGH-PERFORMANCE LIQUID CHROMATOGRAPHY ON OCTADECYLSILANE SUPPORTS
Hawke, David; Yuan, Pau Miau; Shively, John E.
Immunol. Div. City of Hope Res. Inst. Duarte USA US CA 91010

Anal. Biochem.; (82) P 302-11; Vol 120; No 2

16-82-0009 (0009)
HIGH-PRESSURE LIQUID CHROMATOGRAPHIC PROCEDURE FOR THE DETERMINATION OF .EPSILON.-(.GAMMA.-GLUTAMYL)LYSINE IN PROTEINS
Griffin, M.; Wilson, J.; Lorand, L.
Dep. Biochem., Mol., Cell Biol. Northwestern Univ. Evanston USA US IL 60201
Anal. Biochem.; (82) P 406-13; Vol 124; No 2

16-82-0010 (0010)
ADAPTATION OF REVERSED-PHASE HIGH-PERFORMANCE LIQUID CHROMATOGRAPHY FOR THE ISOLATION AND SEQUENCE ANALYSIS OF PEPTIDES FROM PLASMA AMYLOID P-COMPONENT
Anderson, Jacqueline K.; Mole, John E.
Med. Sch. Univ. Massachusetts Worcester USA US MA 01605
Anal. Biochem.; (82) P 413-21; Vol 123; No 2

16-82-0011 (0011)
PURIFICATION OF GLYCOPEPTIDES OF HUMAN CERULOPLASMIN AND IMMUNOGLOBULIN D BY HIGH-PRESSURE LIQUID CHROMATOGRAPHY
Tetaert, Daniel; Takahashi, Nobuhiro; Putnam, Frank W.
Dep. Biol. Indiana Univ. Bloomington USA US IN 47405
Anal. Biochem.; (82) P 430-7; Vol 123; No 2

16-82-0012 (0012)
IDENTIFICATION OF ENZYMICALLY PRODUCED PEPTIDE FRAGMENTS BY LIQUID CHROMATOGRAPHY AND MASS SPECTROMETRY
Baranowski, Robert; Westenfelder, Christof; Currie, Bruce L.
Abraham Lincoln Sch. Med. Univ. Illinois Chicago USA US IL 60680
Anal. Biochem.; (82) P 97-102; Vol 121; No 1

16-82-0013 (0013)
APPLICATION OF HIGH-PERFORMANCE LIQUID CHROMATOGRAPHY TO THE SEPARATION AND ISOLATION OF BIOPOLYMERS
Unger, K. K.; Roumeliotis, P.
Inst. Anorg. Chem. Anal. Chem. Johannes Gutenberg-Univ. Mainz Fed. Rep. Ger. DE 6500
Anal. Chem. Symp. Ser.; (82) P 455-70; Vol 9; No Affinity Chromatogr. Relat. Tech.

16-82-0014 (0014)
CHEMICAL STRUCTURE OF RAT LIVER CYTOCHROME B5. ISOLATION OF PEPTIDES BY HIGH-PRESSURE LIQUID CHROMATOGRAPHY
Ozols, Juris; Heinemann, Francis S.
Health Cent. Univ. Connecticut Farmington USA US CT 06032
Biochim. Biophys. Acta; (82) P 163-73; Vol 704; No 1

16-82-0015 (0015)
SEPARATION OF BASIC, HYDROPHILIC PEPTIDES BY REVERSED-PHASE ION-PAIR CHROMATOGRAPHY. I. ANALYTICAL APPLICATIONS WITH PARTICULAR REFERENCE TO A CLASS OF SERINE PEPTIDE SUBSTRATES OF CYCLIC AMP-STIMULATED PROTEIN KINASE
Fransson, B.; Ragnarsson, U.; Zetterqvist, O.
Biomed. Cent. Univ. Uppsala Uppsala Swed. SE S-751 23
J. Chromatogr.; (82) P 165-71; Vol 240; No 1

16-82-0016 (0016)
APPLICATION OF HIGH-PERFORMANCE LIQUID CHROMATOGRAPHY TO THE SEPARATION AND ISOLATION OF BIOPOLYMERS
Unger, K. K.; Roumeliotis, P.
Inst. Anorg. Chem. Anal. Chem. Johannes Gutenberg-Univ. Mainz Fed. Rep. Ger. DE 6500

Anal. Chem. Symp. Ser.; (82) P 455-70; Vol 9; No Affinity Chromatogr. Relat. Tech.

16-82-0017 (0017)
HPLC AND RIA OF CHOLECYSTOKININ PEPTIDES IN THE VERTEBRATE NEURAL RETINA
Eskay, Robert L.; Beinfeld, Margery C.
Lab. Clin. Sci. Natl. Inst. Mental Health Bethesda USA US MD 20205
Brain Res.; (82) P 315-18; Vol 246; No 2

16-82-0018 (0018)
HPLC SEPARATION OF BITTER PEPTIDES FROM CHEDDAR CHEESE
Champion, H. M.; Stanley, D. W.
Dep. Food Sci. Univ. Guelph Guelph Can. CA ON N1G 2W1
Can. Inst. Food Sci. Technol. J.; (82) P 283-8; Vol 15; No 4

16-82-0019 (0019)
THE POTENCY OF LC AS AN ANALYTICAL TOOL IN PEPTIDE CHEMISTRY
Goehring, Walter; Gemeiner, Manfred; Moroder, Luis; Nyfeler, Rolf; Wuensch, Erich
Dep. Peptide Chem. Max Planck Inst. Biochem. Martinsried Fed. Rep. Ger. DE
Chromatogr. Sci.; (82) P 13-29; Vol 20; No Biol./Biomed. Appl. Liq. Chromatogr. 4

16-82-0020 (0020)
REVERSE-PHASE LIQUID CHROMATOGRAPHY OF NEUROPEPTIDES
Verhoef, J.; Witter, A.
Med. Fac. Univ. Utrecht Utrecht Neth. NL
Chromatogr. Sci.; (82) P 57-70; Vol 20; No Biol./Biomed. Appl. Liq. Chromatogr. 4

16-82-0021 (0021)
HPLC STUDIES OF PEPTIDES. THE EFFECT OF THE MOBILE PHASE SYSTEM IN REVERSED-PHASE CHROMATOGRAPHY
Hu, Zhenyuan; Xia, Lijun; Xu, Jiacheng; Cheng, Linlin
Shanghai Inst. Org. Chem. Acad. Sin. Shanghai Peop. Rep. China CN
Chromatographia; (82) P 5-8; Vol 15; No 1

16-82-0022 (0022)
SYNTHESIS AND HPLC OF DELTA-INDUCING PEPTIDE (DSIP) AND ITS ANALOGS
Xu, Jie Cheng; Chen, Ling Ling; Hu, Zheng Yuan; Xia, Li Jun; Huang, Jia Xin
Shanghai Inst. Org. Chem. Acad. Sinica Shangai Peop. Rep. China CN
CONFERENCE; (82), Chem. Nat. Prod., Proc. Sino-Am. Symp.; P 268-70; Avail.: Wang, Yu; Date: 80; Publisher: Sci. Press, Beijing, Peop. Rep. China

16-82-0023 (0023)
PROTEIN MICROSEQUENCING BY HPLC PEPTIDE PURIFICATION AND SOLID PHASE EDMAN DEGRADATION
L'Italien, James J.; Laursen, Richard A.
Dep. Intern. Med. Yale Univ. New Haven USA US CT
CONFERENCE; (82), Methods Protein Sequence Anal., (Proc. Int. Conf.), 4th; P 383-99; Avail.: Elzinga, Marshall; Date: 81; Publisher: Humana, Clifton, N. J.

16-82-0024 (0024)
LIQUID CHROMATOGRAPHY : THE EFFECT OF PEPTIDE STRUCTURE ON RETENTION, SEPARATION AND DETERMINATION OF ENANTIOMERIC AMINO ACIDS AND PEPTIDES, AND PREPARATIVE LIQUID CHROMATOGRAPH USING A RADIALLY COMPRESSED XAD-4 CARTRIDGE
Cahill, William Robert, Jr.
Univ. Iowa, Iowa City, USA, IA
1981. 154 pp. Avail. Univ. Microfilms Int., Order No.DA 8209968
Diss. Abstr. Int. B 1982 P 4403;(82); Vol 42; No 11

16-82-0025 (0025)
SIMPLE PURIFICATION OF ACETONITRILE FOR HIGH-PERFORMANCE LIQUID CHROMATOGRAPHY (HPLC) OF PEPTIDES AND PROTEINS
Braunitzer, Gerhard; Ruecknagel, Peter; Oberthuer, Walter
Abt. Proteinchem. Max-Planck-Inst. Biochem. Martinsried/Muenchen Fed. Rep. Ger. DE D-8033
Hoppe-Seyler's Z. Physiol. Chem.; (82) P 485-6; Vol 363; No 5; In Ger

16-82-0026 (0026)
DETERMINATION OF BRADYKININ, KALLIDIN AND MET-LYS-BRADYKININ BY HIGH PERFORMANCE LIQUID CHROMATOGRAPHY
Geiger, Reinhard; Hell, Ruza; Fritz, Hans
Abt. Klin. Chem. Univ. Muenchen Munich Fed. Rep. Ger. DE D-8000/2
Hoppe-Seyler's Z. Physiol. Chem.; (82) P 527-30; Vol 363; No 5

16-82-0027 (0027)
STUDIES ON SPIN-TRAPPED RADICALS IN .GAMMA.-IRRADIATED AQUEOUS SOLUTIONS OF L-ALANYLGLYCINE AND L-ALANYL-L-ALANINE BY HIGH-PERFORMANCE LIQUID CHROMATOGRAPHY AND ESR SPECTROSCOPY
Moriya, Fumio; Makino, Keisuke; Suzuki, Nobuhiro; Rokushika, Souji ; Hatano, Hiroyuki
Fac. Sci. Kyoto Univ. Kyoto Japan JP 606
J. Am. Chem. Soc.; (82) P 830-6; Vol 104; No 3

16-82-0028 (0028)
SEPARATION OF GLYCOPEPTIDES BY HIGH PERFORMANCE LIQUID CHROMATOGRAPHY
Rosner, Marsha Rich; Robbins, Phillips W.
Dep. Biol. Massachusetts Inst. Technol. Cambridge USA US MA 02139
J. Cell. Biochem.; (82) P 37-47; Vol 18; No 1

16-82-0029 (0029)
HIGH-PERFORMANCE LIQUID CHROMATOGRAPHIC SEPARATION OF AMINO SUGARS AND PEPTIDES WITH METAL ION-MODIFIELD MOBILE PHASES
Dua, Virendra K.; Bush, C.Allen
Dep.Chem., Illinois Inst. Technol., Chicago USA, IL 60 616
J. Chromatogr. (82) P 128-33; Vol 244; No 1

16-82-0030 (0030)
EVALUATION OF STATIONARY AND MOBILE PHASES FOR REVERSED-PHASE HIGH PERFORMANCE LIQUID CHROMATOGRAPHY OF PEPTIDES
Wehr, C. T.; Correia, L.; Abbott, S. R.
Varian Instrum. Group Walnut Creek USA US CA 94598
J. Chromatogr. Sci.; (82) P 114-19; Vol 20; No 3

16-82-0031 (0031)
PURIFICATION OF POLYOXIN D BY REVERSED-PHASE HIGH-PERFORMANCE LIQUID CHROMATOGRAPHY
Shenbagamurthi, P.; Smith, H. A.; Becker, J. M.; Naider, F.
Coll. Staten Island City Univ. New York Staten Island USA US NY 10301
J. Chromatogr.; (82) P 133-7; Vol 245; No 1

16-82-0032 (0032)
THE POTENCY OF LC AS AN ANALYTICAL TOOL IN PEPTIDE CHEMISTRY
Goehring, Walter; Gemeiner, Manfred; Moroder, Luis; Nyfeler, Rolf; Wuensch, Erich
Dep. Peptide Chem. Max Planck Inst. Biochem. Martinsried Fed. Rep. Ger. DE
Chromatogr. Sci.; (82) P 13-29; Vol 20; No Biol./Biomed. Appl. Liq. Chromatogr. 4

16-82-0033 (0033)
SEPARATION OF BASIC, HYDROPHILIC PEPTIDES BY REVERSED-PHASE ION-PAIR CHROMATOGRAPHY. I. ANALYTICAL APPLICATIONS WITH PARTICULAR REFERENCE TO A CLASS OF SERINE PEPTIDE SUBSTRATES OF CYCLIC AMP-STIMULATED PROTEIN KINASE
Fransson, B.; Ragnarsson, U.; Zetterqvist, O.
Biomed. Cent. Univ. Uppsala Uppsala Swed. SE S-751 23
J. Chromatogr.; (82) P 165-71; Vol 240; No 1

16-82-0034 (0034)
ISOCRATIC REVERSED-PHASE HIGH-PERFORMANCE LIQUID CHROMATOGRAPHIC SEPARATION OF UNDERIVATIZED TYROSINE-RELATED PEPTIDES OF THYMOPOIETIN32-36 PENTAPEPTIDE
Tischio, John P.; Hetyei, Nancy
Res. Lab. Ortho Pharm. Corx. Raritan USA NJ 08869
J. Chromatogr.; (82) P 237-43; Vol 236; No 1

16-82-0035 (0035)
USE OF ION-PAIRED, REVERSED-PHASE THIN-LAYER CHROMATOGRAPHY FOR THE ANALYSIS OF PEPTIDES. A SIMPLE PROCEDURE FOR THE MONITORING OF PREPARATIVE REVERSED-PHASE HIGH-PERFORMANCE LIQUID CHROMATOGRAPHY
Poll, D. J.; Knighton, D. R.; Harding, D. R. K.; Hancock, W. S.
Dep. Chem. Biochem. Biophys. Massey Univ. Palmerston North N. Z. NZ
J. Chromatogr.; (82) P 244-8; Vol 236; No 1

16-82-0036 (0036)
ANALYSIS OF .GAMMA.-GLUTAMYL DIPEPTIDES IN F2 HYBRID OF BLACK GRAM AND MUNG BEAN BY HIGH-PERFORMANCE LIQUID CHROMATOGRAPHY
Wu, Shih Hsiung
Inst. Biol. Chem. Acad. Sin. Taipei Taiwan TW
J. Chromatogr.; (82) P 268-70; Vol 245; No 2

16-82-0037 (0037)
EFFECT OF MOLECULAR STRUCTURE AND CONFORMATIONAL CHANGE OF PROLINE-CONTAINING DIPEPTIDES IN REVERSED-PHASE CHROMATOGRAPHY
Melander, Wayne R.; Jacobson, Jana; Horvath, Csaba
Dep. Chem. Eng. Yale Univ. New Haven USA US CT 06520
J. Chromatogr.; (82) P 269-76; Vol 234; No 2

16-82-0038 (0038)
ANALYSIS OF SMALL PEPTIDES IN UREMIC SERUM BY HIGH-PERFORMANCE LIQUID CHROMATOGRAPHY
Mabuchi, Hisao; Nakahashi, Hisamitsu
Dep. Intern. Med. Nephrol. Nishijin Hosp. Kyoto Japan JP 602
J. Chromatogr.; (82) P 292-7; Vol 228

16-82-0039 (0039)
PREDICTION OF PEPTIDE RETENTION TIMES IN REVERSED-PHASE HIGH-PERFORMANCE LIQUID CHROMATOGRAPHY DURING LINEAR GRADIENT ELUTION
Sasagawa, Tatsuru; Okuyama, Tsuneo; Teller, David C.
Dep. Biochem. Univ. Washington Seattle USA US WA 98195
J. Chromatogr.; (82) P 329-40; Vol 240; No 2

16-82-0040 (0040)
SENSITIVE AND RAPID AMINO ACID ANALYSIS OF PEPTIDE HYDROLYZATES BY HIGH-PERFORMANCE LIQUID CHROMATOGRAPHY OF o-PHTHALDIALDEHYDE DERIVATIVES
Burbach, J. Peter H.; Prins, Arie; Lebouille, Jos L. M.; Verhoef, J.; Witter, Albert
Med. Fac. Univ. Utrecht Utrecht Neth. NL 3521 GD
J. Chromatogr.; (82) P 339-43; Vol 237; No 2

16-82-0041 (0041)
AMINO ACIDS AND PEPTIDES. CLXXVIII. CORRELATION BETWEEN THE HYDROPHOBICITY OF SUBSTITUENTS IN THE PHENYLALANINE MOIETY OF OXYTOCIN CARBA ANALOGS AND REVERSED-PHASE-CHROMATOGRAPHIC k' VALUES
Lebl, Michel
Inst. Org. Chem. Biochem. Czech. Acad. Sci. Prague Czech. CS 166 10/6
J. Chromatogr.; (82) P 342-5; Vol 242; No 2

16-82-0042 (0042)
COMPARISON OF THE HIGH-PERFORMANCE LIQUID CHROMATOGRAPHY OF PEPTIDES AND PROTEINS ON 100- AND 300-.ANG. REVERSED-PHASE SUPPORTS
Wilson, Kenneth J.; Van Wieringen, Erika; Klauser, Stefan; Berchtold, Martin W.; Hughes, Graham J.
Biochem. Inst. Univ. Zurich Zurich Switz. CH CH-8028
J. Chromatogr.; (82) P 407-16; Vol 237; No 3

16-82-0043 (0043)
ANALYSIS OF INSULIN PREPARATIONS BY REVERSED-PHASE HIGH-PERFORMANCE LIQUID CHROMATOGRAPHY
Lloyd, L. F.; Corran, P. H.
Chem. Sect. Natl. Inst. Biol. Stand. Control Holly Hill/Hampstead/London UK GB NW3 6RB
J. Chromatogr.; (82) P 445-54; Vol 240; No 2

16-82-0044 (0044)
FACTORS INFLUENCING THE RETENTION OF INSULINS IN REVERSED-PHASE HIGH-PERFORMANCE LIQUID CHROMATOGRAPHIC SYSTEMS
Vigh, Gyula; Varga-Puchony, Zita; Hlavay, Jozsef; Papp-Hites, E.
Inst. Anal. Chem. Veszprem Univ. Chem. Eng. Veszprem Hung. HU
J. Chromatogr.; (82) P 51-9; Vol 236; No 1

16-82-0045 (0045)
SEPARATION OF HUMAN INSULIN AND SOME STRUCTURAL ISOMERS BY HIGH-PERFORMANCE LIQUID CHROMATOGRAPHY
Lloyd, L. F.; Calam, D. H.
Natl. Inst. Biol. Stand. Control London UK GB NW3 6RB
J. Chromatogr.; (82) P 511-14; Vol 237; No 3

16-82-0046 (0046)
RAPID ISOLATION OF NANOGRAM AMOUNTS OF CRUSTACEAN ERYTHROPHORE CONCENTRATING HORMONE FROM INVERTEBRATE NERVE TISSUE BY RP-HPLC
Jaffe, Howard; Loeb, Marcia; Hayes, Dora K.; Holston, Nancy
Agric. Environ. Qual. Inst. ARS Beltsville USA US MD 20705
J. Liq. Chromatogr.; (82) P 1375-90; Vol 5; No 7

16-82-0047 (0047)
PRECOLUMN DERIVATIZATION IN HPLC OF AMINO ACIDS AND PEPTIDES
Szokan, Gyula
Inst. Org. Chem. Eotvos Univ. Budapest Hung. HU H-1088
J. Liq. Chromatogr.; (82) P 1493-8; Vol 5; No 8

16-82-0048 (0048)
IODINATION OF PEPTIDE HORMONES AND PURIFICATION OF IODINATED PEPTIDES BY HPLC
Janaky, T.; Toth, G.; Penke, B.; Kovacs, K.; Laszlo, F. A.
1st Dep. Med. Univ. Med. Sch. Szeged Hung. HU
J. Liq. Chromatogr.; (82) P 1499-507; Vol 5; No 8

16-82-0049 (0049)
SEPARATION OF THE SECRETIN-GLUCAGON FAMILY PEPTIDES BY RP-HPLC IN TRIETHYL-AMMONIUM PHOSPHATE BUFFERS

Fourmy, D.; Pradayrol, L.; Ribet, A.
CHU Ranqueil Toulouse Fr. FR 31054
J. Liq. Chromatogr.; (82) P 2123-34; Vol 5; No 11

16-82-0050 (0050)
PEPTIDE FORMATION IN THE PRESENCE OF METAL ION PROTECTING GROUPS. III. THE SEPARATION OF ((NH3)5Co(III)-) AMINO ACIDS AND PEPTIDES BY REVERSE-PHASE HIGH-PRESSURE LIQUID CHROMATOGRAPHY
Isied, Stephan S.; Lyon, John; Vassilian, Asbed; Worosila, Greg
Dep. Chem. Rutgers, State Univ. New Brunswick USA US NJ 08903
J. Liq. Chromatogr.; (82) P 537-47; Vol 5; No 3

16-82-0051 (0051)
PURIFICATION OF RADIO-IODINATED CHOLECYSTOKININ PEPTIDES BY REVERSE PHASE HPLC
Fourmy, D.; Pradayrol, L.; Antoniotti, H.; Esteve, J. P.; Ribet, A.
Lab. Biol. Pathol. Dig. Cent. Hosp. Univ. Rangveil Toulouse Fr. FR 31054
J. Liq. Chromatogr.; (82) P 757-66; Vol 5; No 4

16-82-0052 (0052)
HIGH-PERFORMANCE LIQUID CHROMATOGRAPHY OF PHARMACOLOGICALLY ACTIVE AMINES AND PEPTIDES IN BIOLOGICAL MATERIALS
Davis, Thomas P.; Schoemaker, Hans; Chen, Andrew; Yamamura, Henry I.
Heatlh Sci. Cent. Univ. rizona Tucson USA US AZ 85724
Life Sci.; (82) P 971-87; Vol 30; No 12

16-82-0053 (0053)
REVERSED PHASE HIGH-PRESSURE LIQUID CHROMATOGRAPHY FOR THE IDENTIFICATION AND PURIFICATION OF NEUROPEPTIDES
Moore, Graham J.
Fac. Med. Univ. Calgary Calgary Can. CA AB T2N 1N4
Life Sci.; (82) P 995-1002; Vol 30; No 12

16-82-0054 (0054)
CHARACTERIZATION OF CORTICOTROPIN-LIKE IMMUNOREACTIVE PEPTIDES IN RAT BRAIN USING HIGH PERFORMANCE LIQUID CHROMATOGRAPHY
Smith, A.; Keith, A. B.; Edwardson, J. A.; Biggins, J. A.; McDermott, J. R.
MRC Neuroendocrinol. Unit Newcastle Gen. Hosp. Newcastle upon Tyne UK GB NE4
Neurosci. Lett.; (82) P 133-8; Vol 30; No 2

16-82-0055 (0055)
SEPARATION OF NEUROPEPTIDES BY HIGH-PERFORMANCE LIQUID CHROMATOGRAPHY AND ITS APPLICATION TO THE ANALYSIS OF PEPTIDES IN THE RAT PITUITARY NEUROINTERMEDIATE LOBE
Hashimoto, Kozo; Ohno, Norihito; Hattori, Teruhiko; Takahara, Jiro ; Ofuji, Tadashi
Sch. Med. Okayama Univ. Okayama Japan JP
Nippon Naibumpi Gakkai Zasshi; (82) P 167-76; Vol 58; No 3; In Japan

16-82-0056 (0056)
ANALYSIS OF .GAMMA.-GLUTAMYL DIPEPTIDES IN BEAN EXTRACTS BY HIGH-PERFORMANCE LIQUID CHROMATOGRAPHY
Wu, Shih Hsiung; Wang, Kung Tsung; Tsou, Samson C. S.
Inst. Biol. Chem. Acad. Sinica Taipei Taiwan TW
Proc. Natl. Sci. Counc., Repub. China, Part A; (82) P 34-8; Vol 6 ; No 1

16-82-0057 (0057)
A STUDY OF THE METABOLISM OF DES-TYROSINE-.GAMMA.-ENDORPHIN USING AN IMPROVED HPLC SYSTEM
Schoemaker, Hans; Chen, Andrew; Davis, Thomas P.; Yamamura, Henry I.
Health Sci. Cent. Univ. Arizona Tucson USA US AZ 85724

Psychopharmacol. Bull.; (82) P 144-8; Vol 18; No 4

16-82-0058 (0058)
CHARACTERIZATION OF SUBSTANCE P-LIKE IMMUNOREACTIVITY IN PERIPHERAL SENSORY NERVES AND ENTERIC NERVES BY HIGH-PRESSURE LIQUID CHROMATOGRAPHY AND RADIOIMMUNOASSAY
Murphy, R.; Furness, J. B.; Beardsley, A. M.; Costa, M.
Sch. Med. Flinders Univ. Bedford Park Australia AU 5042
Regul. Pept.; (82) P 203-12; Vol 4; No 4

16-82-0059 (0059)
CORTICOTROPIN-RELATED PEPTIDES IN THE INTERMEDIATE LOBE OF THE RODENT PITUITARY GLAND: CHARACTERIZATION BY HIGH PERFORMANCE LIQUID CHROMATOGRAPHY AND RADIOIMMUNOASSAY
Al-Noaemi, M. C.; Biggins, J. A.; Edwardson, J. A.; McDermott, J. R.; Smith,
MRC Neuroendocrinol. Unit Newcastle Gen. Hosp. Newcastle upon Tyne UK GB NE4 6BE
Regul.Pept.; (82) P 351-9; Vol 3; No 5-6

16-82-0060 (0060)
DETERMINATION OF METHIONINE-ENKEPHALIN, NOREPINEPHRINE, DOPAMINE, 3,4-DIHYDROXYPHENYLACETIC ACID (DOPAC) AND 3-METHOXY-4-HYDROXYPHENYLACETIC ACID (HVA) IN BRAIN BY HIGH-PRESSURE LIQUID CHROMATOGRAPHY WITH ELECTROCHEMICAL DETECTOR
Nabeshima, Toshitaka; Hiramatsu, Masayuki; Noma, Shigeki; Ukai, Makoto; Amano, Manabu; Kameyama, Tsutomu
Fac. Pharm. Sci. Meijo Univ. Nagoya Japan JP 468
Res. Commun. Chem. Pathol. Pharmacol.; (82) P 421-42; Vol 35; No 3

16-82-0061 (0061)
USE OF HIGH-PRESSURE LIQUID CHROMATOGRAPHY IN PEPTIDE SYNTHESES
Sakakibara, Shumpei
Tanpakushitsu Kenkyu. Shoreikai Minoo Japan bP
Tanpakushitsu Kakusan Koso; (82) P 1938-46; Vol 27; No 12; In Japan

16-82-0062 (0062)
DETERMINATION OF GLUTATHIONE AND D-PENICILLAMINE USING HIGH-PERFORMANCE LIQUID CHROMATOGRAPHY WITH A VOLTAMMETRIC DETECTOR
Inoue, Kenji; Otake, Tateo; Noda, Katsumi; Kyogoku, Kazuaki; Uchida, Shoji
Res. Cent. Taisho Pharm. Co., Ltd. Saitama Japan JP 330
Yakugaku Zasshi; (82) P 659-65; Vol 102; No 7; In Japan

CHAPTER 7.17
Plant Extracts

--- 1982 ---

17-82-0001 (0001)
LIQUID PHASE CHROMATOGRAPHY OF JASMINE CONCRETES. ELUANT ANALYSIS BY UV-VISIBLE SPECTROPHOTOMETER WITH PARALLEL DETECTION
Linossier, J. L.; Pevron, L.; Rocca, J. L.
Cent. Rech. Saint-Fons Rhone-Poulenc Rech. Saint-Fons Fr. FR 69190
Analusis; (82) P 310-18; Vol 10; No 7; In Fr

17-82-0002 (0002)
SEPARATION OF o-PHTHALDIALDEHYDE DERIVATIVES OF FREE AMINO ACIDS FROM PLANT TISSUES BY ISOCRATIC REVERSE-PHASE HIGH-PERFORMANCE LIQUID CHROMATOGRAPHY
Martin, F.; Maudinas, B.; Gadal, P.
Unite Microbiol. Sols INRA Seichamps Fr. FR
Ann. Bot. (London); (82) P 401-6; Vol 50; No 3

17-82-0003 (0003)
ISOLATION OF GINSENOSIDES RB1, RBR, RC, RD, RE, RF, AND RG1 FROM GINSENG ROOT BY HIGH-PERFORMANCE LIQUID CHROMATOGRAPHY
Paik, Nam Ho; Park, Man Ki; Choi, Kana Ju; Cho, Yuna Hvun
Coll. Pharm. Seoul Natl. Univ. Seoul S. Korea KR 151
Arch. Pharmacal Res.; (82) P 7-12; Vol 5; No 1

17-82-0004 (0004)
HISTAMINE IN COTTON PLANT AND DUST DETERMINED BY HIGH-PERFORMANCE LIQUID CHROMATOGRAPHY
Wall, J. H.; Muller, L. L.; Berni, R. J.
South. Rea. Res. Cent. USDA New Orleans USA US LA 70179
ACS Svmp. Ser.; (82) P 301-11; Vol 189; No Cotton Dust

17-82-0005 (0005)
RETENTION BEHAVIOR OF PANAX GINSENG SAPOGENINS IN SILICA GEL LIQUID-SOLID CHROMATOGRAPHY
Kunihiro, Kazuo; Yamaauchi, Hirovuki; Hara, Shoii
Cent. Res. Lab. Wakunaaa Pharm. Co., Ltd. Hiroshima Japan JP
Bunseki Kaaaku; (82) P 83-7; Vol 31; No 2; In Japan

17-82-0006 (0006)
ISOLATION AND CHARACTERIZATION OF A VITAMIN D3 COMPOUND FROM A PLANT BY LIQUID CHROMATOGRAPHY
Rambeck, Walter A.; Zucker, H.
Vet. Fac. Univ. Munich Munich Fed. Rep. Ger. DE
Chromatoar. Sci.; (82) P 319-26; Vol 20; No Biol./Biomed. Appl. Liq. Chromatoar. 4

17-82-0007 (0007)
PROLINE METABOLISM THROUGH GABA-SHUNT IN BARLEY ROOTS MONITORED WITH A HPLC-RADIOANALYZER
Mori, S.
Dep. Aaric. Chem. Univ. Tokvo Tokvo Japan JP
CONFERENCE; Plant Nutr., Proc. Int. Plant Nutr. Colloq., 9th; (82) P 400-5; Vol 2; Avail.: Scaife, A; Publisher: CAB, Farnham Roval, UK.

17-82-0008 (0008)
PROLINE METABOLISM THROUGH GABA-SHUNT IN BARLEY ROOTS MONITORED WITH A HPLC-RADIOANALYZER
Mori, S.
Dep. Aaric. Chem. Univ. Tokvo Tokvo Japan JP
CONFERENCE;(82), Plant Nutr., Proc. Int. Plant Nutr. Colloq., 9th; P 400-5; Vol 2,; Avail.: Scaife, A; Publisher: CAB, Farnham Roval, UK.

17-82-0009 (0009)
QUALITY CONTROL OF VALERIAN PREPARATIONS. 2. VALERENIC ACID AND VALERENAL AS CHARACTERISTIC COMPONENTS OF VALERIANA OFFICINALIS. DETERMINATION BY HPLC
Haensel, Rudolf; Schulz, Jutta
Inst. Pharmakogn. Phytochem. Freie Univ. Berlin Berlin Fed. Rep. Ger. DE 1000
Dtsch. Apoth.-Ztg.; (82) P 215-19; Vol 122; No 5; In Ger

17-82-0010 (0010)
COMPARISON OF CHLOROPHYLL A MEASUREMENTS BY FLUOROMETRIC, SPECTROPHOTOMETRIC AND HIGH PRESSURE LIQUID CHROMATOGRAPHIC METHODS IN AQUATIC ENVIRONMENTS
Jacobsen, Timothy R.
Inst. Ecol. Univ. Georgia Athens USA US GA 30602
Ergeb. Limnol.; (82) P 35-45; Vol 16

17-82-0011 (0011)
HPLC DETERMINATION OF COUMARINIC CONSTITUENTS IN LEAVES OF FICUS CARICA
Innocenti, G.; Bettero, A.; Caporale, G.
Ist. Chim. Farm. Univ. Padova Padua Italy IT
Farmaco, Ed. Sci.; (82) P 475-85; Vol 37; No 7; In Ital

17-82-0012 (0012)
THE USE OF HIGH PRESSURE LIQUID CHROMATOGRAPHY WITH FLUORIMETRIC DETECTION FOR CHLOROPHYLL A DETERMINATION IN NATURAL EXTRACTS OF CHLOROPIGMENTS AND THEIR DEGRADATION PRODUCTS
Goevens, L.; Post, E.; Dehpirs, F.; Vandenhoudt, A.; Baevens, W.
Lab. Anal. Scheikd. Vrije Univ. Brussel Brussels Belg. BE B-1050
Int. J. Environ. Anal. Chem.; (82) P 51-63; Vol 12; No 1

17-82-0013 (0013)
A HIGH-PERFORMANCE LIQUID CHROMATOGRAPHIC METHOD FOR QUANTITATION OF DIMBOA AND MBOA IN MAIZE PLANT EXTRACT
Gutierrez, Carmen; Guerrero, Ana; Castanera, P.; Torres, J. V.
Inst. Nac. Invest. Agrar. CRIDA Madrid Spain ES
J. Agric. Food Chem.; (82) P 1258-60; Vol 30; No 6

17-82-0014 (0014)
TRACE ANALYSIS OF NATURAL CAPSAICINOIDS IN ANIMAL FEED, HUMAN URINE, AND WASTEWATER BY HIGH-PRESSURE LIQUID CHROMATOGRAPHY
Johnson, Eric K.; Thompson, Harold C., Jr.; Bowman, Malcolm C.
Dep. Health Hum. Serv. Food and Drug Adm. Jefferson USA AR 72079
J. Agric. Food Chem.; (82) P 344-9; Vol 30; No 2

17-82-0015 (0015)
TRACE ANALYSIS OF NATURAL CAPSAICINOIDS IN ANIMAL FEED, HUMAN URINE, AND WASTEWATER BY HIGH-PRESSURE LIQUID CHROMATOGRAPHY
Johnson, Eric K.; Thompson, Harold C., Jr.; Bowman, Malcolm C.
Dep. Health Hum. Serv. Food and Drug Adm. Jefferson USA AR 72079
J. Agric. Food Chem.; (82) P 3X4-9; Vol 30; No 2

17-82-0016 (0016)
IDENTIFICATION OF FLAVONOID CHEMICAL MARKERS IN ROSES AND THEIR HIGH-PRESSURE LIQUID CHROMATOGRAPHIC RESOLUTION AND QUANTITATION FOR CULTIVAR IDENTIFICATION
Asen, S.
Florist Nursery Crops Lab. Agric. Res. Serv. Beltsville USA US MD 20705
J. Am. Soc. Hortic. Sci.; (82) P 744-50; Vol 107; No 5

17-82-0017 (0017)
DETERMINATION OF INDIVIDUAL AND TOTAL CARDENOLIDES IN CONVALLARIA MAJALIS L. BY HIGH-PERFORMANCE LIQUID CHROMATOGRAPHY
Jurenitsch, J.; Kopp, B.; Bambera-Kubelka, E.; Kubelka, W.

Inst. Pharmakoa., Univ. Wien, Vienna Austria A-1090
J. Chromatoar. (82) P 235-42; Vol 240; No 1; In Ger

17-82-0018 (0018)
HIGH-PERFORMANCE LIQUID CHROMATOGRAPHY OF BACTERIOCHLOROPHYLL B AND ITS DERIVATIVES AS AN AID FOR STRUCTURE ANALYSIS
Steiner, R.; Wieschhoff, H.; Scheer, H.
Bot. Inst. Univ. Mvenchen Munich Fed. Rep. Ger. DE D-8000/19
J. Chromatoar.; (82) P 127-34; Vol 242; No 1

17-82-0019 (0019)
GENERAL SCHEME OF ANALYSIS OF PHENOLIC COMPOUNDS IN PLANT EXTRACTS BY REVERSED-PHASE HIGH-PERFORMANCE LIQUID CHROMATOGRAPHY
Villeneuve, Francis; Abravan11, Gerard; Moutounet, Michel; Alibert, Gilbert
Cent. Phvsiol. Vea. Univ. Paul Sabatier Toulouse Fr. FR 31062
J. Chromatoar.; (82) P 131-40; Vol 234; No 1

17-82-0020 (0020)
HIGH-PERFORMANCE LIQUID CHROMATOGRAPHIC ANALYSIY OF FLAVONOIDS FROM PROPOLIS
Bankova, V. S.; Popov, S. S.; Marekov, N.PL.
Cent. Phvtochem. Inst. Ora. Chem. Sofia Bula. BG 1113
J. Chromatoar.; (82) P 135-43; Vol 242; No 1

17-82-0021 (0021)
QUANTITATIVE HIGH-PERFORMANCE LIQUID CHROMATOGRAPHIC METHOD FOR THE ESTIMATION OF HECOGENIN AND TIGOGENIN IN THE LEAVES AND SAPOGENIN CONCENTRATES OF AGAVE SISALANA
Sitton, Dov; Blunden, Gerald; Cripps, Alan L.
Sch. Pharm. Portsmouth Polvtech. Portsmouth/Hants. UK GB P01 2DZ
J. Chromatoar.; (82) P 138-40; Vol 245; No 1

17-82-0022 (0022)
REVERSED- AND NORMAL-PHASE HIGH-PERFORMANCE LIQUID CHROMATOGRAPHY OF GIBBERELLIN METHOXYCOUMARYL ESTERS
Crozier, Alan; Zaerr, Joe B.; Morris, Rov O.
Dep. Bot. Univ. Glasaow Glasaow UK GB G12 8QQ
J. Chromatoar.; (82) P 157-66; Vol 238; No 1

17-82-0023 (0023)
DETERMINATION OF HOST-SELECTIVE PHYTOTOXINS FROM ALTERNARIA ALTERNATA F. SP. LYCOPERSICI AS THEIR MALEYL DERIVATIVES BY HIGH-PERFORMANCE LIQUID CHROMATOGRAPHY
Siler, D. J.; Gilchrist, D. G.
Dep. Plant Pathol. Univ. California Davis USA US CA 95616
J. Chromatoar.; (82) P 167-73; Vol 238; No 1

17-82-0024 (0024)
NEW LIQUID CHROMATOGRAPHIC APPROACHES FOR FREE AMINO ACID ANALYSIS IN PLANTS AND INSECTS
Barcelon, Maria de los Anaeles
Aaric. Res. Educ. Cent. Univ. Florida Fort Lauderdale USA US FL 33314
J. Chromatoar.; (82) P 175-82; Vol 238; No 1

17-82-0025 (0025)
SEPARATION OF FARNESOL ISOMERS BY LIQUID CHROMATOGRAPHY
Sato, Hisava; Kaaevu, Akira; Mivashita, Kivoshi; Tanaka, Yasuvuki
Fac. Technol. Tokvo Univ. Aaric. Technol. Tokvo Japan JP 184
J. Chromatoar.; (82) P 178-82; Vol 237; No 1

17-82-0026 (0026)
COMPARISON OF STYRENE-DIVINYLBENZENE COPOLYMER AND C18 SILICA REVERSED-PHASE PACKINGS FOR HIGH-PERFORMANCE LIQUID CHROMATOGRAPHY OF GIBBERELLINS
Grevson, Richard l.; Patch, Andrea M.
Dep. Plant Sci. Univ. West. Ontario London Can. CA ON N6A 5B7
J. Chromatogr.; (82) P 349-52; Vol 242; No 2

17-82-0027 (0027)
HIGH-PERFORMANCE LIQUID CHROMATOGRAPHIC ANALYSIS OF DIGITALIS LEAF EXTRACTS. II. QUANTITATIVE ANALYSIS
Wichtl, M.; Wichtl-Bleier, W.; Manakudidiojo, M.
Inst. Pharm. Biol. Philipps-Univ. Marburg Fed. Rep. Ger. DE D-3550
J. Chromatogr.; (82) P 359-65; Vol 247; No 2; In Ger

17-82-0028 (0028)
QUANTITATIVE ANALYSIS OF THE ISOFLAVONE PHYTO-ESTROGENS GENISTEIN, FORMONONETIN AND BIOCHANIN A, IN SUBTERRANEAN CLOVER LEAVES BY HIGH-PERFORMANCE LIQUID CHROMATOGRAPHY
Patroni, Joseph J.; Collins, W. J.; Stern, W. R.
Dep. Org. Chem. Univ. West. Australia Nedlands Australia AU 6009
J. Chromatogr.; (82) P 366-8; Vol 247; No 2

17-82-0029 (0029)
STEVIA REBAUDIANA. III. HIGH-PERFORMANCE LIQUID CHROMATOGRAPHIC SEPARATION AND QUANTITATION OF REBAUDIOSIDES B, D, AND E, DULCOSIDE A, AND STEVIOLBIOSIDE
Ahmed, M. S.; Dobberstein, R. H.
Stevia Co., Inc. Arlington Heights USA US IL 60005
J. Chromatogr.; (82) P 373-6; Vol 245; No 3

17-82-0030 (0030)
SEPARATION AND QUANTITATION OF ESTROGENIC ISOFLAVONES FROM CLOVERS BY HIGH-PERFORMANCE LIQUID CHROMATOGRAPHY
Nicollier, G. F.; Thompson, A. C.
Cent. Alluvial Plains Stud. Delta State Univ. Cleveland USA US MS 38773
J. Chromatogr.; (82) P 399-402; Vol 249; No 2

17-82-0031 (0031)
ANALYTICAL AND PREPARATIVE SEPARATION OF WITHANOLIDES FROM CRUDE EXTRACTS OF ACNISTUS BREVIFLORUS LEAVES BY HIGH-PERFORMANCE LIQUID CHROMATOGRAPHY
Burton, Gerardo; Veleiro, Adriana S.; Gros, Eduardo G.
Fac. Cienc. Exactas Nat. Univ. Buenos Aires Buenos Aires Argent. AR 1428
J. Chromatogr.; (82) P 472-5; Vol 248; No 3

17-82-0032 (0032)
HIGH-PERFORMANCE LIQUID CHROMATOGRAPHY SEPARATION OF SOYBEAN ISOFLAVONES AND THEIR GLUCOSIDES
Eldridge, A. C.
Northern Reg. Res. Cent. Agric. Res. Sci. Educ. Adm. Peoria USA US IL 61604
J. Chromatogr.; (82) P 494-6; Vol 234; No 2

17-82-0033 (0033)
STEVIA REBAUDIANA. II. HIGH-PERFORMANCE LIQUID CPROMATOGRAPHIC SEPARATION AND QUANTITATION OF STEVIOSIDE, REBAUDIOSIDE A AND REBAUDIOSIDE C
Ahmed, M. S.; Dobberstein, R. H.
Stevia Co., Inc. Arlington Heights USA US IL 60005
J. Chromatogr.; (82) P 523-6; Vol 236; No 2

17-82-0034 (0034)
DETERMINATION OF EMETINE AND CEPHAELINE IN IPECAC ROOTS BY HIGH-PERFORMANCE LIQUID CHROMATOGRAPHY

Sahu, N. P.; Mahato, S. B.
Indian Inst. Chem. Biol. Calcutta India IN 700032
J. Chromatogr.; (82) P 525-9; Vol 238; No 2

17-82-0035 (0035)
HIGH-PERFORMANCE LIQUID CHROMATOGRAPHY OF CHROMENES AND BENZOFURANS FROM THE GENUS ENCELIA (ASTERACEAE)
Proksch, Peter; Rodriguez, Floy
Dep. Ecol. Univ. California Irvine USA US CA 92717
J. Chromatogr.; (82) P 543-6; Vol 240; No 2

17-82-0036 (0036)
CONFIGURATION OF THE OLEFINIC BONDS IN THE HETEROOLEFINIC SIDE-CHAINS OF JAPANESE LACQUER URUSHIOL. SEPARATION AND IDENTIFICATION OF COMPONENTS OF DIMETHYLURUSHIOL BY MEANS OF REDUCTIVE OZONOLYSIS AND HIGH-PERFORMANCE LIQUID CHROMATOGRAPHY
Yamauchi, Yoshio; Oshima, Ryuichi; Kumanotani, Ju
Inst. Ind. Sci. Univ. Tokyo Tokyo Japan JP 106
J. Chromatogr.; (82) P 71-84; Vol 243; No 1

17-82-0037 (0037)
DETERMINATION OF CHROMIUM IN ORCHARD LEAVES BY REVERSED-PHASE HIGH-PERFORMANCE LIQUID CHROMATOGRAPHY
Willett, James D.; Knight, Margaret M.
Dep. Chem. Univ. Idaho Moscow USA US ID 83843
J. Chromatogr.; (82) P 99-105; Vol 237; No 1

17-82-0038 (0038)
HPLC ANALYSES OF AMINO ACIDS IN PLANT MATERIALS AS DNP-DERIVATIVES BY USE OF METHYL BENZOATE AS THE INTERNAL STANDARD
Kozukue, Nobuyuki; Kozukue, Etsuko; Terai, Hirofumi; Mizuno, Susumu
Dep. Home Econ. Kenmei Jr. Coll. Himeji City Japan JP
J. Food Sci.; (82) P 1584-6; Vol 47; No 5

17-82-0039 (0039)
HIGH PERFORMANCE LIQUID CHROMATOGRAPHY OF PLANT PIGMENTS - A REVIEW
Schwartz, S. J.; Von Elbe, J. H.
Dep. Food Sci. Univ. Wisconsin Madison USA US WI 53706
J. Liq. Chromatogr.; (82) P 43-73; Vol 5; No Suppl. 1

17-82-0040 (0040)
LIQUID CHROMATOGRAPHIC DETERMINATION OF PHENOLIC ACIDS OF VEGETABLE ORIGIN
Roston, Daryl A.; Kissinger, Peter T.
Dep. Chem. Purdue Univ. West Lafayette USA US IN 47907
J. Liq. Chromatogr.; (82) P 75-103; Vol 5; No Suppl. 1

17-82-0041 (0041)
APPLICATION OF HIGH PERFORMANCE LIQUID CHROMATOGRAPHY IN THE CHARACTERIZATION OF IRON STRESS RESPONSE
Johnson, Gordon V.
Dep. Biol. Univ. New Mexico Albuquerque USA US NM 87131
J. Plant Nutr.; (82) P 499-514; Vol 5; No 4-7

17-82-0042 (0042)
CHARACTERIZATION OF EXTRACTS FROM NORWAY SPRUCE (PICEA ABIES) BY HIGH PRESSURE LIQUID CHROMATOGRAPHY
Pulkkinen, Erkki; Vaisanen, Sirkka
Dep. Chem. Univ. Oulu Oulu Finland FI SF-90570/57
Pap. Puu; (82) P 72-5, 82; Vol 64; No 2

17-82-0043 (0043)
A STUDY OF THE FATE OF (14C)MALEIC HYDRAZIDE IN WHITE ASH AND BLACK LOCUST SEEDLINGS BY HIGH-PERFORMANCE LIQUID CHROMATOGRAPHY
Mischke, Charles F.; Domir, Subhash C.
Dep. Bot. Bacteriol. Ohio Weslevan Univ. Delaware USA US OH 43015
Pestic. Sci.; (82) P 304-8; Vol 13; No 3

17-82-0044 (0044)
SEPARATION AND QUANTITATIVE DETERMINATION OF THE PRIMARY PUNGENT COMPONENTS OF ZINGIBERIS RHIZOMA WITH COMBINED TLC/HPLC
Steineaaer, E.; Stucki, K.
Pharm. Inst. Univ. Bern Bern Switz. CH
Pharm. Acta Helv.; (82) P 66-71; Vol 57; No 3; In Ger

17-82-0045 (0045)
PRECISION AND ACCURACY OF INDOLE-3-ACETIC ACID ANALYSES PERFORMED WITH THE 2-METHYLINDOLO-.ALPHA.-PYRONE FLUORESCENCE ASSAY AND WITH A HIGH-PERFORMANCE LIQUID CHROMATOGRAPHY TECHNIQUE WITH SPECTROFLUOROMETRIC DETECTION, EXEMPLIFIED ON PINE TISSUE (PINUS SYLVESTRIS L)
Sandbera, Goeran; Dunbera, Arne
Dep. For. Genet. Plant Phvsiol. Swed. Univ. Aaric. Sci. Umea Swed. SE S-901 83
Phvsiol. Plant.; (82) P 315-22; Vol 55; No 3

17-82-0046 (0046)
HIGH-PERFORMANCE LIQUID CHROMATOGRAPHIC IDENTIFICATION AND QUANTITATION OF CYCLIC ADENOSINE 3':5'-MONOPHOSPHATE IN HIGHER (PHASEOLUS VULGARIS L.) AND LOWER (CHLORELLA SP.) PLANTS
Van Onckelen, H. A.; Dupon, M.; De Greef, J. A.
Dep. Biol. Univ. Instellina Antwerpen Wilriik Bela. BE B-2610
Phvsiol. Plant.; (82) P 93-7; Vol 55; No 2

17-82-0047 (0047)
ANALYSIS OF POLYAMINES IN HIGHER PLANTS BY HIGH PERFORMANCE LIQUID CHROMATOGRAPHY
Flores, Hector E.; Galston, Arthur W.
Dep. Biol. Yale Univ. New Haven USA US CT 06511
Plant Phvsiol.; (82) P 701-6; Vol 69; No 3

17-82-0048 (0048)
TLC- AND HPLC-ANALYSIS OF ELEUTHEROCOCCUS PREPARATIONS
Waaner, H.; Heur, Y. H.; Obermeier, A.; Tittel, G.; Bladt, S.
Inst. Pharm. Biol. Univ. Muenchen Munich Fed. Rep. Ger. DE D-8000/2
Planta Med.; (82) P 193-8; Vol 44; No 4; In Ger

17-82-0049 (0049)
DETERMINATION OF FLAVONOIDS IN SEEDS OF ZIZYPHUS VULGARIS VAR. SPINOSUS BY HIGH PERFORMANCE LIQUID CHROMATOGRAPHY
Shin, Kuk Hvun; Kim, Hee Yona; Woo, Won Sick
Nat. Prod. Res. Inst. Seoul Natl. Univ. Seoul S. Korea KR
Planta Med.; (82) P 94-6; Vol 44; No 2

17-82-0050 (0050)
ANALYSIS AND ISOLATION OF PHENOLICS FROM PLANTS BY HPLC: POSSIBILITIES AND LIMITS
Tittel, G.; Waaner, H.
Inst. Pharm. Biol. Univ. Muenchen Munich Fed. Rep. Ger. DE D-8000/2
Stud. Ora. Chem. (Amsterdam); (82) P 299-310; Vol 11; No Flavonoids Bioflavonoids

17-82-0051 (0051)
DETERMINATION OF FLUE-CURED TOBACCO PIGMENTS BY HIGH-PERFORMANCE LIQUID CHROMATOGRAPHY
Court, William A.; Hendel, John G.
Chem. Lab. Agric. Canada Delhi Can. CA ON N4B 2W9
Tob. ant.; (82) P 68-70; Vol 184; No 18

17-82-0052 (0052)
AN IMPROVED EXTRACTION-HPLC METHOD FOR TOBACCO POLYPHENOLS
Snook, M. E.; Chortyk, O. T.
Tobacco Saf. Res. Unit Agric. Res. Serv. Athens USA US GA 30613
Tob. Int.; (82) P 111-15; Vol 184; No 5

17-82-0053 (0053)
A SELECTIVE AND RAPID METHOD OF CHLOROPHYLLS A AND B SEPARATIVE DETERMINATION BY HIGH-PRESSURE LIQUID CHROMATOGRAPHY
Bessiere, J.; Montiel, A.
Serv. Controle Paris Fr. FR 75014
Water Res.; (82) P 987-93; Vol 16; No 6; In Fr

17-82-0054 (0054)
A BRIEF REVIEW ON THE APPLICATION OF CLASSICAL LIQUID CHROMATOGRAPHY IN THE SEPARATION OF NATURAL PRODUCTS
Yao, Qianyuan; Ji,PDezhen
Shandong Inst. Trad. Chinese Med. Peop. Rep. China CN
Zhongcaoyao; (82) P 41-7; Vol 13; No 5; In Ch

CHAPTER 7.18

Polymers—Plastics—Antioxidants

--- 1982 ---

18-82-0001 (0001)
ANALYSIS OF PHENOL-FORMALDEHYDE RESINS BY HIGH-PRESSURE LIQUID CHROMATOGRAPHY
Much, H.; Pasch, H.
Zentralinst. Org. Chem. DAW Berlin Ger. Dem. Rep. DD DDR-1199
Acta Polym.; (82) P 366-9; Vol 33; No 6; In Ger

18-82-0002 (0002)
APPLICATION OF HIGH-PERFORMANCE LIQUID CHROMATOGRAPHY TO THE SEPARATION AND ISOLATION OF BIOPOLYMERS
Unger, K. K.; Roumeliotis, P.
Inst. Anorg. Chem. Anal. Chem. Johannes Gutenberg-Univ. Mainz Fed. Rep. Ger. DE 6500
Anal. Chem. Symp. Ser.; (82) P 455-70; Vol 9; No Affinity Chromatogr. Relat. Tech.

18-82-0003 (0003)
NONAQUEOUS REVERSED-PHASE LIQUID CHROMATOGRAPHIC FRACTIONATION OF POLYSTYRENE
Armstrong, Daniel W.; Bui, K. H.
Dep. Chem. Georgetown Univ. Washington USA US DC 20057
Anal. Chem.; (82) P 706-8; Vol 54; No 4

18-82-0004 (0004)
ANALYSIS OF RUBBER AND PLASTIC CHEMICALS BY LC/SPECTROSCOPY
Pausch, Jerry B.
B. F. Goodrich Res. Dev. Cent. Brecksville USA US OH 44141
Anal. Chem.; (82) P 89A-90A, 92A, 95A, 96A; Vol 54; No 1

18-82-0005 (0005)
CHARACTERIZATION OF NOVOLAKS AND RESOLS BY HPLC
Werner, W.; Barber, O.
Ber. Anal. Dynamit Nobel A.-G. Troisdorf Fed. Rep. Ger. DE 5210
Chromatographia; (82) P 101-6; Vol 15; No 2; In Ger

18-82-0006 (0006)
REVERSED PHASE HIGH-PERFORMANCE LIQUID-CHROMATOGRAPHIC SEPARATION OF GALLIC ESTER ANTIOXIDANTS
Varga-Puchony, Zita; Hites-Papp, Mrs. E.; Vigh, G.
Dep. Anal. Chem. Univ. Chem. Eng. Veszprem Hung. HU
Hung. J. Ind. Chem.; (82) P 43-50; Vol 10; No 1

18-82-0007 (0007)
ANALYSIS OF AMIDE SLIP AGENTS IN POLYETHYLENES AND POLY(ETHYLENE-VINYLACETATE) COPOLYMERS AS FILMS AND RESIN GRANULES BY HPLC
Edwards, D. J. H.
Mardon Flexible Packag. Ltd. Bath UK GB BA3 4AA
HRC CC, J. High Resolut. Chromatogr. Chromatogr. Commun.; (82) P 161-3; Vol 5; No 3

18-82-0008 (0008)
SIMULTANEOUS USE OF FLUORESCENCE, ULTRAVIOLET, AND ELECTROCHEMICAL DETECTORS IN HIGH-PERFORMANCE LIQUID CHROMATOGRAPHY - SEPARATION AND IDENTIFICATION OF PHENOLIC ANTIOXIDANTS AND RELATED COMPOUNDS
Masoud, A. N.; Cha, Young Nam
Dep. Environ. Health Sci. Johns Hopkins Univ. Baltimore USA US MD 21205
HRC CC, J. High Resolut. Chromatogr. Chromatogr. Commun.; (82) P 299-305; Vol 5; No 6

18-82-0009 (0009)
HIGH PERFORMANCE LIQUID CHROMATOGRAPHY. A RELIABLE TECHNIQUE FOR EPOXY RESIN PREPREG ANALYSIS
Haanauer, Garv L.; Dunn, David A.
Polym. Res. Div. U. S. Army Mater. Mech. Res. Cent. Watertown USA US MA 02172
Ind. Eng. Chem. Prod. Res. Dev.; (82) P 68-73; Vol 21; No 1

18-82-0010 (0010)
HIGH-PERFORMANCE ADSORPTION CHROMATOGRAPHY OF POLYMERS: SEPARATION OF STYRENE-ACRYLONITRILE COPOLYMERS ON CN SILICA
Danielewicz, M.; Kubin, M.; Vozka, S.
Inst. Macromol. Chem. Czech. Acad. Sci. Prague Czech. CS 162 03/6
J. Appl. Polym. Sci.; (82) P 3629-31; Vol 27; No 9

18-82-0011 (0011)
HIGH-PERFORMANCE LIQUID CHROMATOGRAPHY OF 6-CAPROLACTAM AND ITS CYCLIC OLIGOMERS PRESENT IN POLYAMIDE 6
Krainik, Vaclav; Bozek, Peter; Kondelikova, Jaroslava; Kralicek, Jaroslav
State Hosp., Bratislava Czech., 821 24
J. Chromatogr. (82) P539-42; Vol 240; No 2

18-82-0012 (0012)
RETENTION BEHAVIOR OF ALKYLENE OXIDES IN REVERSED-PHASE HIGH-PERFORMANCE LIQUID CHROMATOGRAPHY
Kudoh, Megumu; Konami, Sumiyo; Fudano, Shintaro; Yamaguchi, Shinichiro
Wakayama Res. Lab. Kao Soap Co. Ltd. Wakayama Japan JP 640-91E
J. Chromatogr.; (82) P 209-13; Vol 234; No 1

18-82-0013 (0013)
LIQUID CHROMATOGRAPHY OF SOME POLYURETHANE POLYOLS
Guise, G. Bruce; Smith, Geoffrey C.
Div. Text. Ind. CSIRO Belmont Australia AU 3216
J. Chromatogr.; (82) P 369-73; Vol 247; No 2

18-82-0014 (0014)
ANALYTICAL CHARACTERIZATION OF BISPHENOL A EPOXY RESINS BY HIGH-PERFORMANCE LIQUID CHROMATOGRAPHY. II
Eppert, Guenter; Liebscher, Gert
VEB Leuna-Werke "Walter Ulbricht" Leuna Ger. Dem. Rep. DD 4220
J. Chromatogr.; (82) P 399-407; Vol 238; No 2; In Ger

18-82-0015 (0015)
SIMULTANEOUS DETERMINATION OF THE ANTIOXIDANT, THE CROSSLINKING-AGENT AND DECOMPOSITION PRODUCTS IN POLYETHYLENE BY REVERSE-PHASE HPLC
Duval, Michel; Giguere, Yves
Inst. Rech. Hydro-Quebec Varennes Can. CA PQ J0L 2P0
J. Liq. Chromatogr.; (82) P 1847-54; Vol 5; No 10

18-82-0016 (0016)
OLIGOMER ANALYSIS BY HPLC AND GPC
Curtis, Michael A.; Rogers, L. B.
Dep. Chem. Univ. Georgia Athens USA US GA 30602
J. Liq. Chromatogr.; (82) P 319-29; Vol 5; No Suppl. 2

18-82-0017 (0017)
DETERMINATION OF UV ABSORBING POLYOLEFIN ADDITIVES BY GRADIENT AND ISOCRATIC NORMAL-PHASE HIGH-PERFORMANCE LIQUID CHROMATOGRAPHY
Schabron, J. F.; Smith, V. J.; Ware, J. L.
Phillips Res. Cent. Phillips Pet. Co. Bartlesville USA US OK 74004
J. Liq. Chromatogr.; (82) P 613-24; Vol 5; No 4

18-82-0018 (0018)
DETERMINATION OF ETHYLENE OXIDE OLIGOMER DISTRIBUTIONS IN ALCOHOL ETHOXYLATES BY HPLC USING A ROTATING DISK-FLAME IONIZATION DETECTOR
McClure, J. D.
Westhollow Res. Cent. Shell Dev. Co. Houston USA US TX 77001
JAOCS, J. Am. Oil Chem. Soc.; (82) P 364-73; Vol 59; No 8

18-82-0019 (0019)
PARTICLE SIZE MEASUREMENTS FOR POLYMER LATEXES BY LIQUID CHROMATOGRAPHY USING POROUS SILICA PACKINGS
Kato, Tadaya; Takami, Hajime; Takahashi, Akira
Fac. Eng. Mie Univ. Tsu Japan JP 514
Kobunshi Ronbunshu; (82) P 487-92; Vol 39; No 8; In Japan

18-82-0020 (0020)
HPLC DETERMINATION OF 2-MERCAPTOBENZOTHIAZOLE IN RUBBER BABY BOTTLE NIPPLES
Blosczyk, Guenter; Doemling, Hans Juergen
Landesuntersuchungsamt Gesundheitswesen Nordbayern Erlangen Fed. Rep. Ger. DE 8520
Lebensmittelchem. Gerichtl. Chem.; (82) P 90; Vol 36; No 4; In Ger

18-82-0021 (0021)
THE CHARACTERIZATION OF OLIGOMER AND POLYMER BY HIGH PERFORMANCE LIQUID CHROMATOGRAPHY
Oguri Hiroyuki
Nippon Paints K. K. Osaka Japan JP
Shikizai Kyokaishi; (82) P 389-98; Vol 55; No 6; In Japan

18-82-0022 (0022)
IDENTIFICATION OF POLYISOBUTYLENE BY DERIVATIVE DIFFERENTIAL THERMAL ANALYSIS AND LIQUID CHROMATOGRAPHY
Babaitseva, L. T.; Tsvetkova, V. N.; Kamaev, A. V.
Vses. Nauchno-Issled. Inst. Sudob. Eksp. Moscow USSR SU
Zavod. Lab.; (82) P 17-19; Vol 48; No 1; In Russ

CHAPTER 7.19

Polyaromatic Hydrocarbons

--- 1982 ---

19-82-0001 (0001)
RESOLUTION OF OPTICAL ISOMERS BY CHIRAL HIGH-PERFORMANCE LIQUID CHROMATOGRAPHY: SEPARATION OF DIHYDRODIOLS AND TETRAHYDRODIOLS OF BENZO(A)PYRENE AND BENZ(A)ANTHRACENE
Weems, Henri B.; Yavg, Shen K.
Sch. Med. Uniformed Serv. Univ. of the Health Sci. Bethesda USA US MD 20814
Anal. Biochem.; (82) P 156-61; Vol 125; No 1

19-82-0002 (0002)
ON-LINE MULTIDIMENSIONAL LIQUID CHROMATOGRAPHIC DETERMINATION OF POLYNUCLEAR AROMATIC HYDROCARBONS IN COMPLEX SAMPLES
Sonnefeld, W. J.; Zoller, W. H.; May, W. E.; Wise, S. A.
Dep. Chem. Univ. Maryland College Park USA MD 20742
Anal. Chem.; (82) 723-7; Vol 54; No 4

19-82-0003 (0003)
LIQUID CHROMATOGRAPHIC SEPARATION OF AROMATIC HYDROCARBONS WITH CHEMICALLY BONDED (2,4-DINITROANILINOPROPYL)SILICA
Grizzle, Patrick L.; Thomson, Jane S.
Bartlesville Energy Technol. Cent. U. S. Dep. Energy Bartlesville USA US OK 74003
Anal. Chem.; (82) P 1071-8; Vol 54; No 7

19-82-0004 (0004)
ISOLATION OF POLYCYCLIC AROMATIC HYDROCARBONS IN AIR PARTICULATE MATTER BY LIQUID CHROMATOGRAPHY
Romanowski, T.; Funcke, W.; Koenig, J.; Balfanz, E.
Fraunhofer-Inst. Toxikol. Aerosolforsch. Muenster-Roxel Fed. Rep. Ger. DE D-4400
Anal. Chem.; (82) P 1285-7; Vol 54; No 8

19-82-0005 (0005)
HYDROCARBON CLASS FRACTIONATION WITH BONDED-PHASE LIQUID CHROMATOGRAPHY
Miller, Ronald
Perkin-Elmer Corp. Norwalk USA US CT 06856
Anal. Chem.; (82) P 1742-6; Vol 54; No 11

19-82-0006 (0006)
TIME-RESOLVED FLUORIMETRY WITH A SUB-NANOSECOND DYE LASER SOURCE FOR THE DETERMINATION OF POLYNUCLEAR AROMATIC HYDROCARBONS AFTER SEPARATION BY HIGH-PERFORMANCE LIQUID CHROMATOGRAPHY
Imasaka, Totaro; Ishibashi, Kouichi; Ishibashi, Nobuhiko
Fac. Eng. Kyushu Univ. Hakozaki Japan JP
Anal. Chim. Acta; (82) P 1-12; Vol 142

19-82-0007 (0007)
DETERMINATION OF POLYNUCLEAR AROMATIC HYDROCARBONS BY A SPIKED MOBILE PHASE FLUORESCENCE HPLC DETECTOR
Su, Syang Y.; Lai, E. P. C.; Winefordner, J. D.
Dep. Chem. Univ. Florida Gainesville USA US FL 32611
Anal. Lett.; (82) P 439-50; Vol 15; No A5

19-82-0008 (0008)
A FLUOROMETRIC-HPLC ASSAY FOR QUANTITATING THE BINDING OF BENZO(A)PYRENE METABOLITES TO DNA
Rahn, R. O.; Chang, S. S.; Holland, J. M.; Shugart, L. R.
Biol. Div. Oak Ridge Natl. Lab. Oak Ridge USA US TN 37830
Biochem. Biophys. Res. Commun.; (82) P 262-8; Vol 109; No 1

19-82-0009 (0009)
ESTIMATING THE AQUEOUS SOLUBILITY OF AROMATIC HYDROCARBONS BY HIGH PERFORMANCE LIQUID CHROMATOGRAPHY
Whitehouse, Brian G.
Dep. Oceanogr. Dalhousie Univ. Halifax Can. CA NS B3H 4J1
Chemosphere; (82) P 689-99; Vol 11; No 8

19-82-0010 (0010)
FINGERPRINTING OF ENVIRONMENTAL PAH BY HIGH SPEED HPLC
Albaiges, Joan; Grimalt, Joan
Environ. Chem. Unit Inst. Bio-Org. Chem. Barcelona Spain ES
Chromatogr. Newsl.; (82) P 8-11; Vol 10; No 1

19-82-0011 (0011)
ANALYSIS OF MULTIRING AROMATIC HYDROCARBONS BY HPLC USING A REVERSE-PHASE RADIAL COMPRESSION COLUMN SYSTEM
Alexander, Robert; Cumbers, Kimberley M.; Kagi, Robert I.
Sch. Appl. Chem. Western Australian Inst. Technol. Bentley Australia AU 6102
Int. J. Environ. Anal. Chem.; (82) P 161-7; Vol 12; No 3-4

19-82-0012 (0012)
AROMATIC HYDROCARBON METABOLITES IN FISH: AUTOMATED EXTRACTION AND HIGH-PERFORMANCE LIQUID-CHROMATOGRAPHIC SEPARATION INTO CONJUGATE AND NONCONJUGATE FRACTIONS
Krahn, Margaret M.; Collier, Tracy K.; Malins, Donald C.
Environ. Conservation Div. Northwest and Alaska Fis. Cent. Seattle USA US WA 98112
J. Chromatogr. (82) P 441-52; Vol 236; No 2

19-82-0013 (0013)
HIGH-PERFORMANCE LIQUID CHROMATOGRAPHY WITH REDUCTIVE ELECTROCHEMICAL DETECTION OF MUTAGENIC NITRO-SUBSTITUTED POLYNUCLEAR AROMATIC HYDROCARBONS IN DIESEL EXHAUSTS
Rappaport, S. M.; Jin, Z. L.; Xu, X. B.
Sch. Public Health Univ. California Berkeley USA US CA 94720
J. Chromatogr.; (82) P 145-54; Vol 240; No 1

19-82-0014 (0014)
THIN-LAYER CHROMATOGRAPHIC AND HIGH-PERFORMANCE LIQUID CHROMATOGRAPHIC SEPARATION OF METABOLITES OF THE WEAK CARCINOGEN, 7-METHYLBENZ(C)ACRIDINE
Boux, L. J.; Ireland, C. M.; Wright, D. J.; Holder, G. M.; Ryan, A. J.
Dep. Pharm. Univ. Sydney Sydney Australia AU 2006
J. Chromatogr.; (82) P 149-57; Vol 227; No 1

19-82-0015 (0015)
REVERSED-PHASE THIN-LAYER CHROMATOGRAPHY OF POLYNUCLEAR AROMATIC HYDROCARBONS AND CHLORINATED BIPHENYLS. RELATIONSHIP WITH HYDROPHOBICITY AS MEASURED BY AQUEOUS SOLUBILITY AND OCTANOL-WATER PARTITION COEFFICIENT
Bruggeman, W. A.; Van der Steen, J.; Hutzinger, O.
Lab. Environ. Toxicol. Chem. Univ. Amsterdam Amsterdam Neth. NL 1018 WV
J. Chromatogr.; (82) P 335-46; Vol 238; No 2

19-82-0016 (0016)
POLYAROMATIC HYDROCARBONS AS HIGH-PERFORMANCE LIQUID CHROMATOGRAPHIC CALIBRATION STANDARDS FOR THE LOW-LEVEL DETERMINATION OF CHLORINATED DIBENZO-p-DIOXINS AND CHLORINATED DIBENZOFURANS IN BIOLOGICAL SAMPLES
Ryan, John J.; Pilon, Jean C.
Food Div. Health Pro4. Branch Ottawa Can. CA ON K1A 0L2
J. Chromatogr.; (82) P 409-15; Vol 248; No 3

19-82-0017 (0017)
COMPARISON OF MOLECULAR CONNECTIVITY AND A CHROMATOGRAPHIC CORRELATION FACTOR IN REVERSED-PHASE HIGH-PERFORMANCE LIQUID CHROMATOGRAPHY FOR POLYCYCLIC AROMATIC HYDROCARBONS
Hurtubise, Robert J.; Allen, Todd W.; Silver, Howard F.
Chem. Dep. Univ. Wyoming Laramie USA US WY 82071
J. Chromatogr.; (82) P 517-22; Vol 235; No 2

19-82-0018 (0018)
PAH ANALYSES USING HPLC
Dorner, Wolf G.
Inzell Fed. Rep. Ger. DE 8221
LaborPraxis; (82) P 854, 856, 858; Vol 6; No 7-8; In Ger

19-82-0019 (0019)
RAPID DETERMINATION OF AROMATIC HYDROCARBONS IN SOLVENT NAPHTHA BY HPLC
Galtieri, A.; Mascherpa, A.
Stn. Sper. Combust. San Donato Milanese Italy IT
Riv. Combust.; (82) P 181-3; Vol 36; No 6; In Ital

19-82-0020 (0020)
HIGH-PRESSURE LIQUID CHROMATOGRAPHIC METHOD FOR ROUTINE ANALYSIS OF POLYCYCLIC AROMATIC HYDROCARBONS IN AIRBORNE PARTICULATES
Suga, Kuniko; Ishiguro, Tatsukichi; Matsushita, Hidetsuru
Tokyo Metrop. Res. Inst. Environ. Prot. Tokyo Japan JP 100
Taiki Osen Gakkaishi; (82) P 117-25; Vol 17; No 2; In Japan

19-82-0021 (0021)
ROUTINE HPLC OF POLYNUCLEAR AROMATIC HYDROCARBONS
O'Donnell, C.
Water Res. Div. St. Martin's House Dublin Ire. IE
TECHNICAL REPORT; Comm. Eur. Communities, (Rep.) EUR; (82) P 174-7; No EUR 7623, Anal. Org. Micropollut. Water

19-82-0022 (0022)
DETERMINATION OF METABOLITES OF AROMATIC HYDROCARBONS BY HIGH-PERFORMANCE LIQUID CHROMATOGRAPHY
Bergert, K. D.; Zehner, C.; Voigt, H.; Maehrlein, W.
Arbeitshyg. Zent. Niederdorf Ger. Dem. Rep. DD DDR-9154
Z. Gesamte Hyg. Ihre Grenzgeb.; (82) P 521-4; Vol 28; No 8; In Ger

CHAPTER 7.20
Porphyrins

--- 1982 ---

20-82-0001 (0001)
HIGH-PERFORMANCE LIQUID CHROMATOGRAPHIC ANALYSIS OF PORPHYRINS AND THEIR ISOMERS WITH RADIAL COMPRESSION COLUMNS
Seubert, Andreas; Seubert, Sigrid
Dep. Dermatol. Univ. Goettingen Goettingen Fed. Rep. Ger. DE D-3400
Anal. Biochem.; (82) P 303-7; Vol 124; No 2

20-82-0002 (0002)
DETERMINATION OF URINARY COPROPORPHYRINS BY HIGH PRESSURE LIQUID CHROMATOGRAPHY. SIGNIFICANCE FOR THE PREVENTION OF LEAD POISONING
Pekari, K.; Arsac, F.; Esmaelzadeh, A.; Boudene, C.
Lab. Biochem. Inst. Occup. Health Helsinki Finland FI 00370/37
Arch. Mal. Prof. Med. Trav. Secur. Soc.; (82) P 449-54; Vol 43; No 6; In Fr

20-82-0003 (0003)
ANALYSIS AND SEPARATION OF NATURAL AND SYNTHETIC MIXTURES OF UROPORPHYRINS BY HIGH-PRESSURE LIQUID CHROMATOGRAPHY
Jackson, Anthony H.; Rao, K. R. Nagaraja; Smith, Sydney G.
Dep. Chem. Univ. Coll. Cardiff Cardiff UK GB CF1 1XL
Biochem. J.; (82) P 515-17; Vol 203; No 2

20-82-0004 (0004)
A RAPID AND QUANTITATIVE HIGH PERFORMANCE LIQUID CHROMATOGRAPHIC METHOD FOR ASSAYING BILIRUBIN AND ITS CONJUGATES IN BILE
Gordon, Ellen R.; Goresky, Carl A.
Med. Clin. McGill Univ. Montreal Can. CA PQ H3G 1A4
Can. J. Biochem.; (82) P 1050-7; Vol 60; No 11

20-82-0005 (0005)
HIGH-PRESSURE LIQUID CHROMATOGRAPHY OF PETROPORPHYRINS: EVALUATION AS A GEOCHEMICAL FINGERPRINTING METHOD BY PRINCIPAL COMPONENTS ANALYSIS
Hohn, M. E.; Hajibrahim, S. K.; Eglinton, G.
Sch. Chem. Univ. Bristol Bristol UK GB BS8 1TS
Chem. Geol.; (82) P 229-37; Vol 37; No 3-4

20-82-0006 (0006)
HIGH-PERFORMANCE LIQUID CHROMATOGRAPHIC AND MASS SPECTROMETRIC ANALYSES OF PORPHYRINS FROM DEEP-SEA SEDIMENTS
Quirke, J. M. E.; Eglinton, G.; Palmer, S. E.; Baker, E. W.
Org. Geochem. Unit Univ. Bristol Bristol UK GB BS8 1TS
Chem. Geol.; (82) P 69-85; Vol 35; No 1-2

20-82-0007 (0007)
THE SEPARATION OF BILIRUBIN SPECIES IN PATHOLOGICAL SERA BY HPLC
Lauff, John J.; Kasper, Mary E.; Ambrose, Robert T.
Res. Lab. Eastman Kodak Co. Rochester USA US NY 14650
CONFERENCE; Proc. - Int. Congr. Clin. Chem., 11th ; (82) P 995-1000; Avail.: Kaiser, Erich; Gabl, Franz; Mueller, Mathias M; Date: 81; Publisher: de Gruyter, Berlin, Fed. Rep. Ger.

20-82-0008 (0008)
IDENTIFICATION OF THE PRODUCT OF HEME DEGRADATION CATALYZED BY THE HEME OXYGENASE SYSTEM AS BILIVERDIN IX.ALPHA. BY REVERSED-PHASE HIGH-PERFORMANCE LIQUID CHROMATOGRAPHY
Noguchi, Masato; Yoshida, Tadashi; Kikuchi, Goro
Sch. Med. Tohoku Univ. Sendai Japan JP 980
J. Biochem. (Tokyo); (82) P 1479-83; Vol 91; No 5

20-82-0009 (0009)
SAMPLE PROCESSING FOR HIGH-PERFORMANCE LIQUID CHROMATOGRAPHY OF ECDYSTEROIDS
Lafont, Rene; Pennetier, Jean Louis; Andrianjafintrimo, Mex;
Claret, Jacques; Modde, Jean Francois; Blais, Catherine
Lab. Zool. Ec. Norm. Super. Paris Fr. FR F-75230/05
J. Chromatogr.; (82) P 137-49; Vol 236; No 1

20-82-0010 (0010)
DEVELOPMENT AND UTILIZATION OF A PROCEDURE FOR MEASURING URINARY PORPHYRINS BY HIGH-PERFORMANCE LIQUID CHROMATOGRAPHY
Hill, Robert H., Jr.; Bailey, Sandra L.; Needham, Larry L.
Cent. Dis. Control U. S. Dep. Health Hum. Serv. Atlanta USA US GA 30333
J. Chromatogr.; (82) P 251-60; Vol 232; No 2

20-82-0011 (0011)
SEPARATION AND ANALYSIS OF HEMATOPORPHYRIN DERIVATIVE COMPONENTS BY HIGH-PERFORMANCE LIQUID CHROMATOGRAPHY
Cadby, Peter A.; Dimitriadis, Eugene; Grant, Hamish G.; Ward, A. David; Forbes, Ian J.
Dep. Org. Chem. Univ. Adelaide Adelaide Australia AU 5001
J. Chromatogr.; (82) P 273-81; Vol 231; No 2

20-82-0012 (0012)
THE DETERMINATION OF UROPORPHYRINOGEN DECARBOXYLASE IN TISSUES BY HIGH-PERFORMANCE LIQUID CHROMATOGRAPHY COUPLED TO FLUORESCENCE DETECTION
Cantoni, L.; Ruggieri, R.; Dal Fiume, D.; Rizzardini, M.
Ist. Ric. Farmacol. "Mario Negri" Milan Italy IT 20157
J. Chromatogr.; (82) P 311-18; Vol 229; No 2

20-82-0013 (0013)
HIGH-PERFORMANCE LIQUID CHROMATOGRAPHIC DIFFERENTIATION OF URINARY FREE PORPHYRINS
Gaetani, E.; Laureri, C. F.; Vitto, M.; Rocchi, E.; Gibertini, P.; Farina, F.; Cassanelli, M.; Ventura, E.
Ist. Chim. Farm. Tossicol. Univ. Parma Parma Italy IT
J. Chromatogr.; (82) P 425-32; Vol 231; No 2

20-82-0014 (0014)
SEPARATION AND QUANTITATION OF URINARY PORPHYRINS BY HIGH-PERFORMANCE LIQUID CHROMATOGRAPHY
Johansson, B.; Nilsson, B.
Malmoe Gen. Hosp. Univ. Lund Malmoe Swed. SE 214 01
J. Chromatogr.; (82) P 439-44; Vol 229; No 2

20-82-0015 (0015)
ISOCRATIC HIGH-PERFORMANCE LIQUID CHROMATOGRAPHY OF BILE PIGMENTS
Cole, K. D.; Little, G. H.
Health Sci. Cent. Texas Tech Univ. Lubbock USA 79430
J. Chromatogr.; (82) P 503-9; Vol 227; No 2.

20-82-0016 (0016)
HIGH-PERFORMANCE LIQUID CHROMATOGRAPHY OF BILE PIGMENTS
Lim, C. K.
Div. Clin. Chem. MRC Clin. Res. Cent. Harrow/Middlesex UK GB HA1 3UJ
J. Liq. Chromatogr.; (82) P 305-18; Vol 5; No Suppl. 2

CHAPTER 7.21
Proteins

--- 1982 ---

21-82-0001 (0001)
HIGH-PERFORMANCE LIQUID CHROMATOGRAPHY AND ITS APPLICATION TO PROTEIN CHEMISTRY
Hearn, Milton T. W.
St. Vincent's Sch. Med. Res. Fitzroy Australia A
Adv. Chromatogr. (N. Y.); (82) P 1-82; Vol 20

21-82-0002 (0002)
HPLC OF PEPTIDES AND PROTEINS
Hearn, T. W.; Regnier, F. E.; Wehr, C. T.
St. Vincent's Sch. Med. Res. Australia AU
Am. Lab. (Fairfield, Conn.); (82) P 18, 20, 23-4, 27-30, 35-6, 38-9; Vol 14; No 10

21-82-0003 (0003)
NOVEL SEPARATION OF PROTEIN AND LYMPHOCYTES BY LIQUID GEL CHROMATOGRAPHY DURING EXPOSURE TO RADIOFREQUENCY RADIATION
Liburdy, Robert P.
Radiat. Sci. Div. USAF Sch. Aerosp. Med. Brooks AFB USA US TX 78235
Anal. Biochem.; (82) P 148-58; Vol 122; No 1

21-82-0004 (0004)
ASSESSMENT OF CISPLATIN REACTIVITY WITH PEPTIDES AND PROTEINS USING REVERSE-PHASE HIGH-PERFORMANCE LIQUID CHROMATOGRAPHY AND FLAMELESS ATOMIC ABSORPTION SPECTROSCOPY
Riley, Christopher M.; Sternson, Larry A.; Repta, Arnold J.
Dep. Pharm. Chem. Univ. Kansas Lawrence USA US KS 66045
Anal. Biochem.; (82) P 167-79; Vol 124; No 1

21-82-0005 (0005)
HIGH-PERFORMANCE LIQUID CHROMATOGRAPHY OF PROTEINS: PURIFICATION OF THE ACIDIC ISOZYME OF ADENYLOSUCCINATE SYNTHETASE FROM RAT LIVER
Rudolph, Frederick B.; Clark, Sandra W.
Dep. Biochem. Rice Univ. Houston USA US TX 77001
Anal. Biochem.; (82) P 193-7; Vol 127; No 1

21-82-0006 (0006)
APPLICATION OF HIGH-PERFORMANCE LIQUID CHROMATOGRAPHY TO COMPETITIVE LABELING STUDIES: THE CHEMICAL PROPERTIES OF FUNCTIONAL GROUPS OF GLUCAGON
Cockle, Stephen A.; Kaplan, Harvey; Hefford, Mary A.; Young, N. Martin
Div. Biol. Sci. Natl. Res. Counc. Canada Ottawa Can. CA ON K1A 0R6
Anal. Biochem.; (82) P 210-16; Vol 125; No 1

21-82-0007 (0007)
REVERSE-PHASE HIGH-PERFORMANCE LIQUID CHROMATOGRAPHY OF ESCHERICHIA COLI RIBOSOMAL SMALL SUBUNIT PROTEINS
Kerlavage, Anthony R.; Kahan, Lawrence; Cooperman, Barry S.
Dep. Chem. Univ. Pennsylvania Philadelphia USA US PA 19104
Anal. Biochem.; (82) P 342-8; Vol 123; No 2

21-82-0008 (0008)
CHARACTERIZATION OF INTACT AND TRYPSIN-DIGESTED BIOSYNTHETIC HUMAN GROWTH HORMONE BY HIGH-PRESSURE LIQUID CHROMATOGRAPHY
Kohr, William J.; Keck, Rodney; Harkins, Richard N.
Dep. Protein Biochem. Genentech, Inc. South San Francisco USA US CA 94080
Anal. Biochem.; (82) P 348-59; Vol 122; No 2

21-82-0009 (0009)
ANALYSIS OF PEPTIC FRAGMENTATION OF HUMAN IMMUNOGLOBULIN G USING HIGH-PERFORMANCE LIQUID CHROMATOGRAPHY
Tomono, Tsugikazu; Suzuki, Tohru; Tokunaga, Eiichi
Plasma Fractionation Dep. Japanese Red Cross Cent. Blood Cent Tokyo Japan JP 150
Anal. Biochem.; (82) P 394-401; Vol 123; No 2

21-82-0010 (0010)
APPLICATION OF HIGH-PERFORMANCE LIQUID CHROMATOGRAPHY TO THE SEPARATION AND ISOLATION OF BIOPOLYMERS
Unger, K. K.; Roumeliotis, P.
Inst. Anorg. Chem. Anal. Chem. Johannes Gutenberg-Univ. Mainz Fed. Rep. Ger. DE 6500
Anal. Chem. Symp. Ser.; (82) P 455-70; Vol 9; No Affinity Chromatogr. Relat. Tech.

21-82-0011 (0011)
METABOLISM OF PARATHYROID HORMONE BY KUPFFER CELLS: ANALYSIS BY REVERSE-PHASE HIGH-PERFORMANCE LIQUID CHROMATOGRAPHY
Bringhurst, F. Richard; Segre, Gino V.; Lampman, Gary W.; Potts, John T., Jr.
Dep. Med. Massachusetts Gen. Hosp. Boston USA US MA 02114
Biochemistry; (82) P 4252-8; Vol 21; No 18

21-82-0012 (0012)
ISOLATION OF NEURONAL PARVALBUMIN BY HIGH-PERFORMANCE LIQUID CHROMATOGRAPHY. CHARACTERIZATION AND COMPARISON WITH MUSCLE PARVALBUMIN
Berchtold, Martin W.; Wilson, Kenneth J.; Heizmann, Claus W.
Sch. Vet. Med. Univ. Zurich Zurich Switz. CH CH-8057
Biochemistry; (82) P 6552-7; Vol 21; No 25

21-82-0013 (0013)
A HIGH PERFORMANCE LIQUID CHROMATOGRAPHY ASSAY FOR THE RAPID ANALYSIS OF THE SUBUNIT CONTENT OF CONCANAVALIN A
Sitrin, Robert D.; Antell, Laura; Griswold, Don E.; Bender, Paul E.; Greig, Russell G.; Poste, George
Res. Dev. Div. Smith Kline and French Lab. Philadelphia USA US PA 19101
Biochim. Biophys. Acta; (82) P 175-8; Vol 717; No 1

21-82-0014 (0014)
ISOLATION OF DENATURED PROTEINS AND PEPTIDES BY HIGH-PERFORMANCE LIQUID CHROMATOGRAPHY. EFFECT OF DIFFERENT PERFLUORINATED ACIDS, COLUMN LENGTH AND LARGE-PORE SUPPORTS
Mahoney, Walter C.
Dep. Genet. Univ. Washington Seattle USA US WA 98195
Biochim. Biophys. Acta; (82) P 284-9; Vol 704; No 2

21-82-0015 (0015)
SEPARATION OF PROTEIN MIXTURES BY HIGH-PERFORMANCE LIQUID CHROMATOGRAPHY
Shimada, Kenji; Sakaguchi, Takeichi; Ikeda, Mitsuo
Niigata Coll. Pharm. Niigata Japan JP 950-21
Chem. Pharm. Bull.; (82) P 1370-6; Vol 30; No 4

21-82-0016 (0016)
PROTEIN PURIFICATION AND ANALYSIS BY HPLC
Pfeifer, Robert E.; Skea, William M.; Waraska, John; Cohen, Charles; Burnworth, Larry
Waters Assoc. Milford USA US MA
Chromatogr. Sci.; (82) P 43-55; Vol 20; No Biol./Biomed. Appl. Liq. Chromatogr. 4

21-82-0017 (0017)
LIQUID-CHROMATOGRAPHIC MEASUREMENT OF ELASTIN
Covault, H. Patrick; Lubrano, Tina; Dietz, Albert A.; Rubinstein, Herbert M.
Res. Serv. VA Hosp. Hines USA US IL 60141
Clin. Chem. (Winston-Salem, N. C.); (82) P 1465-8; Vol 28; No 7

21-82-0018 (0018)
SEPARATION OF HUMAN PRO .ALPHA.1(I) AND PRO .ALPHA.2(I) PROCOLLAGEN CHAINS BY REVERSED-PHASE HIGH-PERFORMANCE LIQUID CHROMATOGRAPHY
Van der Rest, Michel; Stolle, Catherine A.; Prockop, Darwin J.; Fietzek, Peter P.
Dep. Biochem. Rutgers Med. Sch. Piscataway USA US NJ 08854
Collagen Relat. Res.: Clin. Exp.; (82) P 281-5; Vol 2; No 4

21-82-0019 (0019)
A COMPREHENSIVE APPROACH TO THE STUDY OF COLLAGEN PRIMARY STRUCTURE BASED ON HIGH-PERFORMANCE LIQUID CHROMATOGRAPHY
Van der Rest, Michel; Fietzek, Peter P.
Rutgers Med. Sch. Coll. Med. Dent. Piscataway USA US NJ 08854
Eur. J. Biochem.; (82) P 491-6; Vol 125; No 3

21-82-0020 (0020)
FURTHER STUDIES ON THE QUANTITATION OF THE HEMOGLOBINS A, S, C, AND F IN NEWBORN BABIES WITH DIFFERENT HEMOGLOBINOPATHIES USING HIGH PRESSURE LIQUID CHROMATOGRAPHY
Gardiner, M. B.; Carver, J.; Abraham, B. L.; Wilson, J. B.; Huisman, T. H. J.
Compr. Sickle Cell Cent. Med. Coll. Georgia Augusta USA US GA 30912
Hemoglobin; (82) P 1-13; Vol 6; No 1

21-82-0021 (0021)
SIMPLE PURIFICATION OF ACETONITRILE FOR HIGH-PERFORMANCE LIQUID CHROMATOGRAPHY (HPLC) OF PEPTIDES AND PROTEINS
Braunitzer, Gerhard; Ruecknagel, Peter; Oberthuer, Walter
Abt. Proteinchem. Max-Planck-Inst. Biochem. Martinsried/Muenchen Fed. Rep. Ger. DE D-8033
Hoppe-Seyler's Z. Physiol. Chem.; (82) P 485-6; Vol 363; No 5; In Ger

21-82-0022 (0022)
HETEROGENEITY OF HUMAN SERUM HIGH-DENSITY LIPOPROTEINS ON HIGH-PERFORMANCE LIQUID CHROMATOGRAPHY
Okazaki, Mitsuyo; Hagiwara, Noriko; Hara, Ichiro
Dep. Gen. Educ. Tokyo Med. Dent. Univ. Ichikawa Japan JP 272
J. Biochem. (Tokyo); (82) P 517-24; Vol 92; No 2

21-82-0023 (0023)
IODINATION OF .BETA.-MELANOTROPIN. TIME COURSE ANALYSIS OF REACTION MIXTURES BY HIGH PRESSURE LIQUID CHROMATOGRAPHY AND CHARACTERIZATION OF BIOLOGICALLY ACTIVE MONO- AND DIIODO-.BETA.-MELANOTROPIN
Lambert, Drew T.; Stachelek, Conrad; Varga, Janos M.; Lerner, Aaron B.
Sch. Med. Yale Univ. New Haven USA US CT 06510
J. Biol. Chem.; (82) P 8211-15; Vol 257; No 14

21-82-0024 (0024)
MOLECULAR SIEVING DURING REVERSED-PHASE HIGH-PERFORMANCE LIQUID CHROMATOGRAPHY OF PROTEINS
Van der Zee, R.; Welling, G. W.
Lab. Med. Microbiol. Rijksuniv. Groningen Groningen Neth. NL 9713 EZ
J. Chromatogr.; (82) P 134-6; Vol 244; No 1

21-82-0025 (0025)
SEPARATION OF ACIDIC PROTEINS FROM MINERALIZED TISSUES BY REVERSED-PHASE HIGH-PERFORMANCE LIQUID CHROMATOGRAPHY
Weiner, Stephen
Isot. Dep. Weizmann Inst. Sci. Rehovot Israel IL 76 100
J. Chromatogr.; (82) P 148-54;P Vol 245; No 1

21-82-0026 (0026)
HIGH-PERFORMANCE LIQUID CHROMATOGRAPHY OF PROTEINS: ANALYTICAL APPLICATIONS
Buchholz, K.; Goedelmann, B.; Molnar, I.
DECHEMA-Inst. Frankfurt/Main Fed. Rep. Ger. DE D-6000/97
J. Chromatogr.; (82) P 193-202; Vol 238; No 1

21-82-0027 (0027)
MECHANISM OF PROTEIN RETENTION IN REVERSED-PHASE HIGH-PERFORMANCE LIQUID CHROMATOGRAPHY
Barford, R. A.; Sliwinski, B. J.; Breyer, A. C.; Rothbart, H. L.
East. Reg. Res. Cent. Philadelphia USA US PA 19118
J. Chromatogr.; (82) P 281-8; Vol 235; No 2

21-82-0028 (0028)
COMPARISON OF THE HIGH-PERFORMANCE LIQUID CHROMATOGRAPHY OF PEPTIDES AND PROTEINS ON 100- AND 300-.ANG. REVERSED-PHASE SUPPORTS
Wilson, Kenneth J.; Van Wieringen, Erika; Klauser, Stefan; Berchtold, Martin W.; Hughes, Graham J.
Biochem. Inst. Univ. Zurich Zurich Switz. CH CH-8028
J. Chromatogr.; (82) P 407-16; Vol 237; No 3

21-82-0029 (0029)
TWO-DIMENSIONAL SEPARATION SYSTEM FOR ANALYSIS OF PROTEINS EMPLOYING ISOELECTRIC FOCUSING AND HIGH-PERFORMANCE LIQUID CHROMATOGRAPHY
Kojima, Kiyotsugu; Manabe, Takashi; Okuyama, Tsuneo; Tomono, Tsugikazu; Suzuki, Tohru; Tokunaga, Eiichi
Fac. Sci. Tokyo Metrop. Univ. Tokyo Japan JP 158
J. Chromatogr.; (82) P 565-70; Vol 239

21-82-0030 (0030)
CROSS-LINKING OF STROMA-FREE HEMOGLOBIN MONITORED BY HIGH-PERFORMANCE LIQUID CHROMATOGRAPHY
Pietta, Piergiorgio; Calatroni, Alma; Palazzini, Gabriele; Agostoni, Angelo
Ist. Chim. Org. Anal. Milan Italy IT 20133
J. Chromatogr.; (82) P 65-71; Vol 246; No 1

21-82-0031 (0031)
ADSORPTION CHROMATOGRAPHY OF PROTEINS ON SILICONIZED POROUS GLASS
Mizutani, Takaharu; Narihara, Takahiko
Fac. Pharm. Sci. Nagoya City Univ. Nagoya Japan JP 467
J. Chromatogr.; (82) P 755-60; Vol 239

21-82-0032 (0032)
REVERSE-PHASE HPLC OF PROTEINS: EFFECTS OF VARIOUS BONDED PHASES
Lewis, Randolph V.; DeWald, Daryll
Dep. Biochem. Univ. Wyoming Laramie USA US WY 82071
J. Liq. Chromatogr.; (82) P 1367-74; Vol 5; No 7

21-82-0033 (0033)
FRACTIONATION BY HIGH-PERFORMANCE LIQUID CHROMATOGRAPHY OF MICROSOMAL CYTOCHROME P-450 INDUCED BY HEXACHLOROBIPHENYL ISOMERS
Kohli, Krishan K.; Hernandez, Oscar; McKinney, James D.
Natl. Inst. Environ. Health Sci. Research Triangle Park USA US NC 27709

J. Liq. Chromatogr.; (82) P 367-77; Vol 5; No 2

21-82-0034 (0034)
HPLC ANALYSIS OF NEUROPHYSIN PROTEINS FROM NEUROSECRETORY GRANULES
Schwandt, P.; Richter, W. O.; Reuschel-Janetschek, E.
Klinik. Grosshadern Univ. Munich Munich Fed. Rep. Ger. DE
J. Neurochem.; (82) P 357-9; Vol 39; No 2

21-82-0035 (0035)
RADIOCHEMICAL ANALYSIS OF TECHNETIUM-99M-LABELED-HUMAN SERUM ALBUMIN WITH HIGH-PRESSURE LIQUID CHROMATOGRAPHY: CONCISE COMMUNICATION
Vallabhajosula, Shankar; Goldsmith, Stanley J.; Pollina, Robert; Lipszyc, Helena
Mount Sinai Med. Cent. New York USA US NY 10029
J. Nucl. Med.; (82) P 326-9; Vol 23; No 4

21-82-0036 (0036)
PEPTIDE MAPPING OF 125I-LABELED MEMBRANE PROTEIN O2 INFLUENZA VIRUSES BY REVERSE-PHASE HIGH-PERFORMANCE LIQUID CHROMATOGRAPHY
Darveau, A.; Seidah, N. G.; Chretien, M.; Lecomte, J.
Cent. Rech. Virol. Inst. Armand-Frappier Laval Can. CA PQ H7N 4Z3
J. Virol. Methods;(82) P 77-85; Vol 4; No 2

21-82-0037 (0037)
APPLICATION OF HIGH-PERFORMANCE LIQUID CHROMATOGRAPHY TO THE PREPARATION OF ENCEPHALITOGENIC MYELIN BASIC PROTEIN
Hinman, Channing L.; Rauch, Helene C.; Pfeifer, Robert F.
Sch. Med. Wayne State Univ. Detroit USA US MI 48201
Life Sci.; (82) P 989-93; Vol 30; No 12

21-82-0038 (0038)
POTENTIAL BARRIER CHROMATOGRAPHY: AN HPLC METHOD FOR PROTEIN SEPARATIONS
Ruckenstein, E.; Srinivasan, R.
Dep. Chem. Eng. State Univ. New York Buffalo USA US NY 14260
Sep. Sci. Technol.; (82) P 763-83; Vol 17; No 6

21-82-0039 (0039)
APPLICATION OF HIGH PERFORMANCE LIQUID CHROMATOGRAPHY TO PROTEIN CHEMISTRY
Sasagawa, Tatsuru; Titani, Koiti
Dep. Biochem. Univ. Washington Seattle USA US WA 98195
Tanpakushitsu Kakusan Koso; (82) P 1056-68; Vol 27; No 7; In Japan

21-82-0040 (0040)
MOLECULAR SIEVE CHROMATOGRAPHY OF SOME PLASMA PROTEIN PRODUCTS BY HIGH PERFORMANCE LIQUID CHROMATOGRAPHY: COMPARISON OF QUANTITATION BY UV ABSORPTION AND LOWRY PROTEIN
Bertolini, Maurice J.
Cutter Lab. Berkeley USA US CA
Vox Sang.; (82) P 87-90; Vol 43; No 2

21-82-0041 (0041)
HIGH PERFORMANCE LIQUID CHROMATOGRAPHY OF PROTEIN. IV. SEPARATION OF INSULIN FROM DIFFERENT SPECIES (EQUINE, PORCINE, BOVINE AND OVINE) BY REVERSED PHASE HIGH PERFORMANCEP LIQUID CHROMATOGRAPHY
Asakawa, Naoki; Tsuno, Masanori; Saeki, Yasuharu; Matsuda, Muneaki ; Hattori, Teiichi; Ueyama, Masaaki; Shinoda, Aishin; Miyake, Yasuo
Res. Lab. Pharm. Dev. Eisai Co., Ltd. Gifu Japan JP 483
Yakugauu Zasshi; (82) P 43-8; Vol 102; No 1; In Japan

CHAPTER 7.22

Radiolabeled Compounds

--- 1982 ---

22-82-0001 (0001)
PURIFICATION OF 11C-LABELED RADIOPHARMACEUTICALS BY LC
Berger, G.; Maziere, M.; Godot, J. J.; Prenant, C.; Comar, D.
Dep. Biol. Hop. Orsay Orsay Fr. FR
Chromatogr. Sci.; (82) P 223-30; Vol 20; No Biol./Biomed. Appl. Liq. Chromatogr. 4

22-82-0002 (0002)
SEPARATION OF ISOTOPICALLY LABELED MOLECULES BY MICRO HIGH-PERFORMANCE LIQUID CHROMATOGRAPHY WITH A PACKED MICRO GLASS CAPILLARY COLUMN
Jinno, Kiyokatsu; Hirata, Yukio; Hiyoshi, Yoshihiko
Sch. Mater. Sci. Toyohashi Univ. Technol. Toyohashi Japan JP 440
HRC CC, J. High Resolut. Chromatogr. Chromatogr. Commun.; (82) P 102-3; Vol 5; No 2

22-82-0003 (0003)
CHARACTERIZATION OF THE RABBIT RENAL RECEPTOR FOR NATIVE PARATHYROID HORMONE EMPLOYING A RADIOLIGAND PURIFIED BY REVERSED-PHASE LIQUID CHROMATOGRAPHY
Kremer, Richard; Bennett, Hugh P. J.; Mitchell, Jane; Goltzman, David
Dep. Med. McGill Univ. Montreal Can. CA PQ H3A 1A1
J. Biol. Chem.; (82) P 14048-54; Vol 257; No 23

22-82-0004 (0004)
QUANTITATION OF RADIOLABELED COMPOUNDS ELUTING FROM THE HPLC SYSTEM
Kessler, Michael J.
Med. Sch. Univ. Texas Houston USA US TX 77030
J. Chromatogr. Sci.; (82) P 523-7; Vol 20; No 11

22-82-0005 (0005)
PURIFICATION OF TRITIUM-LABELED 8-METHOXYPSORALEN BY HIGH PERFORMANCE LIQUID CHROMATOGRAPHY
Adriaens, P.; De Schrijver, F.; Van Boven, M.; Roelandts, R.; Roets, E.; Hoogmartens, J.
Dep. Chem. Kathol. Univ. Leuven Louvain Belg. BE
J. Labelled Compd. Radiopharm.; (82) P 463-8; Vol 19; No 3

22-82-0006 (0006)
IODINATION OF PEPTIDE HORMONES AND PURIFICATION OF IODINATED PEPTIDES BY HPLC
Janaky, T.; Toth, G.; Penke, B.; Kovacs, K.; Laszlo, F. A.
1st Dep. Med. Univ. Med. Sch. Szeged Hung. HU
J. Liq. Chromatogr.; (82) P 1499-507; Vol 5; No 8

22-82-0007 (0007)
PURIFICATION OF RADIO-IODINATED CHOLECYSTOKININ PEPTIDES BY REVERSE PHASE HPLC
Fourmy, D.; Pradayrol, L.; Antoniotti, H.; Esteve, J. P.; Ribet, A.
Lab. Biol. Pathol. Dig. Cent. Hosp. Univ. Rangveil Toulouse Fr. FR 31054
J. Liq. Chromatogr.; (82) P 757-66; Vol 5; No 4

22-82-0008 (0008)
RADIOCHEMICAL ANALYSIS OF TECHNETIUM-99M-LABELED-HUMAN SERUM ALBUMIN WITH HIGH-PRESSURE LIQUID CHROMATOGRAPHY: CONCISE COMMUNICATION
Vallabhajosula, Shankar; Goldsmith, Stanley J.; Pollina, Robert; Lipszyc, Helena
Mount Sinai Med. Cent. New York USA US NY 10029
J. Nucl. Med.; (82) P 326-9; Vol 23; No 4

22-82-0009 (0009)
A HIGH PERFORMANCE LIQUID CHROMATOGRAPHIC METHOD FOR DETERMINATION OF THE

RADIOCHEMICAL PURITY OF IODINE-123-LABELED IODOHIPPURIC ACID INJECTION
Millar, A. M.
Dep. Pharm. Med. Phys. R. Infirm. Edinburgh UK GB EH3 9YW
J. Pharm. Pharmacol.; (82) P 14-17; Vol 34; No 1

22-82-0010 (0010)
DETERMINATION OF THE RADIOCHEMICAL PURITY OF HIPPURAN BY HPLC
Pfeiffer, G.; Beranek, I.
Swiss Fed. Inst. React. Res. Wuerenlingen Switz. CH
Prog. Radiopharmacol.; (82) P 27-30; Vol 3; No Proc. Eur. Symp. Radiopharmacol., 3rd

22-82-0011 (0011)
REMOTE-CONTROLLED INJECTION OF RADIOACTIVE SOLUTIONS INTO A HIGH-PERFORMANCE LIQUID CHROMATOGRAPH
Vandewalle, T.; Vandecasteele, C.; Van Haver, D.
Inst. Nucl. Sci. Rijksuniv. Gent Gent Belg. BE B-9000
Radiochem. Radioanal. Lett.; (82) P 29-32; Vol 51; No 1

CHAPTER 7.23
Steroids

--- 1982 ---

23-82-0001 (0001)
APPLICATION OF HIGH-PERFORMANCE LIQUID CHROMATOGRAPHIC SEPARATION OF FREE STEROLS TO THE SCREENING OF YEAST STEROL MUTANTS
Rodriguez, R. J.; Parks, L. W.
Dep. Microbiol. Oregon State Univ. Corvallis USA US OK 97331
Anal. Biochem.; (82) P 200-4; Vol 119; No 1

23-82-0002 (0002)
QUANTITATION OF MULTIPLE STEROIDS IN HUMAN SERUM AND URINE BY COMBINATION OF AUTOMATIC HPLC AND RIA
Kage, A.; Fenner, A.; Weber, B.; Schoeneshoefer, M.
Inst. Clin. Chem. Biochem. FU Berlin Berlin Fed. Rep. Ger. DE D-1000/45
Anal. Chem. Symp. Ser.; (82) P 327-32; Vol 10; No Adv. Steroid Anal.

23-82-0003 (0003)
CRITERIA FOR THE EVALUATION OF THE PURITY OF STEROIDS FOR PHARMACEUTICAL PREPARATIONS BY MEANS OF HPLC
Cavina, G.; Moretti, G.; Alimenti, R.; Gallinella, B.; Porra, R.; Valvo, L.
Ist. Super. Sanita Fellow Ist. Super. Sanita Rome Italy IT
Anal. Chem. Symp. Ser.; (82) P 343-60; Vol 10; No Adv. Steroid Anal.

23-82-0004 (0004)
HPLC STEROID ANALYSES: GENERIC COLUMN DESCRIPTION AND VARIABLE SELECTIVITY
Kirschbaum, J.; Clay, R.; Poet, R.
Squibb Inst. Med. Res. New Brunswick USA US NJ 08903
Anal. Chem. Symp. Ser.; (82) P 361-6; Vol 10; No Adv. Steroid Anal.

23-82-0005 (0005)
A NEW POLAROGRAPHIC RESPONSE FOR ETHYNYLESTRADIOL AND ITS APPLICATION TO ELECTROCHEMICAL DETECTION ASSOCIATED WITH REVERSE-PHASE LIQUID CHROMATOGRAPHY AND IN CATHODIC STRIPPING VOLTAMMETRY
Bond, A. M.; Heritage, I. D.; Briggs, M. H.
Div. Chem. Phys. Sci. Deakin Univ. Waurn Ponds Australia AU 3217
Anal. Chim. Acta; (82) P 35-45; Vol 138

23-82-0006 (0006)
ANALYSIS OF CLINICALLY IMPORTANT BILE PIGMENTS BY HIGH-PERFORMANCE LIQUID CHROMATOGRAPHY
Bull, Rosalind V. A.; Lim, C. K.
Div. Clin. Chem. Clin. Res. Cent. Harrow UK GB HA1 3UJ
Anal. Proc. (London); (82) P 186-9; Vol 19; No 4

23-82-0007 (0007)
LC-MS DETECTION OF STEROIDS AND DRUGS AT LOW AND MEDIUM RESOLUTION
Baty, J. D.; Willis, R. G.
Dep. Biochem. Med. Ninewells Hosp. Med. Sch. Dundee UK GB DD1 9SY
Anal. Proc. (London); (82) P 251-3; Vol 19; No 5

23-82-0008 (0008)
STUDY OF EXPERIMENTAL CONDITIONS FOR RAPID PREGNANCY URINARY ESTRIOL EVALUATION BY HIGH-PERFORMANCE LIQUID CHROMATOGRAPHY
Nadjem, B.; Abdeddaim, K.; Guermouche, M. H.
Inst. Chim. USTHB Alger Algeria DZ
Analusis; (82) P 83-6; Vol 10; No 2; In Fr

23-82-0009 (0009)
FETOMATERNAL RELATIONSHIPS OF SERUM BILE ACID PATTERN ESTIMATED BY HIGH-PRESSURE LIQUID CHROMATOGRAPHY

Itoh, Susumu; Onishi, Shoju; Isobe, Kenichi; Manabe, Masahiro; Inukai, Kazuhisa
Dep. Pediatr. Kagawa Med. Sch. Kagawa Japan JP 760-07
Biochem. J.; (82) P 141-5; Vol 204; No 1

23-82-0010 (0010)
SOLVENT OPTIMIZATION IN NORMAL-PHASE LIQUID CHROMATOGRAPHY OF SELECTED STEROIDS
Antle, P. E.
Anal. Instrum. Div. E. I. du Pont de Nemours and Co., Inc. Wilmington USA US DE 19898
Chromatographia; (82) P 277-81; Vol 1; No 5

23-82-0011 (0011)
DETERMINATION OF C21-STEROIDS IN MIXTURES BY HIGH PERFORMANCE LIQUID CHROMATOGRAPHY
Kim, Choong Young
Sch. Med. Kyungpook Natl. Univ. Taegu S. Korea KR
Chungang Uihak; (82) P 121-5;P Vol 42; No 2; In Korean

23-82-0012 (0012)
RADIOIMMUNOASSAY AND LIQUID-CHROMATOGRAPHIC ANALYSIS FOR FREE CORTISOL IN URINE COMPARED WITH ISOTOPE DILUTION-MASS SPECTROMETRY
Lantto, Olle
Karolinska Inst. Huddinge Univ. Hosp. Huddinge Swed. SE S-141 86
Clin. Chem. (Winston-Salem, N. C.); (82) P 1129-32; Vol 28; No 5

23-82-0013 (0013)
TWO-CYCLE LIQUID-CHROMATOGRAPHIC QUANTITATION OF CORTISOL IN URINE
Nakamura, Jiro; Yakata, Minoru
Sch. Med. Niigata Univ. Niigata Japan JP 951
Clin. Chem. (Winston-Salem, N. C.); (82) P 1497-500; Vol 28; No 7

23-82-0014 (0014)
MICROMETHOD FOR LIQUID-CHROMATOGRAPHIC DETERMINATION OF CHOLESTEROL IN LIPEMIC SERUMS
Webb, Lawrence E.; Lengle, Ernest; Izzo, Rodger; Woods, Eugene
Travenol. Lab., Inc. Morton Grove USA US IL 60053
Clin. Chem. (Winston-Salem, N. C.); (82) P 1769-72; Vol 28; No 8

23-82-0015 (0015)
ANALYSIS FOR CORTISOL IN SERUM BY LIQUID CHROMATOGRAPHY
Caldarella, Antoinette M.; Reardon, George E.; Canalis, Ernesto
Med. Cent. Saint Francis Hosp. Hartford USA US CT 06105
Clin. Chem. (Winston-Salem, N. C.); (82) P 538-43; Vol 28; No 3

23-82-0016 (0016)
DIRECT DETERMINATION OF THE LINOLEATE/OLEATE RATIO IN SERUM CHOLESTEROL ESTERS BY LIQUID CHROMATOGRAPHY
Bernert, John T., Jr.; Akins, James R.; Miller, Dayton T.
Clin. Chem. Div. Cent. Environ. Health Atlanta USA US GA 30333
Clin. Chem. (Winston-Salem, N. C.); (82) P 676-80; Vol 28; No 4, Pt. 1

23-82-0017 (0017)
ANALYSIS OF STEROIDS FROM NORMAL AND TUMOR TISSUE BY HPLC
Kessler, Michael J.
Health Sci. Cent. Univ. Texas Houston USA US TX 77025
Clin. Chim. Acta; (82) P 21-30; Vol 125; No 1

23-82-0018 (0018)
A SIMPLE REVERSE-PHASE HIGH PRESSURE LIQUID CHROMATOGRAPHIC DETERMINATION OF CONJUGATED BILE ACIDS IN SERUM AND BILE USING A NOVEL RADIAL COMPRESSION SEPARATION SYSTEM
Ruben, Annie T.; Van Berge-Henegouwen, Gerard P.
Lab. Med. Res. Munic. Hosp. Arnhem Neth. NL 6815 AD
Clin. Chim. Acta; (82) P 41-50; Vol 119; No 1-2

23-82-0019 (0019)
LIQUID CHROMATOGRAPHIC ASSAY OF URINARY FREE CORTISOL
Shihabi, Zakariya K.; Andrews, Rosalyn I.; Scaro, Judy
Bowman Gray Sch. Med. Wake Forest Univ. Winston-Salem USA US NC 27103
Clin. Chim. Acta; (82) P 75-83; Vol 124; No 1

23-82-0020 (0020)
COMPARISON OF TLC, HPLC AND GC IN ANALYSIS OF BILE ACIDS
Levitt, Robert E.; Touchstone, Joseph C.; Soloway, Roger D.
Univ. Pennsylvania Philadelphia USA US PA
CONFERENCE; Adv. Thin Layer Chromatogr., (Proc. Bienn. Symp.), 2nd; (82) P 241-7; Avail.: Touchstone, Joseph C; Date: 80; Publisher: Wiley, New York, N. Y.

23-82-0021 (0021)
STUDIES ON THE PERFLUOROAMIDE-BONDED PHASE AS HPLC PACKING. III. APPLICATION OF PERFLUOROAMIDE-BONDED SILICA IN THE SEPARATION OF CHOLANIC ACIDS AND ESTERS
Li, Wenhua; Wang, Jinhua
Shanghai Inst. Org. Chem. Acad. Sin. Shanghai Peop. Rep. China CN
Huaxue Xuebao; (82) P 377-80; Vol 40; No 4; In Ch

23-82-0022 (0022)
A STABILITY INDICATING HIGH-PERFORMANCE LIQUID CHROMATOGRAPHY DETERMINATION OF TRIPLE CORTICOID INTEGRATED SYSTEM IN A CREAM
Shek, Efraim; Bragonje, Joan; Benjamin, Eric J.; Sutherland, Mark J.; Gluck, John A. P.
Inst. Pharm. Sci. Syntex Res. Palo Alto USA US CA 94304
Int. J. Pharm.; (82) P 257-69; Vol 11; No 3

23-82-0023 (0023)
CHARACTERIZATION OF THE RABBIT RENAL RECEPTOR FOR NATIVE PARATHYROID HORMONE EMPLOYING A RADIOLIGAND PURIFIED BY REVERSED-PHASE LIQUID CHROMATOGRAPHY
Kremer, Richard; Bennett, Hugh P. J.; Mitchell, Jane; Goltzman, David
Dep. Med. McGill Univ. Montreal Can. CA PQ H3A 1A1
J. Biol. Chem.; (82) P 14048-54; Vol 257; No 23

23-82-0024 (0024)
DETERMINATION OF 17-HYDROXYCORTICOSTEROIDS IN URINE BY FLUORESCENCE HIGH-PERFORMANCE LIQUID CHROMATOGRAPHY USING DNS-HYDRAZINE AS A PRECOLUMN LABELING REAGENT
Kawasaki, Takao; Maeda, Masako; Tsuji, Akio
Sch. Pharm. Sci. Showa Univ. Tokyo Japan JP
J. Chromatogr.; (82) P 1-11; Vol 232; No 1

23-82-0025 (0025)
SEPARATION OF SOME NATURAL AND SYNTHETIC CORTICOSTEROIDS IN BIOLOGICAL FLUIDS AND TISSUES BY HIGH-PERFORMANCE LIQUID CHROMATOGRAPHY
Althaus, PZelda R.; Rowland, Jon M.; Freeman, James P.; Slikker, William, Jr.
Dep. Health Hum. Serv. Natl. Cent. Toxicol. Res. Jefferson USA US AR 72079
J. Chromatogr.; (82) P 11-23; Vol 227; No 1

23-82-0026 (0026)
STUDIES ON STEROIDS. CLXXIX. DETERMINATION OF ESTRIOL 16- AND 17-GLUCURONIDE IN BIOLOGICAL FLUIDS BY HIGH-PERFORMANCE LIQUID CHROMATOGRAPHY WITH ELECTROCHEMICAL DETECTION
Shimada, Kazutake; Xie, Fuming; Nambara, Toshio
Pharm. Inst. Tohoku Univ. Sendai Japan JP 980
J. Chromatogr.; (82) P 13-18; Vol 232; No 1

23-82-0027 (0027)
QUANTITATIVE HIGH-PERFORMANCE LIQUID CHROMATOGRAPHIC METHOD FOR THE ESTIMATION OF HECOGENIN AND TIGOGENIN IN THE LEAVES AND SAPOGENIN CONCENTRATES OF AGAVE SISALANA
Sitton, Dov; Blunden, Gerald; Cripps, Alan L.
Sch. Pharm. Portsmouth Polytech. Portsmouth/Hants. UK GB P01 2DZ
J. Chromatogr.; (82) P 138-40; Vol 245; No 1

23-82-0028 (0028)
HIGH-PERFORMANCE LIQUID CHROMATOGRAPHY OF BILE ACIDS WITH A REVERSED-PHASE RADIAL COMPRESSION COLUMN
Reid, A. D.; Baker, P. R.
Ninewells Hosp. Med. Sch. Univ. Dep. Surg. Dundee UK GB DD1 9SY
J. Chromatogr.; (82) P 149-56; Vol 247; No 1

23-82-0029 (0029)
COMPARISON OF ADSORPTION AND REVERSED-PHASE PARTITION HIGH-PERFORMANCE LIQUID CHROMATOGRAPHY FOR THE SEPARATION OF ANDROGENS
Lin, Jiann Tsyh; Heftmann, Erich
West. Reg. Res. Cent. U. S. Dep. Agric. Berkeley USA US CA 94710
J. Chromatogr.; (82) P 215-24; Vol 237; No 2

23-82-0030 (0030)
ANALYSIS OF PIPERAZINE ESTRONE SULFATE IN TABLETS BY ION-PAIR HIGH-PERFORMANCE LIQUID CHROMATOGRAPHY
Carignan, G.; Lodge, B. A.; Skakum, W.
Health Protect. Branch Bur. Drug Res. Ottawa Can. CA ON K1A 0L2
J. Chromatogr.; (82) P 240-3; Vol 234; No 1

23-82-0031 (0031)
HIGH-PERFORMANCE LIQUID CHROMATOGRAPHIC DETERMINATION OF PHOSPHATE ESTERS OF DEXAMETHASONE AND PREDNISOLONE AND THEIR SULFITE ADDUCTS
Dijkstra, J.; Dekker, D.
Hosp. Venlo-Tegelen Venlo Neth. NL 5914 BA
J. Chromatogr.; (82) P 247-9; Vol 238; No 1

23-82-0032 (0032)
SEPARATION OF STEROID CONJUGATES BY HIGH-PERFORMANCE LIQUID CHROMATOGRAPHY
Van der Wal, S.; Huber, J. F. K.
Tech. Instruments Corp. Tarrytown USA US NY 10591
J. Chromatogr.; (82) P 289-310; Vol 251; No 3

23-82-0033 (0033)
DETERMINATION OF METHYLPREDNISOLONE IN CENTRAL NERVOUS TISSUE AND PLASMA USING NORMAL-PHASE HIGH-PERFORMANCE LIQUID CHROMATOGRAPHY
McGinley, P. A.; Braughler, J. M.; Hall, E. D.
Coll. Med. Northeast. Ohio Univ. Rootstown USA US OH 44272
J. Chromatogr.; (82) P 29-35; Vol 230; No 1

23-82-0034 (0034)
RAPID METHOD FOR THE SIMULTANEOUS ANALYSIS OF HYDROCORTISONE AND CLIOQUINOL IN TOPICAL PREPARATIONS BY HIGH-PERFORMANCE LIQUID CHROMATOGRAPHY

Phoon, Khye Wang; Stubley, C.
Dep. Pharm. Chem. Univ. Bradford Bradford UK GB BD7 1DP
J. Chromatogr.; (82) P 297-303; Vol 246; No 2

23-82-0035 (0035)
HIGH-PERFORMANCE LIQUID CHROMATOGRAPHY OF ECDYSTEROIDS AND ECDYSTEROID-22-PHOSPHATES
Isaac, R. Elwyn; Milner, Nicholas P.; Rees, Huw H.
Dep. Biochem. Univ. Liverpool Liverpool UK GB L69 3BX
J. Chromatogr.; (82) P 317-22; Vol 246; No 2

23-82-0036 (0036)
DETERMINATION OF PLASMA LEVELS OF SPIRORENONE, A NEW ALDOSTERONE ANTAGONIST, AND ONE OF ITS METABOLITES BY HIGH-PERFORMANCE LIQUID CHROMATOGRAPHY
Krause, Werner; Jakobs, Uwe
Dep. Pharmacokinet. Schering Berlin/Bergkamen Berlin Fed. Rep. Ger. DE 1000/65
J. Chromatogr.; (82) P 37-45; Vol 230; No 1

23-82-0037 (0037)
DETERMINATION OF MEDROXYPROGESTERONE ACETATE IN PLASMA BY HIGH-PERFORMANCE LIQUID CHROMATOGRAPHY
Milano, G.; Carle, G.; Renee, N.; Boublil, J. L.; Namer, M.
Cent. Antoine-Lacassagne Nice Fr. FR 06054
J. Chromatogr.; (82) P 413-17; Vol 232; No 2

23-82-0038 (0038)
RAPID AND DIRECT METHOD FOR THE DETERMINATION OF ECDYSTEROID CONJUGATES BY LIQUID CHROMATOGRAPHY
Scalia, S.; Morgan, E. D.
Dep. Chem. Univ. Keele Keele/Staffordshire UK GB ST5 5BG
J. Chromatogr.; (82) P 457-64; Vol 238; No 2

23-82-0039 (0039)
RAPID, SPECIFIC METHOD FOR DIETHYLSTILBESTROL ANALYSIS USING AN IN-LINE PHOTOCHEMICAL REACTOR WITH HIGH-PERFORMANCE LIQUID CHROMATOGRAPHY AND FLUORESCENCE DETECTION
Williams, A. T. Rhys; Winfield, S. A.; Belloli, R. C.
Perkin-Elmer Ltd. Buckinghamshire UK GB HP9 1QA
J. Chromatogr.; (82) P 461-70; Vol 235; No 2

23-82-0040 (0040)
DETERMINATION OF URINARY DEHYDROEPIANDROSTERONE SULFATE BY COMBINED HIGH-PERFORMANCE LIQUID CHROMATOGRAPHY AND RADIOIMMUNOASSAY
Causon, Roger C.; Collins, Susan L.; Fry, Denys E.
Epsom Hosp. Lab. Epsom/Surrey UK GB KT19 8PB
J. Chromatogr.; (82) P 485-91; Vol 227; No 2

23-82-0041 (0041)
HIGH-PERFORMANCE LIQUID CHROMATOGRAPHIC SEPARATION OF THE TWO ESTROGEN ISOMERS OF ESTRADIOL WITH ELECTROCHEMICAL DETECTION
Prescott, William R., Jr.; Boyd, Brenda K.; Seaton, John F.
Milton S. Hershey Med. Cent. Pennsylvania State Univ. Hershey USA US PA 17033
J. Chromatogr.; (82) P 513-16; Vol 234; No 2

23-82-0042 (0042)
DETERMINATION OF PREDNISONE AND PREDNISOLONE IN HUMAN SERUM BY HIGH-PERFORMANCE LIQUID CHROMATOGRAPHY - ESPECIALLY ON IMPAIRED CONVERSION OF CORTICOSTEROIDS IN PATIENTS WITH CHRONIC LIVER DISEASE
Ui, Tadahiro; Mitsunaga, Masayuki; Tanaka, Teruji; Horiguchi, Masaharu
Sch. Med. Jikei Univ. Komae Japan JP

J. Chromatogr.; (82) P 711-16; Vol 239

23-82-0043 (0043)
SEPARATION AND DETERMINATION OF BILE ACIDS BY HIGH-PERFORMANCE LIQUID CHROMATOGRAPHY USING IMMOBILIZED 3.ALPHA.-HYDROXYSTEROID DEHYDROGENASE AND AN ELECTROCHEMICAL DETECTOR
Kamada, Satoru; Maeda, Masako; Tsuji, Akio; Umezawa, Yoshio; Kurahashi, Tatsuo
Sch. Pharm. Sci. Showa Univ. Tokyo Japan JP
J. Chromatogr.; (82) P 773-83; Vol 239

23-82-0044 (0044)
SELECTIVE EFFECTS OF MOBILE AND STATIONARY PHASES IN REVERSED-PHASE HIGH-PERFORMANCE LIQUID CHROMATOGRAPHY OF ECDYSTEROIDS
Wilson, I. D.; Bielby, C. R.; Morgan, E. D.
Dep. Chem. Univ. Keele Keele/Staffordshire UK GB ST5 5BG
J. Chromatogr.; (82) P 97-102; Vol 238; No 1

23-82-0045 (0045)
STEROID ANALYSIS BY HIGH-PERFORMANCE LIQUID CHROMATOGRAPHY
Heftmann, Erich; Lin, Jiann Tsyh
WRRC ARS Berkeley USA US CA 94710
J. Liq. Chromatogr.; (82) P 121-73; Vol 5; No Suppl. 1

23-82-0046 (0046)
A RAPID HIGH PERFORMANCE LIQUID CHROMATOGRAPHY SYSTEM FOR THE SEPARATION OF GONADAL STEROIDS
Kessler, Michael J.
Health Sci. Cent. Univ. Texas Houston USA US TX 77025
J. Liq. Chromatogr.; (82) P 125-39; Vol 5; No 1

23-82-0047 (0047)
STUDIES ON STEROIDS. CLXXVIII. SEPARATION OF ESTROGEN GLUCURONIDES BY HIGH-PERFORMANCE LIQUID CHROMATOGRAPHY
Shimada, Kazutake; Kaji, Masanori; Xie, Fuming; Nambara, Toshio
Pharm. anst. Tohoku Univ. Sendai Japan JP
J. Liq. Chromatogr.; (82) P 1763-70; Vol 5; No 9

23-82-0048 (0048)
RAPID SEPARATION OF TESTOSTERONE AND ITS MICROSOMAL METABOLITES BY REVERSE-PHASE HIGH PERFORMANCE LIQUID CHROMATOGRAPHY
Newton, J. F.; Braselton, W. E., Jr.; Lepper, L. F.; McCormack, K. M.; Hook, J. B.
Cent. Environ. Toxicol. Michigan State Univ. East Lansing USA US MI 48824
J. Liq. Chromatogr.; (82) P 563-70; Vol 5; No 3

23-82-0049 (0049)
SIMULTANEOUS DETERMINATION OF HYDROCORTISONE AND BENZYL ALCOHOL IN PHARMACEUTICAL FORMULATIONS BY REVERSED-PHASE HIGH-PRESSURE LIQUID CHROMATOGRAPHY
Rego, Albert; Nelson, Brett
Anal. Dev. Group Allergan Pharm., Inc. Irvine USA US CA 92713
J. Pharm. Sci.; (82) P 1219-23; Vol 71; No 11

23-82-0050 (0050)
QUANTITATIVE ANALYSIS OF ETHYNODIOL DIACETATE AND ETHINYL ESTRADIOL/MESTRANOL IN ORAL CONTRACEPTIVE TABLETS BY HIGH-PERFORMANCE LIQUID CHROMATOGRAPHY
Carignan, G.; Lodge, B. A.; Skakum, W.
Bur. Drug Res. Health and Welfare Canada Ottawa Can. CA ON K1A 0L2
J. Pharm. Sci.; (82) P 264-6; Vol 71; No 2

23-82-0051 (0051)
QUANTITATION OF HYDROXYPROGESTERONE CAPROATE, MEDROXYPROGESTERONE ACETATE, AND PROGESTERONE BY REVERSED-PHASE HIGH-PRESSURE LIQUID CHROMATOGRAPHY
Das Gupta, V.
Coll. Pharm. Univ. Houston Houston USA US TX 77030
J. Pharm. Sci.; (82) P 294-7; Vol 71; No 3

23-82-0052 (0052)
HIGH-PERFORMANCE LIQUID CHROMATOGRAPHIC ANALYSIS OF HYDROCORTISONE DRUG SUBSTANCE, TABLETS, AND ENEMA
Walters, Milda J.; Dunbar, Walter E.
Dep. Health Hum. Serv. FDA Detroit USA US MI 48207
J. Pharm. Sci.; (82) P 446-51; Vol 71; No 4

23-82-0053 (0053)
ORAL HYDROCORTISONE PHARMACOKINETICS: A COMPARISON OF FLUORESCENCE AND ULTRAVIOLET HIGH-PRESSURE LIQUID CHROMATOGRAPHIC ASSAYS FOR HYDROCORTISONE IN PLASMA
Toothaker, Roger D.; Sundaresan, Gloria M.; Hunt, John P.; Goehl, Thomas J.; Rotenberg, Keith S.; Prasad, Vadlamani K.; Craig, William A.; Welling, Peter
Sch. Pharm. Univ. Wisconsin Madison USA US WI 53706
J. Pharm. Sci.; (82) P 573-6; Vol 71; No 5

23-82-0054 (0054)
DETERMINATION OF ETHINYLESTRADIOL IN SOLID DOSAGE FORMS BY HIGH-PERFORMANCE LIQUID CHROMATOGRAPHY
Strusiak, Suzanne H.; Hoogerheide, John G.; Gardner, Maureen S.
Schering Corp. Bloomfield USA US NJ 07003
J. Pharm. Sci.; (82) P 636-40; Vol 71; No 6

23-82-0055 (0055)
SIMULTANEOUS DETERMINATION OF CORTICOSTERONE, HYDROCORTISONE, AND DEXAMETHASONE IN DOG PLASMA USING HIGH-PERFORMANCE LIQUID CHROMATOGRAPHY
Alvinerie, M.; Toutain, P. L.
Stn. Pharmacol. Toxicol. Toulouse Fr. FR 31300
J. Pharm. Sci.; (82) P 816-18; Vol 71; No 7

23-82-0056 (0056)
DETERMINATION OF URINARY FREE CORTISOL AND CORTISONE BY SEQUENTIAL THIN-LAYER AND HIGH-PERFORMANCE LIQUID CHROMATOGRAPHY
Lewbart, Marvin L.; Elverson, Robert A.
Dep. Med. Crozer-Chester Med. Cent. Chester USA US PA 19013
J. Steroid Biochem.; (82) P 185-90; Vol 17; No 2

23-82-0057 (0057)
SEPARATION OF STEROIDAL ESTROGENS AND THEIR MAJOR UNCONJUGATED METABOLITES BY HIGH-PERFORMANCE LIQUID CHROMATOGRAPHY
Aten, R. F.; Eisenfeld, A. J.; MacLusky, N. J.; Hochberg, R. B.
Dep. Obstet. Gynecol. Yale Univ. Sch. Med. New Haven USA US CT 05610
J. Steroid Biochem.; (82) P 447-9; Vol 16; No 3

23-82-0058 (0058)
DETERMINATION OF BILE ACIDS IN BLOOD BY HIGH PERFORMANCE LIQUID CHROMATOGRAPHY USING A METHANOL SYSTEM
Okuda, Hiroaki; Hashimoto, Etsuko; Sujino, Hideko; Takeda, Keiko; Fujino, Nobuyuki; Motoike, Yoji; Nakanishi, Toshimi; Obata, Hiroshi; Matsuhara, Kenichi
Tokyo Women's Med. Coll. Japan JP
Kanzo; (82) P 89; Vol 23; No 1; In Japan

23-82-0059 (0059)
DETERMINATION OF CORTISOL IN BLOOD AND URINE - COMPARISON BETWEEN ENZYME IMMUNOASSAY AND HIGH-PERFORMANCE LIQUID CHROMATOGRAPHY
Nakamura, Jiro; Ogata, Minoru
Med. Sch. Univ. Niigata Niigata Japan JP
Niigata Igakkai Zasshi; (82) P 90-3; Vol 96; No 2; In Japan

23-82-0060 (0060)
TWO FORMS OF MURINE EPIDERMAL GROWTH FACTOR: RAPID SEPARATION BY USING REVERSE-PHASE HPLC
Burgess, Antony W.; Knesel, Jaromir; Sparrow, Lindsay G.; Nicola, Nicos A.; Nice, Edouard C.
Ludwig Inst. Cancer Res. Post Off. R. Melbourne Hosp. Victoria Australia AU 3050
Proc. Natl. Acad. Sci. U. S. A.; (82) P 5753-7; Vol 79; No 19

23-82-0061 (0061)
SIMULTANEOUS DETERMINATION OF PREDNISOLONE AND CORTISOL IN SERUM BY HPLC AND BY ISOTOPE DILUTION-MASS SPECTROMETRY
Oest, L.; Falk, O.; Lantto, O.; Bjoerkhem, I.
Karolinska Inst. Huddinge Univ. Hosp. Huddinge Swed. SE S-141 86
Scand. J. Clin. Lab. Invest.; (82) P 181-7; Vol 42; No 2

23-82-0062 (0062)
ANALYSIS OF PROGESTERONE AND 20.ALPHA.-DIHYDROPROGESTERONE IN PLASMA AND OVARY OF RATS BY HPLC
Guan, Ruiqin; Zhou, Zukun; Li, Wujun
Hubei Inst. Pharm. Eng. Peop. Rep. China CN
Shengzhi Yu Biyun; (82) P 56-8; Vol 2; No 3; In Ch

23-82-0063 (0063)
HIGH-PERFORMANCE LIQUID CHROMATOGRAPHY OF STEROID METABOLITES IN THE PREGNENOLONE AND PROGESTERONE PATHWAYS
Kessler, Michael J.
Houston-Med. Sch. Univ. Texas Houston USA US TX 77030
Steroids; (82) P 21-32; Vol 39; No 1

23-82-0064 (0064)
CHOLECALCIFEROL SULFATE IDENTIFICATION IN HUMAN MILK BY HPLC
Le Boulch, N.; Cancela, L.; Miravet, L.
INSERM Paris Fr. FR 75010
Steroids; (82) P 391-8; Vol 39; No 4

23-82-0065 (0065)
HIGH-PERFORMANCE LIQUID CHROMATOGRAPHY AND RADIOIMMUNOASSAY OF RAT PLASMA CORTICOSTERONE
Hofreiter, B. T.; Allen, J. P.; Mizera, A. C.; Powers, C. D.; Masi, A. M.
Peoria Sch. Med. Univ. Illinois Peoria USA US IL 61656
Steroids; (82) P 547-55; Vol 39; No 5

23-82-0066 (0066)
QUANTITATION OF PROGESTERONE AND TESTOSTERONE PROPIONATE IN A HORMONE INJECTION MIXTURE BY HIGH-PERFORMANCE LIQUID CHROMATOGRAPHY (HPLC)
Hu, Jiazhi; Ye, Zhenming
Canton Inst. Drug Control Canton Peop. Rep. China CN
Yaowu Fenxi Zazhi; (82) P 169-71; Vol 2; No 3; In Ch

CHAPTER 7.24
Surfactants

--- 1982 ---

24-82-0001 (0001)
DETERMINATION OF ANIONIC SURFACTANTS BY HIGH-PERFORMANCE LIQUID CHROMATOGRAPHY II. DETERMINATION OF SODIUM ALKYL SULFATES AS 3,5-DINITROBENZOATE DERIVATIVES
Utsunomiya, Akiko; Ikeda, Tatsuo; Takamatsu, Kazuyuki; Naito, Syoji
Kanagawa Prefect. Public Health Lab. Yokohama Japan JP
Bunseki Kagaku; (82) P 115-20; Vol 31; No 3; In Japan

24-82-0002 (0002)
DETERMINATION OF ALKYLBENZENESULFONATE IN WATER BY HIGH-PERFORMANCE LIQUID CHROMATOGRAPHY
Matsueda, Takahiko; Osaki, Yasuhiko; Shigee, Shinya
Fukuoka Environ. Res. Cent. Fukuoka Japan JP
Bunseki Kagaku; (82) P 59-63; Vol 31; No 2; In Japan

24-82-0003 (0003)
DETERMINATION OF TRACES OF SODIUM ALKYLBENZENESULFONATE BY HIGH-PERFORMANCE LIQUID CHROMATOGRAPHY. APPLICATION TO WATER
Saito, Toshihide; Higashi, Kunishige; Hagiwara, Kazuyoshi
Gov. Ind. Res. Inst. Ikeda Japan JP 563
Fresenius' Z. Anal. Chem.; (82) P 21-3; Vol 313; No 1

24-82-0004 (0004)
HIGH-PERFORMANCE LIQUID CHROMATOGRAPHIC SEPARATION AND SELECTIVE DETECTION OF ANIONIC SURFACTANTS. APPLICATION TO COMMERCIAL FORMULATION AND WATER SAMPLE
Smedes, F.; Kraak, J.C.; Werkhoven-Goewie, C.F.; Brinkman, U.A.T.; Frei, R.W.
Lab. Anal. Chrm., Univ. Amsterdam, Neth. Amsterdam
J. Chromatogr. (82) P 123-32; Vol 247; No 1

24-82-0005 (0005)
QUANTITATION OF ALKYLSULFONATES USING UV DETECTOR SENSITIVE "ION-PAIR" REAGENTS IN REVERSED-PHASE LIQUID CHROMATOGRAPHY
Sachok, Bohdan; Deming, Stanley N.; Bidlingmeyer, Brian A.
Dep. Chem. Univ. Houston Houston USA US TX
J. Liq. Chromatogr.; (82) P 389-402; Vol 5; No 3

24-82-0006 (0006)
HIGH PERFORMANCE LIQUID CHROMATOGRAPHIC ANALYSIS OF INTACT AND PARTIALLY BIODEGRADED LINEAR ALKYLBENZENE SULFONATES
Linder, D. E.; Allen, M. C.
Res. Dev. Dep. Conoco Inc. Ponca City USA US OK
JAOCS, J. Am. Oil Chem. Soc.; (82) P 152-5; Vol 59; No 3

24-82-0007 (0007)
SEPARATION OF SURFACTANT MIXTURES AND THEIR HOMOLOGS BY HIGH PERFORMANCE LIQUID CHROMATOGRAPHY
Nakamura, K.; Morikawa, Y.
Shiseido Lab. Yokohama Japan JP 223
JAOCS, J. Am. Oil Chem. Soc.; (82) P 64-8; Vol 59; No 1

24-82-0008 (0008)
SURFACTANTS AND HIGH-PERFORMANCE LIQUID CHROMATOGRAPHY
Hayano, Shigeo
Inst. Ind. Sci Univ. Tokyo Tokyo Japan JP
Yukagaku; (82) P 744-7; Vol 31; No 10; In Japan

CHAPTER 7.25
Toxins

--- 1982 ---

25-82-0001 (0001)
ANALYSIS OF AFLATOXINS IN MILK AND MILK PRODUCTS BY HIGH-PERFORMANCE LIQUID CHROMATOGRAPHY
Goto, Tetsuhisa; Manabe, Masaru Matsuura, Shinji
Natl. Food Res. Inst. Minist. Agric. Ibaraki Japan JP 305
Agric. Biol. Chem.; (82) P 801-2; Vol 46; No 3

25-82-0002 (0002)
HPLC SEPARATION AND COMPARATIVE TOXICITY OF SAXITOXIN AND ITS REACTION PRODUCTS
Rubinson, Kenneth A.
Univ. Chem. Lab. Cambridge UK GB CB2 1EW
Biochim. Biophys. Acta; (82) P 315-20; Vol 687; No 2

25-82-0003 (0003)
THE ROLE OF LC IN THE ELUCIDATION OF THE MECHANISM OF HYDROLYSIS OF AN ORGAN SPECIFIC CYTOTOXIN, PRECOCENE I EPOXIDE, AND SOME INACTIVE ANALOGS
Hamnett, Anthony F.; Pratt, Grahame E.; Ottridge, Anthony P.; Alexander, B. Alan J.
Agric. Res. Counc. Univ. Sussex Sussex UK GB
Chromatogr. Sci.; (82) P 111-33; Vol 20; No Biol./Biomed. Appl. Liq. Chromatogr. 4

25-82-0004 (0004)
HIGH-PERFORMANCE LIQUID CHROMATOGRAPHY OF THE STAPHYLOCOCCAL ENTEROTOXINS
Williams, Robert Roy
Oregon State Univ. Corvallis USA US OR
From: Diss. Abstr. Int. B 1982, 42(10), 3961.
DISSERTATION; (82) P 102 pp.; Avail.:Univ. Microfilms Int.,Order No. DA8207495

25-82-0005 (0005)
HIGH-PERFORMANCE LIQUID CHROMATOGRAPHIC DETERMINATION OF .BETA.-EXOTOXIN PRODUCED BY THE BACTERIUM BACILLUS THURINGIENSIS
Oehler, Delbert D.; Gingrich, Richard E.; Haufler, Maurice
U. S. Livestock Insects Lab. Agric. Res. Serv. Kerrville USA US TX 78028
J. Agric. Food Chem.; (82) P 407-8; Vol 30; No 2

25-82-0006 (0006)
DETECTING AN ALGALTOXIN BY HIGH-PRESSURE LIQUID CHROMATOGRAPHY
Wong, See Hoi; Hindin, Ervin
Dep. Civ. Struct. Eng. Hong Kong Polytech. Hung Hom Hong Kong HK
J. Am. Water Works Assoc.; (82) P 528-9; Vol 74; No 10

25-82-0007 (0007)
HIGH-PERFORMANCE LIQUID CHROMATOGRAPHIC QUANTITATION OF AFLATOXIN METABOLITES IN ANIMAL TISSUES
Gregory, Jesse F.,III; Manley, Doris B.
Food Sci. Hum. Nutr. Dep., Univ. Florida, Gainesville USA, FL 32 611
J. Assoc. Off. Anal. Chem. (82) P 869-75; Vol 65; No 4

25-82-0008 (0008)
HIGH-PERFORMANCE LIQUID CHROMATOGRAPHIC DETERMINATION AND CLEARANCE TIME OF AFLATOXIN RESIDUES IN SWINE TISSUES
Miller, Doris M.; Wilson, David M.; Wyatt, Roger D.; McKinney, Jerry K.; Crowell, Wayne A.; Stuart, Barry P.
Coll. Vet. Med. Univ. Georgia Athens USA US GA 30602
J. Assoc. Off. Anal. Chem.; (82) P 1-4; Vol 65; No 1

25-82-0009 (0009)
COMPARISON OF RAPID HIGH-PRESSURE LIQUID CHROMATOGRAPHIC AND CB METHODS FOR DETERMINATION OF AFLATOXINS IN CORN AND PEANUTS
DeVries, Jonathan W.; Chang, Henry L.
Medallion Lab. James Ford Bell Techn. Cent. Minneapolis USA US MN 55427
J. Assoc. Off. Anal. Chem.; (82) P 206-9; Vol 65; No 2

25-82-0010 (0010)
HIGH-PRESSURE LIQUID CHROMATOGRAPHIC DETERMINATION OF AFLATOXINS BY USING RADIAL COMPRESSION SEPARATION
Shepherd, Eric C.; Phillips, Timothy D.; Heidelbaugh, Norman D.; Hayes, A. Wallace
Dep. Vet. Public Health Texas A and M Univ. College Station USA US TX 77843
J. Assoc. Off. Anal. Chem.; (82) P 665-71; Vol 65; No 3

25-82-0011 (0011)
LIQUID CHROMATOGRAPHIC DETERMINATION OF AFLATOXINS IN ARTIFICIALLY CONTAMINATED COCOA BEANS
Hurst, W. Jeffrey; Lenovich, Lawrence M.; Martin, Robert A., Jr.
Tech. Cent. Hershey Foods Corp. Hershey USA US PA 17033
J. Assoc. Off. Anal. Chem.; (82) P 888-91; Vol 65; No 4

25-82-0012 (0012)
DETERMINATION OF HOST-SELECTIVE PHYTOTOXINS FROM ALTERNARIA ALTERNATA F. SP. LYCOPERSICI AS THEIR MALEYL DERIVATIVES BY HIGH-PERFORMANCE LIQUID CHROMATOGRAPHY
Siler, D. J.; Gilchrist, D. G.
Dep. Plant Pathol. Univ. California Davis USA US CA 95616
J. Chromatogr.; (82) P 167-73; Vol 238; No 1

25-82-0013 (0013)
SEPARATION OF FOUR CARDIOTOXINS OF TAIWAN COBRA (NAJA NAJA ATRA) BY REVERSED-PHASE HIGH-PERFORMANCE LIQUID CHROMATOGRAPHY
Wu, Shih Siung; Tseng, Min Jen; Wang, Kung Tsung
Inst. Biol. Chem. Acad. Sin. Taipei Taiwan TW
J. Chromatogr.; (82) P 369-73; Vol 242; No 2

25-82-0014 (0014)
HIGH-PERFORMANCE LIQUID CHROMATOGRAPHY AND THIN-LAYER CHROMATOGRAPHY OF PENITREMS A-F, TREMORGENIC MYCOTOXINS FROM PENICILLIUM CRUSTOSUM
Maes, Catherine M.; Steyn, Pieter S.; Van Heerden, Fanie R.
Counc. Sci. Ind. Res. Natl. Chem. Res. Lab. Pretoria S. Afr. ZA 0001
J. Chromatogr.; (82) P 489-93; Vol 234; No 2

CHAPTER 7.26
Various Organics

--- 1982 ---

26-82-0001 (0001)
EXAMINATION OF THE VOLATILE CARBONYL FRACTION OF FRESH TOMATOES AND TOMATO PREPARATIONS. II. HIGH-PERFORMANCE LIQUID CHROMATOGRAPHIC SEPARATIONS
Varga-Puchony, Zita; Vigh, Gyula; Hlavay, Jozsef; Petro-Turza, Marta; Szarfoldi-Szalma, Ilona
Inst. Anal. Chem. Univ. Chem. Eng. Veszprem Hung. HU H-8200
Acta Aliment.; (82) P 103-15; Vol 11; No 1

26-82-0002 (0002)
DETERMINATION OF URINARY OXALATE BY HIGH-PERFORMANCE LIQUID CHROMATOGRAPHY
Hughes, H.; Hagen, L.; Sutton, R. A. L.
Clin. Invest. Unit Shaughnessy Hosp. Vancouver Can. CA BC
Anal. Biochem.; (82) P 1-3; Vol 119; No 1

26-82-0003 (0003)
RESOLUTION OF OPTICAL ISOMERS BY CHIRAL HIGH-PERFORMANCE LIQUID CHROMATOGRAPHY: SEPARATION OF DIHYDRODIOLS AND TETRAHYDRODIOLS OF BENZO(A)PYRENE AND BENZ(A)ANTHRACENE
Weems, Henri B.; Yavg, Shen K.
Sch. Med. Uniformed Serv. Univ. of the Health Sci. Bethesda USA US MD 20814
Anal. Biochem.; (82) P 156-61; Vol 125; No 1

26-82-0004 (0004)
PURIFICATION BY REVERSE-PHASE HIGH-PERFORMANCE LIQUID CHROMATOGRAPHY OF AN EPIDERMAL GROWTH FACTOR-DEPENDENT TRANSFORMING GROWTH FACTOR
Anzano, Mario A.; Roberts, Anita B.; Smith, Joseph M.; Lamb, Lois C.; Sporn, Michael B.
Lab. Chemoprev. Natl. Cancer Inst. Bethesda USA US MD 20205
Anal. Biochem.; (82) P 217-24; Vol 125; No 1

26-82-0005 (0005)
FLUORESCENT/ULTRAVIOLET ABSORBING ESTER DERIVATIVE FORMATION AND ANALYSIS OF EICOSANOIDS BY HIGH-PRESSURE LIQUID CHROMATOGRAPHY
Watkins, W. David; Peterson, Myron B.
Anesth. Serv. Massachusetts Gen. Hosp. Boston USA US MA 02114
Anal. Biochem.; (82) P 30-40; Vol 125; No 1

26-82-0006 (0006)
FURTHER PURIFICATION OF EPIDERMAL GROWTH FACTOR BY HIGH-PERFORMANCE LIQUID CHROMATOGRAPHY
Matrisian, Lynn M.; Larsen, Brent R.; Finch, Joanne S.; Magun, Bruce E.
Arizona Health Sci. Cent. Univ. Arizona Tucson USA US AZ 85724
Anal. Biochem.; (82) P 339-51; Vol 125; No 2

26-82-0007 (0007)
HIGH-PERFORMANCE LIQUID CHROMATOGRAPHIC METHOD FOR THE DETERMINATION OF PHYTATE
Graf, Ernst; Dintzis, Frederick R.
North. Reg. Res. Cent. Agric. Res. Serv. Peoria USA US IL 61604
Anal. Biochem.; (82) P 413-17; Vol 119; No 2

26-82-0008 (0008)
REVERSED-PHASE LIQUID CHROMATOGRAPHY FOR DETERMINATION OF BETA-HYDROXY OXIMES AND RELATED COMPOUNDS FROM COPPER SOLVENT EXTRACTION
Stone, L.C.; Andersen, T.N.; Parr, J.C.; Metcalf, D.D.
Process. Technol., Kennecott Miner. Co., Salt Lake City USA, UT 84147
Anal. Chem. (82) P 1965-8; Vol 54; No 12

26-82-0009 (0009)
DETERMINATION OF ENVIRONMENTAL PHENOLS BY LIQUID CHROMATOGRAPHY / ELECTROCHEMISTRY
Shoup, Ronald E.; Mayer, Ginny S.
Res. Lab. Bioanal. Syst. Inc. West Lafayette USA US IN 47906
Anal. Chem.; (82) P 1164-9; Vol 54; No 7

26-82-0010 (0010)
LIQUID CHROMATOGRAPHIC SEPARATIONS OF DIHYDROXYBENZENES WITH SILICA, DIAMINE, AND COBALT COMPLEX-BONDED PHASE COLUMNS
Chang, C. Allen; Tu, Cheng Fan
Dep. Chem. Univ. Texas El Paso USA US TX 79968
Anal. Chem.; (82) P 1179-82; Vol 54; No 7

26-82-0011 (0011)
SEPARATION OF ALIPHATIC ALCOHOLS BY PAIRED ION LIQUID CHROMATOGRAPHY
Gnanasambandan, T.; Freiser, H.
Dep. Chem. Univ. Arizona Tucson USA US AZ 85721
Anal. Chem.; (82) P 1282-5; Vol 54; No 8

26-82-0012 (0012)
AIR SAMPLING AND LIQUID CHROMATOGRAPHIC DETERMINATION OF ETHYLENIMINE
Morales, Raul; Stampfer, J. F.; Hermes, R. E.
Los Alamos Natl. Lab. Univ. California Los Alamos USA US NM 87545
Anal. Chem.; (82) P 1340-4; Vol 54; No 8

26-82-0013 (0013)
HYDROCARBON CLASS FRACTIONATION WITH BONDED-PHASE LIQUID CHROMATOGRAPHY
Miller, Ronald
Perkin-Elmer Corp. Norwalk USA US CT 06856
Anal. Chem.; (82) P 1742-6; Vol 54; No 11

26-82-0014 (0014)
ISOLATION AND IDENTIFICATION OF BENZENE METABOLITES IN VITRO WITH LIQUID CHROMATOGRAPHY/ELECTROCHEMISTRY
Roston, Daryl A.; Kissinger, Peter T.
Dep. Chem. Purdue Univ. West Lafayette PUSA US IN 47907
Anal. Chem.; (82) P 1798-802; Vol 54; No 11

26-82-0015 (0015)
DETERMINATION OF TETRAHYDROISOQUINOLINES BY REVERSED-PHASE LIQUID CHROMATOGRAPHY WITH GRADIENT ELUTION AND AMPEROMETRIC DETECTION
St. Claire, R. L., III; Ansari, G. A. S.; Abell, Creed W.
Med. Branch Univ. Texas Galveston USA US TX 77550
Anal. Chem.; (82) P 186-9; Vol 54; No 2

26-82-0016 (0016)
PHOTOREDUCTION-FLUORESCENCE DETECTION OF ALIPHATIC ALCOHOLS, ALDEHYDES, AND ETHERS INLIQUID CHROMATOGRAPHY
Gandelman, Mitchell S.; Birks, John W.
Dep. Chem. Univ. Colorado Boulder USA US CO 80309
Anal. Chem.; (82) P 2131-3; Vol 54; No 12

26-82-0017 (0017)
DETERMINATION OF ALKYL NITRATE ADDITIVES IN DIESEL FUEL BY LIQUID CHROMATOGRAPHY WITH INFRARED SPECTROMETRIC DETECTION
Schabron, J. F.; Fuller, M. P.
Phillips Petrol. Co. Bartlesville USA US OK 74004
Anal. Chem.; (82) P 2599-601; Vol 54; No 14

26-82-0018 (0018)
SEPARATION OF INORGANIC AND ORGANIC ANIONS ON REVERSED-PHASE LIQUID CHROMATOGRAPHY COLUMNS
Skelly, Norman E.
Anal. Lab. Dow Chem. Co. Midland USA MI 48640
Anal. Chem.; (82) P 712-15; Vol 54; No 4

26-82-0019 (0019)
SEPARATION AND DETERMINATION OF 4-(1,1,3,3-TETRAMETHYLBUTYL)PHENYL DIHYDROGEN PHOSPHATE AND DI-(4-(1,1,3,3-TETRAMETHYLBUTYL)PHENYL) HYDROGEN PHOSPHATE BY HIGH-PERFORMANCE LIQUID CHROMATOGRAPHY
Lin, Cheng Hsien; Chen, Haw Jan; Ting, Gann
Inst. Nucl. Energy Res. At. Energy Counc. Lungtan Taiwan TW
Anal. Chim. Acta; (82) P 319-23; Vol 140; No 1

26-82-0020 (0020)
HPLC IDENTIFICATION AND SEPARATION OF PHENOLIC COMPOUNDS DERIVATIZED WITH DIAZOTIZED SULFANILIC ACID. STRUCTURAL EFFECTS ON RETENTION TIMES
Baiocchi, C.; Gennaro, M. C.; Campi, E.; Mentasti, E.; Aruga, R.
Ist. Chim. Anal. Univ. Torino Turin Italy IT 10125
Anal. Lett.; (82) P 1539-48; Vol 15; No A19

26-82-0021 (0021)
DETERMINATION OF THE MOTHPROOFING AGENT MITIN FF ON WOOL TEXTILES BY HIGH-PERFORMANCE LIQUID CHROMATOGRAPHY
Mayfield, R. J.
Div. Tex. Ind. CSIRO Belmont Australia AU 3216
Analyst (London); (82) P 324-7; Vol 107; No 1272

26-82-0022 (0022)
DETERMINATION OF ORGANIC CONTAMINANTS IN ULTRA-PURE WATER BY REVERSED-PHASE HIGH-PERFORMANCE LIQUID CHROMATOGRAPHY WITH ULTRAVIOLET DETECTION
Reust, Juerg B.; Meyer, Veronika R.
Inst. Anorg. Anal. Phys. Chem. Univ. Berne Berne Switz. CH CH-3012
Analyst (London); (82) P 673-9; Vol 107; No 1275

26-82-0023 (0023)
RAPID SCREENING TECHNIQUE UTILIZING HIGH-PERFORMANCE LIQUID CHROMATOGRAPHY FOR ASSESSING ACRYLAMIDE CONTAMINATION IN EFFLUENTS
Brown, Leslie; Rhead, Michael M.; Bancroft, Keith C. C.
Dep. Environ. Sci. Plymouth Polytech. Plymouth UK GB PL4 8AA
Analyst (London); (82) P 749-54; Vol 107; No 1276

26-82-0024 (0024)
SIMPLIFIED EXTRACTION PROCEDURE FOR THE DETERMINATION OF BRODIFACOUM IN WHEAT BAITS BY HIGH-PERFORMANCE LIQUID CHROMATOGRAPHY
Perez, R. L.
Roche-Maag Ltd. Bankstown Australia AU 2200
Analyst (London); (82) P 968-71; Vol 107; No 1277

26-82-0025 (0025)
LIQUID CHROMATOGRAPHIC DETERMINATION OF DIMETHYL FORMAMIDE, METHYLENE BISPHENYL ISOCYANATE AND METHYLENE BISPHENYL AMINE IN AIR SAMPLES
Lipski, Krzysztof
Inst. Occup. Med. Lodz Pol. PL 90-950
Ann. Occup. Hyg.; (82) P 1-4; Vol 25; No 1

26-82-0026 (0026)
HIGH-PERFORMANCE LIQUID CHROMATOGRAPHY USING POLYMER GELS FOR DIRECT ANALYSES AND CHARACTERIZATIONS OF SOLUBLE ORGANIC POLLUTANTS IN WATER

Urano, Kohei; Furyune, Hiroyuki; Hayashi, Koji
Dep. Saf. Environ. Eng. Yokohama Natl. Univ. Yokohama Japan JP 240
Bull. Chem. Soc. Jpn.; (82) P 2248-53; Vol 55; No 7

26-82-0027 (0027)
MICRO HIGH-PERFORMANCE LIQUID CHROMATOGRAPHY OF MUTAGEN, ETHYL n-ETHYL-N-NITROSOCARBAMATE, FORMED FROM URETHANE AND NITRITE
Kohashi, Kazuya; Kawazoe, Yuichi; Horiuchi, Tadao; Ohkura, Yosuke
Fac. Pharm. Sci. Kyushu Univ. Fukuoka Japan JP 812
Bunseki Kagaku; (82) P E41-E44; Vol 31; No 1

26-82-0028 (0028)
DETERMINATION OF URINARY OXALATE BY REVERSED-PHASE ION-PAIR "HIGH-PERFORMANCE" LIQUID CHROMATOGRAPHY
Larsson, Lasse; Libert, Bo; Asperud, Monica
Dep. Clin. Chem. Univ. Linkoeping Linkoeping Swed. SE S-581 85
Clin. Chem. (Winston-Salem, N. C.); (82) P 2272-4; Vol 28; No 11

26-82-0029 (0029)
LIQUID-CHROMATOGRAPHIC DETERMINATION OF ETHYLENE GLYCOL IN PLASMA
Gupta, Ram N.; Eng, Francis; Gupta, Mohan L.
Clin. Chem. Lab. St. Joseph's Hosp. Hamilton Can. CA ON L8N 1Y4
Clin. Chem. (Winston-Salem, N. C.); (82) P 32-3; Vol 28; No 1

26-82-0030 (0030)
HIGH-PRESSURE LIQUID CHROMATOGRAPHIC ANALYSIS OF PHENOLIC AND AMINE STABILIZERS
Juergens, U.; Doemling, H. J.
Landesuntersuchungsamt Gesundheitswesen Nordbayern Erlangen Fed. Rep. Ger. DE D-8520
Dtsch. Lebensm.-Rundsch.; (82) P 49-51; Vol 78; No 2; In Ger

26-82-0031 (0031)
HIGH PERFORMANCE LIQUID CHROMATOGRAPHIC DETERMINATION OF CAFFEINE AND SODIUM BENZOATE
Kozu, Tadashi
Sci. Invest. Res. Lab. Nagano Pref. Police Headquarters Nagano Japan JP
Eisei Kagaku; (82) P 111-13; Vol 28; No 2; In Japan

26-82-0032 (0032)
DYE DETERMINATION IN LIGHT FUEL OILS BY HIGH PRESSURE LIQUID CHROMATOGRAPHY
Peltzer, Joerg
Wiss. Abt. Zolltech. Pruef. Lehranst. Muenchen Munich Fed. Rep. Ger. DE D-8000/2
Erdoel Kohle, Erdgas, Petrochem.; (82) P 312; Vol 35; No 6; In Ger

26-82-0033 (0033)
DETERMINATION OF SULFHYDRYL COMPOUNDS IN PHYSIOLOGICAL FLUIDS BY LIQUID CHROMATOGRAPHY WITH POLAROGRAPHIC DETECTION
Zygmunt, B.; Hanekamp, H. B.; Bos, P.; Frei, R. W.
Dep. Anal. Chem. Free Univ. Amsterdam Neth. NL NL-1081 HV
Fresenius' Z. Anal. Chem.; (82) P 197-200; Vol 311; No 3

26-82-0034 (0034)
DETERMINATION OF PERCHLOROETHYLENE (PCE) IN CARCASSE MEAL USING HPLC
Fankel, Rudolf; Slad, Ilse
Landwirtschaftlich-Chem. Bundesversuchsanst. Vienna Austria AT A-1020
Fresenius' Z. Anal. Chem.; (82) P 47; Vol 313; No 1; In Ger

26-82-0035 (0035)
METHOD FOR DETERMINATION OF FORMALDEHYDE IN AIR IN THE PPT-RANGE BY HPLC AFTER EXTRACTION AS 2,4-DINITROPHENYLHYDRAZONE
Tuss, H.; Neitzert, V.; Seiler, W.; Neeb, R.
Max-Planck-Inst. Chem. Mainz Fed. Rep. Ger. DE D-6500
Fresenius' Z. Anal. Chem.; (82) P 613-17; Vol 312; No 7

26-82-0036 (0036)
DETERMINATION OF REDUCED BIOPTERINS BY HIGH PRESSURE LIQUID CHROMATOGRAPHY AND SUBSEQUENT ELECTROCHEMICAL DETECTION
Braeutigam, Matthias; Dreesen, Rainer; Herken, Hans
Pharmakol. Inst. Freien Univ. Berlin Fed. Rep. Ger. DE
Hoppe-Seyler's Z. Physiol. Chem.; (82) P 341-3; Vol 363; No 3

26-82-0037 (0037)
HPLC GRADIENT ELUTION OF DINITROPHENYLHYDRAZONES OF ALDEHYDES AND KETONES FROM AQUEOUS SAMPLES
Liebezeit, G.
Sonderforschungsber. 95 Univ. Kiel Kiel Fed. Rep. Ger. DE 2300
HRC CC, J. High Resolut. Chromatogr. Chromatogr. Commun.; (82) P 215-16; Vol 5; No 4

26-82-0038 (0038)
HIGH-PERFORMANCE LIQUID CHROMATOGRAPHY FOR THE QUANTITATIVE DETERMINATION OF THE URINARY METABOLITES OF TOLUENE, XYLENE, AND STYRENE
Poggi, G.; Giusiani, M.; Palagi, U.; Paggiaro, P. L.; Loi, A. M.; Dazzi, F.; Siclari, C.; Baschieri, L.
Ist. Med. Legale Assicurazioni Pisa Italy IT
Int. Arch. Occup. Environ. Health; (82) P 25-31; Vol 50; No 1

26-82-0039 (0039)
DETERMINATION OF 2,4-(DICHLOROPHENOXY)ACETIC ACID AND OF DICHLOBENIL IN WATER BY HIGH-PERFORMANCE LIQUID CHROMATOGRAPHY
Connick, William J., Jr.; Simoneaux, Jacqueline M.
South. Reg. Res. Cent. Sci. Educ. Adm. New Orleans USA US LA 70179
J. Agric. Food Chem.; (82) P 258-60; Vol 30; No 2

26-82-0040 (0040)
ANALYSIS OF FORMALDEHYDE IN SHRIMP BY HIGH-PRESSURE LIQUID CHROMATOGRAPHY
Radford, Terence; Dalsis, David E.
Coca-Cola Co. Atlanta USA US GA 30301
J. Agric. Food Chem.; (82) P 600-2; Vol 30; No 3

26-82-0041 (0041)
DETERMINATION OF VOLATILE ALDEHYDES IN MEAT AS 2,4-DINITROPHENYLHYDRAZONES USING REVERSED-PHASE HIGH-PERFORMANCE LIQUID CHROMATOGRAPHY
Reindl, Bertram; Stan, Hans Juergen
Inst. Sozialmed. Epidemiol. Bundesgesundheitsamt Berlin Fed. Rep. Ger. DE 1000/33
J. Agric. Food Chem.; (82) P 849-54; Vol 30; No 5

26-82-0042 (0042)
RESOLUTION OF CHIRAL OLEFINIC HYDROCARBONS AND SULFOXIDES BY HIGH-PERFORMANCE LIQUID CHROMATOGRAPHY VIA DIASTEREOMERIC PLATINUM COMPLEXES
Goldman, M.; Kustanovich, Z.; Weinstein, S.; Tishbee, A.; Gil-Av, E.
Dep. Org. Chem. Weizmann Inst. Sci. Rehovot Israel IL
J. Am. Chem. Soc.; (82) P 1093-5; Vol 104; No 4

26-82-0043 (0043)
METABOLISM AND DISTRIBUTION OF TWO 14C-BENZIDINE-CONGENER-BASED DYES IN RATS

AS DETERMINED BY GC, HPLC, AND RADIOASSAYS
Bowman, Malcolm C.; Oller, William L.; Nony, Charles R.; Rowland, Kenneth L.; Billedeau, Stanley M.; Lowry, Larry K.
Dep. Health Human Serv. Natl. Cent. Toxicol. Res. Jefferson USA US AR 72079
J. Anal. Toxicol.; (82) P 164-74; Vol 6; No 4

26-82-0044 (0044)
LIQUID-CHROMATOGRAPHIC DETERMINATION OF RATE CONSTANTS FOR THE CELLULOSE-DIMETHYLOLDIHYDROXYETHYLENEUREA REACTION
Beck, Keith R.; Pasad, Dilip M.
Dep. Consum. Sci. Retailing Purdue Univ. West Lafayette USA US IN 47907
J. Appl. Polym. Sci.; (82) P 1131-8; Vol 27; No 4

26-82-0045 (0045)
REVERSE-PHASE LIQUID CHROMATOGRAPHIC DETERMINATION OF CHLOROPHACINONE AND DIPHACINONE IN BAIT FORMULATIONS
Bennett, Brian R.; Grimes, Gregory S.
Biochem. Dep. Purdue Univ. West Lafayette USA US IN 47907
J. Assoc. Off. Anal. Chem.; (82) P 927-9; Vol 65; No 4

26-82-0046 (0046)
RAPID DETERMINATION OF CAPSAICIN, CN AND CS IN TEAR GAS BY HPLC
Krebs, J.; Prime, R. J.; Leung, K.
Chem. Sect. Cent. Forensic Sci. Toronto Can. CA ON M7A 2G8
J. Can. Soc. Forensic Sci.; (82) P 29-33; Vol 15; No 1

26-82-0047 (0047)
SEPARATION OF DESULFOGLUCOSINOLATES BY REVERSED-PHASE HIGH-PERFORMANCE LIQUID CHROMATOGRAPHY
Minchinton, I.; Sang, J.; Burke, D.; Truscott, R.J.W.
Div. Agric. Chem., Dep. Agric., Melbourne, Austrialia, 3002
J. Chromatogr. (82) P 141-8; Vol 247; No 1

26-82-0048 (0048)
ON-LINE HPLC/GC TECHNIQUES FOR THE ANALYSES OF BIOMASS DERIVED HYDROCARBONS
Luzbetak, Dori J.; Hoffmann, Joseph J.
Off. Arid Lands Stud. Univ. Arizona Tucson USA US AZ 85719
J. Chromatogr. Sci.; (82) P 132-5; Vol 20; No 3

26-82-0049 (0049)
DETERMINATION OF PHENOLIC COMPOUNDS BY HPLC
Chao, Grace Kung Jou; Suatoni, Joseph C.
Refin. Products Div. Anal. Technol. Dep. Gulf Sci. Technol. Co. Pittsburgh USA US PA 15230
J. Chromatogr. Sci.; (82) P 436-40; Vol 20; No 9

26-82-0050 (0050)
A COMPARISON OF THREE DIFFERENT HIGH PERFORMANCE LIQUID CHROMATOGRAPHY SYSTEMS FOR THE DETERMINATION OF ALDEHYDES AND KETONES IN DIESEL EXHAUST. PART I
Creech, G.; Johnson, R. T.; Stof6er, J. O.
Univ. Missouri Rolla USA US MO 65401
J. Chromatogr. Sci.; (82) P 67-72; Vol 20; No 2

26-82-0051 (0051)
LIQUID CHROMATOGRAPHY OF ADAMANTANES ON CARBON ADSORBENTS
Prusova, Dana; Colin, Henri; Guiochon, Georges
Lab. Synthetic Fuels Inst. Chem. Technol. Prague Czech. CS 16628/6
J. Chromatogr.; (82) P 1-11; Vol 234; No 1

26-82-0052 (0052)
UNIVERSAL,MULTICHANNEL ULTRAVIOLET DETECTION IN THE PURITY ANALYSIS OF 2-ETHYL 3-(4-HYDROXYBENZOYL)INDOLIZINE BY HIGH-PERFORMANCE LIQUID CHROMATOGRAPHY
Drenth, B. F. H.; Ghijsen, R. T.; De Zeeuw, R. A.
Dep. Toxicol. State Univ. Groningen Neth. NL 9713 AW
J. Chromatogr.; (82) P 113-20; Vol 238; No 1

26-82-0053 (0053)
ANALYSIS OF INDOLE DERIVATIVES BY REVERSED-PHASE HIGH-PERFORMANCE LIQUID CHROMATOGRAPHY
Jensen, Einar
Inst. Biol. Geol. Univ. Tromso Tromsoe Norway NO N-9001
J. Chromatogr.; (82) P 126-32 Vol 246; No 1

26-82-0054 (0054)
A SIMPLE LIQUID CHROMATOGRAPHIC METHOD FOR THE ANALYSIS OF PENTA- AND TETRACHLOROPHENOLS IN URINE OF EXPOSED WORKERS
Pekari, Kaija; Aitio, Antero
Dep. Toxicol. Ind. Hyg. Inst. Occup. Health Helsinki Finland FI 00370/37
J. Chromatogr.; (82) P 129-36; Vol 232; No 1

26-82-0055 (0055)
GENERAL SCHEME OF ANALYSIS OF PHENOLIC COMPOUNDS IN PLANT EXTRACTS BY REVERSED-PHASE HIGH-PERFORMANCE LIQUID CHROMATOGRAPHY
Villeneuve, Francis; Abravan1l, Gerard; Moutounet, Michel; Alibert, Gilbert
Cent. Physiol. Veg. Univ. Paul Sabatier Toulouse Fr. FR 31062
J. Chromatogr.; (82) P 131-40; Vol 234; No 1

26-82-0056 (0056)
SIMULTANEOUS DETERMINATION OF MYOCARDIAL ADENINE NUCLEOTIDES AND CREATINE PHOSPHATE BY HIGH-PERFORMANCE LIQUID CHROMATOGRAPHY
Harmsen, E.; De Tombe, P. P.; De Jong, J. W
Thoraxcent. Erasmus Univ. Rotterdam Rotterdam Neth. NL
J. Chromatogr.; (82) P 131-6; Vol 230; No 1

26-82-0057 (0057)
DETERMINATION OF URINARY HIPPURIC ACID AND o-CRESOL, AS INDICES OF TOLUENE EXPOSURE, BY LIQUID CHROMATOGRAPHY ON DYNAMICALLY MODIFIED SILICA
Hansen, Steen Honore; Doessing, Martin
Dep. Org. Chem. R. Dan. Sch. Pharm. Copenhagen Den. DK DK-2100
J. Chromatogr.; (82) P 141-8; Vol 229; No 1

26-82-0058 (0058)
THIN-LAYER CHROMATOGRAPHIC AND HIGH-PERFORMANCE LIQUID CHROMATOGRAPHIC SEPARATION OF METABOLITES OF THE WEAK CARCINOGEN, 7-METHYLBENZ(C)ACRIDINE
Boux, L. J.; Ireland, C. M.; Wright, D. J.; Holder, G. M.; Ryan, A. J.
Dep. Pharm. Univ. Sydney Sydney Australia AU 2006
J. Chromatogr.; (82) P 149-57; Vol 227; No 1

26-82-0059 (0059)
HIGH-PERFORMANCE LIQUID CHROMATOGRAPHIC DETERMINATION OF SPIROHYDANTOIN MUSTARD IN A CLINICALLY ACCEPTABLE FORMULATION OF FAT EMULSION
Rose, Lucy M.; James, Ruby H.; Duncan, M. Susan; Struck, Robert F.
South. Res. Inst. Birmingham USA US AL 35255
J. Chromatogr.; (82) P 152-6; Vol 246; No 1

26-82-0060 (0060)
SEPARATION OF INDOLE METABOLITES FROM URINE WITH AN ODS TYPE RESIN BY HIGH-PERFORMANCE LIQUID CHROMATOGRAPHY
Yamaguchi, Tomoko; Yokota, Kazuko; Uematsu, Fumie

Dep. Biochem. Tokyo Women's Med. Coll. Tokyo Japqn JP
J. Chromatogr.; (82) P 166-72; Vol 231; No 1

26-82-0061 (0061)
RETENTION BEHAVIOR OF ACETYL-INDANDIONES IN REVERSED-PHASE HIGH-PERFORMANCE LIQUID CHROMATOGRAPHY
Vigh, Gyula; Varga-Puchony, Zita; Bartha, Akos; Balogh, Sandor
Inst. Anal. Chem. Univ. Chem. Eng. Veszprem Hung. HU
J. Chromatogr.; (82) P 169-76; Vol 241; No 1

26-82-0062 (0062)
DETERMINATION OF MIMOSINE AND 3-HYDROXY-4(1H)-PYRIDONE IN LEUCAENA, AVIAN EXCRETA AND SERUM USING REVERSED-PHASE ION-PAIR HIGH-PERFORMANCE LIQUID CHROMATOGRAPHY
Acamovic, T.; D'Mello, J. P. F.; Fraser, K. W.
Dep. Agric. Biochem. Edinburgh Sch. Agric. Edinburgh UK GB EH9 3JG
J. Chromatogr.; (82) P 169-79; Vol 236; No 1

26-82-0063 (0063)
RETENTION BEHAVIOR OF ALKYLENE OXIDES IN REVERSED-PHASE HIGH-PERFORMANCE LIQUID CHROMATOGRAPHY
Kudoh, Megumu; Konami, Sumiyo; Fudano, Shintaro; Yamaguchi, Shinichiro
Wakayama Res. Lab. Kao Soap Co. Ltd. Wakayama Japan JP 640-91E
J. Chromatogr.; (82) P 209-13; Vol 234; No 1

26-82-0064 (0064)
PHOTOOXYGENATION-CHEMILUMINESCENCE HIGH-PERFORMANCE LIQUID CHROMATOGRAPHIC DETECTOR FOR THE DETERMINATION OF ALIPHATIC ALCOHOLS, ALDEHYDES, ETHERS AND SACCHARIDES
Gandelman, Mitchell S.; Birks, John W.
Dep. Chem. Univ. Colorado Boulder USA US Co 80309
J. Chromatogr.; (82) P 21-31; Vol 242; No 1

26-82-0065 (0065)
DNS DERIVATIZATION OF ANABOLIC AGENTS WITH HIGH-PERFORMANCE LIQUID CHROMATOGRAPHIC SEPARATION AND FLUORESCENCE DETECTION
Williams, A. T. Rhys; Winfield, S. A.; Belloli, R. C.
Perkin-Elmer Ltd. Beaconsfield/Buckinghamshire UK GB HP9 1QA
J. Chromatogr.; (82) P 224-9; Vol 240; No 1

26-82-0066 (0066)
HIGH-BACK-PRESSURE LIQUID CHROMATOGRAPHY. IV. STUDIES ON THE RETENTION BEHAVIOR OF AROMATIC COMPOUNDS
Takeuchi, Toyohide; Ishii, Daido
Fac. Eng. Nagoya Univ. Nagoya Japan JP 464
J. Chromatogr.; (82) P 23-32; Vol 244; No 1

26-82-0067 (0067)
INVESTIGATION OF N-ALKYLBENZAMIDES BY REVERSED-PHASE LIQUID CHROMATOGRAPHY. V. CHARACTERISTICS OF SOME TERTIARY ALKYLBENZAMIDES
P Wells, Martha J. M.; Clark, C. Randall
Sch. Pharm. Auburn Univ. USA US AL 36849
J. Chromatogr.; (82) P 231-40; Vol 244; No 2

26-82-0068 (0068)
LIQUID CHROMATOGRAPHY STUDY OF BROMINATED ANILINES AND INVESTIGATION OF PRODUCT FORMATION IN THE BROMINATION REACTION. I. ANILINES WITHOUT RING SUBSTITUENTS OR WITH ALKYL GROUPS IN THE ORTHO- AND PARA-POSITIONS
Truedsson, L. A.; Smith, B. E. F.
Chem. Cent. Lund Inst. Technol. Lund Swed. SE S-220 07/7

J. Chromatogr.; (82) P 25-46; Vol 234; No 1

26-82-0069 (0069)
REVERSED-PHASE HIGH-PERFORMANCE LIQUID CHROMATOGRAPHY OF PROCYANIDINS AND OTHER PHENOLICS IN FRESH AND OXIDIZING APPLE JUICES USING A PH SHIFT TECHNIQUE
Lea, Andrew G. H.
Long Ashton Res. Stn. Univ. Bristol Long Ashton/Bristol UK GB BS18 9AF
J. Chromatogr.; (82) P 253-7; Vol 238; No 1

26-82-0070 (0070)
ANALYSIS OF ETHYLENETHIOUREA IN BEER BY HIGH-PERFORMANCE LIQUID CHROMATOGRAPHY
Massey, R. C.; Key, P. E.; McWeeny, D. J.
Food Lab. Minist. Agric. Fish. Food Norwich UK GB NR2 4SX
J. Chromatogr.; (82) P 254-6; Vol 240; No 1

26-82-0071 (0071)
SEPARATION OF DICHLORO- AND CHLOROFLUOROBENZOPHENONE ISOMERS BY HIGH-PERFORMANCE LIQUID CHROMATOGRAPHY
Bishara, R. H.; Smith, S. L.
Lilly Res. Lab. Eli Lilly and Co. Indianapolis USA IN 46285
J. Chromatogr.; (82) P 261-3; Vol 234; No 1

26-82-0072 (0072)
INVESTIGATION OF n-ALKYLBENZAMIDES BY REVERSED-PHASE LIQUID CHROMATOGRAPHY. I. THE STUDY OF A HOMOLOGOUS SERIES OF n-ALKYLBENZAMIDES USING THE SOLVOPHOBIC THEORY AND MOLECULAR CONNECTIVITY
Wells, Martha J. M.; Clark C. Randall
Sch. Pharm. Auburn Univ. Auburn USA US AL 36849
J. Chromatogr.; (82) P 263-77; Vol 243; No 2

26-82-0073 (0073)
INVESTIGATION OF N-ALKYLBENZAMIDES BY REVERSED-PHASE LIQUID CHROMATOGRAPHY. I. ISOCRATIC ELUTION CHARACTERISTICS OF THE C1-C5 n-ALKYLBENZAMIDES
Wells, Martha J. M.; Clark, C. Randall
Sch. Pharm. Auburn Univ. Auburn USA AL 36849
J. Chromatogr.; (82) P 31-41; Vol 235; No 1

26-82-0074 (0074)
HIGH-PERFORMANCE LIQUID CHROMATOGRAPHIC DETERMINATION OF PYRIDAZINONES IN WASTEWATERS. II
Lehotay, J.; Matisova, E.; Garaj, J.; Violova, A.
Fac. Chem. Technol. Slovak Tech. Univ. Bratislava Czech. CS 812 37
J. Chromatogr.; (82) P 323-6; Vol 246; No 2

26-82-0075 (0075)
HIGH-PERFORMANCE LIQUID CHROMATOGRAPHIC SEPARATION OF THE CHLORINATION PRODUCTS OF TEREPHTHALAMIDE
Lee, Kuen Shan; Hsio, Kuo Liang; Su, Tsung Tsan; Kuo, Kung Tu
Union Ind. Res. Lab. Ind. Technol. Res. Inst. Hsinchu Taiwan TW
J. Chromatogr.; (82) P 324-9; Vol 237; No 2

26-82-0076 (0076)
LIQUID CHROMATOGRAPHY OF CHLORINATED BIPHENYLOLS
Hernandez, Adan; Walton, Harold F.
Coop. Inst. Res. Environ. Sci. Univ. Colorado Boulder USA US CO 80309
J. Chromatogr.; (82) P 346-8; Vol 242; No 2

26-82-0077 (0077)
PREPARATIVE-SCALE SEPARATION OF ALKENE GEOMETRIC ISOMERS BY LIQUID CHROMATOGRAPHY

Evershed, R. P.; Morgan, E. D.; Thompson, Lorna D.
Dep. Chem. Univ. Keele Keele/Staffordshire UK GB ST5 5BG
J. Chromatogr.; (82) P 350-4; Vol 237; No 2

26-82-0078 (0078)
DETERMINATION OF EDTA IN WATER BY HIGH-PERFORMANCE LIQUID CHROMATOGRAPHY
Harmsen, Joop; Van den Toorn, Antoni
Inst. Land Water Manage. Res. Wageningen Neth. NL 6700 AA
J. Chromatogr.; (82) P 379-84; Vol 249; No 2 In Eng

26-82-0079 (0079)
DETERMINATION OF DIMETHYL SULFATE IN AIR BY REVERSED-PHASE LIQUID CHROMATOGRAPHY
Williams, R. G.
Upjohn Co. Kalamazoo USA US MI 49001
J. Chromatogr.; (82) P 381-4; Vol 245; No 3

26-82-0080 (0080)
SEPARATION OF CHLORINATED PHENOLS BY REVERSED-PHASE HIGH-PERFORMANCE LIQUID CHROMATOGRAPHY AT AN ALKALINE pH
McLeod, H. A.; Laver, G.
Food Res. Div. Health Prot. Branch Ottawa Can. CA ON K1A 0L2
J. Chromatogr.; (82) P 385-90; Vol 244; No 2

26-82-0081 (0081)
HIGH-PERFORMANCE LIQUID CHROMATOGRAPHY OF SOME HYDROXY-SUBSTITUTED AROMATIC COMPOUNDS
Chiavari, Giuseppe; Concialini, Vittorio; Vitali, Patrizia
Ist. Chim. "G. Ciamician" Univ. Bologna Bologna Italy IT 40126
J. Chromatogr.; (82) P 385-92; Vol 249; No 2

26-82-0082 (0082)
SEPARATION AND QUANTITATION OF POLYNITRATE ESTERS IN PHARMACEUTICAL PREPARATIONS BY REVERSED-PHASE HIGH-PERFORMANCE LIQUID-CHROMATOGRAPHY
Cavazzutti, C.; Gagliardi, L.; Amato, A.; Gattavecchia, E.; Tonelli, D.
Ist. Super. Sanita Rome Italy IT
J. Chromatogr.; (82) P 391-5; Vol 244; No 2

26-82-0083 (0083)
SEPARATION AND QUANTIFICATION OF BASIC HYDROXYCINNAMIC AMIDES AND HYDROXYCINNAMIC ACIDS BY REVERSED-PHASE HIGH-PERFORMANCE LIQUID CHROMATOGRAPHY
Ponchet, Michel; Martin-Tanguy, Josette; Poupet, Alain; Marais, Antoine; Beck, Daniel
Stn. Bot. Pathol. Inst. Natl. Rech. Agron. Antibes Fr. FR 06602
J. Chromatogr.; (82) P 397-404; Vol 240; No 2

26-82-0084 (0084)
HIGH-PERFORMANCE LIQUID CHROMATOGRAPHY OF NAPHTHYLURETHANES WITH FLUORESCENCE DETECTION
Wintersteiger, R.; Wenninger-Weinzierl, G.; Pacha, W.
Inst. Pharm. Chem. Univ. Graz Graz Austria AT A-8010
J. Chromatogr.; (82) P 399-406; Vol 237; No 3

26-82-0085 (0085)
HIGH-PERFORMANCE LIQUID-CHROMATOGRAPHIC DETERMINATION OF 2,4- AND 2,6-TOLUENE-DIAMINE IN AQUEOUS EXTRACTS
Snyder, Roger C.; Breder, Charles V.
Div. Chem. Phys. Food and Drug Adm. Washington USA US DC 20204
J. Chromatogr.; (82) P 429-40; Vol 236; No 2

26-82-0086 (0086)
SYSTEMS FOR THE SEPARATION OF METABOLITES OF THE CARCINOGEN, N-2-FLUORENYL-ACETAMIDE, BY HIGH-PERFORMANCE LIQUID CHROMATOGRAPHY
Ryzewski, C. N.; Malejka-Giganti, D.
Dep. Lab. Med. Pathol. Univ. Minnesota Minneapolis USA US MN 55417
J. Chromatogr.; (82) P 447-56; Vol 237; No 3

26-82-0087 (0087)
HIGH-PERFORMANCE LIQUID CHROMATOGRAPHIC METHOD FOR THE DETERMINATION OF CHLORHEXIDINE
Huston, Carolyn E.; Wainwright, Paul; Cooke, Michael; Simpson, Rosemary
Dep. Chem. Univ. Bristol Bristol UK GB BS8 1TS
J. Chromatogr.; (82) P 457-64; Vol 237; No 3

26-82-0088 (0088)
SEPARATION OF CIS- AND TRANS-3,3'-BIS(TRIMETHYLAMMONIOMETHYL)AZOBENZENE SALTS (BIS Q SALTS) BY REVERSED-PHASE HIGH-PERFORMANCE LIQUID CHROMATOGRAPHY
Delcour, Anne; Cash, Derek J.; Erlanger, Bernard F.; Hess, George P.
Sect. Biochem., Mol. Cell Biol. Cornell Univ. Ithaca USA US NY
J. Chromatogr.; (82) P 461-8; Vol 248; No 3

26-82-0089 (0089)
LIQUID CHROMATOGRAPHY STUDY OF BROMINATED ANILINES AND INVESTIGATION OF PRODUCT FORMATION IN THE BROMINATION REACTION. II. ANILINES WITH ALKYL GROUPS IN THE META-POSITION
Truedsson, L. A.
Chem. Cent. Lund Inst. Technol. Lund Swed. SE S-220 07/7
J. Chromatogr.; (82) P 47-56; Vol 234; No 1

26-82-0090 (0090)
SEPARATION OF SATURATED, MONOUNSATURATED AND DIUNSATURATED ALDEHYDES AS 2,4-DINITROPHENYLHYDRAZONES USING HIGH-PERFORMANCE LIQUID CHROMATOGRAPHY AT INCREASED TEMPERATURE
Reindl, B.; Stan, H. J.
Inst. Sozialmed. Epidemiol. Bundesgesundheitsamt Berlin Fed. Rep. Ger. E 1000/33
J. Chromatogr.; (82) P 481-8; Vol 235; No 2

26-82-0091 (0091)
ANALYSIS OF UNQUENCHED REACTION MIXTURES OF CHLORINE DIOXIDE AND PHENOLS BY REVERSED-PHASE HIGH-PERFORMANCE LIQUID CHROMATOGRAPHY
Brueggemann, Ernst; Waj'n, J. Edmund; Wade, Clarence W. R.; Burrows, Elizabeth P.
U.S. Army Medical Bioeng. Res. Dev. Lab. Frederick USA US MD 21701
J. Chromatogr.; (82) P 484-8; Vol 237; No 3

26-82-0092 (0092)
HIGH-PERFORMANCE LIQUID CHROMATOGRAPHIC ANALYSIS OF THE IN VIVO METABOLITES OF (14C)-PYRIDINE
Shaker, M. S.; Crooks, P. A.; Damani, L. A.
Dep. Pharm. Univ. Manchester Manchester UK GB ML3 9PL
J. Chromatogr.; (82) P 489-95; Vol 237; No 3

26-82-0093 (0093)
ANALYSIS OF ISOFLAVONES IN BENGAL GRAM BY HIGH-PERFORMANCE LIQUID CHROMATOGRAPHY
Dziedzic, S. Z.; Dick, J.
Dep. Food Sci. Univ. Reading Reading UK GB RG1 5AQ
J. Chromatogr.; (82) P 497-9; Vol 234; No 2

26-82-0094 (0094)
ANALYSIS OF OLIVETOL IN RABBIT SERUM BY HIGH-PERFORMANCE LIQUID CHROMATOGRAPHY
Rauls, D.; Penney, L. L.
Dep. Clin. Invest. William Beaumont Army Med. Cent. El Paso USA US TX 79920
J. Chromatogr.; (82) P 500-2; Vol 234; No 2

26-82-0095 (0095)
DETERMINATION OF 2,4,7-TRINITRO-9-FLUORENONE IN WORKPLACE ENVIRONMENTAL SAMPLES USING HIGH-PERFORMANCE LIQUID CHROMATOGRAPHY
Seymour, Martha J.
Div. Phys. Sci. Eng. Natl. Inst. Occup. Saf. Health Cincinnati USA US OH 45226
J. Chromatogr.; (82) P 530-4; Vol 236; No 2

26-82-0096 (0096)
HIGH-PERFORMANCE LIQUID CHROMATOGRAPHY OF 6-CAPROLACTAM AND ITS CYCLIC OLIGOMERS PRESENT IN POLYAMIDE 6
Krajnik, Vaclav; Bozek, Peter; Kondelikova, Jaroslava; Kralicek, Jaroslav
State Hosp. Bratislava Czech. CS 821 24
J. Chromatogr.; (82) P 539-42; Vol 240; No 2

26-82-0097 (0097)
SILICA GEL LIQUID-LIQUID CHROMATOGRAPHY USING AQUEOUS BINARY PHASE SYSTEMS. HIGH-EFFICIENCY EXTRACTION AND RESOLUTION OF PHENOLS AND CARBOXYLIC ACIDS
Hara, Shoji; Dobashi, Yasuo; Oka, Kitaro
Tokyo Coll. Pharm. Hachioji Japan JP 192-03
J. Chromatogr.; (82) P 677-85; Vol 239

26-82-0098 (0098)
DETERMINATION OF ANTHRANILATE, 3-HYDROXYANTHRANILATE AND 2,3-DIHYDROXYBENZOATE IN THE ENZYME ASSAY OF ANTHRANILATE HYDROXYLASE BY THIN-LAYER AND HIGH-PERFORMANCE LIQUID CHROMATOGRAPHY
Schott, Hans Henning; Langenbach-Schmidt, Doris; Zinke, Karen
Biochem. Inst. Justus Liebig-Univ. Giessen Giessen Fed. Rep. Ger. DE D-6300
J. Chromatogr.; (82) P 85-91; Vol 245; No 1

26-82-0099 (0099)
HIGH-PERFORMANCE LIQUID CHROMATOGRAPHY OF PRENYLQUINONES, PRENYLVITAMINS AND PRENOLS
Prenzel, Ursula; Lichtenthaler, Hartmut K.
Bot. Inst. Univ. Karlsruhe Karlsruhe Fed. Rep. Ger. DE D-7500
J. Chromatogr.; (82) P 9-19; Vol 242; No 1

26-82-0100 (0100)
EXAMINATION OF BALL PEN INK BY HIGH-PRESSURE LIQUID CHROMATOGRAPHY
Lyter, A. H., III
Bur. Alcohol, Tobacco Firearms Rockville USA US MD
J. Forensic Sci.; (82) P 154-60; Vol 27; No 1

26-82-0101 (0101)
SEPARATION OF BACTERIAL MENAQUINONES BY HPLC USING REVERSE PHASE (RS18) AND A SILVER LOADED ION EXCHANGER AS STATIONARY PHASES
Kroppenstedt, Reiner M.
Dtsch. Sammlung von Mikroorg. Ges. Biotechnol. Forsch. m.b.H. Braunschweig Darmstadt Fed. Rep. Ger. DE D-6100
J. Liq. Chromatogr.; (82) P 2359-67; Vol 5; No 12

26-82-0102 (0102)
HPLC ANALYSIS OF THE ISOMERIC THIOETHER METABOLITES OF STYRENE OXIDE
Hernandez, O.; Yagen, B.; Cox, R. H.; Bend, J. R.; McKinney, J. D.
Natl. Inst. Environ. Health Sci. Research Triangle Park USA US NC 27709

J. Liq. Chromatogr.; (82) P 345-65; Vol 5; No 2

26-82-0103 (0103)
LIQUID CHROMATOGRAPHIC DETERMINATION OF AZIDE AS THE 3,5-DINITROBENZOYL DERIVATIVE
Swarin, Stephen J.; Waldo, Richard A.
Anal. Chem. Dep. Gen. Motors Res. Lab. Warren USA US MI 48090
J. Liq. Chromatogr.; (82) P 597-604; Vol 5; No 4

26-82-0104 (0104)
DETERMINATION OF CARBONYL 2,4-DINITROPHENYLHYDRAZONES BY LIQUID CHROMATOGRAPHY/ ELECTROCHEMISTRY
Jacobs, W. A.; Kissinger, P. T.
Dep. Chem. Purdue Univ. West Lafayette USA US IN 47907
J. Liq. Chromatogr.; (82) P 669-76; Vol 5; No 4

26-82-0105 (0105)
AN HPLC METHOD FOR THE ANALYSIS OF GLYCEROL AND OTHER GLYCOLS IN SOAP
George, E. D.; Acquaro, J. A.
Bradford Lab. Orig. Bradford Soap Works, Inc. West Warwick USA US RI 02893
J. Liq. Chromatogr.; (82) P 927-38; Vol 5; No 5

26-82-0106 (0106)
A NOTE ON THE SEPARATION OF NATURAL MIXTURES OF BACTERIAL METAQUINONES USING REVERSED-PHASE HIGH-PERFORMANCE LIQUID CHROMATOGRAPHY
Collins, M. D.
Dep. Microbiol., Natl. Inst. Res.Dairy.,Shinfield/Reading UK RG2 9AT
J.Appl. Bacteriol. (82) P 457-60; Vol 52; No 3

26-82-0107 (0107)
HIGH-PERFORMANCE LIQUID CHROMATOGRAPHIC DETERMINATION OF HYDROLYZED ACID ANHYDRIDES
Mitsubishi Electric Corp. Japan JP
Jpn. Kokai Tokkyo Koho; (021982) P 4 pp.; In Japan; Pat. No.: 82 30947; Int. Class: G01N-031/08; Doc. Code: A2; Appl.: JP 80/108569 X00804

26-82-0108 (0108)
CONJUGATED DIENES OF CRUDE SOY OIL: DETECTION BY UV SPECTROPHOTOMETRY AND SEPARATION BY HPLC
Brown, H. G.; Snyder, H. E.
Dep. Hortic. Food Sci. Univ. Arkansas Fayetteville USA US AR 72701
JAOCS, J. Am. Oil Chem. Soc.; (82) P 280-3; Vol 59; No 7

26-82-0109 (0109)
SEPARATION OF PIGMENTS BY CENTRIFUGAL LIQUID CHROMATOGRAPHY
Utsugi, Hiroshi; Endo, Atsushi; Suzuki, Noboru; Umemura, Mitsuhiko
Coll. Eng. Utsunomiya Univ. Utsunomiya Japan JP
Kagaku Kogaku; (82) P 241-3; Vol 46; No 4; In Japan

26-82-0110 (0110)
DETERMINATION OF THE CONTENT OF ANTIFREEZE ADDITIVES BY A METHOD OF INDICATOR-LIQUID CHROMATOGRAPHY
Zrelov, V. N.; Krasnaya, L. V.; Piskunov, V. A.; Popleteev, S. I.
USSR SU
Khim. Tekhnol. Topl. Masel; (82) P 37-8; No 6; In Russ

26-82-0111 (0111)
HPLC DETERMINATION OF 2-MERCAPTOBENZOTHIAZOLE IN RUBBER BABY BOTTLE NIPPLES
Blosczyk, Guenter; Doemling, Hans Juergen
Landesuntersuchungsamt Gesundheitswesen Nordbayern Erlangen Fed. Rep. Ger. DE

8520
Lebensmittelchem. Gerichtl. Chem.; (82) P 90; Vol 36; No 4; In Ger

26-82-0112 (0112)
HIGH-PERFORMANCE LIQUID CHROMATOGRAPHY OF POLYTETRAMETHYLENE ETHER GLYCOLS
Andrews, G. D.; Vatvars, A.; Pruckmayr, G.
Cent. Res. Dev. Dep. E. I. du Pont de Nemours and Co., Inc. Wilmington USA US DE 19898
Macromolecules; (82) P 1580-3; Vol 15; No 6

26-82-0113 (0113)
HPLC/GC DETERMINATION OF AETT IN COSMETICS
Rooselaar, J.; Liem, D. H.
Keuringsdienst Van Waren Enschede Neth. NL 7500 AT
Mitt. Geb. Lebensmittelunters. Hyg.; (82) P 468-74; Vol 73; No 4

26-82-0114 (0114)
NEW QUANTITATIVE METHOD FOR THE DETERMINATION OF UV STABILITY IN VARIOUS LC + DYE MIXTURES
Veghely, T.
Ind. Res. Inst. Electron. Budapest Hung. HU
Mol. Cryst. Liq. Cryst.; (82) P 35-50; Vol 80; No 1-4

26-82-0115 (0115)
LIQUID CHROMATOGRAPHIC ANALYSIS OF MIXTURES OF HIGHER ALKYLNAPHTHALENES
Aratskova, A. A.; Ageev, A. N.; Anpilogov, Yu. E.; Alekseeva, T. V.; Lipshtein, A. R.; Margulis, M. A.; Yashin, Ya. I.
Vses. Nauchno-Issled. Inst. Org. Sint. Moscow USSR SU
Neftekhimiya; (82) P 309-11; Vol 22; No 3; In Russ

26-82-0116 (0116)
IMPROVING THE RELIABILITY OF A METHOD OF LIQUID ADSORPTION CHROMATOGRAPHY ANALYSIS OF OIL-ALKYLPHENOL-CALCIUM ALKYLSALICYLATE ADDITIVES
Kulikov, A. A.; Sukhoverkhov, V. D.
Novokuibyshev. Neftepererab. Zavod Novokuibyshev USSR SU
Neftepererab. Neftekhim. (Moscow); (82) P 19-20; No 7; In Russ

26-82-0117 (0117)
A STUDY OF THE FATE OF (14C)MALEIC HYDRAZIDE IN WHITE ASH AND BLACK LOCUST SEEDLINGS BY HIGH-PERFORMANCE LIQUID CHROMATOGRAPHY
Mischke, Charles F.; Domir, Subhash C.
Dep. Bot. Bacteriol. Ohio Wesleyan Univ. Delaware USA US OH 43015
Pestic. Sci.; (82) P 304-8; Vol 13; No 3

26-82-0118 (0118)
DETERMINATION OF ANTHRACYCLINES IN FERMENTATION BROTHS BY HPLC
Alemanni, A.; Breme, U.; Vigevani, A.
Anal. Dep. Farmitalia Carlo Erba S.p.A. Milan Italy IT 20146
Process Biochem.; (82) P 9-12; Vol 17; No 3

26-82-0119 (0119)
HPLC SEPARATION OF PATULIN FROM 5-(HYDROXYMETHYL)FURFUROL
Ruggieri, G.; Ruggeri, P.
Ist. Merceol. Univ. Napoli Naples Italy IT
Rass. Chim.; (82) P 121-4; Vol 34; No 3; In Ital

26-82-0120 (0120)
DETERMINATION OF DIPHENYLAMINE AND VARIOUS NITRATED DIPHENYLAMINES BY REVERSE-PHASE HPLC
Richardson, B. R.

Mason and Hanger-Silas Mason Co., Inc. Amarillo USA US TX
From: Energy Res. Abstr. 1982, 7(17), Abstr. No. 44598. TECHNICAL REPORT; Report; (82) P 14 pp.; No MHSMP-82-11; Order No. DE82008769,; Avail.: NTIS

26-82-0121 (0121)
DETERMINATION OF OLIVE OIL PHENOLIC COMPOUNDS BY GLC AND HPLC: HYPOTHETICAL ROLE OF TYROSOL IN DETERMINING VIRGIN OLIVE OIL IN MIXTURES WITH REFINED OILS
Solinas, M.; Cichelli, A.
Ist. Sper. Elaiotec. Pescara Italy IT
Riv. Soc. Ital. Sci. Aliment.; (82) P 223-30; Vol 11; No 4; In Ital

26-82-0122 (0122)
HPLC ANALYSIS IN THE PROCESS OF ENZYMIC SYNTHESIS OF CEPHALEXIN
Wang, Yumei; Wang, Zhenxiang; Yue, Huaai; Jiao, Qinghua; Deng, Liru
Inst. Biophys. Chinese Acad. Sci. Beijing Peop. Rep. China CN
Shengwu Huaxue Yu Shengwu Wuli Xuebao; (82) P 547-52; Vol 14; No 6; In Ch

26-82-0123 (0123)
DETERMINATION OF NITROFURANS IN CULTURED FISHES BY HIGH-PERFORMANCE LIQUID CHROMATOGRAPHY
Nagata, Tomoko; Niyamoto, Fumio; Saeki, Masanobu
Public Health Lab., Chiban, Japan, 280
Shokuhin Eisagaku Zasshi (82) P 278-82; Vol 23; No 3; In Japan

26-82-0124 (0124)
ANALYSIS AND ISOLATION OF PHENOLICS FROM PLANTS BY HPLC: POSSIBILITIES AND LIMITS
Tittel, G.; Wagner, H.
Inst. Pharm. Biol. Univ. Muenchen Munich Fed. Rep. Ger. DE D-8000/2
Stud. Org. Chem. (Amsterdam); (82) P 299-310; Vol 11; No Flavonoids Bioflavonoids

26-82-0125 (0125)
AN IMPROVED EXTRACTION-HPLC METHOD FOR TOBACCO POLYPHENOLS
Snook, M. E.; Chortyk, O. T.
Tobacco Saf. Res. Unit Agric. Res. Serv. Athens USA US GA 30613
Tob. Int.; (82) P 111-15; Vol 184; No 5

26-82-0126 (0126)
IDENTIFICATION OF 1-METHYLTHIOUREA AS THE METABOLITE OF ETHYLENETHIOUREA IN RATS BY HIGH-PERFORMANCE LIQUID CHROMATOGRAPHY
Kobayashi, Hiroko; Kaneda, Masahiro; Teramoto, Shoji
Chem. Div. Inst. Environ. Toxicol. Tokyo Japan JP 187
Toxicol. Lett.; (82) P 109-14; Vol 12; No 2-3

26-82-0127 (0127)
EVALUATION OF A MASS DETECTOR FOR HPLC DETERMINATION OF ORGANIC COMPOUNDS IN WATER
Connor, K. J.; Waggott, A.
Stevenage Lab. Water Res. Cent. Stevenage/Herts. UK GB SG1 1TH
TECHNICAL REPORT; Comm. Eur. Communities, (Rep.) EUR; (82) P 178-87; No EUR 7623, Anal. Org. Micropollut. Water

26-82-0128 (0128)
DETERMINATION OF PHENOLS IN WATER BY HPLC
Sonneborn, M.; Pabel, E.; Schwabe, R.
Fed. Health Office Inst. Water-, Soil Air Hyg. Berlin Fed. Rep. Ger. DE
TECHNICAL REPORT; Comm. Eur. Communities, (Rep.) EUR; (82) P 335-8; No EUR 7623, Anal. Org. Micropollut. Water

26-82-0129 (0129)
RETENTION BEHAVIOR OF SOME 4'-ALKANOYLBENZO-15-CROWN-5 DERIVATIVES IN REVERSED-PHASE HIGH-PERFORMANCE LIQUID CHROMATOGRAPHY (HPLC)
Stein, Johann; Meerbote, Michael; Beger, Joerg
Sekt. Chem. Bergakad. Freiberg Freiberg Ger. Dem. Rep. DD DDR-9200
Z. Chem.; (82) P 146-7; Vol 22; No 4; In Ger

26-82-0130 (0130)
DETERMINATION OF METABOLITES OF AROMATIC HYDROCARBONS BY HIGH-PERFORMANCE LIQUID CHROMATOGRAPHY
Bergert, K. D.; Zehner, C.; Voigt, H.; Maehrlein, W.
Arbeitshyg. Zent. Niederdorf Ger. Dem. Rep. DD DDR-9154
Z. Gesamte Hyg. dhre Grenzgeb.; (82) P 521-4; Vol 28; No 8; In Ger

26-82-0131 (0131)
DETERMINATION OF QUININE IN SOFT DRINKS AND DRUGS BY HIGH-PERFORMANCE LIQUID CHROMATOGRAPHY WITH ELECTROCHEMICAL DETECTION
Kral, Karl; Sontag, Gerhard
Inst. Anal. Chem. Univ. Wien Vienna Austria AT A-1090
Z. Lebensm.-Unters. Forsch.; (82) P 22-4; Vol 175; No 1; In Ger

26-82-0132 (0132)
SENSITIVE DETERMINATION OF p-NITROPHENOL IN BLOOD AND URINE BY HPLC AFTER E 605 INTOXICATION
Michalke, P.
Inst. Rechtsmed. Univ. Duesseldorf Duesseldorf Fed. Rep. Ger. DE D-4000
Z. Rechtsmed.; (82) P 195-202; Vol 88; No 3; In Ger

CHAPTER 7.27
Pharmaceuticals

7.27.1
Generalities

--- 1982 ---

27-82-0001 (0001)
THE USE OF HIGH PRESSURE LIQUID CHROMATOGRAPHY IN UNDERGRADUATE PHARMACY EXPERIMENTS
Houglum, Joel E.; Cascella, Peter J.
Coll. Pharm. South Dakota State Univ. Brookings USA US SD 57007
Am. J. Pharm. Educ.; (82) P 32-4; Vol 46; No 1

27-82-0002 (0002)
CRITERIA FOR THE EVALUATION OF THE PURITY OF STEROIDS FOR PHARMACEUTICAL PREPARATIONS BY MEANS OF HPLC
Cavina, G.; Moretti, G.; Alimenti, R.; Gallinella, B.; Porra, R.; Valvo, L.
Ist. Super. Sanita Fellow Ist. Super. Sanita Rome Italy IT
Anal. Chem. Symp. Ser.; (82) P 343-60; Vol 10; No Adv. Steroid Anal.

27-82-0003 (0003)
DETERMINATION OF SULFA DRUGS IN BIOLOGICAL FLUIDS BY LIQUID CHROMATOGRAPHY/MASS SPECTROMETRY/MASS SPECTROMETRY
Henion, Jack D.; Thomson, Bruce A.; Dawson, Peter H.
Coll. Vet. Med. Cornell Univ. Ithaca USA US NY 14853
Anal. Chem.; (82) P 451-6; Vol 54; No 3

27-82-0004 (0004)
APPLICATION OF HIGH-PERFORMANCE LIQUID CHROMATOGRAPHY IN THE BIOLOGICAL ANALYSIS OF DRUGS
Schill, Goeran
Dep. Anal. Pharm. Chem. Univ. Uppsala Uppsala Swed. SE S-751 23
Anal. Proc. (London); (82) P 163-6; Vol 19; No 4

27-82-0005 (0005)
LIQUID CHROMATOGRAPHY IN DRUG PRE-FORMULATION AND FORMULATION SCIENCE
Van Rooij, H. H.; Tomlinson, E.
Subfac. Pharm. Univ. Amsterdam Amsterdam Neth. NL 1018 TV
Anal. Proc. (London); (82) P 174-9; Vol 19; No 4

27-82-0006 (0006)
SOME APPLICATIONS OF LIQUID CHROMATOGRAPHY TO THE ANALYSIS OF MEDICINAL PRODUCTS
Hopkins, R. G.
Med. Prod. Sub-Div. Lab. Gov. Chem. London UK GB SE1 9NQ
Anal. Proc. (London); (82) P 193-5; Vol 19; No 4

27-82-0007 (0007)
LC-MS DETECTION OF STEROIDS AND DRUGS AT LOW AND MEDIUM RESOLUTION
Baty, J. D.; Willis, R. G.
Dep. Biochem. Med. Ninewells Hosp. Med. Sch. Dundee UK GB DD1 9SY
Anal. Proc. (London); (82) P 251-3; Vol 19; No 5

27-82-0008 (0008)
ASSAY METHODS OF VITAMIN A. 3. ASSAY METHODS OF VITAMIN A IN PHARMACEUTICAL PREPARATIONS, FOODS AND FEEDS BY HIGH-PERFORMANCE LIQUID CHROMATOGRAPHY
Kastui, Goichiro
Eisai Co. Ltd. Tokyo Japan JP 112
Bitamin; (82) P 151-6; Vol 56; No 3; In Japan

27-82-0009 (0009)
THE INCREASING IMPORTANCE AND CAPACITY OF LIQUID CHROMATOGRAPHY IN DRUG MONITORING AND ANALYTICAL TOXICOLOGY
Uges, Donald R. A.; Bouma, Piet

Dep. Pharm. State Univ. Hosp. Groningen Neth. NL
Chromatogr. Sci.; (82) P 145-60; Vol 20; No Biol./Biomed. Appl. Liq.
Chromatogr. 4

27-82-0010 (0010)
HPLC WITH UV-VIS SPECTROPHOTOMETRIC DETECTION FOR BLOOD AND TISSUE DRUG ASSAYS
Desager, J. P.; Sclavons, M.
Lab. Pharmacother. Univ. Cathol. Louvain Brussels Belg. BE B-1200
Chromatographia; (82) P 451-2; Vol 15; No 7

27-82-0011 (0011)
ON-LINE DIRECT LIQUID INTRODUCTION INTERFACE FOR MICRO-LIQUID CHROMATOGRAPHY/MASS SPECTROMETRY: APPLICATION TO DRUG ANALYSIS
Eckers, Christine; Skrabalak, Dale S.; Henion, Jack
New York State Coll. Vet. Med. Cornell Univ. Ithaca USA US NY 14853
Clin. Chem. (Winston-Salem, N. C.); (82) P 1882-6; Vol 28; No 9

27-82-0012 (0012)
SCREENING FOR UNKNOWN COMPOUNDS BY HPLC USING A MULTICHANNEL DIODE ARRAY UV-VIS SPECTROPHOTOMETER: APPLICATION TO DRUG METABOLISM
Ghijsen, R. T.; Drenth, B. H. F.; Overzet, F.; De Zeeuw, R. A.
Dep. Toxicol. State Univ. Groningen Groningen Neth. NL 9713 AW
HRC CC, J. High Resolut. Chromatogr. Chromatogr. Commun.; (82) P 192-8; Vol 5; No 4

27-82-0013 (0013)
METHODOLOGY FOR ISOLATING DRUG METABOLITES FROM BIOLOGICAL FLUIDS BY HPLC
Overzet, F.
Dep. Toxicol. State Univ. Groningen Groningen Neth. NL 9713 AW
HRC CC, J. High Resolut. Chromatogr. Chromatogr. Commun.; (82) P 604-15; Vol 5; No 11

27-82-0014 (0014)
OFF-LINE COMBINATION OF LIQUID CHROMATOGRAPHY AND FIELD DESORPTION MASS SPECTROMETRY: PRINCIPLES AND ENVIRONMENTAL, MEDICAL, AND PHARMACEUTICAL APPLICATIONS
Schulten, H. R.
Inst. Chem. Univ. Bonn Bonn Fed. Rep. Ger. DE 5300/1
J. Chromatogr.; (82) P 105-28; Vol 251; No 2

27-82-0015 (0015)
LIQUID CHROMATOGRAPHY-MASS SPECTROMETRY IN THE PHARMACEUTICAL INDUSTRY: OBJECTIVES AND NEEDS
Erni, F.
Pharm. Dep. Sandoz Ltd. Basel Switz. CH
J. Chromatogr.; (82) P 141-51; Vol 251; No 2

27-82-0016 (0016)
HIGH-PERFORMANCE LIQUID CHROMATOGRAPHIC ANALYSIS OF BASIC DRUGS ON SILICA COLUMNS USING NONAQUEOUS IONIC ELUENTS
Flanagan, R. J.; Storey, G. C. A.; Bhamra, R. K.; Jane, I.
Poisons Univ. Guy's Hosp. London UK GB SE1 9RT
J. Chromatogr.; (82) P 15-37; Vol 247; No 1

27-82-0017 (0017)
SEPARATION AND QUANTITATION OF POLYNITRATE ESTERS IN PHARMACEUTICAL PREPARATIONS BY REVERSED-PHASE HIGH-PERFORMANCE LIQUID-CHROMATOGRAPHY
Cavazzutti, C.; Gagliardi, L.; Amato, A.; Gattavecchia, E.; Tonelli, D.
Ist. Super. Sanita Rome Italy IT
J. Chromatogr.; (82) P 391-5; Vol 244; No 2

27-82-0018 (0018)
HIGH PRESSURE LIQUID CHROMATOGRAPHIC DETERMINATION OF PARABENS IN PHARMACEU-TICAL PREPARATIONS CONTAINING HYDROXYQUINOLINES
Padmanabhan, G. R.; Smith, J.; Mellish, N.; Fogel, G.
Pharm. Div. Ciba-Geigy Corp. Suffern USA US NY 10901
J. Liq. Chromatogr.; (82) P 1357-66; Vol 5; No 7

27-82-0019 (0019)
DETERMINATION OF AMINE INGREDIENTS IN COUGH-COLD LIQUIDS BY REVERSED-PHASE ION-PAIR HIGH-PERFORMANCE LIQUID CHROMATOGRAPHY
Halstead, G. W.
Control Anal. Res. Dev. Upjohn Co. Kalamazoo USA US MI 49001
J. Pharm. Sci.; (82) P 1108-12; Vol 71; No 10

27-82-0020 (0020)
RELATION BETWEEN HPLC PARAMETER log k' AND BIOLOGICAL ACTIVITY OF SULFA DRUGS
Zhang, Shengqiang; An, Dengkui; Wu, Rujin; Sheng, Longsheng
Dep. Pharm. Anal. Nanjing Coll. Pharm. Nanjing Peop. Rep. China CN
Nanjing Yaoxueyuan Xuebao; (82) P 1-9; No 2; In Ch

27-82-0021 (0021)
QUANTITATIVE CALIFORNIUM-252 PLASMA DESORPTION MASS SPECTROMETRY FOR PHARMACEUTICALS. A NEW APPROACH TO COUPLING LIQUID CHROMATOGRAPHY WITH MASS SPECTROMETRY
Jungclas, Hartmut; Danigel, Harald; Schmidt, Lothar
Kernchem. Philipps-Univ. Marburg Fed. Rep. Ger. DE D-3550
Org. Mass Spectrom.; (82) P 86-90; Vol 17; No 2

27-82-0022 (0022)
THERAPEUTIC DRUG MONITORING BY HIGH-PERFORMANCE LIQUID CHROMATOGRAPHY
Goto, Junichi
Pharm. Inst. Tohoku Univ. Sendai Japan JP
Rinsho Byori; (82) P 384-9; Vol 30; No 4; In Japan

27-82-0023 (0023)
APPLICATION OF HIGH-PERFORMANCE LIQUID CHROMATOGRAPHY IN PHARMACEUTICAL ANALYSIS
Yang, Zhongyuan; Zhou, Yuanyao; Xu, Ruzheng
Wuhan Inst. Drug Control Wuhan Peop. Rep. China CN
Yaowu Fenxi Zazhi; (82) P 182-92; Vol 2; No 3; In Ch

27-82-0024 (0024)
DETERMINATION OF QUININE IN SOFT DRINKS AND DRUGS BY HIGH-PERFORMANCE LIQUID CHROMATOGRAPHY WITH ELECTROCHEMICAL DETECTION
Kral, Karl; Sontag, Gerhard
Inst. Anal. Chem. Univ. Wien Vienna Austria AT A-1090
Z. Lebensm.-Unters. Forsch.; (82) P 22-4; Vol 175; No 1; In Ger

7.27.2
Alkaloids and Drugs of Abuse

--- 1982 ---

27-82-0001 (0001)
A STABILITY STUDY OF AQUEOUS SOLUTIONS OF DIAMORPHINE AND MORPHINE USING HIGH-PERFORMANCE LIQUID CHROMATOGRAPHY
Beaumont, I. M.
Pharm. Dep. East Birmingham Hosp. Birmingham UK GB 9
Anal. Proc. (London); (82) P 128-31; Vol 19; No 3

27-82-0002 (0002)
STUDIES OF ERGOT ALKALOIDS USING HIGH-PERFORMANCE LIQUID CHROMATOGRAPHY-MASS SPECTROMETRY AND MASS SPECTROMETRY-MASS SPECTROMETRY
Eckers, Christine; Games, David E.; Mallen, David N. B.; Swann, Brian P.
Dep. Chem. Univ. Coll. Cardiff Cardiff UK GB CF1 1XL
Anal. Proc. (London); (82) P 133-7; Vol 19; No 3

27-82-0003 (0003)
ACETYLPROCAINE IN HEROIN PREPARATIONS. DETECTION AND QUANTIFICATION BY HIGH-PERFORMANCE LIQUID CHROMATOGRAPHY
Bernhauer, Dietmar; Fuchs, Ernst F.
Hess. Landeskrim. Wiesbaden Fed. Rep. Ger. DE 6200
Arch. Kriminol.; (82) P 25-30; Vol 169; No 1-2; In Ger

27-82-0004 (0004)
STUDIES OF ERGOT ALKALOIDS USING HIGH-PERFORMANCE LIQUID CHROMATOGRAPHY-MASS SPECTROMETRY AND B/E LINKED SCANS
Eckers, Christine; Games, David E.; Mallen, David N. B. Swann, Brian P.
Dep. Chem. Univ. Coll. Cardiff Cardiff UK GB CF1 1XL
Biomed. Mass Spectrom.; (82) P 162-73; Vol 9; No 4

27-82-0005 (0005)
REVERSE PHASE HIGH-PERFORMANCE LIQUID CHROMATOGRAPHIC METHOD FOR DETERMINATION OF CAFFEINE IN DECAFFEINATED COFFEE
Ferrara, Antonio; Reale, Patrizia; Grazia Calaminici, Maria; Iaccarino, Tommasina
Lab. Chim. Prov. Caserta Italy IT 81100
Boll. Chim. Unione Ital. Lab. Prov., Parte Sci.; (82) P 55-60; Vol 33; No S1; In Ital

27-82-0006 (0006)
LIQUID CHROMATOGRAPHY OF CODEINE IN PLASMA WITH FLUORESCENCE DETECTION
Tna, Irene W.; Fass, Marian; Debban, Jennifer A.; Matin, Shaikh B.
Dep. Anal. Metab. Chem. Syntex Res. Palo Alto USA US CA 94304
Clin. Chem. (Winston-Salem, N. C.); (82) P 1137-9; Vol 28; No 5

27-82-0007 (0007)
SMOKING-INDUCED CHANGES IN NICOTINE DISPOSITION: APPLICATION OF A NEW HPLC ASSAY FOR NICOTINE AND ITS METABOLITES
Kyerematen, Gabriel A.; Damiano, Michael D.; Dvorchik, Barry H.; Vesell, Elliot S.
Coll. Med. Pennsylvania State Univ. Hershey USA US PA 17033
Clin. Pharmacol. Ther. (St. Louis); (82) P 769-80; Vol 32; No 6

27-82-0008 (0008)
DETERMINATION OF METHYL AND ETHYL CAFFEATE IN VEGETABLES BY HPLC WITH ELECTROCHEMICAL DETECTION
Sontag, G.; Schaefers, F. I.; Herrmann, K.
Inst. Anal. Chem. Univ. Vienna Vienna Austria AT
CONFERENCE; (82); Recent Dev. Food Anal., Proc. Eur. Conf. Food Chem., 1st; P 41-6; Avail.: Baltes, Werner; Czedik-Eysenberg, Peter B.; Pfannhauser, Werner;

Date: 81; Publisher: Verlag Chem., Weinheim, Fed. Rep. Ger.

27-82-0009 (0009)
QUALITY CONTROL OF VALERIAN PREPARATIONS. 2. VALERENIC ACID AND VALERENAL AS CHARACTERISTIC COMPONENTS OF VALERIANA OFFICINALIS. DETERMINATION BY HPLC
Haensel, Rudolf; Schulz, Jutta
Inst. Pharmakogn. Phytochem. Freie Univ. Berlin Berlin Fed. Rep. Ger. DE 1000
Dtsch. Apoth.-Ztg.; (82) P 215-19; Vol 122; No 5; In Ger

27-82-0010 (0010)
HIGH PERFORMANCE LIQUID CHROMATOGRAPHIC DETERMINATION OF CAFFEINE AND SODIUM BENZOATE
Kozu, Tadashi
Sci. Invest. Res. Lab. Nagano Pref. Police Headquarters Nagano Japan JP
Eisei Kagaku; (82) P 111-13; Vol 28; No 2; In Japan

27-82-0011 (0011)
STABILITY OF COCAINE HYDROCHLORIDE SOLUTIONS AT VARIOUS pH VALUES AS DETERMINED BY HIGH-PRESSURE LIQUID CHROMATOGRAPHY
Das Gupta, V.
Coll. Pharm. Univ. Houston Houston USA US TX 77030
Int. J. Pharm.; (82) P 249-57; Vol 10; No 3

27-82-0012 (0012)
LIQUID CHROMATOGRAPHIC IDENTIFICATION OF CIS- AND TRANS-CINNAMOYLCOCAINE IN ILLICIT COCAINE
Noggle, F. Taylor, Jr.; Clark, C. Randall
Alabama Dep. Forensic Sci. Auburn USA US AL 36830
J. Assoc. Off. Anal. Chem.; (82) P 756-61; Vol 65; No 3

27-82-0013 (0013)
DETERMINATION OF MORPHINE IN CEREBROSPINAL FLUID AND PLASMA BY HIGH-PERFORMANCE LIQUID CHROMATOGRAPHY WITH ELECTROCHEMICAL DETECTION
Todd, Richard D.; Muldoon, Sheila M.; Watson, Robert L.
Dep. Anesthesiol. Uniformed Serv. Univ. Health Sci. Bethesda USA US MD 20014
J. Chromatogr.; (82) P 101-10; Vol 232; No 1

27-82-0014 (0014)
HIGH-PERFORMANCE LIQUID CHROMATOGRAPHIC SEPARATION OF POTATO GLYCOALKALOIDS USING A RADIALLY COMPRESSED AMINO COLUMN
Bushway, R. J.
Dep. Food Sci. Univ. Maine Orono USA US ME 04469
J. Chromatogr.; (82) P 180-3; Vol 247; No 1

27-82-0015 (0015)
ANALYSIS OF POPPY STRAW CONCENTRATE BY HIGH-PERFORMANCE LIQUID CHROMATOGRAPHY
Pettitt, B. C., Jr.; Damon, C. E.
U. S. Customs Serv. Washington USA US DC 20229
J. Chromatogr.; (82) P 189-92; Vol 242; No 1

27-82-0016 (0016)
READY SEPARATION OF ERGOCORNINE, .ALPHA.- AND .BETA.-ERGOCRYPTINE BY HIGH-PERFORMANCE LIQUID CHROMATOGRAPHY
Herenyi, Bulcsu; Gorog, Sandor
Sci. Dep. Gedeon Richter Ltd. Budapest Hung. HU H-1475
J. Chromatogr.; (82) P 250-2; Vol 238; No 1

27-82-0017 (0017)
HIGH-PERFORMANCE LIQUID CHROMATOGRAPHY OF EBURNANE ALKALOIDS. III. APPLICATION OF DIFFERENT PHASE SYSTEMS

Gazdag, M.; Szepesi, G.; Csomor, K.
Chem. Works Gedeon Richter Ltd. Budapest Hung. HU H-1475
J. Chromatogr.; (82) P 315-22; Vol 243; No 2

27-82-0018 (0018)
ANALYSIS OF INDOLE ALKALOIDS IN NORWEGIAN PSILOCYBE SEMILANCEATA USING HIGH-PERFORMANCE LIQUID CHROMATOGRAPHY AND MASS SPECTROMETRY
Christiansen, A. L.; Rasmussen, K. E.
Inst. Pharm. Univ. Oslo Oslo Norway NO
J. Chromatogr.; (82) P 357-64; Vol 244; No 2

27-82-0019 (0019)
SEPARATION OF CINCHONA ALKALOIDS BY HIGH-PERFORMANCE LIQUID CHROMATOGRAPHY
Hobson-Frohock, A.; Edwards, W. T. E.
Food Res. dnst. ARC Norwich UK GB NR4 7UA
J. Chromatogr.; (82) P 369-72; Vol 249; No 2

27-82-0020 (0020)
LIQUID CHROMATOGRAPHIC DETERMINATION OF CAFFEINE IN BIOLOGICAL FLUIDS
Haughey, David B.; Greenberg, Richard; Schaal, Stephen F.; Lima, John J.
Coll. Pharm. Ohio State Univ. Columbus USA US OH 43210
J. Chromatogr.; (82) P 387-95; Vol 229; No 2

27-82-0021 (0021)
HIGH-PERFORMANCE LIQUID CHROMATOGRAPHY DETECTION OF MORPHINE BY FLUORESCENCE AFTER POSTCOLUMN DERIVATIZATION
Nelson, Peter E.; Nolan, Susan L.; Bedford, Keith R.
Dep. Sci. Ind. Res. Univ. Auckland Auckland N. Z. NZ
J. Chromatogr.; (82) P 407-14; Vol 234; No 2

27-82-0022 (0022)
DETERMINATION OF 11-BROMOVINCAMINE IN HUMAN PLASMA BY HIGH-PERFORMANCE LIQUID CHROMATOGRAPHY
Brodie, R. R.; Chasseaud, L. F.
Dep. Metab. Pharmacokinet. Huntingdon Res. Cent. Huntingdon UK GB PE18 6ES
J. Chromatogr.; (82) P 413-17; Vol 228

27-82-0023 (0023)
DETERMINATION OF MORPHINE, MORPHINE-3-GLUCURONIDE AND (TENTATIVELY) MORPHINE-6-GLUCURONIDE IN PLASMA AND URINE USING ION-PAIR HIGH-PERFORMANCE LIQUID CHROMATOGRAPHY
Svensson, Jan Olof; Rane, Anders; Saewe, Juliette; Sjoeqvist, Folke
Karolinska Inst. Huddinge Univ. Hosp. Huddinge Swed. SE S-141 86
J. Chromatogr.; (82) P 427-32; Vol 230; No 2

27-82-0024 (0024)
SEPARATION OF FOUR ISOMERS OF LYSERGIC ACID .ALPHA.-HYDROXYETHYLAMIDE BY LIQUID CHROMATOGRAPHY AND THEIR SPECTROSCOPIC IDENTIFICATION
Flieger, M.; Sedmera, P.; Vokoun, J.; Ricicova, A.; Rehacek, Z.
Inst. Microbiol. Czech. Acad. Sci. Prague Czech. CS
J. Chromatogr.; (82) P 453-9; Vol 236; No 2

27-82-0025 (0025)
ANALYSIS OF OLIVETOL IN RABBIT SERUM BY HIGH-PERFORMANCE LIQUID CHROMATOGRAPHY
Rauls, D.; Penney, L. L.
Dep. Clin. Invest. William Beaumont Army Med. Cent. El Paso USA US TX 79920
J. Chromatogr.; (82) P 500-2; Vol 234; No 2

27-82-0026 (0026)
RAPID EXTRACTION AND MEASUREMENT OF MORPHINE AND OPIATE ANTAGONISTS FROM RAT

BRAIN USING HIGH-PERFORMANCE LIQUID CHROMATOGRAPHY AND ELECTROCHEMICAL DETECTION
Raffa, Robert B.; O'Neill, John J.; Tallarida, Ronald J.
Sch. Med. Temple Univ. Philadelphia USA US PA 19140
J. Chromatogr.; (82) P 515-19; Vol 238; No 2

27-82-0027 (0027)
DETERMINATION OF MORPHINE, DIAMORPHINE AND THEIR DEGRADATION PRODUCTS IN PHARMACEUTICAL PREPARATIONS BY REVERSED-PHASE HIGH-PERFORMANCE LIQUID CHROMATOGRAPHY
Beaumont, I.; Deeks, T.
Pharm. Dep. East Birmingham Hosp. Birmingham UK GB B9 5ST
J. Chromatogr.; (82) P 520-4; Vol 238; No 2

27-82-0028 (0028)
IMPROVED HIGH-PERFORMANCE LIQUID CHROMATOGRAPHIC ASSAY FOR THEOPHYLLINE IN PLASMA AND SALIVA IN THE PRESENCE OF CAFFEINE AND ITS METABOLITES AND COMPARISONS WITH THREE OTHER ASSAYS
Muir, Keith T.; Kunitani, Michael; Riegelman, Sidney
Sch. Pharm. Univ. California San Francisco USA US CA 94143
J. Chromatogr.; (82) P 73-82; Vol 231; No 1

27-82-0029 (0029)
SEPARATION OF HETEROYOHIMBINE AND OXINDOLE ALKALOIDS BY REVERSED-PHASE HIGH-PERFORMANCE LIQUID CHROMATOGRAPHY
Phillipson, J. David; Supavita, Narong; Anderson, Linda A.
Sch. Pharm. Univ. London London UK GB WC1N 1AX
J. Chromatogr.; (82) P 91-8; Vol 244; No 1

27-82-0030 (0030)
A NEW HIGH PERFORMANCE LIQUID CHROMATOGRAPHY (HPLC) METHOD FOR THE QUANTITATION OF STRYCHNINE IN URINE AND TISSUE EXTRACTS
Egloff, T.; Niederwieser, A.; Pfister, K.; Otten, A.; Steinmann, B.; Steiner, W.; Gitzelmann, R.
Dep. Pediatr. Univ. Zurich Zurich Switz. CH
J. Clin. Chem. Clin. Biochem.; (82) P 203-6; Vol 20; No 4

27-82-0031 (0031)
HIGH-PERFORMANCE LIQUID CHROMATOGRAPHIC ANALYSIS OF HEROIN BY REVERSE-PHASE ION-PAIR CHROMATOGRAPHY
Lurie, I. S.; Sottolano, S. M.; Blasof, Stanley
Special Test. Res. Lab. Drug Enforcement Adm. McLean USA US VA 22102
J. Forensic Sci.; (82) P 519-26; Vol 27; No 3

27-82-0032 (0032)
HIGH-PERFORMANCE LIQUID CHROMATOGRAPHIC DETERMINATION OF THE METABOLITES OF THE POTATO GLYCOALKALOIDS, .ALPHA.-CHACONINE AND .ALPHA.-SOLANINE, IN POTATO TUBERS AND POTATO PRODUCTS
Bushway, R. J.
Dep. Food Sci. Univ. Maine Orono USA US ME 04469
J. Liq. Chromatogr.; (82) P 1313-22; Vol 5; No 7

27-82-0033 (0033)
THE HPLC ANALYSIS OF CAFFEINE AND THEOBROMINE IN ANIMAL DIETS
Hurst, W. Jeffrey; Martin, Robert A., Jr.
Tech. Cent. Hershey Food Corp. Hershey USA US PA 17033
J. Liq. Chromatogr.; (82) P 585-9; Vol 5; No 3

27-82-0034 (0034)
BELLADONNA ALKALOIDS AND PHENOBARBITAL COMBINATION PHARMACEUTICALS ANALYSIS.

I: HIGH-PERFORMANCE LIQUID CHROMATOGRAPHIC DETERMINATIONS OF HYOSCYAMINE-ATROPINE AND SCOPOLAMINE
Pennington, Linda J.; Schmidt, Walter F.
FDA Atlanta USA US GA 30309
J. Pharm. Sci.; (82) P 951-3; Vol 71; No 8

27-82-0035 (0035)
BELLADONNA ALKALOIDS AND PHENOBARBITAL COMBINATION PHARMACEUTICALS ANALYSIS. II: HIGH-PERFORMANCE LIQUID CHROMATOGRAPHIC DETERMINATION OF PHENOBARBITAL
Schmidt, Walter F.; Pennington, Linda J.
FDA Atlanta USA US GA 30309
J. Pharm. Sci.; (82) P 954-5; Vol 71; No 8

27-82-0036 (0036)
A SENSITIVE PROCEDURE FOR DETERMINATION OF MORPHINE IN MOUSE WHOLE BLOOD BY HIGH-PERFORMANCE LIQUID CHROMATOGRAPHY WITH ELECTROCHEMICAL DETECTION
Ishikawa, Koichi; McGaugh, James L.; Shibanoki, Shinji; Kubo, Taizo
Sch. Biol. Sci. Univ. California Irvine USA US CA 92717
Jpn. J. Pharmacol.; (82) P 969-71; Vol 32; No 5

27-82-0037 (0037)
HIGH-PRESSURE LIQUID CHROMATOGRAPHY OF PARACETAMOL, ACETYLSALICYLIC ACID, AND CODEINE PHOSPHATE
Stuber, B.; Mueller, K. H.
Dr. Spirig A. G. Olten Switz. CH CH-4601
Pharm. Acta Helv.; (82) P 181; Vol 57; No 7; In Ger

27-82-0038 (0038)
STABILITY STUDY OF AQUEOUS SOLUTIONS OF DIAMORPHINE AND MORPHINE USING HPLC
Beaumont, I. M.
Pharm. Dep. East Birmingham Hosp. Birmingham UK GB B9 5ST
Pharm. J.; (82) P 39-41; Vol 229; No 6183

27-82-0039 (0039)
ANALYSIS OF MORPHINE IN SERUM BY HIGH PERFORMANCE LIQUID CHROMATOGRAPHY WITH AMPEROMETRIC DETECTION
Vandenberghe, H.; MacLeod, S. M.; Chinyanga, H.; Soldin, S. J.
Dep. Clin. Biochem. Univ. Toronto Toronto Can. CA ON
Ther. Drug Monit.; (82) P 307-14; Vol 4; No 3

27-82-0040 (0040)
HPLC DETERMINATION OF STRYCHNINE IN STRYCHNOS PIERSIANA AND ITS PREPARATION
Li, Haisheng; Zhou, Jingyuan
Inst. Drug Control Tianjin Peop. Rep. China CN
Yaowu Fenxi Zazhi; (82) P 326-30; Vol 2; No 6; In Ch

7.27.3
Antibiotics

--- 1982 ---

27-82-0001 (0001)
HIGH-PERFORMANCE LIQUID CHROMATOGRAPHY AS AN AID TO THE RATIONAL USE OF ANTIMICROBIAL AGENTS
Watson, Ian D.; Platt, David J.; McIntosh, Stuart J.; Stewart, Michael J.; Cohen, Howard N.
Dep. Biochem. R. Infirm. Glasgow UK GB G4 0SF
Anal. Proc. (London); (82) 180-3; Vol 19; No 4

27-82-0002 (0002)
HIGH-PERFORMANCE LIQUID CHROMATOGRAPHY OF POLYMYXIN B SULFATE AND GRAMICIDIN
Fisher, B. V.; Raja, R. B.
Cent. Anal. Lab. (Chem.) Wellcome Found. Ltd. Dartford UK GB DA1 5AH
Anal. Proc. (London); (82) P 137-40; Vol 19; No 3

27-82-0003 (0003)
PERFORMANCE CHARACTERISTICS OF BIOASSAY, RADIOENZYMATIC ASSAY, HOMOGENEOUS ENZYME IMMUNOASSAY, AND HIGH-PERFORMANCE LIQUID CHROMATOGRAPHIC DETERMINATION OF SERUM GENTAMICIN
Delaney, Collene J.; Opheim, Kent E.; Smith, Arnold L.; Plorde, James J.
Sch. Med. Univ. Washington Seattle USA US WA 98105
Antimicrob. Agents Chemother.; (82) P 19-25; Vol 21; No 1

27-82-0004 (0004)
HIGH-PERFORMANCE LIQUID CHROMATOGRAPHIC DETERMINATION OF PIPERACILLIN IN PLASMA
Brisson, A. M.; Fourtillan, J. B.
Lab. Pharm. Chim. Unites Enseign. Rech. Med. Pharm. Poitiers Fr. FR 86034
Antimicrob. Agents Chemother.; (82) P 664-5; Vol 21; No 4

27-82-0005 (0005)
HIGH-PERFORMANCE LIQUID CHROMATOGRAPHIC DETERMINATION OF ANTIMICROBIAL ADDITIVES
Lazzarini, Giuseppe
Lab. Chim. Ig. Profil. Cremona Italy IT 26100
Boll. Chim. Unione Ital. Lab. Prov., Parte Sci.; (82) P 27-32; Vol 33; No S1; In Ital

27-82-0006 (0006)
DETERMINATION OF RIFAMPICIN IN SERUM BY HIGH-PERFORMANCE LIQUID CHROMATOGRAPHY
Kubo, Hiroaki; Kinoshita, Toshio; Kobayashi, Yoshie; Tokunaga, Ken
Sch. Pharm. Sci. Kitasato Univ. Tokyo Japan JP
Bunseki Kagaku; (82) P 175-9; Vol 31; No 4; In Japan

27-82-0007 (0007)
MICRO DETERMINATION OF TOBRAMYCIN IN SERUM BY HIGH-PERFORMANCE LIQUID CHROMATOGRAPHY
Kubo, Hiroaki; Kobayashi, Yoshie; Kinoshita, Toshio; Nishikawa, Takashi
Sch. Pharm. Sci. Kitasato Univ. Tokyo Japan JP 108
Bunseki Kagaku; (82) P E263-E268; Vol 31; No 8

27-82-0008 (0008)
MEASUREMENT OF THEOPHYLLINE, MEXILETINE AND A NEW CEPHALOSPORIN IN PLASMA BY LC (LIQUID CHROMATOGRAPHY)
Dresse, Al E.; Delapierre-Gabriel, D.
Dep. Pharmacol. Univ. Liege Liege Belg. BE
Chromatogr. Sci.; (82) P 161-72; Vol 20; No Biol./Biomed. Appl. Liq. Chromatogr. 4

27-82-0009 (0009)
THE ASSAY OF ANTIBIOTICS AND THEIR METABOLITES IN BODY FLUIDS BY REVERSE-PHASE LC WITH SPECIAL REFERENCE TO CEPHALOSPORINS AND CHLORAMPHENICOL
White, Les O.; Reeves, David S.
Dep. Microbiol. Southmead Hosp. Bristol UK GB
Chromatogr. Sci.; (82) P 185-95; Vol 20; No Biol./Biomed. Appl. Liq. Chromatogr. 4

27-82-0010 (0010)
ANALYSES OF TETRACYLINE ANTIBIOTICS BY REVERSED-PHASE HIGH-PERFORMANCE LIQUID CHROMATOGRAPHY
Aszalos, A.; Haneke, C.; Hayden, M. J.; Crawford, J.
Natl Cent. Antibiot. Anal. Food Drug Adm. Washington USA US DC 20204
Chromatographia; (82) P 367-73; Vol 15; No 6

27-82-0011 (0011)
HIGH-PERFORMANCE LIQUID CHROMATOGRAPHY ANALYSIS OF SOME BRANDS OF GENTAMICIN USED CLINICALLY
White, L. O.; Reeves, D. S.
Dep. Microbiol. Southmead Hosp. Bristol UK GBBS10 5NB
CONFERENCE; (82), Curr. Chemother. Immunother., Proc. Int. Congr. Chemother., 12th; P 270-1 ; Vol 1,; Avail.: Periti, Piero; Gialdroni Grassi, Giuliana; Date: 81; Publisher: Am. Soc. Microbiol., Washington, D. C.

27-82-0012 (0012)
STUDIES OF SOME ANTIMICROBIAL AGENTS. I. DETERMINATION OF PENICILLINASE-RESISTANT PENICILLINS IN BIOLOGICAL FLUIDS BY HIGH PRESSURE LIQUID CHROMATOGRAPHY. II. MICROBIAL PHYSIOLOGY AND GENETICS OF MERCURY AND ORGANOMERCURIAL RESISTANCE IN ANAEROBIC BACTERIA
Rudrik, James Thomas
Wayne State Univ. Detroit USA US MI
From: Diss. Abstr. Int. B 1982, 43(2), 346.
DISSERTATION; (82) P 182 pp.; Avail.: Univ. Microfilms Int., Order No. DA8216164

27-82-0013 (0013)
PLASMA CONCENTRATIONS OF CARMINOMYCIN AND CARMINOMYCINOL IN MAN, MEASURED BY HIGH PRESSURE LIQUID CHROMATOGRAPHY
Lankelma, J.; Penders, P. G. M.; McVie, J. G.; Leyva, A.; Ten Bokkel-Huinink, W. W.; De Planque, M. M.; Pinedo, H. M.
Div. Exp. Chemother. Netherlands Cancer Inst. Amsterdam Neth. NL
Eur. J. Cancer Clin. Oncol.; (82) P 363-7; Vol 18; No 4

27-82-0014 (0014)
DETECTION AND DETERMINATION OF CHLORAMPHENICOL BY HPLC AND GC/MS
Kutter, R.; Jahr, D.; Stritzinger, H.
Landesuntersuchungsamt Gesundheitswesen Suedbayern Munich Fed. Rep. Ger. DE
Fleischwirtschaft; (82) P 515-16; Vol 62; No 4; In Ger

27-82-0015 (0015)
A NEW CORRELATION STUDY OF HPLC CHROMATOGRAM WITH ANTIFUNGAL ACTIVITY OF MACROLIDE POLYENES
Swami, M. B.; Nanda, R. K.
Hindustan Antibiotics Ltd. Poona City India IN 411 018
Hind. Antibiot. Bull.; (82) P 1-8; Vol 24; No 1-2

27-82-0016 (0016)
DETERMINATION OF CYCLOSPORIN A IN BLOOD AND PLASMA BY COLUMN SWITCHING HIGH-PERFORMANCE LIQUID CHROMATOGRAPHY AFTER RAPID SAMPLE PREPARATION
Nussbaumer, K.; Niederberger, W.; Keller, H. P.

Div. Pharm. Res. Dev. Sandoz Ltd. Basel Switz. CH CH-4002
HRC CC, J. High Resolut. Chromatogr. Chromatogr. Commun.; (82) P 424-7; Vol 5; No 8

27-82-0017 (0017)
A HIGH-PERFORMANCE LIQUID CHROMATOGRAPHIC METHOD FOR THE DETERMINATION AND CONTROL OF THE COMPOSITION OF GENTAMICIN SULFATE
Kraisintu, K.; Parfitt, R. T.; Rowman, M. G.
Sch. Pharm. Pharmacol. Univ. Bath Bath UK GB BA2 7AY
Int. J. Pharm.; (82) P 67-75; Vol 10; No 1

27-82-0018 (0018)
HIGH PERFORMANCE LIQUID CHROMATOGRAPHIC ASSAY OF CYCLIZATION ACTIVITY IN CELL-FREE SYSTEMS FROM STREPTOMYCES CLAVULIGERUS
Jensen, Susan E.; Westlake, Donald W. S.; Wolfe, Saul
Dep. Microbiol. Univ. Alberta Edmonton Can. CA AB T6G 2E9
J. Antibiot.; (82) P 1026-32; Vol 35; No 8

27-82-0019 (0019)
HIGH PERFORMANCE LIQUID CHROMATOGRAPHY (HPLC) OF NATURAL PRODUCTS. V. THE USE OF HPLC IN THE CELL-FREE BIOSYNTHETIC CONVERSION OF .ALPHA.-AMINOADIPYL-CYSTEINYL-VALINE (LLD) INTO ISOPENICILLIN N
Neuss, N.; Berry, D. M.; Kupka, J.; Demain, A. L.; Queener, S. W.; Duckworth, D. C.; Huckstep, L. L.
Dep. Nutr. Food Sci. Massachusetts Inst. Technol Cambridge USA US MA 02139
J. Antibiot.; (82) P 580-4; Vol 35; No 5

27-82-0020 (0020)
SERUM GENTAMICIN, NETILMICIN AND TOBRAMYCIN ASSAYS BY HIGH PERFORMANCE LIQUID CHROMATOGRAPHY
Marples, J.; Oates, M.D.G.
Dep. Pathol., R.Albert Edward Infirm., Wigan/Lancs., UK WN1 2NN
J. Antimicrob. Chemother. (82) P 311-18; Vol 10; No 4

27-82-0021 (0021)
LIQUID CHROMATOGRAPHIC DETERMINATION OF NATAMYCIN IN CHEESE AT RESIDUE LEVELS
Tuinstra, Louis G. M. Th.; Traag, Wim A.
State Inst. Qual. Control Agric. Prod. Wageningen Neth. NL 6708 PD
J. Assoc. Off. Anal. Chem.; (82) P 820-2; Vol 65; No 4

27-82-0022 (0022)
DETERMINATION OF AMDINOCILLIN IN PLASMA AND URINE BY HIGH-PERFORMANCE LIQUID CHROMATOGRAPHY
Lee, Teh Lo; Brooks, Marvin A.
Dep. Pharmacokinet. Biopharm. Hoffman-La Roche Inc. Nutley USA US NJ 07110
J. Chromatogr.; (82) P 137-48; Vol 227; No 1

27-82-0023 (0023)
A VERY PRECISE HIGH-PERFORMANCE LIQUID CHROMATOGRAPHIC PROCEDURE FOR THE DETERMINATION OF CEFMENOXIME, A NEW CEPHALOSPORIN ANTIBIOTIC, IN PLASMA
Granneman, G. Richard; Sennello, Lawrence T.
Drug Metab. Dep. Abbott Lab. North Chicago USA US IL 60064
J. Chromatogr.; (82) P 149-57; Vol 229; No 1

27-82-0024 (0024)
HIGH-PERFORMANCE LIQUID CHROMATOGRAPHIC DETERMINATION OF CLOXACILLIN IN PHARMACEUTICAL DOSAGE FORMS
Lauriault, G.; LeBelle, M. J.; Vilim, A.
Bur. Drug Res. Health and Welfare Canada Ottawa Can. CA ON K1A OL2
J. Chromatogr.; (82) P 157-60; Vol 246; No 1

27-82-0025 (0025)
HIGH-PERFORMANCE LIQUID CHROMATOGRAPHIC RESOLUTION AND QUANTIFICATION OF A DILACTONIC ANTIBIOTIC MIXTURE (ANTIMYCIN A)
Abidi, S. L.
Natl. Fish. Res. Lab. U. S. Fish Wildl. Serv. La Crosse USA US WI 54601
J. Chromatogr.; (82) P 187-200; Vol 234; No 1

27-82-0026 (0026)
HIGH-PERFORMANCE LIQUID CHROMATOGRAPHIC METHOD FOR THE DETERMINATION OF NOVOBIOCIN
Tsuji, Kiyoshi; Rahn, Paul D.; Kane, Michael P.
Control Anal. Res. Dev. Upjohn Co. Kalamazoo USA US MI 49001
J. Chromatogr.; (82) P 205-14; Vol 235; No 1

27-82-0027 (0027)
DETERMINATION OF 4'-EPIDOXORUBICIN AND ITS 13-DIHYDRO DERIVATIVE IN HUMAN PLASMA BY HIGH-PERFORMANCE LIQUID CHROMATOGRAPHY WITH FLUORESCENCE DETECTION
Moro, Enzo; Jannuzzo, Maria G.; Ranghieri, Massimo; Stegnjaich, Silvano; Valzelli, Guido
Pharmacokinet. Dep. Farmitalia Carlo Erba Milan Italy IT 20159
J. Chromatogr.; (82) P 207-11; Vol 230; No 1

27-82-0028 (0028)
DETERMINATION OF GENTAMICIN COMPONENTS C1A, C2 AND C1 IN PLASMA AND URINE BY HIGH-PERFORMANCE LIQUID CHROMATOGRAPHY
D'Souza, J.; Ogilvie, R. I.
Dep. Pharmacol. Ther. McGill Univ. Montreal Can. CA PQ H3G 1A4
J. Chromatogr.; (82) P 212-18; Vol 232; No 1

27-82-0029 (0029)
IMPROVED HIGH-PERFORMANCE LIQUID CHROMATOGRAPHIC METHOD FOR SIMULTANEOUS DETERMINATION OF NEAMINE, NEOMYCIN B AND NEOMYCIN C IN NEOMYCIN SULFATE
Helboe, Per; Kryger, Sonja
Drug Stand. Lab. Natl. Board Health Broenshoej Den. DK DK-2700
J. Chromatogr.; (82) P 215-20; Vol 235; No 1

27-82-0030 (0030)
MICRO DETERMINATION OF GENTAMICIN IN SERUM BY HIGH-PERFORMANCE LIQUID CHROMATOGRAPHY
Kubo, Hiroaki; Kinoshita, Toshio; Kobayashi, Yoshie; Tokunaga, Ken
Sch. Pharm. Sci. Kitasato Univ. Tokyo Japan JP 108
J. Chromatogr.; (82) P 244-8; Vol 227; No 1

27-82-0031 (0031)
SEPARATION OF THE SIDE-CHAIN DIASTEREOISOMERS OF PENICILLINS BY HIGH-PERFORMANCE LIQUID CHROMATOGRAPHY
Hoogmartens, J.; Roets, E.; Janssen, G.; Vanderhaeghe, H.
Inst. Farm. Wet. Kathol. Univ. Leuven Louvain Belg. BE 3000
J. Chromatogr.; (82) P 299-309; Vol 244; No 2

27-82-0032 (0032)
RAPID, RELIABLE AND SENSITIVE ASSAY FOR WARFARIN USING NORMAL-PHASE HIGH-PERFORMANCE LIQUID CHROMATOGRAPHY
Tasker, R. A. R.; Nakatsu, K.
Dep. Pharmacol. Queen's Univ. Kingston Can. CA ON K7L 3N6
J. Chromatogr.; (82) P 346-9; Vol 228

27-82-0033 (0033)
INFLUENCE OF TEMPERATURE ON THE STABILITY OF SOLID TETRACYCLINE HYDROCHLORIDE, MEASURED BY HIGH-PERFORMANCE LIQUID CHROMATOGRAPHY

Dihuidi, Kabala; Roets, E.; Hoogmartens, J.; Vanderhaeghe, H.
Lab. Farm. Chem. Katholieke Univ. Leuven Louvain Belg. BE B-3000
J. Chromatogr.; (82) P 350-5; Vol 246; No 2

27-82-0034 (0034)
RAPID DETERMINATION OF TETRACYCLINE AND LUMECYCLINE IN HUMAN PLASMA AND URINE USING HIGH-PERFORMANCE LIQUID CHROMATOGRAPHY
Hermansson, Joergen
Dep. Drugs Natl. Board Health Welfare Uppsala Swed. SE S-751 25
J. Chromatogr.; (82) P 385-93; Vol 232; No 2

27-82-0035 (0035)
IMPROVED PROCEDURE FOR THE DETERMINATION OF THE UREIDOPENICILLINS AZLOCILLIN AND MEZLOCILLIN IN PLASMA BY HIGH-PERFORMANCE LIQUID CHROMATOGRAPHY
Hildebrandt, R.; Gundert-Remy, U.
Abt. Klin. Pharmakol. Med. Universitaetsklin. Heidelberg Fed. Rep. Ger. DE D-6900
J. Chromatogr.; (82) P 409-12; Vol 228

27-82-0036 (0036)
SIMULTANEOUS DETERMINATION OF CHLORAMPHENICOL AND CHLORAMPHENICOL SUCCINATE IN PLASMA USING HIGH-PERFORMANCE LIQUID CHROMATOGRAPHY
Velagapudi, Raja; Smith, Robert V.; Ludden, Thomas M.; Sagraves, Rosalie
Coll. Pharm. Univ. Texas Austin USA US TX 78712
J. Chromatogr.; (82) P 423-8; Vol 228

27-82-0037 (0037)
SIMULTANEOUS DETERMINATION OF THE CATIONIC AND ANIONIC PARTS IN REPOSITORY PENICILLINS BY HIGH-PERFORMANCE LIQUID CHROMATOGRAPHY
Nachtmann, F.; Gstrein, K.
Biochem. G.m.b.H. Kundl Austria AT A-6250
J. Chromatogr.; (82) P 461-8; Vol 236; No 2

27-82-0038 (0038)
VANCOMYCIN QUANTITATION BY HIGH-PERFORMANCE LIQUID CHROMATOGRAPHY IN HUMAN SERUM
McClain, John Bruce Lundy; Bongiovanni, Rodolpho; Brown, Scott
Infect. Dis. Serv. Walter Reed Army Med. Cent. Washington USA US DC 20012
J. Chromatogr.; (82) P 463-6; Vol 231; No 2

27-82-0039 (0039)
QUANTITATIVE REVERSED-PHASE HIGH-PERFORMANCE LIQUID CHROMATOGRAPHIC ANALYSIS OF AMPICILLIN
Margosis, Michel
Natl. Cent. Antibiotic Anal. Food and Drug Adm. Washington USA US DC 20204
J. Chromatogr.; (82) P 469-80; Vol 236; No 2

27-82-0040 (0040)
PAIRED-ION HIGH-PERFORMANCE LIQUID-CHROMATOGRAPHIC DETERMINATION OF THE STABILITY OF NOVOBIOCIN IN MASTITIS PRODUCTS STERILIZED BY COBALT-60 IRRADIATION
Hornish, R. E.
Upjohn Co. Kalamazoo USA US MI 49001
J. Chromatogr.; (82) P 481-8; Vol 236; No 2

27-82-0041 (0041)
MEASUREMENT OF KETOCONAZOLE, A NEW ANTIFUNGAL AGENT, BY HIGH-PERFORMANCE LIQUID CHROMATOGRAPHY
Swezey, Sarah F.; Giacomini, Kathleen M.; Abang, Anthony; Brass, Corstiaan; Stevens, David A.; Blaschke, Terrence F.

Med. Sch. Stanford Univ. Stanford USA US CA 94305
J. Chromatogr.; (82) P 510-15; Vol 227; No 2

27-82-0042 (0042)
LIQUID CHROMATOGRAPHY WITH CROWN ETHER-CONTAINING MOBILE PHASES. II. RETENTION BEHAVIOR OF .BETA.-LACTAM ANTIBIOTICS IN REVERSED-PHASE HIGH-PERFORMANCE LIQUID CHROMATOGRAPHY
Nakagawa, Terumichi; Shibukawa, Akimasa; Uno, Toyozo
Fac. Pharm. Sci. Kyoto Univ. Kyoto Japan JP 606
J. Chromatogr.; (82) P 695-706; Vol 239

27-82-0043 (0043)
SEPARATION AND DETECTION OF MONOPYRROLE AND BIPYRROLE PRECURSORS OF PRODIGIOSIN FROM SERRATIA MARCESCENS BY A COMBINED METHOD OF HIGH-PERFORMANCE LIQUID CHROMATOGRAPHY AND SYNTROPHIC PIGMENT SYNTHESIS
Feng, J. S.; Qian, H. L.; Tsang, J. C.
Dep. Chem. Illinois State Univ. Bloomington USA US IL
J. Liq. Chromatogr.; (82) P 1329-40; Vol 5; No 7

27-82-0044 (0044)
APPLICATIONS OF A MODIFIED "ISOHYDRIC SOLVENT SYSTEM" IN HPLC ON SILICA GEL FOR THE ANALYSIS OF THE MACROLIDE ANTIBIOTICS TURIMYCINS AND SPIRAMYCINS
Bens, G. A.; Crombez, E.; De Moerloose, P.
Dep. Pharm. Chem. Drug Qual. Control State Univ. Ghent Ghent Belg. BE B-9000
J. Liq. Chromatogr.; (82) P 1449-65; Vol 5; No 8

27-82-0045 (0045)
DETERMINATION OF TETRACYCLINE AND RELATED COMPOUNDS BY HIGH-PERFORMANCE LIQUID CHROMATOGRAPHY
Hon, J. Y. C.; Murray, L. R.; Walker, B.
Antibiot. Sect. Natl. Biol. Stand. Lab. Canberra Australia AU 2601
J. Liq. Chromatogr.; (82) P 1973-90; Vol 5; No 10

27-82-0046 (0046)
PRECISE HIGH-PERFORMANCE LIQUID CHROMATOGRAPHIC PROCEDURE FOR THE DETERMINATION OF CEFSULODIN, A NEW ANTIPSEUDOMONAL CEPHALOSPORIN ANTIBIOTIC, IN PLASMA
Granneman, G. Richard; Sennello, Lawrence T.
Drug Metab. Dep. Abbott Lab. North Chicago USA US IL 60064
J. Pharm. Sci.; (82) P 1112-15; Vol 71; No 10

27-82-0047 (0047)
HIGH-PRESSURE LIQUID CHROMATOGRAPHIC ASSAYS FOR TICARCILLIN IN SERUM AND URINE
Kwan, R. H.; MacLeod, S. M.; Spino, M.; Teare, F. W.
Fac. Pharm. Univ. Toronto Toronto Can. CA ON M5S 1A1
J. Pharm. Sci.; (82) P 1118-21; Vol 71; No 10

27-82-0048 (0048)
DETERMINATION OF BACMECILLINAM, AN AMDINOCILLIN PRODRUG, IN HUMAN AND CANINE WHOLE BLOOD BY REVERSED-PHASE LIQUID CHROMATOGRAPHY
Westerlund, Douglas; Pettersson, Bodil; Carlqvist, Jan
Dev. Lab., Anal. Chem. Astra Laekemedel A.B. Soedertaelje Swed. SE S-151 85
J. Pharm. Sci.; (82) P 1148-51; Vol 71; No 10

27-82-0049 (0049)
IMPROVED HIGH-PRESSURE LIQUID CHROMATOGRAPHIC METHOD FOR THE ANALYSIS OF ERYTHROMYCIN IN SOLID DOSAGE FORMS
Tsuji, Kiyoshi; Kane, Michael P.
Control Anal. Res. Dev. Upjohn Co. Kalamazoo USA US MI 49001
J. Pharm. Sci.; (82) P 1160-4; Vol 71; No 10

27-82-0050 (0050)
REVERSED-PHASE ION-PAIR CHROMATOGRAPHY OF TETRACYCLINE, TETRACYCLINE ANALOGS, AND THEIR POTENTIAL IMPURITIES
Hermansson, Joergen; Andersson, Mats
Div. Pharm. Natl. Board Health Welfare Uppsala Swed.
J. Pharm. Sci.; (82) P 222-9; Vol 71; No 2

27-82-0051 (0051)
SEPARATION OF PENICILLIN AND ITS MAJOR DEGRADATION PRODUCTS BY ION-PAIR REVERSED-PHASE HIGH-PRESSURE LIQUID CHROMATOGRAPHY
Ghebre-Sellassie, Isaac; Hem, Stanley L.; Knevel, Adelbert M.
Sch. Pharm. Pharm. Sci. Purdue Univ. West Lafayette USA US PIN 47907
J. Pharm. Sci.; (82) P 351-3; Vol 71; No 3

27-82-0052 (0052)
HIGH-PRESSURE LIQUID CHROMATOGRAPHIC ASSAY OF CLOXACILLIN IN SERUM AND URINE
Teare, F. W.; Kwan, R. H.; Spino, M.; MacLeod, S. M.
Fac. Pharm. Univ. Toronto Toronto Can. CA ON M5S 1A1
J. Pharm. Sci.; (82) P 938-41; Vol 71; No 8

27-82-0053 (0053)
HIGH-PRESSURE LIQUID CHROMATOGRAPHY OF DIBENZYLETHYLENEDIAMINE IN BENZATHINE PENICILLINS
Stuber, B.
Spirig A.-G. Olten Switz. CH CH-4601
Pharm. Acta Helv.; (82) P 293-4; Vol 57; No 10-11; In Ger

27-82-0054 (0054)
THE DETERMINATION OF AMINOGLYCOSIDE ANTIBIOTICS IN SERUM: A COMPARISON OF A HIGH PERFORMANCE LIQUID CHROMATOGRAPHIC METHOD WITH A MICROBIOLOGICAL ASSAY
Barends, D. M.; Rutgers, A.; Van Klingeren, B.; Hulshoff, A.
Dep. Anal. Pharm. Univ. Utrecht, Pharmaceutical Laboratory Utrecht Neth. NL 3511 GH
Pharm. Weekbl., Sci. Ed.; (82) P 104-11; Vol 4; No 4

27-82-0055 (0055)
COMPARISON OF AN HPLC METHOD WITH A RIA, EMIT AND FIA METHOD FOR THE ASSAY OF SERUM GENTAMICIN WITH EXTENSIVE STATISTICAL EVALUATION
Hospes, W.; Boskma, R. J.; Brouwers, J. R. B. J.
Dep. Clin. Pharm. Gen. Hosp. De Tjongerschans' Heerenveen Neth. NL 8441 PW
Pharm. Weekbl., Sci. Ed.; (82) P 32-7; Vol 4; No 2

27-82-0056 (0056)
HPLC SEPARATION OF PATULIN FROM 5-(HYDROXYMETHYL)FURFUROL
Ruggieri, G.; Ruggeri, P.
Ist. Merceol. Univ. Napoli Naples Italy IT
Rass. Chim.; (82) P 121-4; Vol 34; No 3; In Ital

27-82-0057 (0057)
HIGH-PRESSURE LIQUID CHROMATOGRAPHIC DETERMINATION OF CYCLOSPORIN A IN HUMAN PLASMA
Leyland-Jones, Brian; Clark, Ann; Kreis, Willi; Dinsmore, Robert; O'Reilly, Richard; Young, Charles W.
Clin. Pharmacol. Lab. Mem. Sloan-Kettering Cancer Cent. New York USA US NY 10021
Res. Commun. Chem. Pathol. Pharmacol.; (82) P 431-44; Vol 37; No 3

7.27.4
Various Pharmaceuticals

--- 1982 ---

27-82-0001 (0001)
ALKYLATED 2- AND 4-THIOURACILS. SYNTHESES AND HPLC SEPARATIONS
Gacek, Michel; Undheim, Kjell
Dep. Chem. Univ. Oslo Oslo Norway NO
Acta Chem. Scand., Ser. B; (82) P 15-18; Vol B36; No 1

27-82-0002 (0002)
DETERMINATION OF SERUM OXPRENOLOL BY HIGH PERFORMANCE LIQUID CHROMATOGRAPHY
Plavsic, Franjo
Fac. Med. Clin. Hosp. Cent. Zagreb Yugoslavia YU
Acta Pharm. Jugosl.; (82) P 137-40; Vol 32; No 2

27-82-0003 (0003)
ISOLATION AND DETERMINATION OF THE DIASTEREOMERS OF THE GLUCURONIDATED 5-(4-HYDROXYPHENYL)-5-PHENYLHYDANTOIN IN HUMAN URINE USING HPLC
Hermansson, Joergen; Iversen, Tommy; Lindquist, Ulf
Dep. Drugs Natl. Board Health Welfare Uppsala Swed. SE S-751 25
Acta Pharm. Suec.; (82) P 199-210; Vol 19; No 3

27-82-0004 (0004)
DISPARATE SERUM THIORIDAZINE CONCENTRATIONS: LIQUID CHROMATOGRAPHY VERSUS RADIORECEPTOR ASSAY
Linnoila, Markku; Rosenblatt, Jack E.; Jeste, Dilip; Skinner, Travis; Potter, William Z.; Wyatt, Richard Jed
Clin. Psychobiol. Branch Natl. Inst. Mental Health Bethesda USA US MD
Acta Pharmacol. Toxicol.; (82) P 25-9; Vol 50; No 1

27-82-0005 (0005)
COLORIMETRIC OR HPLC METHOD FOR ACETYLATOR PHENOTYPING WITH SULFADIMIDINE?
Olsen, Harald
Dep. Clin. Pharmacol. Univ. Hosp. Tromso Norway NO N-9012
Acta Pharmacol. Toxicol.; (82) P 75-7; Vol 50; No 1

27-82-0006 (0006)
METRONIDAZOLE INTERFERENCE WITH PROCAINAMIDE HPLC ASSAY
Gannon, Richard H.; Phillips, Lynn R.
Dep. Pharm. Serv. Yale-New Haven Hosp. New Haven USA US CT
Am. J. Hosp. Pharm.; (82) P 1966-7; Vol 39; No 11

27-82-0007 (0007)
RAPID HIGH-PRESSURE LIQUID CHROMATOGRAPHIC DETERMINATION OF METHOTREXATE AND ITS METABOLITES 7-HYDROXYMETHOTREXATE AND 2,4-DIAMINO-N10-METHYLPTEROIC ACID IN BIOLOGICAL FLUIDS
Breithaupt, H.; Kuenzlen, E.; Goebel, G.
Med. Klin. Justus Liebig-Univ. Giessen Fed. Rep. Ger. DE 6300
Anal. Biochem.; (82) P 103-13; Vol 121; No 1

27-82-0008 (0008)
QUANTITATIVE DETERMINATION OF ALLANTOIN IN BIOLOGICAL FLUIDS BY REVERSED-PHASE HIGH-PRESSURE LIQUID CHROMATOGRAPHY
Tiemeyer, W.; Giesecke, D.
Tieraerztl. Fak. Univ. Muenchen Munich Fed. Rep. Ger. DE D-8000/22
Anal. Biochem.; (82) P 11-13; Vol 123; No 1

27-82-0009 (0009)
DETERMINATION OF ACETAMINOPHEN METABOLITES IN URINE BY LIQUID CHROMATOGRAPHY/ELECTROCHEMISTRY
Hamilton, Marta; Kissinger, Peter T.

Dep. Chem. Purdue Univ. West Lafayette USA US IN 47907
Anal. Biochem.; (82) P 143-8; Vol 125; No 1

27-82-0010 (0010)
ASSESSMENT OF CISPLATIN REACTIVITY WITH PEPTIDES AND PROTEINS USING REVERSE-PHASE HIGH-PERFORMANCE LIQUID CHROMATOGRAPHY AND FLAMELESS ATOMIC ABSORPTION SPECTROSCOPY
Riley, Christopher M.; Sternson, Larry A.; Repta, Arnold J.
Dep. Pharm. Chem. Univ. Kansas Lawrence USA US KS 66045
Anal. Biochem.; (82) P 167-79; Vol 124; No 1

27-82-0011 (0011)
URIDINE DIPHOSPHO-N-ACETYLGALACTOSAMINE 4-SULFATE: ISOLATION AND HIGH-PERFORMANCE LIQUID CHROMATOGRAPHY ASSAY OF ITS REACTION WITH ARYLSULFATASE B
Fluharty, Arvan L.; Glick, Jeffrey A.; Samaan, Georgy F.; Kihara, Hayato
Sch. Med. Univ. California Los Angeles Los Angeles USA US CA
Anal. Biochem.; (82) P 310-14; Vol 121; No 2

27-82-0012 (0012)
USE OF REVERSE-PHASE HIGH-PERFORMANCE LIQUID CHROMATOGRAPHY IN STRUCTURAL STUDIES OF NEUROPHYSINS, PHOTOLABELED DERIVATIVES, AND BIOSYNTHETIC PRECURSORS
Abercrombie, David M.; Hough, Christopher J.; Seeman, Jonathan R.; Brownstein, Michael J.; Gainer, Harold; Russell, James T.; Chaiken, Irwin M.
Natl. Inst. Arthritis, Diabetes Dig. Kid. Dis. Univ. Maryland College Park USA US MD 20742
Anal. Biochem.; (82) P 395-405; Vol 125; No 2

27-82-0013 (0013)
CRITERIA FOR THE EVALUATION OF THE PURITY OF STEROIDS FOR PHARMACEUTICAL PREPARATIONS BY MEANS OF HPLC
Cavina, G.; Moretti, G.; Alimenti, R.; Gallinella, B.; Porra, R.; Valvo, L.
Ist. Super. Sanita Fellow Ist. Super. Sanita Rome Italy IT
Anal. Chem. Symp. Ser.; (82) P 343-60; Vol 10; No Adv. Steroid Anal.

27-82-0014 (0014)
LIQUID CHROMATOGRAPHIC DETERMINATION OF INTACT LEUKOTRIENE A4
Wynalda, M. A.; Morton, D. R.; Kelly, R. C.; Fitzpatrick, F. A.
Pharm. Res. Dev. Upjohn Co. Kalamazoo USA US MI 49001
Anal. Chem.; (82) P 1079-82; Vol 54; No 7

27-82-0015 (0015)
LIQUID CHROMATOGRAPHIC DETERMINATION OF GUANADREL IN LABORATORY ANIMAL DIET AS THE FLUORESCENT ACETYLACETONE DERIVATIVE
Bombardt, Paul A.; Adams, Wade J.
Upjohn Co. Kalamazoo USA US MI 49001
Anal. Chem.; (82) P 1087-90; Vol 54; No 7

27-82-0016 (0016)
PRECOLUMNS OF AMBERLITE XAD-2 FOR DIRECT INJECTION LIQUID CHROMATOGRAPHIC DETERMINATION OF METHAQUALONE IN BLOOD PLASMA
Hux, Robert A.; Mohammed, Hussain Y.; Cantwell, Frederick F.
Dep. Chem. Univ. Alberta Edmonton Can. CA AB T6G 2G2
Anal. Chem.; (82) P 113-17; Vol 54; No 1

27-82-0017 (0017)
DETERMINATION OF GLYBURIDE IN HUMAN SERUM BY LIQUID CHROMATOGRAPHY WITH FLUORESCENCE DETECTION
Adams, W. J.; Skinner, G. S.; Bombardt, P. A.; Courtney, M.; Brewer, J. E.
Pharm. Res. Dev. Upjohn Co. Kalamazoo USA US MI 49001
Anal. Chem.; (82) P 1287-91; Vol 54; No 8

27-82-0018 (0018)
OPTIMIZATION OF A PAIRED-ION REVERSED-PHASE LIQUID CHROMATOGRAPHIC SEPARATION OF SYNTHETIC PHENCYCLIDINE MIXTURES
Jones, Louis A.; Beaver, Rodney W.; Schmoeger, T. L.
Dep. Chem. North Carolina Rtate Univ. Raleigh USA US NC 27650
Anal. Chem.; (82) P 182-6; Vol 54; No 2

27-82-0019 (0019)
ELECTROCHEMICAL, FLUORESCENCE, AND LIQUID CHROMATOGRAPHIC STUDIES ON THE OXIDATION OF SOTERENOL
Miner, David J.; Kissinger, Peter T.
Dep. Chem. Purdue Univ. West Lafayette USA US IN 47907
Anal. Chim. Acta; (82) P 231-5; Vol 138

27-82-0020 (0020)
HPLC METHOD FOR DETERMINATION OF DIAZEPAM INJECTION
Smith, F. Mauriece; Nuessle, Noel O.
Sch. Pharm. Univ. Missouri Kansas City USA US MO 64110
Anal. Lett.; (82) P 363-71; Vol 15; No B4

27-82-0021 (0021)
THIAZIDES. XII. A SIMPLE HPLC METHOD FOR DETERMINATION OF THIAZIDES IN URINE
Shah, Vinod P.; Lee, Jennifer; Prasad, Vadlamani K.
Div. Biopharm. FDA Washington USA US DC 20204
Anal. Lett.; (82) P 529-36; Vol 15; No B6

27-82-0022 (0022)
DETERMINATION OF BENZODIAZEPINES IN SERUM BY HIGH PERFORMANCE LIQUID CHROMATOGRAPHY USING RADIALLY COMPRESSED COLUMNS AND AN AQUEOUS METHANOLIC MOBILE PHASE
Kinberger, Bengt; Wahrgren, Peter
Dep. Clin. Chem. Cent. Hosp. Halmstad Swed. SE S-301 85
Anal. Lett.; (82) P 549-57; Vol 15; No B6

27-82-0023 (0023)
HIGH-PERFORMANCE LIQUID CHROMATOGRAPHY OF AMINOPYRIDINES
Scott, D. K.; Irwin, W. J.
Pharm. Dep. East Birmingham Hosp. Birmingham UK GB 9
Anal. Proc. (London); (82) P 123-5; Vol 19; No 3

27-82-0024 (0024)
HIGH-PERFORMANCE LIQUID CHROMATOGRAPHY IN THE MEASUREMENT OF ANTINEOPLASTIC DRUGS
Stewart, M. J.; Farid, Y. Y. Z.; Watson, I. D.
Dep. Biochem. R. Infirm. Glasgow UK GB G4 0SF
Anal. Proc. (London); (82) P 179-80; Vol 19; No 4

27-82-0025 (0025)
SOME APPLICATIONS OF LIQUID CHROMATOGRAPHY TO THE ANALYSIS OF MEDICINAL PRODUCTS
Hopkins, R. G.
Med. Prod. Sub-Div. Lab. Gov. Chem. London UK GB SE1 9NQ
Anal. Proc. (London); (82) P 193-5; Vol 19; No 4

27-82-0026 (0026)
OPTIMAL OPERATING CONDITIONS FOR THE SEPARATION OF SALICYLIC ACID AND ACETYL-SALICYLIC ACID BY HPLC. APPLICATION TO TRACES ANALYSIS OF SALICYLIC ACID IN ASPIRIN
Menouer, M.; Ghernati, H. M.; Bouabdallah, F.; Guermouche, M. H.
Lab. Biophys. Univ. Alger Alger Algeria DZ

Analusis; (82) P 172-6; Vol 10; No 4; In Fr

27-82-0027 (0027)
DETERMINATION OF TIAMULIN HYDROGEN FUMARATE IN ANIMAL FEEDS USING HIGH-PERFORMANCE LIQUID CHROMATOGRAPHY
Howard, D.; Cowen, T.
Int. Dev. Lab. E. R. Squibb and Sons Moreton UK GB L46 1QW
Analyst (London); (82) P 319-23; Vol 107; No 1272

27-82-0028 (0028)
STABILITY INDICATING ASSAY FOR DIPYRONE. PART II. SEPARATION AND QUANTITATIVE DETERMINATION OF DIPYRONE AND ITS DEGRADATION PRODUCTS BY HIGH-PERFORMANCE LIQUID CHROMATOGRAPHY
Eddine, N. Hussam; Bressolle, F.; Mandrou, B.; Fabre, H.
Lab. Chim. Anal. Toxicol. Fac. Pharm. Montpellier Fr. FR 34060
Analyst (London); (82) P 67-70; Vol 107; No 1270

27-82-0029 (0029)
A RAPID HIGH-PERFORMANCE LIQUID CHROMATOGRAPHIC METHOD FOR QUANTITATION OF 5-FLUOROURACIL IN PLASMA AFTER CONTINUOUS INTRAVENOUS INFUSION
Sampson, D. C.; Fox, R. M.; Tattersall, M. H. N.; Hensley, W. J.
Dep. Biochem. R. Prince Alfred Hosp. Camperdown Australia AU 2050
Ann. Clin. Biochem.; (82) P 125-8; Vol 19; No 2

27-82-0030 (0030)
HIGH-PERFORMANCE LIQUID CHROMATOGRAPHIC METHOD FOR DETERMINATION OF CEFMETAZOLE IN HUMAN SERUM
Sekine, Minoru; Sasahara, Kunihiro; Kojima, Toshimasa; Morioka, Tadashi
Prod. Dev. Lab. Sankyo Co. Ltd. Tokyo Japan JP
Antimicrob. Agents Chemother.; (82) P 740-3; Vol 21; No 5

27-82-0031 (0031)
DETERMINATION OF NORFLOXACIN, A NEW NALIDIXIC ACID ANALOG, IN HUMAN SERUM AND URINE BY HIGH-PERFORMANCE LIQUID CHROMATOGRAPHY
Boppana, Venkata K.; Swanson, Brian N.
Jefferson Med. Coll. Thomas Jefferson Univ. Philadelphia USA US PA 19107
Antimicrob. Agents Chemother.; (82) P 808-10; Vol 21; No 5

27-82-0032 (0032)
IDENTIFICATION OF METABOLISM PATHWAYS OF ANTICANCER DRUGS BY HIGH-PRESSURE LIQUID CHROMATOGRAPHY IN COMBINATION WITH FIELD DESORPTION MASS SPECTROMETRY
Przybylski, M.
Inst. Org. Chem. Univ. Mainz Mainz Fed. Rep. Ger. DE D-6500
Arzneim. Forsch.; (82) P 995-1012; No 9

27-82-0033 (0033)
QUANTITATIVE DETERMINATION OF SUPROFEN IN HUMAN PLASMA AND URINE BY FULLY AUTOMATED HIGH-PERFORMANCE LIQUID CHROMATOGRAPHY
Mueller, H.; Zulliger, Hans W.; Doerig, B. A.; Egli, R. A.
Pharm. Res. Dev. Dep. Cilag A.-G. Ltd. Schaffhausen Switz. CH CH-8201
Arzneim.-Forsch.; (82) P 257-60; Vol 32; No 3

27-82-0034 (0034)
HUMAN PHARMACOKINETICS OF DIFFERENT I.V. DOSES OF AZAPROPAZONE (HPLC-DETERMINATION)
Gikalov, I.; Kaufmann, R.; Schuster, O.
Med. Abt. Firma Siegfried A.-G. Zofingen Switz. CH CH-4800
Arzneim.-Forsch.; (82) P 423-6; Vol 32; No 4; In Ger

27-82-0035 (0035)
HPLC OF NEOPTERIN PHOSPHATES
Blau, N.; Bosshard, R.; Redweik, U.; Niederwieser, A.
Dep. Pediatr. Univ. Zurich Zurich Switz. CH CH-8032
Biochem. Clin. Aspects Pteridines; (82) P 103-16; Vol 1; No Cancer, Immunol., Metab. Dis.

27-82-0036 (0036)
PTERIDINES AS TUMOR MARKERS? ATTEMPT OF EVALUATION BY HPLC AND A RADIO-IMMUNOASSAY FOR NEOPTERIN
Rokos, Katharina; Rokos, Hartmut; Frisius, Hildegard
Res. Lab. Henning Berlin G.m.b.H. Berlin Fed. Rep. Ger. DE D-1000/42
Biochem. Clin. Aspects Pteridines; (82) P 117-30; Vol 1; No Cancer, Immunol., Metab. Dis.

27-82-0037 (0037)
AUTOMATED ROUTINE ESTIMATION OF NEOPTERIN IN HUMAN URINE BY HPLC ON REVERSED PHASE
Fuchs, Dietmar; Hausen, Arno; Reibnegger, Gilbert; Wachter, Helmut
Inst. Med. Chem. Biochem. Univ. Innsbruck Innsbruck Austria AT A-6020
Biochem. Clin. Aspects Pteridines; (82) P 67-79; Vol 1; No Cancer, Immunol., Metab. Dis.

27-82-0038 (0038)
AUTOMATIC HPLC OF PTERINS WITH OR WITHOUT COLUMN SWITCHING
Niederwieser, A.; Staudenmann, W.; Wetzel, E.
Dep. Pediatr. Univ. Zurich Zurich Switz. CH CH-8032
Biochem. Clin. Aspects Pteridines; (82) P 81-102; Vol 1; No Cancer, Immunol., Metab. Dis.

27-82-0039 (0039)
IDENTIFICATION AND QUANTITATION OF METHOTREXATE AND METHOTREXATE METABOLITES IN CLINICAL HIGH-DOSE THERAPY BY HIGH-PRESSURE LIQUID CHROMATOGRAPHY AND FIELD DESORPTION MASS SPECTROMETRY
Przybylski, Michael; Preiss, Joachim; Dennebaum, Reiner; Fischer, Joseph
Inst. Org. Chem. Univ. Mainz Mainz Fed. Rep. Ger. DE
Biomed. Mass Spectrom.; (82) P 22-32; Vol 9; No 1

27-82-0040 (0040)
HPLC DETERMINATION OF BENZTHIAZIDE IN BIOLOGIC MATERIAL
Meyer, Marvin C.; Hwang, Philip; Straughn, Arthur B.; Rotenberg, Keith
Coll. Pharm. Univ. Tennessee Memphis USA US TN
Biopharm. Drug Dispos.; (82) P 1-9; Vol 3; No 1

27-82-0041 (0041)
DETERMINATION OF BLOOD LEVELS OF DILAZEP BY HPLC
De Giuli, G.; Ottolenghi, L.; Signorini, G.
Lab. Controllo Schering S.p.A. Milan Italy IT
Boll. Chim. Farm.; (82) P 10-15; Vol 121; No 1; In Ital

27-82-0042 (0042)
HPLC DETERMINATION OF CEFOXITIN AND STUDIES ON BILIARY EXCRETION
Pierini, N.; Caprioli, A.; Fiocca, F.; Basoli, A.; Chirletti, P.; Panicucci, M.
Ist. Super. Sanita Rome Italy IT
Boll. Chim. Farm.; (82) P 387-401; Vol 121; No 8; In Ital

27-82-0043 (0043)
HPLC DETERMINATION OF AMINOPHENAZONE AND CHLORTHENOXAZINE IN A PHARMACEUTICAL FORM

Maurich, V.; Sciortino, T.; Solinas, D.; Vio, L.
Ist. Tec. Farm. Univ. Trieste Trieste Italy IT
Boll. Chim. Farm.; (82) P 483-9; Vol 121; No 9; In Ital

27-82-0044 (0044)
HPLC STUDY ON SOME CEPHALOSPORINS
Quercia, V.; De Sena, C.; Iela, G.; Pagnozzi, G.
Lab. Chim. Farm. Ist. Super. Sanita Rome Italy IT
Boll. Chim. Farm.; (82) P 65-70; Vol 121; No 2; In Ital

27-82-0045 (0045)
HIGH PERFORMANCE LIQUID CHROMATOGRAPHIC MEASUREMENT OF AMIODARONE AND ITS DESETHYL METABOLITE IN PLASMA
Flanagan, R. J.; Harris, L; Holt, D. W.; McKenna, W. J.; Rowland, E.; Storey, G. C. A.
Poisons Unit Guy's Hosp. London UK GB SE1 9RT
Br. J. Clin. Pharmacol.; (82) P 281P-280P; Vol 13; No 2

27-82-0046 (0046)
SENSITIVE HIGH-PERFORMANCE LIQUID CHROMATOGRAPHIC ASSAY METHODS FOR MONITORING METHOTRIMEPRAZINE AND DIMETHOTHIAZINE LEVELS IN PLASMA
Holt, Jacquelyn E.; Kaye, C. M.; Sankey, M. G.
Clin. Assay Unit May and Baker Ltd. Dagenham/Essex UK GB RM10 7XS
Br. J. Clin. Pharmacol.; (82) P 282P; Vol 13; No 2

27-82-0047 (0047)
A SIMPLE AND SENSITIVE HPLC METHOD FOR THE ASSAY OF PROCHLORPERAZINE IN PLASMA
Sankey, M. G.; Holt, Jacquelyn E.; Kaye, C. M.
Clin. Assay Unit May and Baker Ltd. Dagenham/Essex UK GB RM10 7XS
Br. J. Clin. Pharmacol.; (82) P 578-80; Vol 13; No 4

27-82-0048 (0048)
HIGH PRESSURE LIQUID CHROMATOGRAPHIC METHOD FOR THE DETERMINATION OF MELPHALAN IN PLASMA
Woodhouse, K. W.; Henderson, D. B.
Wolfson Unit Clin. Pharmacol. Univ. Newcastle upon Tyne Newcastle upon Tyne UK GB
Br. J. Clin. Pharmacol.; (82) P 605P; Vol 13; No 4

27-82-0049 (0049)
PROPRANOLOL IN BREAST MILK. AN INVESTIGATION BY HIGH PERFORMANCE LIQUID CHROMATOGRAPHY
Brodrick, A.; Casey, M.
North Yorkshire Area Health Author. UK GB
Br. J. Pharm. Pract.; (82) P 23-4; Vol 3; No 11

27-82-0050 (0050)
DETERMINATION OF INDOMETHACIN IN SERUM BY HIGH-PERFORMANCE LIQUID CHROMATOGRAPHY
Kubo, Hiroaki; Kobayashi, Yoshie; Kinoshita, Toshio
Sch. Pharm. Sci. Kitasato Univ. Tokyo Japan JP
Bunseki Kagaku; (82) P 462-6; Vol 31; No 8; In Japan

27-82-0051 (0051)
A RAPID HIGH-PERFORMANCE LIQUID CHROMATOGRAPHIC METHOD FOR SIMULTANEOUSLY DETERMINING THE CONCENTRATIONS OF TFM AND BAYER 73 IN WATER DURING LAMPRICIDE TREATMENTS
Dawson, V. K.
U. S. Fish Wildlife Serv. La Crosse USA US WI 54601
Can. J. Fish. Aquat. Sci.; (82) P 778-82; Vol 39; No 5

27-82-0052 (0052)
METABOLISM OF DAUNORUBICIN IN SENSITIVE AND RESISTANT EHRLICH ASCITES TUMOR CELLS. DETERMINATION BY HIGH-PRESSURE LIQUID CHROMATOGRAPHY
Londos-Gagliardi, Danielle; Baurain, Roger; Robert, Jacques; Aubel-Sadron, Genevieve
Cent. Biophys. Mol. CNRS Orleans Fr. FR F-45045
Cancer Chemother. Pharmacol.; (82) P 45-8; Vol 9; No 1

27-82-0053 (0053)
SIMULTANEOUS DETERMINATION OF BERBERINE, PALMATINE AND COPTISINE IN CRUDE DRUGS AND ORIENTAL PHARMACEUTICAL PREPARATIONS BY ION-PAIR HIGH-PERFORMANCE LIQUID CHROMATOGRAPHY
Misaki, Tetsuo; Sagara, Kazuhiko; Ojima, Mitsuharu; Kakizawa, Sadao; Oshima, Toshiyuki; Yoshizawa, Hiroshi
Res. Lab. daisho Pharm. Co., Ltd. Omiya Japan JP 330
Chem. Pharm. Bull.; (82) P 354-7; V71 30; No 1

27-82-0054 (0054)
HIGH-PERFORMANCE LIQUID CHROMATOGRAPHIC ASSAY OF CEFSULODIN, CEFOTIAM AND CEFMENOXIME IN SERUM AND URINE
Itakura, Koichi; Mitani, Masayoshi; Aoki, Isamu; Usui, Yoshiro
Cent. Res. Div. Takeda Chem. Ind., Ltd. Osaka Japan JP 532
Chem. Pharm. Bull.; (82) P 622-7; Vol 30; No 2

27-82-0055 (0055)
COMPARISON OF THE CONVENTIONAL METHODS AND HIGH-PERFORMANCE LIQUID CHROMATOGRAPHY FOR THE DETERMINATION OF TRIMETHOPRIM, SULFAMETHOXAZOLE AND ITS METABOLITE IN SERUM
Essers, L.; Korte, H.
Inst. Med. Microbiol. Univ. Essen Essen Fed. Rep. Ger. DE D-4300
Chemotherapy (Basel); (82) P 247-52; Vol 28; No 4

27-82-0056 (0056)
HPLC ANALYSIS OF CLAVULANIC ACID AND PHARMACOKINETIC INVESTIGATION ON URINARY EXCRETION OF BRL25000 (CLAVULANIC ACID-AMOXICILLIN) IN MAN
Uno, Toyozo; Haginaka, Jun; Nakagawa, Terumichi
Fac. Pharm. Sci. Kyoto Univ. Kyoto Japan JP 606
Chemotherapy (Tokyo); (82) P 118-24; Vol 30; No Suppl. 2; In Japan

27-82-0057 (0057)
PHARMACOKINETIC STUDIES ON CEPHALOTHIN USING HIGH-PERFORMANCE LIQUID CHROMATOGRAPHY: HEPATIC CLEARANCE OF CEPHALOTHIN IN GUINEA PIGS WITH ACUTE LIVER IMPAIRMENT
Watanabe, Michio
Inst. Med. Sci. Tokyo Univ. Tokyo Japan JP 10
Chemotherapy (Tokyo); (82) P 163-71; Vol 30; No 2; In Japan

27-82-0058 (0058)
MEASUREMENT OF THEOPHYLLINE, MEXILETINE AND A NEW CEPHALOSPORIN IN PLASMA BY LC (LIQUID CHROMATOGRAPHY)
Dresse, Al E.; Delapierre-Gabriel, D.
Dep. Pharmacol. Univ. Liege Liege Belg. BE
Chromatogr. Sci.; (82) P 161-72; Vol 20; No Biol./Biomed. Appl. Liq. Chromatogr. 4

27-82-0059 (0059)
APPLICATION OF LC (LIQUID CHROMATOGRAPHY) IN THE INVESTIGATION OF THE ANTIARRHYTHMIC EFFICACY OF BRETYLIUM
Carr, Ronald Keith; Woosler, R. L.
Dep. Pharmacol. Vanderbilt Univ. Nashville USA US TN

Chromatogr. Sci.; (82) P 197-206; Vol 20; No Biol./Biomed. Appl. Liq. Chromatogr. 4

27-82-0060 (0060)
DETERMINATION OF CHLORAMPHENICOL AND ITS MONOSUCCINATE ESTER IN PIGLET PLASMA USING HPLC
Wiese, B.; Martin, K.; Hermansson, J.
Fac. Vet. Med. Swedish Univ. Agric. Sci. Uppsala Swed. SE S-751 23
Chromatographia; (82) P 737-42; Vol 15; No 12

27-82-0061 (0061)
HIGH-PERFORMANCE LIQUID CHROMATOGRAPHIC STUDY OF THE ACCELERATED DEGRADATION OF CGP-9000
Cadorniga, R.; Saiz-Vadillo, M. C.; Perotas van Herckenrode, H.
Univ. Complutense Madrid Spain ES
Cienc. Ind. Farm.; (82) P 47-55; Vol 1; No 2; In Span

27-82-0062 (0062)
IMPROVED ISOCRATIC LIQUID-CHROMATOGRAPHIC SIMULTANEOUS MEASUREMENT OF PHENYTOIN, PHENOBARBITAL, PRIMIDONE, CARBAMAZEPINE, ETHOSUXIMIDE, AND N-DESMETHYLMETHSUXIMIDE IN SERUM
Szabo, George K.; Browne, Thomas R.
Sch. Med. Boston Univ. Boston USA US MA 02130
Clin. Chem. (Winston-Salem, N. C.); (82) P 100-4; Vol 28; No 1

27-82-0063 (0063)
IMPROVED LIQUID-CHROMATOGRAPHY OF ASPIRIN, SALICYLATE, AND SALICYLURIC ACID IN PLASMA, WITH A MODIFICATION FOR DETERMINING ASPIRIN METABOLITES IN URINE
Buskin, Jean N.; Upton, Robert A.; Williams, Roger L.
Sch. Pharm. Univ. California San Francisco USA US CA 94143
Clin. Chem. (Winston-Salem, N. C.); (82) P 1200-3; Vol 28; No 5

27-82-0064 (0064)
LIQUID-CHROMATOGRAPHIC MEASUREMENT OF NITRAZEPAM IN PLASMA
Kelly, Helen; Huggett, Anthony; Dawling, Sheila
Poisons Unit Guy's Hosp. London UK GB
Clin. Chem. (Winston-Salem, N. C.); (82) P 1478-81; Vol 28; No 7

27-82-0065 (0065)
RESULTS BY FLUORESCENT IMMUNOASSAY FOR PHENYTOIN COMPARED WITH THOSE BY ENZYME IMMUNOASSAY, LIQUID CHROMATOGRAPHY, AND DISCRETE ANALYSIS (DUPONT ACA)
Gonzalez, G.; Cid-Amador, A.; Steele, B.; Castro, A.
Sch. Med. Univ. Miami Miami USA US FL 33101
Clin. Chem. (Winston-Salem, N. C.); (82) P 1494-6; Vol 28; No 7

27-82-0066 (0066)
IMPROVED LIQUID-CHROMATOGRAPHIC ASSAY OF QUINIDINE AND ITS METABOLITES IN BIOLOGICAL FLUIDS
Rakhit, Ashok; Kunitani, Michael; Holford, Nicholas H. G.; Riegelman, Sidney
Dep. Pharm. Univ. California San Francisco USA US CA 94143
Clin. Chem. (Winston-Salem, N. C.); (82) P 1505-9; Vol 28; No 7

27-82-0067 (0067)
LIQUID-CHROMATOGRAPHIC DETERMINATION OF PENTOBARBITAL IN PLASMA WITH USE OF A RESIN COLUMN AND AN ALKALINE MOBILE PHASE
Gupta, Ram N.; Smith, Paul T.; Eng, Francis
Clin. Chem. Lab. St. Joseph's Hosp. Hamilton Can. CA ON L8N 1Y4
Clin. Chem. (Winston-Salem, N. C.); (82) P 1772-4; Vol 28; No 8

27-82-0068 (0068)
INTERFERENCE IN THE PROPOSED SELECTED METHOD FOR DETERMINATION OF THEOPHYLLINE BY LIQUID CHROMATOGRAPHY. COMMENTS
Jonkman, Jan H. G.; De Zeeuw, Rokus A.; Schoenmaker, Roelof
Lab. Drug Anal. Assen Neth. NL 9400 AC
Clin. Chem. (Winston-Salem, N. C.); (82) P 1987-8; Vol 28; No 9

27-82-0069 (0069)
INTERFERENCE IN THE PROPOSED SELECTED METHOD FOR DETERMINATION OF THEOPHYLLINE BY LIQUID CHROMATOGRAPHY. REPLY TO COMMENTS
Broussard, Larry A.
Med. Lab. Cunningham Pathol. Assoc. Birmingham USA US AL 35256-9899
Clin. Chem. (Winston-Salem, N. C.); (82) P 1988; Vol 28; No 9

27-82-0070 (0070)
LIQUID-CHROMATOGRAPHIC DETERMINATION OF AMITRIPTYLINE AND ITS METABOLITES IN SERUM, WITH ADSORPTION ONTO GLASS MINIMIZED
Edelbroek, Peter M.; De Haas, Ed J. M.; De Wolff, Frederik A.
Lab. Toxicol. Univ. Hosp. Leiden Neth. NL 2333 AA
Clin. Chem. (Winston-Salem, N. C.); (82) P 2143-8; Vol 28; No 10

27-82-0071 (0071)
LIQUID-CHROMATOGRAPHIC DETERMINATION OF AMOXAPINE AND 8-HYDROXYAMOXAPINE IN HUMAN SERUM
Tasset, J. J.; Hassan, F. M.
Med. Cent. Univ. Cincinnati Cincinnati USA US OH 45267
Clin. Chem. (Winston-Salem, N. C.); (82) P 2154-7; Vol 28; No 10

27-82-0072 (0072)
THEOPHYLLINE, DYPHYLLINE, CAFFEINE, ACETAMINOPHEN, SALICYLATE, ACETYLSALICYLATE, PROCAINAMIDE, AND N-ACETYLPROCAINAMIDE DETERMINED IN SERUM WITH A SINGLE LIQUID-CHROMATOGRAPHIC ASSAY
Ou, Ching Nan; Frawley, Vicki L.
Dep. Pathol. Texas Child. Hosp. Houston USA US TX 77030
Clin. Chem. (Winston-Salem, N. C.); (82) P 2157-60; Vol 28; No 10

27-82-0073 (0073)
LIQUID-CHROMATOGRAPHIC DETERMINATION OF CYCLOSPORINE IN SERUM WITH USE OF A RAPID EXTRACTION PROCEDURE
Yee, Gary C.; Gmur, Dennis J.; Kennedy, Michael S.
Clin. Pharmacol. Sect. Fred Hutchinson Cancer Res. Cent. Seattle USA US WA 98104
Clin. Chem. (Winston-Salem, N. C.); (82) P 2269-71; Vol 28; No 11

27-82-0074 (0074)
LIQUID-CHROMATOGRAPHIC PROCEDURE FOR SIMULTANEOUS ANALYSIS FOR EIGHT BENZODIAZEPINES IN SERUM
Lensmeyer, Gary L.; Rajani, Cynthia; Evenson, Merle A.
Clin. Toxicol. Lab. Univ. Wisconsin Hosp. Madison USA US WI 53792
Clin. Chem. (Winston-Salem, N. C.); (82) P 2274-8; Vol 28; No 11

27-82-0075 (0075)
LIQUID-CHROMATOGRAPHIC IDENTIFICATION OF ACETAMINOPHEN IN CEREBROSPINAL FLUID WITH USE OF ELECTROCHEMICAL DETECTION
Walsh, Francis X.; Langlais, Philip J.; Bird, Edward D.
Brain Tissue Resour. Cent. McLean Hosp. Belmont USA US MA 20178
Clin. Chem. (Winston-Salem, N. C.); (82) P 382-3; Vol 28; No 2

27-82-0076 (0076)
QUANTITATIVE LIQUID-CHROMATOGRAPHIC TECHNIQUE FOR THE SIMULTANEOUS ASSAY OF

TRICYCLIC ANTIDEPRESSANT DRUGS IN PLASMA OR SERUM
Sonsalla, Patricia K.; Jennison, Thomas A.; Finkle, Bryan S.
Cent. Hum. Toxicol. Univ. Utah Salt Lake City USA 84112
Clin. Chem. (Winston-Salem, N. C.); (82) P 457-61; Vol 28; No 3

27-82-0077 (0077)
SIMPLE PREPARATION AND LIQUID-CHROMATOGRAPHIC ANALYSIS FOR TRICYCLIC ANTIDEPRESSANTS IN SERUM
Koteel, Prabhakaran; Mullins, Richard E.; Gqdsden, Richard H.
Dep. Lab. Med. Med. Univ. South Carolina Charleston USA US)C 29425
Clin. Chem. (Winston-Salem, N. C.); (82) P 462-6; Vol 28; No 3

27-82-0078 (0078)
LIQUID-CHROMATOGRAPHIC ANALYSIS FOR SERUM THEOPHYLLINE IN LESS THAN 70 SECONDS
Kabra, Pokar M.; Marton, Laurence J.
Sch. Med. Univ. California San Francisco USA US CA 94143
Clin. Chem. (Winston-Salem, N. C.); (82) P 687-9; Vol 28; No 4, Pt. 1

27-82-0079 (0079)
ANALYSIS FOR CARBAMAZEPINE AND PHENYTOIN IN SERUM WITH A HIGH-SPEED LIQUID CHROMATOGRAPHY SYSTEM (PERKIN-ELMER)
Kinberger, Bengt; Holmen, Anders
Dep. Clin. Chem. Cent. Hosp. Halmstad Swed. SE S-301 85
Clin. Chem. (Winston-Salem, N. C.); (82) P 718-19; Vol 28; No 4, Pt. 1

27-82-0080 (0080)
PHARMACOKINETICS OF INTRAVENOUS GLIBENCLAMIDE INVESTIGATED BY A HIGH PERFORMANCE LIQUID CHROMATOGRAPHIC ASSAY
Rogers, H. J.; Spector, R. G.; Morrison, P. J.; Bradbrook, I. D.
Med. Sch. Guy's Hosp. London UK GB SE1 9RT
Diabetologia; (82) P 37-40; Vol 23; No 1

27-82-0081 (0081)
QUALITY CONTROL OF VALERIAN PREPARATIONS. 2. VALERENIC ACID AND VALERENAL AS CHARACTERISTIC COMPONENTS OF VALERIANA OFFICINALIS. DETERMINATION BY HPLC
Haensel, Rudolf; Schulz, Jutta
Inst. Pharmakogn. Phytochem. Freie Univ. Berlin Berlin Fed. Rep. Ger. DE 1000
Dtsch. Apoth.-Ztg.; (82) P 215-19; Vol 122; No 5; In Ger

27-82-0082 (0082)
HPLC ASSAY AND PHARMACOKINETICS OF METHOTREXATE
Chen, Mei Ling
Med. Cent. Univ. Illinois Chicago USA US IL
From: Diss. Abstr. Int. B 1982, 43(4), 1050.
DISSERTATION; (82) P 135 pp.; Avail.: Univ. Microfilms Int., Order No. DA8220701

27-82-0083 (0083)
STUDY ON THE RESORPTION OF CLOXAZEPINE SUCCINATE IN EXPERIMENTAL ANIMALS AND THE USE OF HIGH PERFORMANCE LIQUID CHROMATOGRAPHY (HPLC)
Nakova, G.; Tomov, E.; Georgieva, R.
Nauchnoizsled. Khim.-Farm. Inst. Sofia Bulg. BG
Eksp. Med. Morfol.; (82) P 39-43; Vol 21; No 1; In Bulg

27-82-0084 (0084)
DETERMINATION OF SULFATHIAZOLE IN HONEY FROM MEDICATED HONEY BEE COLONIES BY HIGH-PERFORMANCE LIQUID CHROMATOGRAPHY ON A CYANOAMINO POLAR PHASE
Argauer, Robert J.; Shimanuki, Hachiro; Knox, David A.
ARS USDA Beltsville USA US MD 20705
Environ. Entomol.; (82) P 820-3; Vol 11; No 4

27-82-0085 (0085)
ANTICONVULSANT DRUGS DETERMINED BY EMIT IN SPIKED BRAIN SAMPLES: A CORRELATION WITH HPLC VALUES
Monaco, Francesco; Ramsay, R. Eugene; Vazquez, David
Med. Sch. Sassari Univ. Sassari Italy IT
Epilepsia (N. Y.); (82) P 15-18; Vol 23; No 1

27-82-0086 (0086)
CARBAMAZEPINE DETERMINATION IN SALIVA OF CHILDREN: ENZYME IMMUNOASSAY (EMIT) VERSUS HIGH PRESSURE LIQUID CHROMATOGRAPHY
Paxton, James W.
Sch. Med. Auckland Univ. Auckland N. Z. NZ
Epilepsia (N. Y.); (82) P 185-9; Vol 23; No 2

27-82-0087 (0087)
HIGH-PERFORMANCE REVERSED-PHASE LIQUID CHROMATOGRAPHY (HPLC) OF FAVISM-INDUCING FACTORS IN VICIA FABA L
Lattanzio, V.; Bianco, V. V.; Lafiandra, D.
Cent. Studio Orticoltura CNR Bari Italy IT I-70126
Experientia; (82) P 789-90; Vol 38; No 7

27-82-0088 (0088)
HPLC OF SOME 1,4-BENZODIAZEPINES AND THEIR HYDROLYSIS PRODUCTS, BENZOPHENONES
Kmetec, V.; Kozjek, F.; Japelj, M.
Fac. Nat. Sci. Technol. Univ. Edvard Kardelj Ljubljana Yugoslavia YU
Farm. Vestn. (Ljubljana); (82) P 1-8; Vol 33; No 1; In Slovenian

27-82-0089 (0089)
HPLC DETERMINATION OF THE POTENTIALLY IMMUNOGENIC IMPURITIES AND DECOMPOSITION PRODUCTS IN TABLET FORMULATIONS WITH ACETYLSALICYLIC ACID FROM VARIOUS YUGOSLAVIAN MANUFACTURERS
Pavli, V.; Cinc, M.
Inst. Pharm. Drug Control Ljubljana Yugoslavia YU
Farm. Vestn. (Ljubljana); (82) P 165-78; Vol 33; No 3; In Slovenian

27-82-0090 (0090)
LIQUID CHROMATOGRAPHY OF ANTIDEPRESSANTS
Shcherbina, A. M.
Lvov State Med. Inst.P Lvov USSR SU
Farm. Zh. (Kiev); (82) P 76-7; No 1; In Ukrain

27-82-0091 (0091)
COCCIDOSTATICS IN ANIMAL FEEDS. II. NEW HPLC METHOD FOR THE DETERMINATION OF PYRIMETHAMINE IN MEDICATED FEEDS
Di Simone, L.; Ponti, F.; Settimj, G.; Martillotti, F.
Lab. Chim. Farm. Ist. Super. Sanita Rome Italy IT
Farmaco, Ed. Prat.; (82) P 177-84; Vol 37; No 5; In Ital

27-82-0092 (0092)
HPLC ASSAY OF PYRAZINOIC ACID IN HUMAN PLASMA IN THE PRESENCE OF PYRAZINAMIDE AND OTHER ANTITUBERCULOSIS DRUGS USING AN AUTOMATIC SAMPLER
Ratti, B.; Toselli, A.; Beretta, E.; Bernareggi, A.
Lab. Ric. Gruppo Lepetit S.p.A. Milan Italy IT
Farmaco, Ed. Prat.; (82) P 226-34; Vol 37; No 7

27-82-0093 (0093)
HPLC DETERMINATION OF 2-(7-THEOPHYLLINYLMETHYL)-1,3-DIOXOLANE
Badini, C.; Masera, F.; Franzone, J. S.
Cent. Ric. Ist. Biol. Chemioter. "ABC" S.p.A. Turin Italy IT
Farmaco, Ed. Prat.; (82) P 320-4; Vol 37; No 10; In Ital

27-82-0094 (0094)
THIOCOLCHICOSIDE DETERMINATION BY HPLC IN PHARMACEUTICAL FORMULATIONS
Forni, G. P.
Controllo Qualita Milan Italy IT 20141
Fitoterapia; (82) P 3-8; Vol 53; No 1-2

27-82-0095 (0095)
DETERMINATION OF SULFHYDRYL COMPOUNDS IN PHYSIOLOGICAL FLUIDS BY LIQUID CHROMATOGRAPHY WITH POLAROGRAPHIC DETECTION
Zygmunt, B.; Hanekamp, H. B.; Bos, P.; Frei, R. W.
Dep. Anal. Chem. Free Univ. Amsterdam Neth. NL NL-1081 HV
Fresenius' Z. Anal. Chem.; (82) P 197-200; Vol 311; No 3

27-82-0096 (0096)
HIGH-PERFORMANCE LIQUID-CHROMATOGRAPHIC ANALYSIS OF TRIMETHOPRIM AND SULFAMETHOXAZOLE IN PLASMA
Ali, Syed Laik; Moeller, Helga
Zentrallab. Dtsch. Apoth. Eschborn Fed. Rep. Ger. DE D-6236
Fresenius' Z. Anal. Chem.; (82) P 514-16; Vol 311; No 5

27-82-0097 (0097)
HPLC ASSAY AND PHARMACOKINETICS OF METHOTREXATE
Chen, Mei Ling
Med. Cent., Univ. Illinois, Chicago IL U S A
1982 135 pp. Avail. Univ. Microfilm Int.,Order No. DA 8220701
From Diss. Abstr. Int. B 1982 P 1050; (82), Vol 43, No 4

27-82-0098 (0098)
DETERMINATION OF CARBOQUONE IN PLASMA AND ASCITES BY HIGH PERFORMANCE LIQUID CHROMATOGRAPHY AFTER INTRAVENOUS AND INTRAPERITONEAL ADMINISTRATION IN MAN
Hisaoka, Masafumi; Morioka, Tadashi; Yagita, Akikuni
Prod. Dev. Lab. Sankyo Co., Ltd. Tokyo Japan JP 140
Gann ; (82) P 161-6; Vol 73; No 2

27-82-0099 (0099)
ANALYSIS OF ANTIEPILEPTIC DRUGS BY HPLC AND UV DETECTION WITH PHOTODIODE ARRAYS
Dorner, Wolf G.
Inzell Fed. Rep. Ger. DE 8221
GIT Labor-Med.; (82) P 352-3, 356-7; Vol 5; No 5; In Ger

27-82-0100 (0100)
DETERMINATION OF L-ERYTHRO-TETRAHYDROBIOPTERIN IN BIOLOGICAL TISSUES BY HIGH PRESSURE LIQUID CHROMATOGRAPHY AND ELECTROCHEMICAL DETECTION
Braeutigam, Matthias; Dreesen, Rainer
Pharmakol. Inst. Freie Univ. Berlin Fed. Rep. Ger. DE D-1000/33
Hoppe-Seyler's Z. Physiol. Chem.; (82) P 1203-7; Vol 363; No 10

27-82-0101 (0101)
PROGRESS IN THE DETERMINATION OF ANTIEPILEPTIC DRUGS BY HIGH-PERFORMANCE LIQUID CHROMATOGRAPHY
Miyagawa, Fumio; Maeda, Toshie; Nakamura, Masao; Toida, Toyoji; Arai, Mitsuko
Sch. Med. St. Marianna Univ. Kanagawa Japan JP
Igaku no Ayumi; (82) P 393-400; Vol 121; No 7; In Japan

27-82-0102 (0102)
PHARMACOKINETICS OF PROPYLTHIOURACIL STUDIED BY HIGH-PERFORMANCE LIQUID CHROMATOGRAPHY
Okuno, Akimasa; Taguchi, Tetsuo; Inyaku, Fumie
Dep. Pediatr. Asahikawa Med. Coll. Asahikawa Japan JP

Igaku no Ayumi; (82) P 73-4; Vol 121; No 2; In Japan

27-82-0103 (0103)
ASSAY OF METHYLDOPA AND HYDROCHLOROTHIAZIDE IN COMBINATION DOSAGE FORMS BY HIGH PRESSURE LIQUID CHROMATOGRAPHY
Rao, G. Ramana; Raghuveer, S.; Mohan, K. Rama
Qual. Control Lab. Hyderabad India IN 500 037
Indian Drugs; (82) P 328-30; Vol 19; No 8

27-82-0104 (0104)
THE ANALYSIS OF 4-CHLORO-5-SULFAMOYLANTHRANILIC ACID IN THE BULK MATERIAL AND PHARMACEUTICAL PREPARATIONS OF FUROSEMIDE BY A HIGH-PERFORMANCE LIQUID CHROMATOGRAPHIC METHOD
Rapaka, R. S.; Roth, J.; Prasad, V. K.
Div. Biopharm. Food Drug Adm. Washington USA US DC 20204
Int. J. Pharm.; (82) P 237-47; Vol 11; No 3

27-82-0105 (0105)
HYDROLYSIS AND EPIMERIZATION KINETICS OF PILOCARPINE IN BASIC AQUEOUS SOLUTION AS DETERMINED BY HPLC
Bundgaard, Hans; Hansen, Steen Honore
Dep. Pharm. Chem. R. Dan. Sch. Pharm. Copenhagen Den. DK DK-2100
Int. J. Pharm.; (82) P 281-9; Vol 10; No 4

27-82-0106 (0106)
A SIMPLE HPLC METHOD FOR THE DISSOLUTION STUDIES ON LEVOTHYROXINE SODIUM TABLETS
Rapaka, Rao S.; Roth, Jeri; Brine, G. A.; Prasad, Vadlamani K.
Bur. Drugs Food Drug Adm. Washington USA US DC 20204
Int. J. Pharm.; (82) P 285-94; Vol 12; No 4

27-82-0107 (0107)
MONITORING OF SERUM THEOPHYLLINE BY HIGH-PERFORMANCE LIQUID CHROMATOGRAPHY WITHOUT INTERFERENCE BY VARIOUS COADMINISTERED DRUGS
Frutkoff, Irving W.; Menczel, Jacob; Kidroni, Gideon
Dep. Intern. Med. Hadassah Univ. Hosp., Mt. Scopus Jerusalem Israel IL
Isr. J. Med. Sci.; (82) P 639-41; Vol 18; No 5

27-82-0108 (0108)
TRACE ANALYSIS OF NATURAL CAPSAICINOIDS IN ANIMAL FEED, HUMAN URINE, AND WASTEWATER BY HIGH-PRESSURE LIQUID CHROMATOGRAPHY
Johnson, Eric K.; Thompson, Harold C., Jr.; Bowman, Malcolm C.
Dep. Health Hum. Serv. Food and Drug Adm. Jefferson USA AR 72079
J. Agric. Food Chem.; (82) P 3X4-9; Vol 30; No 2

27-82-0109 (0109)
AN HPLC METHOD FOR THE QUANTITATION OF QUINIDINE AND ITS METABOLITES IN PLASMA: AN APPLICATION TO A QUINIDINE-PHENYTOIN DRUG INTERACTION STUDY
Pershing, Lynn Kennard; Peat, Michael A.; Finkle, Bryan S.
Coll. Med. Univ. Utah Salt Lake City USA US UT 84112
J. Anal. Toxicol.; (82) P 153-6; Vol 6; No 3

27-82-0110 (0110)
DETERMINATION OF LABETALOL IN A POST MORTEM CASE USING HPLC
Pannell, L. K.; Thomson, B. M.; Wilkinson, L. F.
Chem. Div. DSIR Christchurch N. Z. NZ
J. Anal. Toxicol.; (82) P 193-5; Vol 6; No 4

27-82-0111 (0111)
ENZYME IMMUNOASSAY, LIQUID CHROMATOGRAPHY, AND SPECTROFLUOROMETRY COMPARED FOR

THE DETERMINATION OF PROCAINAMIDE AND N-ACETYLPROCAINAMIDE IN SERUM
Pape, Brian E.
Health Sci. Cent. Univ. Missouri Columbia USA US MO 65212
J. Anal. Toxicol.; (82) P 44-8; Vol 6; No 1

27-82-0112 (0112)
HIGH-PERFORMANCE LIQUID CHROMATOGRAPHIC DETERMINATION OF PRIMIDONE IN TABLETS
Roberts, Stanley E.
Winchester Eng. Anal. Cent. Food and Drug Adm. Winchester USA US MA 01890
J. Assoc. Off. Anal. Chem.; (82) P 1063-5; Vol 65; No 5

27-82-0113 (0113)
HIGH PRESSURE LIQUID CHROMATOGRAPHIC DETERMINATION OF BACITRACIN IN PREMIX FEEDS AND FINISHED FEEDS: COLLABORATIVE STUDY
Gallagher, John B.; Love, Paul W.; Knotts, Linda L.
Int. Miner. and Chem. Corp. Terre Haute USA US IN 47808
J. Assoc. Off. Anal. Chem.; (82) P 1178-85; Vol 65; No 5

27-82-0114 (0114)
HIGH-PRESSURE LIQUID CHROMATOGRAPHIC DETERMINATION OF PHYSOSTIGMINE SALICYLATE AND PHYSOSTIGMINE SULFATE IN LIQUIDS AND OINTMENTS: COLLABORATIVE STUDY
Tymes, Norlin W.
Food and Drug Adm. Baltimore USA US MD 21201
J. Assoc. Off. Anal. Chem.; (82) P 132-7; Vol 65; No 1

27-82-0115 (0115)
REVERSE PHASE HIGH-PRESSURE LIQUID CHROMATOGRAPHIC DETERMINATION OF ARPRINOCID IN ANIMAL FEEDS
Thorpe, Virginia A.
Lab. Div. Michigan Dep. Agric. East Lansing USA US MI 48823
J. Assoc. Off. Anal. Chem.; (82) P 43-7; Vol 65; No 1

27-82-0116 (0116)
EVALUATION OF SILICA AND POLAR BONDED COLUMNS FOR LIQUID CHROMATOGRAPHIC ANALYSIS OF TEMEPHOS FORMULATIONS
Mount, Dwight L.; Miles, James W.
Parasit. Dis. Div. Cent. Infect. Dis. Atlanta USA US GA 30333
J. Assoc. Off. Anal. Chem.; (82) P 575-9; Vol 65; No 3

27-82-0117 (0117)
HIGH-PERFORMANCE LIQUID CHROMATOGRAPHIC METHOD FOR DETERMINATION OF TEMEPHOS IN TECHNICAL AND FORMULATED PRODUCTS: COLLABORATIVE STUDY
Miles, James W.; Mount, Dwight L.
Parasit. Dis. Div. Cent. Infect. Dis. Atlanta USA US GA 30333
J. Assoc. Off. Anal. Chem.; (82) P 580-3; Vol 65; No 3

27-82-0118 (0118)
HIGH-PERFORMANCE LIQUID CHROMATOGRAPHIC DETERMINATION OF CLOCAPRAMINE IN FEED AND PLASMA
Geahchan, Aziz; Chambon, Paul; Genoux, Pierre
Lab. Toxicol. Fac. Pharm. Lyon Fr. FR 69373
J. Assoc. Off. Anal. Chem.; (82) P 706-10; Vol 65; No 3

27-82-0119 (0119)
REVERSE PHASE HIGH-PERFORMANCE LIQUID CHROMATOGRAPHIC DETERMINATION OF PHENPROCOUMON IN TABLETS
Schmidt, Walter F.
Food Drug Adm. Atlanta USA US GA 30309
J. Assoc. Off. Anal. Chem.; (82) P 753-5; Vol 65; No 3

27-82-0120 (0120)
HIGH-PRESSURE LIQUID CHROMATOGRAPHY OF ZEARALENONE AND ZEARALENOLS IN RAT URINE AND LIVER
James, Lynda J.; McGirr, Larry G.; Smith, Trevor K.
Coll. Biol. Sci. Univ. Guelph Guelph Can. CA ON N1G 2W1
J. Assoc. Off. Anal. Chem.; (82) P 8-13; Vol 65; No 1

27-82-0121 (0121)
USE OF POST-COLUMN DERIVATIZATION IN LIQUID CHROMATOGRAPHIC DETERMINATION OF SULFAMETHAZINE AND SULFATHIAZOLE IN FEEDS AND FEED PREMIXES
Stringham, Rodger W.; Mundell, Elizabeth C.; Smallidge, Robert L.
Dep. Biochem. Purdue Univ. West Lafayette SA US IN 47907
J. Assoc. Off. Anal. Chem.; (82) P 823-7; Vol 65; No 4

27-82-0122 (0122)
RAPID FORMATION OF POLY-.GAMMA.-GLUTAMYL DERIVATIVES OF METHOTREXATE AND THEIR ASSOCIATION WITH DIHYDROFOLATE REDUCTASE AS ASSESSED BY HIGH PRESSURE LIQUID CHROMATOGRAPHY IN THE EHRLICH ASCITES TUMOR CELL IN VITRO
Fry, David W.; Yalowich, Jack C.; Goldman, I. David
Dep. Med. Med. Coll. Virginia Richmond USA US VA 23298
J. Biol. Chem.; (82) P 1890-6; Vol 257; No 4

27-82-0123 (0123)
RAPID CATABOLISM OF 5-FLUOROURACIL IN FRESHLY ISOLATED RAT HEPATOCYTES AS ANALYZED BY HIGH PERFORMANCE LIQUID CHROMATOGRAPHY
Sommadossi, Jean Pierre; Gewirtz, David A.; Diasio, Robert B.; Aubert, Claude; Cano, Jean Paul; Goldman, I. David
Dep. Med. Med. Coll. Virginia Richmond USA US VA 23298
J. Biol. Chem.; (82) P 8171-6; Vol 257; No 14

27-82-0124 (0124)
PURIFICATION OF SPORIDESMIN BY PREPARATIVE HIGH PRESSURE LIQUID CHROMATOGRAPHY
Ronaldson, John W.
Ruakura Agric. Res. Cent. Hamilton N. Z. NZ
J. Chem. Technol. Biotechnol.; (82) P 556-8; Vol 32; No 4

27-82-0125 (0125)
AN AUTOMATED HPLC METHOD FOR THE ASSAY OF PROPRANOLOL AND ITS BASIC METABOLITES IN PLASMA AND URINE
Lo, Man Wai; Silber, Bernie; Riegelman, Sidney
Sch. Pharm. Univ. California San Francisco USA US CA 94143
J. Chromatogr. Sci.; (82) P 126-31; Vol 20; No 3

27-82-0126 (0126)
HIGH-PERFORMANCE LIQUID CHROMATOGRAPHIC ASSAY OF METHOTREXATE, 7-HYDROXYMETHOTREXATE, 4-DEOXY-4-AMINO-N10-METHYLPTEROIC ACID AND SULFAMETHOXAZOLE IN SERUM, URINE AND CEREBROSPINAL FLUID
Cairnes, David A.; Evans, William E.
Clin. Pharmacokinet. Lab. St. Jude Child. Res. Hosp. Memphis USA US TN 38101
J. Chromatogr.; (82) P 103-10; Vol 231; No 1

27-82-0127 (0127)
SIMULTANEOUS DETERMINATION OF CHLORPROMAZINE AND LEVOMEPROMAZINE IN HUMAN PLASMA AND URINE BY HIGH-PERFORMANCE LIQUID CHROMATOGRAPHY USING ELECTROCHEMICAL DETECTION
Murakami, Kazuo; Murakami, Koji; Ueno, Tadataka; Hijikata, Junko; Shirasawa, Kimio; Muto, Toshitake
Lab. Psychopharmacol. Sagami Hosp. Sagamihara Japan JP
J. Chromatogr.; (82) P 103-12; Vol 227; No 1

27-82-0128 (0128)
DETERMINATION OF PYRIMETHAMINE IN HUMAN PLASMA AFTER ADMINISTRATION OF FANSIDAR OR FANSIDAR-MEFLOQUINE BY MEANS OF HIGH-PERFORMANCE LIQUID CHROMATOGRAPHY WITH FLUORESCENCE DETECTION
Timm, U.; Weidekamm, E.
Biol. Pharm. Res. Dep. F. Hoffmann-La Roche and Co. Basel Switz. CH CH-4002
J. Chromatogr.; (82) P 107-14; Vol 230; No 1

27-82-0129 (0129)
DETERMINATION OF 7-AMINOFLUNITRAZEPAM (RO 20-1815) AND 7-AMINODESMETHYL-METHYLFLUNITRAZEPAM (RO 5-4650) IN PLASMA BY HIGH-PERFORMANCE LIQUID CHROMATOGRAPHY AND FLUORESCENCE DETECTION
Sumirtapura, Y. C.; Aubert, C.; Coassolo, P.; Cano, J. P.
Lab. Pharmacocinet. Toxicocinet. INSERM Marseille Fr. FR F-13385
J. Chromatogr.; (82) P 111-18; Vol 232; No 1

27-82-0130 (0130)
DETERMINATION OF TRIPAMIDE AND ITS METABOLITES IN PLASMA, RED BLOOD CELLS AND URINE BY HIGH-PERFORMANCE LIQUID CHROMATOGRAPHY
Horie, T.; Ohno, T.; Kinoshita, K.
Res. Dev. Div. Eisai Co., Ltd. Tokyo Japan JP
J. Chromatogr.; (82) P 111-19; Vol 231; No 1

27-82-0131 (0131)
DETERMINATION OF (R)- AND (S)-ALPRENOLOL AND (R)- AND (S)-METOPROLOL AS THEIR DIASTEREOMERIC DERIVATIVES IN HUMAN PLASMA BY REVERSED-PHASE LIQUID CHROMATOGRAPHY
Hermansson, Joergen; Von Bahr, Christer
Dep. Drugs Natl. Board Health and Welfare Uppsala Swed. SE S-751 25
J. Chromatogr.; (82) P 113-27; Vol 227; No 1

27-82-0132 (0132)
SEPARATION OF SIX MAJOR PROSTACYCLIN METABOLITES BY HIGH-PERFORMANCE LIQUID CHROMATOGRAPHY
Pieroni, J. P.; Lee, W. H.; Wong, P. Y. K.
Dep. Pharmacol. New York Med. Coll. Valhalla USA US NY 10595
J. Chromatogr.; (82) P 115-20; Vol 230; No 1

27-82-0133 (0133)
SIMULTANEOUS DETERMINATION OF TOLMETIN AND ITS METABOLITE IN BIOLOGICAL FLUIDS BY HIGH-PERFORMANCE LIQUID CHROMATOGRAPHY
Desiraju, Ravi K.; Sedberry, Donald C., Jr.; Ng, Kung Tat
Dep. Drug Metab. McNeil Pharm. Spring House USA US PA 19477
J. Chromatogr.; (82) P 119-28; Vol 232; No 1

27-82-0134 (0134)
METHOD FOR THE SIMPLIFIED ANALYSIS OF DEPROTEINIZED PLASMA AND URINARY ISOPROTERENOL BY HIGH-PERFORMANCE LIQUID CHROMATOGRAPHY
Kishimoto, Yasuro; Ohgitani, Shigeki; Yamatodani, Atsushi; Kuro, Masakazu; Okumura, Fukuichiro
Dep. Cardiovasc. Surg. Natl. Cardiovasc. Cent. Osaka Japan JP 565
J. Chromatogr.; (82) P 121-7; Vol 231; No 1

27-82-0135 (0135)
DETERMINATION OF THE ANTIDEPRESSANT AGENT CITALOPRAM AND METABOLITES IN PLASMA BY LIQUID CHROMATOGRAPHY WITH FLUORESCENCE DETECTION
Oeyehaug, Ellen; Oestensen, Eilif Terje; Salvesen, Bjarne
Agder Coll. Kristiansand Norway NO 4600
J. Chromatogr.; (82) P 129-35; Vol 227; No 1

27-82-0136 (0136)
HIGH-PERFORMANCE LIQUID CHROMATOGRAPHIC DETERMINATION OF PRIFINIUM QUATERNARY AMMONIUM ION IN HUMAN SERUM AND URINE
Tokuma, Yoji; Tamura, Yoko; Noguchi, Hideyo
Res. Lab. Fujisawa Pharm. Co. Ltd. Osaka Japan JP 532
J. Chromatogr.; (82) P 129-36; Vol 231; No 1

27-82-0137 (0137)
SPECIMEN HANDLING AND HIGH-PERFORMANCE LIQUID CHROMATOGRAPHIC DETERMINATION OF FUROSEMIDE
Kerremans, A. L. M.; Tan, Y.; Van Ginneken, C. A. M.; Gribnau, F. W. J.
Div. Clin. Pharmacol. Pharmacokinet. Dep. Pharmacol. Nijmegen Neth. NL 6500 HB
J. Chromatogr.; (82) P 129-39; Vol 229; No 1

27-82-0138 (0138)
DETERMINATION OF ELLIPTICINE IN BIOLOGICAL SAMPLES BY HIGH-PERFORMANCE LIQUID CHROMATOGRAPHY
Bykadi, G.; Flora, K. P.; Cradock, J. C.; Poochikian, G. K.
Div. Cancer Treat. Natl. Cancer Inst. Bethesda USA US MD 20205
J. Chromatogr.; (82) P 137-44; Vol 231; No 1

27-82-0139 (0139)
DETERMINATION OF AMIKACIN IN MICROLITER QUANTITIES OF BIOLOGICAL FLUIDS BY HIGH-PERFORMANCE LIQUID CHROMATOGRAPHY USING 1-FLUORO-2,4-DINITROBENZENE DERIVATIZATION
Wong, Lawrence T.; Beaubien, Arthur R.; Pakuts, Atilla P.
Drug Res. Lab. Frederick G. Banting Build. Ottawa Can. CA ON K1A 0L2
J. Chromatogr.; (82) P 145-54; Vol 231; No 1

27-82-0140 (0140)
DETERMINATION OF HEPTAMINOL IN PHARMACEUTICAL PREPARATIONS BY HIGH-PERFORMANCE LIQUID CHROMATOGRAPHY
Nicolas, A.; Leroy, P.; Moreau, A.; Mirjolet, M.
Lab. Chim. Anal. Toxicol. Fac. Sci. Pharm. Biol. Nancy Fr. FR 54001
J. Chromatogr.; (82) P 148-52; Vol 244; No 1

27-82-0141 (0141)
HIGH-PERFORMANCE LIQUID CHROMATOGRAPHY OF BENZOPHENONE DERIVATIVES FOR THE DETERMINATION OF BENZODIAZEPINES IN CLINICAL EMERGENCIES
Violon, Christiane; Pessemier, Lyotta; Vercruysse, Antoine
Vrije Univ. Brussel Brussels Belg. BE B-1090
J. Chromatogr.; (82) P 157-68; Vol 236; No 1

27-82-0142 (0142)
HIGH-PERFORMANCE LIQUID CHROMATOGRAPHIC DETERMINATION OF CLIOQUINOL AND ITS CONJUGATES IN BIOLOGICAL MATERIALS
Hayakawa, Kazuichi; Kitada, Kumiko; Hamaki, Michiko; Miyazaki, Motoichi
Fac. Pharm. Sci. Kanazawa Univ. Kanazawa Japan JP
J. Chromatogr.; (82) P 159-65; Vol 229; No 1

27-82-0143 (0143)
RAPID HIGH-PERFORMANCE LIQUID CHROMATOGRAPHIC DETERMINATION OF ACETAMINOPHEN IN DOSAGE FORMS, BY USING A TOTALLY AQUEOUS MOBILE PHASE
Gilpin, R. K.; Gaudet, M. H.
IBM Instruments, Inc. Danbury USA US CT 06810
J. Chromatogr.; (82) P 160-4; Vol 248; No 1

27-82-0144 (0144)
DETERMINATION OF BROMOLASALOCID IN PLASMA BY HIGH-PERFORMANCE LIQUID CHROMATOGRAPHY WITH FLUORIMETRIC DETECTION

Brooks, Marvin A.; Stronjny, Norman; Hackman, Martin R.; De Silva, J. Arthur F.
Dep. Pharmacokinet. Biopharmaceut. Hoffmann-La Roche Inc. Nutley USA US NJ 07110
J. Chromatogr.; (82) P 167-78; Vol 229; No 1

27-82-0145 (0145)
DETERMINATION OF SODIUM FLAVODATE IN BODY FLUIDS BY HIGH-PERFORMANCE LIQUID CHROMATOGRAPHY. APPLICATION TO CLINICAL PHARMACOKINETIC STUDIES
Zecca, L.; Guadagni, L.; Bareggi, S. R.
Ist. Farmacol. Milan Italy IT 20129
J. Chromatogr.; (82) P 168-74; Vol 230; No 1

27-82-0146 (0146)
SIMULTANEOUS DETERMINATION OF BENZTROPINE MESYLATE AND BENZOPHENONE BY HIGH-PERFORMANCE LIQUID CHROMATOGRAPHY
Biffar, Stephen; Tibbetts, Donald
Merck, Sharp & Dohme West Point USA US PA 19486
J. Chromatogr.; (82) P 169-73; Vol 237; No 1

27-82-0147 (0147)
LIQUID CHROMATOGRAPHIC SEPARATION AND QUANTITATION OF 2-AMINO-1,3,4-THIADIAZOLE (NSC-4728) FROM HUMAN AND MURINE SERUM
Ackerly, Cynthia C.; Newman, Robert A.; Myers, Carl; McCormack, John J.
Coll. Med. Univ. Vermont Burlington USA US VT 05405
J. Chromatogr.; (82) P 175-80; Vol 230; No 1

27-82-0148 (0148)
HIGH-PERFORMANCE LIQUID CHROMATOGRAPHY-FLUORESCENCE ANALYSIS FOR INDOMETHACIN AND METABOLITES IN BIOLOGICAL FLUIDS
Bernstein, M. S.; Evans, M. A.
Med. Cent. Univ. Illinois Chicago USA US IL 60680
J. Chromatogr.; (82) P 179-87; Vol 229; No 1

27-82-0149 (0149)
HIGH-PERFORMANCE LIQUID CHROMATOGRAPHIC ANALYSIS OF INDAPAMIDE (RHC 2555) IN URINE, PLASMA AND BLOOD
Choi, R. Leslie; Rosenberg, Marc; Grebow, Peter E.; Huntley, Thomas E.
Res. Dev. Div. Revlon Health Care Group Tuckahoe USA US NY 10707
J. Chromatogr.; (82) P 181-7; Vol 230; No 1

27-82-0150 (0150)
RAPID METHOD FOR DETERMINATION OF KETANSERIN, A NOVEL ANTISEROTONIN DRUG, BY HIGH-PERFORMANCE LIQUID CHROMATOGRAPHY
Simon, Verne; Somani, Pitambar
Dep. Pharmacol. Med. Coll. Ohio Toledo USA US OH 43699
J. Chromatogr.; (82) P 186-91; Vol 232; No 1

27-82-0151 (0151)
DETERMINATION OF PENTAZIRIDINOCYCLODIPHOSPHATHIAZENE IN BIOLOGICAL FLUIDS BY HIGH-PERFORMANCE LIQUID CHROMATOGRAPHY
Matsushima, Eiji; Umeno, Yukihiko; Marunaka, Teruyoshi; Akagi, Hiroshi
Res. Lab. Taiho Pharm. Co. Ltd. Tokushima Japan JP 771-01
J. Chromatogr.; (82) P 188-93; Vol 230; No 1

27-82-0152 (0152)
DETERMINATION OF TALINOLOL IN HUMAN PLASMA BY HIGH-PERFORMANCE LIQUID CHROMATOGRAPHY
Poetter, H.; Huelm, M.; Richter, K.
Forschungsabt. 2 VEB Arzneimittelwerk Dresden Radebeul Ger. Dem. Rep. DD

DDR-8122
J. Chromatogr.; (82) P 189-92; Vol 241; No 1; In Ger

27-82-0153 (0153)
MULTICHANNEL DIODE ARRAY UV-VISIBLE SPECTROPHOTOMETER AS DETECTOR IN SCREENING FOR UNKNOWN BUTOPROZINE METABOLITES IN DOG BILE BY HIGH-PERFORMANCE LIQUID CHROMATOGRAPHY
Overzet, Fokko; Ghijsen, Rudy T.; Drenth, Ben F. H.; De Zeeuw, Rokus A.
Dep. Toxicol. State Univ. Groningen Groningen Neth. NL 9713 AW
J. Chromatogr.; (82) P 190-5; Vol 240; No 1

27-82-0154 (0154)
DETERMINATION OF A NOVEL 5-FLUOROURACIL DERIVATIVE IN RAT AND HUMAN PLASMA BY HIGH-PERFORMANCE LIQUID CHROMATOGRAPHY
Takada, Kanji; Yoshikawa, Hiroshi; Muranishi, Shozo
Kyoto Coll. Pharm. Kyoto Japan JP 607
J. Chromatogr.; (82) P 192-6; Vol 232; No 1

27-82-0155 (0155)
SIMULTANEOUS DETERMINATION OF PROMETHAZINE AND TWO OF ITS CIRCULATING METABOLITES BY HIGH-PERFORMANCE LIQUID CHROMATOGRAPHY
Taylor, Glyn; Houston, J. Brian
Dep. Pharm. Univ. Manchester Manchester UK GB M13 9PL
J. Chromatogr.; (82) P 194-8; Vol 230; No 1

27-82-0156 (0156)
LIQUID CHROMATOGRAPHIC DETERMINATION OF CLORAZEPATE DECOMPOSITION RATES
Clark, C. Randall; Ravis, William R.; Dockens, Randy; Barksdale, Jeffrey M.; Arrington, Hugh S.; D'Andrea, George H.
Sch. Pharm. Auburn Univ. Auburn USA US AL 36849
J. Chromatogr.; (82) P 196-201; Vol 240; No 1

27-82-0157 (0157)
HIGH-PERFORMANCE LIQUID CHROMATOGRAPHIC SEPARATION AND PROTON NUCLEAR MAGNETIC RESONANCE IDENTIFICATION OF THE 6-MONO-CIS AND 6,6'-DI-CIS ISOMERS OF RHODOXANTHIN
Englert, Gerhard; Vecchi, Max
Cent. Res. Units F. Hoffmann-La Roche and Co., Ltd. Basel Switz. CH 4002
J. Chromatogr.; (82) P 197-203; Vol 235; No 1

27-82-0158 (0158)
HIGH-PERFORMANCE LIQUID CHROMATOGRAPHIC ASSAY FOR DETERMINATION OF A NEW .BETA.-BLOCKING AGENT FM 24
Lefebvre, M. A.; Julian, B.; Fourtillan, J. B.
Cent. Etud. Rech. Poitiers Fr. FR 86000
J. Chromatogr.; (82) P 199-202; Vol 230; No 1

27-82-0159 (0159)
DETERMINATION OF TIODAZOSIN IN PLASMA AND WHOLE BLOOD BY HIGH-PERFORMANCE LIQUID CHROMATOGRAPHY
Mico, Bruce A.; Baughman, Robert A., Jr.; Benet, Leslie Z.
Sch. Pharm. Univ. California San Francisco USA US CA 94143
J. Chromatogr.; (82) P 203-6; Vol 230; No 1

27-82-0160 (0160)
DETERMINATION OF 2-OXO-PYRROLIDINE-1-ACETAMIDE (PIRACETAM) IN HUMAN PLASMA USING HIGH-PERFORMANCE LIQUID CHROMATOGRAPHY
Rieck, W.; Platt, D.
Inst. Gerontol. Univ. Erlangen-Nuernberg Nuernberg Fed. Rep. Ger. DE
J. Chromatogr.; (82) P 203-6; Vol 232; No 1

27-82-0161 (0161)
HIGH-PERFORMANCE LIQUID CHROMATOGRAPHIC METHOD FOR THE DETERMINATION OF PLASMA AND URINE METAPRAMINE AFTER DANSYLATION
Sommadossi, J. P.; Lemar, M.; Necciari, J.; Sumirtapura, Y.; Cano, J. P.; Gaillot, J.
Lab. Pharmacocinet. INSERM Marseille Fr. FR 13385
J. Chromatogr.; (82) P 205-13; Vol 228

27-82-0162 (0162)
DETERMINATION OF ETOMIDATE IN HUMAN PLASMA BY HIGH-PERFORMANCE LIQUID CHROMATOGRAPHY
Ellis, Edwin Owen; Beck, Peter Richard
Dep. Clin. Biochem. Rotherham Dist. Gen. Hosp. Oakwood/Rotherham UK GB S60 2UD
J. Chromatogr.; (82) P 207-11; Vol 232; No 1

27-82-0163 (0163)
SIMPLE, RAPID AND SENSITIVE REVERSED-PHASE HIGH-PERFORMANCE LIQUID CHROMATOGRAPHIC METHOD FOR THIOPENTAL AND PENTOBARBITAL DETERMINATION IN PLASMA AND BRAIN TISSUE
Shiu, Gerald K.; Nemoto, Edwin M.
Sch. Med. Univ. Pittsburgh Pittsburgh USA US PA 15261
J. Chromatogr.; (82) P 207-12; Vol 227; No 1

27-82-0164 (0164)
DETERMINATION OF TOLAZOLINE IN PLASMA BY HIGH-PERFORMANCE LIQUID CHROMATOGRAPHY
Rovei, V.; Remones, G.; Thenot, J. P.; Morselli, P. L.
Dep. Clin. Res. Synthelabo Paris Fr. FR 75013
J. Chromatogr.; (82) P 210-15; Vol 231; No 1

27-82-0165 (0165)
DETERMINATION OF A NEW ANTIINFLAMMATORY AGENT,1-ISOBUTYL-3,4-DIPHENYLPYRAZOLE-5-ACETIC ACID, BY HIGH-PERFORMANCE LIQUID CHROMATOGRAPHY
Bannier, A.; Brazier, J. L.; Soubeyrand, J.; Comet, F.
Dep. Anal. Chem. Fac. Pharm. Lyon Fr. FR 69373/3
J. Chromatogr.; (82) P 213-18; Vol 227; No 1

27-82-0166 (0166)
DETERMINATION OF METOPROLOL AND .ALPHA.-HYDROXYMETOPROLOL IN PLASMA BY HIGH-PERFORMANCE LIQUID CHROMATOGRAPHY
Pautler, David B.; Jusko, William J.
Sch. Pharm. State Univ. New York Amherst USA US NY 14260
J. Chromatogr.; (82) P 215-22; Vol 228

27-82-0167 (0167)
USE OF HIGH-PERFORMANCE LIQUID CHROMATOGRAPHY-MASS SPECTROMETRY FOR THE STUDY OF THE METABOLISM OF RANITIDINE IN MAN
Martin, L. E.; Oxford, Janet; Tanner, R. J. N.
Biochem. Pharmacol. Dep. Glaxo Group Res. Ltd. Ware/Herts. UK GB
J. Chromatogr.; (82) P 215-24; Vol 251; No 2

27-82-0168 (0168)
HIGH-PERFORMANCE LIQUID CHROMATOGRAPHIC DETERMINATION BY DIPYRIDAMOLE
Rosenfeld, J.; Devereaux, D.; Buchanan, M. R.; Turpie, A. G. G.
Fac. Health Sci. McMaster Univ. Hamilton Can. CA ON L8S 4J9
J. Chromatogr.; (82) P 216-21; Vol 231; No 1

27-82-0169 (0169)
HIGH-PERFORMANCE LIQUID CHROMATOGRAPHIC METHOD FOR THE QUANTITATIVE ANALYSIS OF THE ARYLOXYPROPANOLAMINES PROPRANOLOL, METOPROLOL AND ATENOLOL IN PLASMA

AND TISSUE
Winkler, Horst; Ried, Walter; Lemmer, Bjoern
Cent. Pharmacol. J. W. Goethe-Univ. Frankfurt/Main Fed. Rep. Ger. DE
J. Chromatogr.; (82) P 223-34; Vol 228

27-82-0170 (0170)
HIGH-PERFORMANCE LIQUID CHROMATOGRAPHIC ANALYSIS OF MICONAZOLE IN PLASMA
Sternson, L. A.; Patton, T. F.; King, T. B.
Dep. Pharm. Chem. Univ. Kansas Lawrence USA US KS 66045
J. Chromatogr.; (82) P 223-8; Vol 227; No 1

27-82-0171 (0171)
DETERMINATION OF MICONAZOLE IN HUMAN SALIVA USING HIGH-PERFORMANCE LIQUID CHROMATOGRAPHY
Turner, A.; Warnock, D. W.
Dep. Microbiol. Bristol R. Infirm. Bristol UK GB BS2 8HW
J. Chromatogr.; (82) P 229-32; Vol 227; No 1

27-82-0172 (0172)
DETERMINATION OF IMPURITIES IN PROPRANOLOL HYDROCHLORIDE BY HIGH-PERFORMANCE LIQUID CHROMATOGRAPHY ON DYNAMICALLY MODIFIED SILICA
Helboe, Per
Drug Standardization Lab. Natl. Board Health Broenshoej Den. DK DK-2700
J. Chromatogr.; (82) P 229-38; Vol 245; No 2

27-82-0173 (0173)
DETERMINATION OF FURAZOLIDONE IN SWINE PLASMA USING LIQUID CHROMATOGRAPHY
Veale, Harry S.; Harrington, George W.
Dep. Chem. Temple Univ. Philadelphia USA US PA 19122
J. Chromatogr.; (82) P 230-4; Vol 240; No 1

27-82-0174 (0174)
MEASUREMENT OF HALOPERIDOL IN HUMAN PLASMA USING REVERSED-PHASE HIGH-PERFORMANCE LIQUID CHROMATOGRAPHY
Jatlow, Peter I.; Miller, Ronald; Swigar, Mary
Sch. Med. Yale Univ. New Haven USA US CT 06510
J. Chromatogr.; (82) P 233-8; Vol 227; No 1

27-82-0175 (0175)
RAPID ESTIMATION OF DIFLUNISAL IN PLASMA AND URINE BY HIGH-PERFORMANCE LIQUID CHROMATOGRAPHY AND A COMPARISON WITH A FLUOROMETRIC METHOD
Balali-Mood, Mehdi; King, Ian S.; Prescott, Laurie F.
Univ. Dep. Therapeu. R. Infirm. Edinburgh UK GB EH3 9YW
J. Chromatogr.; (82) P 234-40; Vol 229; No 1

27-82-0176 (0176)
HIGH-PERFORMANCE LIQUID CHROMATOGRAPHIC ANALYSIS OF PENTAZOCINE IN BLOOD AND PLASMA
Anderson, R. D.; Ilett, K. F.; Dusci, L. J.; Hackett, L. P.
Dep. Pharmacol. Univ. West. Australia Nedlands Australia AU 6009
J. Chromatogr.; (82) P 239-43; Vol 227; No 1

27-82-0177 (0177)
DETERMINATION OF CHLOROQUINE AND ITS DESETHYL METABOLITE IN PLASMA, RED BLOOD CELLS AND URINE BY LIQUID CHROMATOGRAPHY
Alvan, Gunnar; Ekman, Lena; Lindstroem, Bjoern
Dep. Clin. Pharmacol. Huddinge Univ. Huddinge Swed. SE S-141 86
J. Chromatogr.; (82) P 241-7; Vol 229; No 1

27-82-0178 (0178)
REVERSED-PHASE ION-PAIR HIGH-PERFORMANCE LIQUID CHROMATOGRAPHIC ASSAY OF 5-FLUOROURACIL, 5'-DEOXY-5-FLUOROURIDINE, THEIR NUCLEOSIDES, MONO-, DI-, AND TRIPHOSPHATE NUCLEOTIDES WITH A MIXTURE OF QUATERNARY AMMONIUM IONS
Au, J. L. S.; Wientjes, M. G.; Luccioni, C. M.; Rustum, Y. M.
Dep. Exp. Therapeut. Roswell Park Mem. Inst. Buffalo USA US NY 14263
J. Chromatogr.; (82) P 245-56; Vol 228

27-82-0179 (0179)
SEPARATION OF ATRAZINE AND SOME OF ITS DEGRADATION PRODUCTS BY HIGH-PERFORMANCE LIQUID CHROMATOGRAPHY
Vermeulen, N. M. J.; Apostolides, Z.; Potgieter, D. J. J.; Nel, P. C.; Smit, N. S. H.
Dep. Biochem. Univ. Pretoria Pretoria S. Afr. ZA 0002
J. Chromatogr.; (82) P 247-53; Vol 240; No 1

27-82-0180 (0180)
DETERMINATION OF CHLOROQUINE AND ITS DE-ETHYLATED METABOLITES IN HUMAN PLASMA BY ION-PAIR HIGH-PERFORMANCE LIQUID CHROMATOGRAPHY
Brown, Nesbitt D.; Poon, Bing T.; Chulay, Jeffrey D.
Div. Biochem. Walter Reed Army Inst. Res. Washington USA US DC 20012
J. Chromatogr.; (82) P 248-54; Vol 229; No 1

27-82-0181 (0181)
IMPROVED HIGH-PERFORMANCE LIQUID CHROMATOGRAPHIC ASSAY FOR CIMETIDINE USING RANITIDINE AS AN INTERNAL STANDARD
Fleitman, J.; Torosian, G.; Perrin, J. H.
Coll. Pharm. Univ. Florida Gainesville USA US FL 32610
J. Chromatogr.; (82) P 255-8; Vol 229; No 1

27-82-0182 (0182)
DETERMINATION OF CEFSULODIN, CEFOTIAM, CEPHALEXIN, CEFOTAXIME, DEACETYLCEFOTA-XIME, CEFUROXIME AND CEFROXADIN IN PLASMA AND URINE BY HIGH-PERFORMANCE LIQUID CHROMATOGRAPHY
Lecaillon, J. B.; Rouan, M. C.; Souppart, C.; Febvre, N.; Juge, F.
Cent. Rech. Biopharmaceut. Ciba-Geigy Rueil-Malmaison Fr. FR 92506
J. Chromatogr.; (82) P 257-67; Vol 228

27-82-0183 (0183)
HIGH-PERFORMANCE LIQUID CHROMATOGRAPHY OF BIFLAVONES FROM GINKGO BILOBA L
Briancon-Scheid, Francoise; Guth, Annelise; Anton, Robert
Lab. Pharmacog. Fac. Pharm. Strasbourg Fr. FR 67048
J. Chromatogr.; (82) P 261-7; Vol 245; No 2

27-82-0184 (0184)
DETERMINATION OF THE .GAMMA.-AMINOBUTYRIC ACID AGONIST 4,5,6,7-TETRAHYDRO-ISOXAZOLO(5,4-C)PYRIDINE-3-OL IN URINE BY HIGH-PERFORMANCE LIQUID CHROMATOGRAPHY
Schultz, Birgit; Hansen, Steen Honore
Dep. Chem. R. Dan. Sch. Pharm. Copenhagen Den. DK DK-2100
J. Chromatogr.; (82) P 279-84; Vol 228

27-82-0185 (0185)
PRECOLUMN DERIVATIZATION OF FLUORESCAMINE AND HIGH-PERFORMANCE LIQUID CHROMATOGRAPHIC ANALYSIS OF DRUGS. APPLICATION TO TOCAINIDE
Sedman, A. J.; Gal, J.
Sch. Med. Univ. Colorado Denver USA US CO 80262
J. Chromatogr.; (82) P 315-26; Vol 232; No 2

27-82-0186 (0186)
DETERMINATION OF SALBUTAMOL IN HUMAN PLASMA WITH BIMODAL HIGH-PERFORMANCE LIQUID CHROMATOGRAPHY AND A ROTATED DISK AMPEROMETRIC DETECTOR
Oosterhuis, B.; Van Boxtel, C. J.
Dep. Med. Pharmacol. Univ. Hosp. Amsterdam Amsterdam Neth. NL 1054 EG
J. Chromatogr.; (82) P 327-34; Vol 232; No 2

27-82-0187 (0187)
HIGH-PERFORMANCE LIQUID CHROMATOGRAPHIC DETERMINATION OF IBUPROFEN AND ITS MAJOR METABOLITES IN BIOLOGICAL FLUIDS
Lockwood, Graham F.; Wagner, John G.
Upjohn Cent. Clin. Pharmacol. Univ. Michigan Ann Arbor USA US MI 48109
J. Chromatogr.; (82) P 335-43; Vol 232; No 2

27-82-0188 (0188)
DETERMINATION OF DIAZEPAM AND ITS PHARMACOLOGICALLY ACTIVE METABOLITES IN BLOOD BY BOND ELUT COLUMN EXTRACTION AND REVERSED-PHASE HIGH-PERFORMANCE LIQUID CHROMATOGRAPHY
Rao, Srinivas N.; Dhar, Amiya K.; Kutt, Henn; Okamoto, Michiko
Med. Coll. Cornell Univ. New York USA US NY 10021
J. Chromatogr.; (82) P 341-8; Vol 231; No 2

27-82-0189 (0189)
AUTOMATED HIGH-PERFORMANCE LIQUID CHROMATOGRAPHIC METHOD FOR THE DETERMINATION OF ANTIPYRINE AND ITS METABOLITES IN URINE. SOME PRELIMINARY RESULTS OBTAINED FROM SMOKERS AND NONSMOKERS
Nakagawa, Akihiko; Nakamura, Kanichi; Ishizaki, Takashi; Chiba, Kan
Anal. Metab. Res. Lab. Sankyo Co., Ltd. Tokyo Japan JP 140
J. Chromatogr.; (82) P 349-60; Vol 231; No 2

27-82-0190 (0190)
RAPID ANALYSIS OF HEXOBARBITAL AND ITS MAIN METABOLITES IN MICE BY HIGH-PERFORMANCE LIQUID CHROMATOGRAPHY
Boecker, R.
Inst. Pharmakol. Toxikol. Univ. Erlangen-Nuernberg Erlangen Fed. Rep. Ger. DE D 8520
J. Chromatogr.; (82) P 350-4; Vol 228

27-82-0191 (0191)
DETERMINATION OF PENBUTOLOL AND ITS HYDROXYLATED METABOLITE IN BIOLOGICAL FLUIDS BY REVERSED-PHASE HIGH-PERFORMANCE LIQUID CHROMATOGRAPHY
Bernard, N.; Cuisinaud, G.; Sassard, J.
Dep. Physiol. Fac. Pharmacy Lyon Fr. FR 69008
J. Chromatogr.; (82) P 355-61; Vol 228

27-82-0192 (0192)
LIQUID CHROMATOGRAPHIC ASSAY OF PHENOTHIAZINE, THIOXANTHENE AND BUTYROPHENONE NEUROLEPTICS AND ANTIHISTAMINES IN BLOOD AND PLASMA WITH CONVENTIONAL AND RADIAL COMPRESSION COLUMNS AND UV AND ELECTROCHEMICAL DETECTION
Curry, Stephen H.; Brown, Edgar A.; Hu, Oliver Y. P.; Perrin, John H.
Coll. Pharm. Univ. Florida Gainesville USA US FL 32610
J. Chromatogr.; (82) P 361-76; Vol 231; No 2

27-82-0193 (0193)
SIMPLE AND RAPID DETERMINATION OF PROPRANOLOL AND ITS ACTIVE METABOLITE, 4-HYDROXYPROPRANOLOL, IN HUMAN PLASMA BY LIQUID CHROMATOGRAPHY WITH FLUORESCENCE DETECTION
Albani, Fiorenzo; Riva, Roberto; Baruzzi, Agostino
Inst. Neurol. Univ. Bologna Bologna Italy IT 40123
J. Chromatogr.; (82) P 362-5; Vol 228

27-82-0194 (0194)
ION-PAIR HIGH-PERFORMANCE LIQUID CHROMATOGRAPHIC ASSAY OF 4-AMINOPYRIDINE IN SERUM
Shinohara, Y.; Miller, R. D.; Castagnoli, N., Jr.
Med. Cent. Univ. California San Francisco USA US CA 94143
J. Chromatogr.; (82) P 363-72; Vol 230; No 2

27-82-0195 (0195)
SIMULTANEOUS DETERMINATION OF EPHEDRINE SULFATE, HYDROXYZINE HYDROCHLORIDE AND THEOPHYLLINE IN TABLETS BY REVERSED-PHASE HIGH-PERFORMANCE LIQUID CHROMATOGRAPHY
Roberts, Stanley E.; Delaney, Michael F.
Winchester Eng. Anal. Cent. U. S. Food Drug Adm. Winchester USA US MA 01890
J. Chromatogr.; (82) P 364-8; Vol 242; No 2

27-82-0196 (0196)
SPECIFIC DETERMINATION OF QUINIDINE AND METABOLITES IN BIOLOGICAL FLUIDS BY REVERSED-PHASE HIGH-PERFORMANCE LIQUID CHROMATOGRAPHY
Leroyer, R.; Jarreau, C.; Pays, M.
Fac. Pharm. Univ. Caen Caen Fr. FR
J. Chromatogr.; (82) P 366-71; Vol 228

27-82-0197 (0197)
DETERMINATION OF RIFAMPICIN, DESACETYLRIFAMPICIN, ISONIAZID AND ACETYLISONIAZID BY HIGH-PERFORMANCE LIQUID CHROMATOGRAPHY: APPLICATION TO HUMAN SERUM EXTRACTS, POLYMORPHONUCLEOCYTES AND ALVEOLAR MACROPHAGES
Guillaumont, M.; Leclercq, M.; Frobert, Y.; Guise, B.; Harf, R.
Inst. Pasteur Lyon Lyon Fr. FR 69365
J. Chromatogr.; (82) P 369-76; Vol 232; No 2

27-82-0198 (0198)
HIGH-PERFORMANCE LIQUID CHROMATOGRAPHY ASSAY FOR FENBUFEN AND TWO SERUM METABOLITES
Fleitman, Jeffrey S.; Schulman, Stephen G.; Perrin, John H.
Coll. Pharm. Univ. Florida Gainesville USA US FL 32610
J. Chromatogr.; (82) P 372-6; Vol 228

27-82-0199 (0199)
QUANTITATION OF 6-MERCAPTOPURINE IN BIOLOGIC FLUIDS USING HIGH-PERFORMANCE LIQUID CHROMATOGRAPHY: A SELECTIVE AND NOVEL PROCEDURE
Narang, Prem K.; Yeager, Russell L.; Chatterji, Dulal C.
Clin. Pharmacokinet. Res. Lab. Pharm. Dev. Serv. Bethesda USA US MD 20205
J. Chromatogr.; (82) P 373-80; Vol 230; No 2

27-82-0200 (0200)
HIGH-PERFORMANCE LIQUID CHROMATOGRAPHIC MEASUREMENT OF AMIODARONE AND DESETHYLAMIODARONE IN PLASMA OR SERUM AT THE CONCENTRATIONS ATTAINED FOLLOWING A SINGLE 400-MG DOSE
Storey, G. C. A.; Holt, D. W.
Poisons Unit Guy's Hosp. London UK GB SEI 9RT
J. Chromatogr.; (82) P 377-80; Vol 245; No 3

27-82-0201 (0201)
HIGH-PERFORMANCE LIQUID CHROMATOGRAPHIC ASSAY FOR MONITORING INDAPAMIDE AND ITS MAJOR METABOLITE IN URINE
Pietta, Piergiorgio; Calatroni, Alma; Rava, Angelo
Ist. Chim. Org. Anal. Milan Italy IT 20133
J. Chromatogr.; (82) P 377-81; Vol 228

27-82-0202 (0202)
SIMULTANEOUS DETERMINATION OF THIORIDAZINE AND ITS S-OXIDIZED AND N-DEMETHYLATED METABOLITES USING HIGH-PERFORMANCE LIQUID CHROMATOGRAPHY ON RADIALLY COMPRESSED SILICA
Kilts, Clinton D.; Patrick, Kennerly S.; Breese, George R.; Mailman, Richard B.
Med. Cent. Duke Univ. Durham USA US NC 27710
J. Chromatogr.; (82) P 377-91; Vol 231; No 2

27-82-0203 (0203)
SEPARATION OF 5.ALPHA.-REDUCED ANDROGENS BY REVERSED-PHASE HIGH-PERFORMANCE LIQUID CHROMATOGRAPHY
Sunde, Arne; Lundmo, Per I.
Inst. Cancer Res. Univ. Trondheim Trondheim Norway NO 7000
J. Chromatogr.; (82) P 381-4; Vol 242; No 2

27-82-0204 (0204)
RAPID ASSAY FOR PLASMA CHLORAMBUCIL AND PHENYL ACETIC MUSTARD USING REVERSED-PHASE LIQUID CHROMATOGRAPHY
Zakaria, Mona; Brown, Phyllis R.
Dep. Chem. Univ. Rhode Island Kingston USA US RI 02881
J. Chromatogr.; (82) P 381-9; Vol 230; No 2

27-82-0205 (0205)
SIMULTANEOUS DETERMINATION OF ALLOPURINOL, OXIPURINOL AND URIC ACID IN HUMAN PLASMA BY HIGH-PERFORMANCE LIQUID CHROMATOGRAPHY
Nissen, Poul
Pharmacol. Dep. DAK-Lab. Copenhagen Den. DK 2300
J. Chromatogr.; (82) P 382-6; Vol 228

27-82-0206 (0206)
ANALYSIS OF BUFLOMEDIL IN MOUSE, RAT AND RABBIT PLASMA BY REVERSED-PHASE ION-PAIR HIGH-PERFORMANCE LIQUID CHROMATOGRAPHY
Thomas, Elizabeth W.
Dep. Drug Metab. Abbott Lab. North Chicago USA US IL 60064
J. Chromatogr.; (82) P 387-91; Vol 228

27-82-0207 (0207)
COMPARISON OF AMINE MODIFIERS USED TO REDUCE PEAK TAILING OF 2-PHENYL-ETHYLAMINE DRUGS IN REVERSED-PHASE HIGH PERFORMANCE LIQUID CHROMATOGRAPHY
Gill, R.; Alexander, S. P.; Moffat, A. C.
Home Office Cent. Res. Establ. Aldermaston/Reading/Berks. UK GB RG7 4PN
J. Chromatogr.; (82) P 39-45; Vol 247; No 1

27-82-0208 (0208)
SIMULTANEOUS DETERMINATION OF AMITRIPTYLINE, NORTRIPTYLINE, AND THEIR RESPECTIVE ISOMERIC 10-HYDROXY METABOLITES IN PLASMA BY LIQUID CHROMATOGRAPHY
Suckow, Raymond F.; Cooper, Thomas B.
Anal. Psychopharmacol. Lab. Rockland Res. Inst. Orangeburg USA US NY 10962
J. Chromatogr.; (82) P 391-400; Vol 230; No 2

27-82-0209 (0209)
SIMULTANEOUS DETERMINATION OF AZATHIOPRINE AND 6-MERCAPTOPURINE IN SERUM BY REVERSED-PHASE HIGH-PERFORMANCE LIQUID CHROMATOGRAPHY
Tsutsumi, Kimiko; Otsuki, Yoshie; Kinoshita, Toshio
Sch. Pharm. Sci. Kitasato Univ. Tokyo Japan JP 108
J. Chromatogr.; (82) P 393-9; Vol 231; No 2

27-82-0210 (0210)
HIGH-PERFORMANCE LIQUID CHROMATOGRAPHIC ANALYSIS OF CIMETIDINE IN SERUM AND

URINE
Guay, David R. P.; Bockbrader, Howard N.; Matzke, Gary R.
Coll. Pharm. Wayne State Univ. Detroit USA US MI
J. Chromatogr.; (82) P 398-403; Vol 228

27-82-0211 (0211)
HIGH-PERFORMANCE LIQUID CHROMATOGRAPHIC DETERMINATION OF PIPOTIAZINE IN HUMAN PLASMA AND URINE
Le Roux, Y.; Gaillot, J.; Bieder, A.
Inst. Biopharm. Rhone-Poulenc Antony Fr. FR F-92160
J. Chromatogr.; (82) P 401-8; Vol 230; No 2

27-82-0212 (0212)
SIMULTANEOUS DETERMINATION OF DANTROLENE AND ITS METABOLITES, 5-HYDROXYDANTROLENE AND NITRO-REDUCED ACETYLATED DANTROLENE (F 490), IN PLASMA AND URINE OF MAN AND DOG BY HIGH-PERFORMANCE LIQUID CHROMATOGRAPHY
Wuis, E. W.; Grutters, A. C. L. M.; Vree, T. B.; Van der Kleyn, E.
Sint Radboud Hosp. Univ. Nijmegen Nijmegen Neth. NL 6526 GA
J. Chromatogr.; (82) P 401-9; Vol 231; No 2

27-82-0213 (0213)
SIMULTANEOUS DETERMINATION OF DANTROLENE AND ITS METABOLITE, 5-HYDROXY-DANTROLENE, IN HUMAN PLASMA BY HIGH-PERFORMANCE LIQUID CHROMATOGRAPHY
Katogi, Yoshio; Tamaki, Noriyasu; Adachi, Mamoru; Terao, Junichi; Mitomi, Mitsuo
Yaizu Plant Yamanouchi Pharm. Co. Yaizu Japan JP
J. Chromatogr.; (82) P 404-8; Vol 228

27-82-0214 (0214)
REGULATION OF SELECTIVITY IN THE SEPARATION OF PAFENOLOL AND POTENTIAL IMPURITIES BY REVERSED-PHASE ION-PAIR CHROMATOGRAPHY
Jansson, Sven Olof; Johansson, Svante
Dep. Anal. Chem. AB Haessle Moelndal Swed. SE S-431 83
J. Chromatogr.; (82) P 41-50; Vol 242; No 1

27-82-0215 (0215)
DETERMINATION OF 11-BROMOVINCAMINE IN HUMAN PLASMA BY HIGH-PERFORMANCE LIQUID CHROMATOGRAPHY
Brodie, R. R.; Chasseaud, L. F.
Dep. Metab. Pharmacokinet. Huntingdon Res. Cent. Huntingdon UK GB PE18 6ES
J. Chromatogr.; (82) P 413-17; Vol 228

27-82-0216 (0216)
DETERMINATION OF MOXALACTAM BY HIGH-PERFORMANCE LIQUID CHROMATOGRAPHY
Aravind, M. K.; Miceli, Joseph N.; Kauffman, Ralph E.
Sch. Med. Wayne State Univ. Detroit USA US MI 48201
J. Chromatogr.; (82) P 418-22; Vol 228

27-82-0217 (0217)
DETERMINATION OF CEFACLOR BY HIGH-PERFORMANCE LIQUID CHROMATOGRAPHY
Nahata, Milap C.
Coll. Pharm. Ohio State Univ. Columbus USA US OH 43210
J. Chromatogr.; (82) P 429-33; Vol 228

27-82-0218 (0218)
SIMULTANEOUS HIGH-PERFORMANCE LIQUID CHROMATOGRAPHIC ASSAY FOR QUINIDINE, DISOPYRAMIDE AND THE MONO-N-DEALKYLATED METABOLITE OF DISOPYRAMIDE
Norris, Ross L. G.; Ahokas, Jorma T.; Ravenscroft, Peter J.
Dep. Clin. Pharmacol. Princess Alexandra Hosp. Brisbane Australia AU
J. Chromatogr.; (82) P 433-7; Vol 230; No 2

27-82-0219 (0219)
HIGH-PERFORMANCE LIQUID CHROMATOGRAPHY ASSAY OF ACEBUTOLOL AND TWO OF ITS METABOLITES IN PLASMA AND URINE
Buskin, Jean N.; Upton, Robert A.; Jones, R. Matthew; Williams, Roger L.
Sch. Pharm. Univ. California San Francisco USA US CA 94143
J. Chromatogr.; (82) P 438-42; Vol 230; No 2

27-82-0220 (0220)
IMPROVED ACETAMINOPHEN ASSAY SENSITIVITY BY MODIFICATION OF A HIGH-PERFORMANCE LIQUID CHROMATOGRAPHY TECHNIQUE
Buskin, Jean N.; Upton, Robert A.; Williams, Roger L.
Sch. Pharm. Univ. California San Francisco USA US CA 94143
J. Chromatogr.; (82) P 443-7; Vol 230; No 2

27-82-0221 (0221)
DETERMINATION OF CAPTOPRIL IN HUMAN BLOOD BY HIGH-PERFORMANCE LIQUID CHROMATOGRAPHY WITH ELECTROCHEMICAL DETECTION
Shimada, Kazutake; Tanaka, Makoto; Nambara, Toshio; Imai, Yutaka; Abe, Keishi; Yoshinaga, Kaoru
Pharm. Inst. Tohoku Univ. Sendai Japan JP 980
J. Chromatogr.; (82) P 445-51; Vol 227; No 2

27-82-0222 (0222)
HIGH-PERFORMANCE LIQUID CHROMATOGRAPHIC ANALYSIS OF PROPAFENONE IN HUMAN PLASMA SAMPLES
Harapat, Sandra R.; Kates, Robert E.
Med. Cent. Stanford Univ. Stanford USA US CA 94305
J. Chromatogr.; (82) P 448-53; Vol 230; No 2

27-82-0223 (0223)
HIGH-PERFORMANCE LIQUID CHROMATOGRAPHIC METHOD FOR THE DETERMINATION OF DIPROPHYLLINE IN HUMAN SERUM
Paterson, N.
Dep. Chem. Pathol. Univ. Aberdeen Foresterhill/Aberdeen UK GB
J. Chromatogr.; (82) P 450-5; Vol 232; No 2

27-82-0224 (0224)
SPECIFIC AND SENSITIVE ASSAY OF CELIPROLOL IN BLOOD, PLASMA AND URINE USING HIGH-PERFORMANCE LIQUID CHROMATOGRAPHY
Buskin, J. N.; Upton, R. A.; Soergel, F.; Williams, R. L.; Lang, E.; Benet, L. Z.
Sch. Pharm. Univ. California San Francisco USA US CA 94143
J. Chromatogr.; (82) P 454-60; Vol 230; No 2

27-82-0225 (0225)
IMPROVED METHOD FOR THE ANALYSIS OF FUROSEMIDE IN PLASMA BY HIGH-PERFORMANCE LIQUID CHROMATOGRAPHY
Rapaka, Rao S.; Roth, Jeri; Viswanathan, C.; Goehl, Thomas J.; Prasad, Vadlamani K.; Cabana, Bernard E.
Div. Biopharm. Food and Drug Adm. Washington USA US DC 20204
J. Chromatogr.; (82) P 463-9; Vol 227; No 2

27-82-0226 (0226)
HIGH-PERFORMANCE LIQUID CHROMATOGRAPHY OF RO 10-9359 (TIGASON) AND ITS METABOLITE RO 10-1670 IN HUMAN PLASMA
Besner, Jean Guy; Meloche, Sylvain; Leclaire, Robert; Band, Pierre ; Mailhot, Sylvie
Fac. Pharm. Univ. Montreal Montreal Can. CA PQ H3C 3J7
J. Chromatogr.; (82) P 467-72; Vol 231; No 2

27-82-0227 (0227)
HIGH-PERFORMANCE LIQUID CHROMATOGRAPHIC METHOD FOR THE DETERMINATION OF DIETHYLAMINE, A METABOLITE OF DISULFIRAM, IN URINE
Neiderhiser, Dewey H.; Fuller, Richard K.
Sch. Med. Case West. Reserve Univ. Cleveland USA US OH 44106
J. Chromatogr.; (82) P 470-4; Vol 229; No 2

27-82-0228 (0228)
DETERMINATION OF PRENALTEROL IN PLASMA BY HIGH-PERFORMANCE LIQUID CHROMATOGRAPHY WITH FLUORESCENCE DETECTION
Oddie, C. J.; Jackman, G. P.; Bobik, A.
Clin. Res. Unit Alfred Hosp. Prahran Australia AU 3181
J. Chromatogr.; (82) P 473-7; Vol 231; No 2

27-82-0229 (0229)
A SIMPLE HIGH-PERFORMANCE LIQUID CHROMATOGRAPHIC PRECOLUMN TECHNIQUE FOR INVESTIGATION OF DRUG METABOLISM IN BIOLOGICAL FLUIDS
Voelter, Wolfgang; Kronbach, Thomas; Zech, Karl; Huber, Reinhard
Physiol.-Chem. Inst. Univ. Tuebingen Tuebingen Fed. Rep. Ger. DE D-7400
J. Chromatogr.; (82) P 475-82; Vol 239

27-82-0230 (0230)
RAPID AND SIMPLE METHOD FOR DETERMINATION OF LORCAINIDE, A NEW ANTIARRHYTHMIC DRUG, AND ITS MAJOR METABOLITE, NORLORCAINIDE, BY HIGH-PERFORMANCE LIQUID CHROMATOGRAPHY
Simon, Verne; Somani, Pitambar
Div. Clin. Pharmacol. Med. Coll. Ohio Toledo USA US OH 43699
J. Chromatogr.; (82) P 478-84; Vol 231; No 2

27-82-0231 (0231)
HIGH-PERFORMANCE LIQUID CHROMATOGRAPHIC ANALYSIS OF IMIDAZOPYRAZOLE (NSC 51143) IN SERUM
Boswell, Garry W.; Cohen, Jordan L.
Sch. Pharm. Univ. South. California Los Angeles USA US CA 90033
J. Chromatogr.; (82) P 485-91; Vol 231; No 2

27-82-0232 (0232)
REVERSED-PHASE HIGH-PERFORMANCE LIQUID CHROMATOGRAPHIC METHOD TO DETERMINE VIDARABINE AND HYPOXANTHINE ARABINOSIDE IN BIOLOGICAL FLUIDS
Bowman, David B.; Kauffman, Ralph E.
Sch. Med. Wayne State Univ. Detroit USA US MI 48201
J. Chromatogr.; (82) P 487-91; Vol 229; No 2

27-82-0233 (0233)
DETERMINATION OF CARBOFURAN AND ITS METABOLITES BY HIGH-PERFORMANCE LIQUID CHROMATOGRAPHY USING ON-LINE TRACE ENRICHMENT
Cramer, Paul H.; Drinkwine, Arbor D.; Going, John E.; Carey, Ann E.
Midwest Res. Inst. Kansas City USA US MO 64110
J. Chromatogr.; (82) P 489-500; Vol 235; No 2

27-82-0234 (0234)
HIGH-PERFORMANCE LIQUID CHROMATOGRAPHIC PROCEDURE FOR SIMULTANEOUS DETERMINATION OF SKF 78729A AND ITS N-ACETYL METABOLITE
Gupta, R. N.; Achong, M. R.; Eng, F.
Dep. Pathol. St. Joseph's Hosp. Hamilton Can. CA ON L8N 1Y4
J. Chromatogr.; (82) P 492-6; Vol 231; No 2

27-82-0235 (0235)
SIMULTANEOUS DETERMINATION OF ACETAMINOPHEN, THEOPHYLLINE AND SALICYLATE IN SERUM BY HIGH-PERFORMANCE LIQUID CHROMATOGRAPHY

Kinberger, Bengt; Holmen, Anders
Dep. Clin. Chem. Cent. Hosp. Halmstad Swed. SE S-301 85
J. Chromatogr.; (82) P 492-7; Vol 229; No 2

27-82-0236 (0236)
FLUOROMETRIC LIQUID CHROMATOGRAPHIC ASSAY OF THE ANTIARRHYTHMIC AGENT FLECAINIDE IN BLOOD PLASMA
De Jong, J. W.; Hegge, J. A. J.; Harmsen, E.; De Tombe, P. P.
Thoraxcent. Erasmus Univ. Rotterdam Neth. NL
J. Chromatogr.; (82) P 498-502; Vol 229; No 2

27-82-0237 (0237)
DETERMINATION OF CLOFAZIMINE IN PLASMA BY HIGH-PERFORMANCE LIQUID CHROMATOGRAPHY
Peters, John H.; Hamme, Karen J.; Gordon, G. Ross
Life Sci. Div. SRI Int. Menlo Park USA US CA 94025
J. Chromatogr.; (82) P 503-8; Vol 229; No 2

27-82-0238 (0238)
HIGH-PERFORMANCE LIQUID CHROMATOGRAPHIC ANALYSIS OF DIGITALIS LEAF EXTRACTS. I. QUALITATIVE ANALYSIS
Wichtl, M.; Mangkudidjojo, M.; Wichtl-Bleier, W.
Inst. Pharm. Biol. Philipps-Univ. Marburg Fed. Rep. Ger. DE D-3550
J. Chromatogr.; (82) P 503-8; Vol 234 No 2; In Ger

27-82-0239 (0239)
DETERMINATION OF THE .GAMMA.-AMINOBUTYRIC ACID AGONIST 4,5,6,7-TETRAHYDRO-ISOXAZOLO(5,4-C)PYRIDIN-3-OL BY HIGH-PERFORMANCE LIQUID CHROMATOGRAPHY USING UV OR ELECTROCHEMICAL DETECTION
Madsen, Soren Munk
H. Lundbeck and Co., A/S Valby Den. DK DK 2500
J. Chromatogr.; (82) P 509-14; Vol 238; No 2

27-82-0240 (0240)
DETERMINATION OF CEFSULODIN SODIUM (D-(-)-SCE-129) BY HIGH-PERFORMANCE LIQUID CHROMATOGRAPHY
Elrod, Lee, Jr.; White, Landy B.; Wimer, David C.; Cox, Robert D.
Anal. Res. Dep. Abbott Lab. North Chicago USA US IL 60064
J. Chromatogr.; (82) P 515-21; Vol 237; No 3

27-82-0241 (0241)
HIGH-PERFORMANCE LIQUID CHROMATOGRAPHIC ANALYSIS OF THE SULFIDE METABOLITE OF SULFINPYRAZONE IN PLASMA
Godbillon, J.; Gauron, S.; Gosset, G.
Cent. Rech. Biopharmaceut. Ciba-Geigy Rueil-Malmaison Fr. FR 92506
J. Chromatogr.; (82) P 516-20; Vol 227; No 2

27-82-0242 (0242)
RAPID HIGH-PERFORMANCE LIQUID CHROMATOGRAPHIC ASSAY OF CINNARIZINE IN HUMAN PLASMA
Nitsche, V.; Mascher, H.
Pharmakol. Untersuchungsges. Vienna Austria AT A-1171
J. Chromatogr.; (82) P 521-5; Vol 227; No 2

27-82-0243 (0243)
DETERMINATION OF EMETINE AND CEPHAELINE IN IPECAC ROOTS BY HIGH-PERFORMANCE LIQUID CHROMATOGRAPHY
Sahu, N. P.; Mahato, S. B.
Indian Inst. Chem. Biol. Calcutta India IN 700032
J. Chromatogr.; (82) P 525-9; Vol 238; No 2

27-82-0244 (0244)
HIGH-PERFORMANCE LIQUID CHROMATOGRAPHIC DETERMINATION OF PLASMA 6-MERCAPTO-PURINE IN CLINICALLY RELEVANT CONCENTRATIONS
De Abreu, Ronney A.; Van Baal, John M.; Schouten, Teunis J.; Schretlen, Egbert D. A. M.; De Bruyn, Chris H. M. M.
Dep. Pediatrics St. Radboud Hosp. Nijmegen Neth. NL
J. Chromatogr.; (82) P 526-33; Vol 227; No 2

27-82-0245 (0245)
HIGH-PERFORMANCE LIQUID CHROMATOGRAPHY OF PHOSFOLAN, MEPHOSFOLAN AND RELATED COMPOUNDS
Abou-Donia, Sherif A.; Abou-Donia, Mohamed B.
Med. Cent. Duke Univ. Durham USA US NC 27710
J. Chromatogr.; (82) P 532-8; Vol 240; No 2

27-82-0246 (0246)
DETERMINATION OF CIPROFIBRATE IN HUMAN PLASMA BY HIGH-PERFORMANCE LIQUID CHROMATOGRAPHY
Park, George B.; Biddlecome, Charles E.; Koblantz, Chester; Edelson, Jerome
Dep. Drug Metab. Disposition Sterling-Winthrop Res. Inst. Rensselaer USA US NY 12144
J. Chromatogr.; (82) P 534-9; Vol 227; No 2

27-82-0247 (0247)
HIGH-PERFORMANCE LIQUID CHROMATOGRAPHIC SEPARATION OF THE DIASTEREOISOMERS OF PROPOXYPHENE. DETERMINATION OF MICROQUANTITIES OF .BETA.-DL-PROPOXYPHENE IN COMMERCIAL PREPARATIONS OF .ALPHA.-D-PROPOXYPHENE
Barkan, Susan; Wainer, I. W.
Div. Drug Chem. Food and Drug Adm. Washington USA US DC 20204
J. Chromatogr.; (82) P 547-51; Vol 240; No 2

27-82-0248 (0248)
SEPARATION AND DETERMINATION OF NAPROXEN ENANTIOMERS IN SERUM BY HIGH-PERFORMANCE LIQUID CHROMATOGRAPHY
Goto, Junichi; Goto, Nobuharu; Nambara, Toshio
Pharm. Inst. Tohoku Univ. Sendai Japan JP 980
J. Chromatogr.; (82) P 559-64; Vol 239

27-82-0249 (0249)
DETERMINATION OF NEOPTERINE IN HUMAN URINE BY REVERSED-PHASE HIGH-PERFORMANCE LIQUID CHROMATOGRAPHY
Hausen, Arno; Fuchs, Dietmar; Koenig, Klaus; Wachter, Helmut
Inst. Med. Chem. Biochem. Univ. Innsbruck Innsbruck Austria AT A-6020
J. Chromatogr.; (82) P 61-70; Vol 227; No 1

27-82-0250 (0250)
INVESTIGATION OF THE EFFECT OF ADDED ORGANIC AMINE ON THE CHROMATOGRAPHY OF TRICYCLIC ANTIDEPRESSANT DRUGS USING REVERSED-PHASE CHROMATOGRAPHY ON OCTADECYLSILICA WITH SODIUM LAURYL SULFATE AS PAIRING ION
Hung, C. T.; Taylor, R. B.; Paterson, N.
Sch. Pharm. Robert Gordon's Inst. Technol. Aberdeen UK GB
J. Chromatogr.; (82) P 61-73; Vol 240; No 1

27-82-0251 (0251)
HIGH-PERFORMANCE LIQUID CHROMATOGRAPHIC ANALYSIS OF THE METABOLISM OF PRIMAQUINE AND THE IDENTIFICATION OF A NEW MAMMALIAN METABOLITE
Baker, John K.; McChesney, James D.; Hufford, Charles D.; Clark, Alice M.
Sch. Pharm. Univ. Mississippi University USA US MS 38677
J. Chromatogr.; (82) P 69-77; Vol 230; No 1

27-82-0252 (0252)
IMPROVED HIGH-PERFORMANCE LIQUID CHROMATOGRAPHIC ASSAY FOR THEOPHYLLINE IN PLASMA AND SALIVA IN THE PRESENCE OF CAFFEINE AND ITS METABOLITES AND COMPARISONS WITH THREE OTHER ASSAYS
Muir, Keith T.; Kunitani, Michael; Riegelman, Sidney
Sch. Pharm. Univ. California San Francisco USA US CA 94143
J. Chromatogr.; (82) P 73-82; Vol 231; No 1

27-82-0253 (0253)
DETERMINATION OF THIABENDAZOLE AND 5-HYDROXYTHIABENDAZOLE IN HUMAN SERUM BY FLUORESCENCE-DETECTED HIGH-PERFORMANCE LIQUID CHROMATOGRAPHY
Watts, Mark T.; Raisys, Vidmantas A.; Bauer, Larry A.
Sch. Med. Univ. Washington Seattle USA US WA 98104
J. Chromatogr.; (82) P 79-86; Vol 230; No 1

27-82-0254 (0254)
REVERSED-PHASE HIGH-PERFORMANCE LIQUID CHROMATOGRAPHY ANALYSIS OF 6-THIOGUANINE APPLICABLE TO PHARMACOLOGIC STUDIES IN HUMANS
Andrews, Paul A.; Egorin, Merrill J.; May, Matthew E.; Bachur, Nicholas R.
Lab. Clin. Biochem. Natl. Cancer Inst. Baltimore USA US MD 21201
J. Chromatogr.; (82) P 83-91; Vol 227; No 1

27-82-0255 (0255)
DETERMINATION OF THE ANTICOAGULANT PHENPROCOUMON IN HUMAN PLASMA AND URINE BY HIGH-PERFORMANCE LIQUID CHROMATOGRAPHY
De Vries, Jan X.; Harenberg, Job; Walter, Erhard; Zimmermann, Rainer; Simon, Michael
Med. Klin. Univ. Heidelberg Heidelberg Fed. Rep. Ger. DE D-6900
J. Chromatogr.; (82) P 83-92; Vol 231; No 1

27-82-0256 (0256)
HIGH-PERFORMANCE LIQUID CHROMATOGRAPHIC DETERMINATION OF INDALPINE, A NEW NONTRICYCLIC ANTIDEPRESSANT, IN HUMAN PLASMA. IDENTIFICATION AND SIMULTANEOUS MEASUREMENT OF ITS MAJOR PLASMA METABOLITE
Jozefczak, C.; Ktorza, N.; Uzan, A.
Pharmuka Lab. Gennevilliers Fr. FR 92231
J. Chromatogr.; (82) P 87-95; Vol 230; No 1

27-82-0257 (0257)
HIGH-PERFORMANCE LIQUID COLUMN AND THIN-LAYER CHROMATOGRAPHIC DETERMINATION OF HUMAN SERUM GLIBENCLAMIDE AT THERAPEUTIC LEVELS
Uihlein, M.; Sistovaris, N.
Hoechst A.G. Frankfurt Fed. Rep. Ger. DE D-6230/80
J. Chromatogr.; (82) P 93-101; Vol 227; No 1

27-82-0258 (0258)
ROUTINE MONITORING OF CARBAMAZEPINE AND CARBAMAZEPINE-10,11-EPOXI- DE IN PLASMA BY HIGH-PERFORMANCE LIQUID CHROMATOGRAPHY USING 10-METHOXYCARBAMAZEPINE AS INTERNAL STANDARD
Elyas, Abdulmajid A.; Ratnaraj, Neville; Goldberg, Valerie D.; Lascelles, Peter T.
Inst. Neurol. Natl. Hosp. Nervous Dis. London UK GB WC1N 3BG
J. Chromatogr.; (82) P 93-101; Vol 231; No 1

27-82-0259 (0259)
DETERMINATION OF MEXILETINE IN BIOLOGICAL FLUIDS BY HIGH-PERFORMANCE LIQUID CHROMATOGRAPHY
Breithaupt, Henning; Wilfling, Michael
Med. Klin. Justus Liebig-Univ. Giessen Fed. Rep. Ger. DE 6300
J. Chromatogr.; (82) P 97-105; Vol 230; No 1

27-82-0260 (0260)
ANALYSIS OF ACETAMINOPHEN METABOLITES IN URINE BY HIGH-PERFORMANCE LIQUID CHROMATOGRAPHY WITH UV AND AMPEROMETRIC DETECTION
Wilson, John M.; Slattery, John T.; Forte, Anthony J.; Nelson, Sidney D.
Dep. Med. Chem. Univ. Washington Seattle USA 98195
J. Chromatogr.; (82)PP 453-62; Vol 227; No 2

27-82-0261 (0261)
PENTAZOCINE TABLET ANALYSIS USING HIGH-PERFORMANCE LIQUID CHROMATOGRAPHY
Wilson, Terry D.
Anal. Chem. Dep. Sterling-Winthrop Res. Inst. Rensselaer USA US NY 12144
J. Chromatogr.; (82)PP 99-109; Vol 243; No 1

27-82-0262 (0262)
SEPARATION OF DOLICHOLS AND POLYPRENOLS BY STRAIGHT-PHASE HIGH-PERFORMANCE LIQUID CHROMATOGRAPHY
Keller, R. Kennedy; Rottler, Grant D.; Adair, W. Lee, Jr.
Coll. Med. Univ. South Florida Tampa USA US FL 33612
J. Chromatogr; (82) P 230-3; Vol 236; No 1

27-82-0263 (0263)
COMPARISON OF A HIGH PRESSURE LIQUID CHROMATOGRAPHIC ANALYSIS AND AN ENZYME IMMUNOASSAY TECHNIQUE FOR QUANTITATION OF DISOPYRAMIDE IN SERUM OR PLASMA
Bryson, Scott M.; Betts, Andrea; Summer, David J.; Whiting, Brian
Stobhill Gen. Hosp. Univ. Glasgow Glasgow UK GB
J. Clin. Hosp. Pharm.; (82) P 119-26; Vol 7; No 2

27-82-0264 (0264)
A CHROMATOGRAPHIC QUALITY CONTROL PROCEDURE BASED ON HPLC FOR 5-HYDROXY-METHYLFURFURAL IN AUTOCLAVED D-GLUCOSE INFUSION FLUIDS
Hung, C. T.; Selkirk, A. B.; Taylor, R. B.
Sch. Pharm. Robert Gordon's Inst. Technol. Schoolhill/Aberdeen UK GB AB9 1FR
J. Clin. Hosp. Pharm.; (82) P 17-23; Vol 7; No 1

27-82-0265 (0265)
A HIGH PRESSURE LIQUID CHROMATOGRAPHIC METHOD FOR THE ANALYSIS OF ZEARALENONE IN CHICKEN BLOOD
Turner, G. Vincent; Phillips, Timothy D.; Heidelbaugh, Norman D.; Russell, Leon H.
Dep. Vet. Public Health Texas A and M Univ. College Station USA US TX 77843
J. Environ. Sci. Health, Part B; (82) P 297-309; Vol B17; No 4

27-82-0266 (0266)
PURIFICATION OF TRITIUM-LABELED 8-METHOXYPSORALEN BY HIGH PERFORMANCE LIQUID CHROMATOGRAPHY
Adriaens, P.; De Schrijver, F.; Van Boven, M.; Roelandts, R.; Roets, E.; Hoogmartens, J.
Dep. Chem. Kathol. Univ. Leuven Louvain Belg. BE
J. Labelled Compd. Radiopharm.; (82) P 463-8; Vol 19; No 3

27-82-0267 (0267)
HIGH-PERFORMANCE LIQUID CHROMATOGRAPHY (HPLC) OF THE NEW ANTINEOPLASTIC 9,10-ANTHRACENEDICARBOXALDEHYDE BIS((4,5-DIHYDRO-1-H-IMIDAZOLE-2-YL) HYDRAZONE) DIHYDROCHLORIDE (CL 216,942; BISANTRENE)
Lu, K.; Savaraj, N.; Huang, M. T.; Moore, D.; Loo, T. L.
M. D. Anderson Hosp. Tumor Inst. Univ. Texas Houston USA US TX 77030
J. Liq. Chromatogr.; (82) P 1323-8; Vol 5; No 7

27-82-0268 (0268)
DETERMINATION OF METHACHOLINE CHLORIDE BY ION-PAIR HIGH-PRESSURE LIQUID

CHROMATOGRAPHY
Woodman, Tracy F.; Johnson, Brian; Marwaha, Raj K.
Med. Sch. Univ. Massachusetts Worcester USA US MA 01605
J. Liq. Chromatogr.; (82) P 1341-8; Vol 5; No 7

27-82-0269 (0269)
HIGH-PRESSURE LIQUID CHROMATOGRAPHY ASSAY FOR DANE SALT POTASSIUM (-)-N-(1-METHOXYCARBONYLPROPENE-2YL)-p-HYDROXYPHENYLGLYCINE
Muhammad, Naseem; Tsai, Peter S.; Lauback, Ronald G.
Bristol Lab. Bristol-Myers Co. Syracuse USA US NY 13201
J. Liq. Chromatogr.; (82) P 1349-55; Vol 5; No 7

27-82-0270 (0270)
ASSAY OF n-PROPYLAJMALINE IN BLOOD PLASMA BY ION-PAIR LIQUID-LIQUID CHROMATOGRAPHY
Grundevik, Inger; Persson, B. A.
AB Haessle Moelndal Swed. SE S-431 83
J. Liq. Chromatogr.; (82) P 141-50; Vol 5; No 1

27-82-0271 (0271)
RAPID HIGH-PERFORMANCE LIQUID CHROMATOGRAPHIC DETERMINATION OF PHYSOSTIGMINE IN PLASMA
Hsieh, John Y. K.; Yang, Ren Kuei; Davis, Kenneth L.
Mount Sinai Sch. Med. City Univ. New York New York USA US NY 10029
J. Liq. Chromatogr.; (82) P 1691-5; Vol 5; No 9

27-82-0272 (0272)
ON LINE DETECTION OF PLASMA BORNE VASOCONSTRICTOR BY THE USE OF HIGH PRESSURE LIQUID CHROMATOGRAPHY
Hazeyama, Y.; Valko, W. J.; Ravert, H. T.; Heindel, N. D.; Moretti, R. L.
Cent. Health Sci. Lehigh Univ. Bethlehem USA US PA 18015
J. Liq. Chromatogr.; (82) P 1749-62; Vol 5; No 9

27-82-0273 (0273)
SIMULTANEOUS DETERMINATION OF FLUNIXIN, PHENYLBUTAZONE, OXYPHENBUTAZONE AND .GAMMA.-HYDROXYPHENYLBUTAZONE IN EQUINE PLASMA BY HIGH-PERFORMANCE LIQUID CHROMATOGRAPHY: WITH APPLICATION TO PHARMACOKINETICS
Hardee, Gregory E.;Lai, Jin Wang,; Moore, James N.
Sch. Pharm. Univ. Georgia Athens USA US GA 3602
J. Liq. Chromatogr.; (82) P 1991-2003; Vol 5; No 10

27-82-0274 (0274)
ANALYSIS OF CIMETIDINE IN BIOLOGICAL FLUIDS BY HIGH PERFORMANCE LIQUID CHROMATOGRAPHY
Apffel, J. A.; Brinkman, U. A. T.; Frei, R. W.
Dep. Anal. Chem. Free Univ. Amsterdam Neth. NL 1081 HV
J. Liq. Chromatogr.; (82) P 2413-22; Vol 5; No 12

27-82-0275 (0275)
DETERMINATION OF ANTICONVULSANTS BY HIGH-PERFORMANCE LIQUID CHROMATOGRAPHY
Kraak, J. C.; Crombeen, J. P.
Lab. Anal. Chem. Univ. Amsterdam Amsterdam Neth. NL
J. Liq. Chromatogr.; (82) P 273-304; Vol 5; No Suppl. 2

27-82-0276 (0276)
A HIGH-PERFORMANCE LIQUID CHROMATOGRAPHIC METHOD FOR THE QUANTITATIVE DETERMINATION OF NAPROXEN AND DES-METHYL-NAPROXEN IN BIOLOGICAL SAMPLES
Van Loenhout, J. W. A.; Van Ginneken, C. A. M.; Ketelaars, H. C. J.; Kimenai, P. M.; Tan, Y.; Gribnau, F. W. J.
Dep. Pharmacol. Univ. Hosp. St. Radboud Nijmegen Neth. NL 6525 GA

J. Liq. Chromatogr.; (82) P 549-61; Vol 5; No 3

27-82-0277 (0277)
HIGH-PRESSURE LIQUID CHROMATOGRAPHIC DETERMINATION OF ISONIAZID AND ACETYLISONIAZID IN HUMAN PLASMA
Holdiness, Mack R.
Sch. Med. Louisiana State Univ. New Orleans USA US LA 70112
J. Liq. Chromatogr.; (82) P 707-14; Vol 5; No 4

27-82-0278 (0278)
A RAPID ISOLATION OF HUMAN CHORIONIC GONADOTROPIN AND ITS SUBUNITS BY REVERSED-PHASE HIGH PERFORMANCE LIQUID CHROMATOGRAPHY
Putterman, Gerald J.; Spear, Mary Beth; Meade-Cobun, Kimberly S.; Widra, Martha; Hixson, Catherine V.
Frederick Cancer Res. Facility Natl. Cancer Inst. Frederick USA US MD 21701
J. Liq. Chromatogr.; (82) P 715-29; Vol 5; No 4

27-82-0279 (0279)
ANALYSIS OF ACETAMINOPHEN AND SALICYLATE BY REVERSED-PHASE HPLC
Ferrell, William J.; Goyette, Gerard W.
Dep. Lab. Med. Children's Hosp. Michigan Detroit USA US MI 48201
J. Liq. Chromatogr.; (82) P 93-6; Vol 5; No 1

27-82-0280 (0280)
THE DETERMINATION OF CAPTOPRIL IN PHYSIOLOGICAL FLUIDS USING HIGH-PERFORMANCE LIQUID CHROMATOGRAPHY WITH ELECTROCHEMICAL DETECTION
Perrett, D.; Drury, P. L.
Dep. Med. St. Bartholomew's Hosp. London UK GB EC1A 7BE
J. Liq. Chromatogr.; (82) P 97-110; Vol 5; No 1

27-82-0281 (0281)
MICROBIAL N-ACETYLATION OF PRIMAQUINE BY TWO STREPTOMYCES SPECIES: TIME COURSE STUDIES AND HPLC ANALYSES
Clark, Alice M.; Evans, Steven L.; Hufford, Charles D.; McChesney, James D.
Sch. Pharm. Univ. Mississippi University USA US MS 38677
J. Nat. Prod.; (82) P 574-81; Vol 45; No 5

27-82-0282 (0282)
ISOCRATIC HIGH-PERFORMANCE LIQUID CHROMATOGRAPHIC METHOD FOR THE DETERMINATION OF TRICYCLIC ANTIDEPRESSANTS AND METABOLITES IN PLASMA
Johnson, S.M.; Chan, C.; Cheng, S.; Shimek, J.L.; Nygard, G.; Khalil,S.K. Wahba
Coll. Pharm., North Dakota State Univ., Fargo, USA ND 58 102
J. Pharm. Sci. (82) P 1027-30; Vol 71; No 9

27-82-0283 (0283)
HIGH-PERFORMANCE LIQUID CHROMATOGRAPHIC ANALYSIS OF DIGITOXIN FORMULATIONS
Desta, Belachew; McErlane, K. M.
Fac. Pharm. Sci. Univ. Br. Columbia Vancouver Can. CA BC V6T 1W5
J. Pharm. Sci.; (82) P 1018-20; Vol 71; No 9

27-82-0284 (0284)
HIGH-PERFORMANCE LIQUID CHROMATOGRAPHIC DETERMINATION OF METOPROLOL IN PLASMA
Rosseel, M. T.; Belpaire, F. M.; Bekaert, I.; Bogaert, M. G.
Med. Sch. Univ. Gent Ghent Belg. BE B-9000
J. Pharm. Sci.; (82) P 114-15; Vol 71; No 1

27-82-0285 (0285)
DETERMINATION OF SULFADIAZINE AND N4-ACETYLSULFADIAZINE IN BIOLOGICAL FLUIDS BY LIQUID CHROMATOGRAPHY ON SILICA GEL WITH AN AQUEOUS BUFFER AS MOBILE PHASE

Westerlund, Douglas; Wijkstroem, Agneta
Res. Dev. Lab. Anal. Chem. Astra Laekemedel A.B. Soedertaelje Swed. SE S-151 85
J. Pharm. Sci.; (82) P 1142-5; Vol 71; No 10

27-82-0286 (0286)
QUALITY CONTROL OF PHENYLBUTAZONE. II: ANALYSIS OF PHENYLBUTAZONE AND ITS DECOMPOSITION PRODUCTS IN DRUGS BY HIGH-PRESSURE LIQUID CHROMATOGRAPHY
Fabre, H.; Mandrou, B.; Eddine, H.
Lab. Chim. Anal. Fac. Pharm. Montpellier Fr. FR 34060
J. Pharm. Sci.; (82) P 120-2; Vol 71; No 1

27-82-0287 (0287)
DETERMINATIONS OF SODIUM FUSIDATE AND FUSIDIC ACID IN DOSAGE FORMS BY HIGH-PERFORMANCE LIQUID CHROMATOGRAPHY AND A MICROBIOLOGICAL METHOD
Hikal, Ahmed H.; Shibl, Atef; El-Hoofy, Sammy
Coll. Pharm. King Saud Univ. Riyadh Riyadh Saudi Arabia SA
J. Pharm. Sci.; (82) P 1297-8; Vol 71; No 11

27-82-0288 (0288)
PREDICTION OF STABILITY IN PHARMACEUTICAL PREPARATIONS. XIX: STABILITY EVALUATION AND BIOANALYSIS OF CLOFIBRIC ACID ESTERS BY HIGH-PRESSURE LIQUID CHROMATOGRAPHY
Garrett, Edward R.; Gardner, Michael R.
Coll. Pharm. Univ. Florida Gainesville USA 32610
J. Pharm. Sci.; (82) P 14-25; Vol 71; No 1

27-82-0289 (0289)
HIGH-PRESSURE LIQUID CHROMATOGRAPHIC ANALYSIS OF INDOCYANINE GREEN
Rappaport, P. L.; Thiessen, J. J.
Fac. Pharm. Univ. Toronto Toronto Can. CA ON M5S 1A1
J. Pharm. Sci.; (82) P 157-61; Vol 71; No 2

27-82-0290 (0290)
ANALYSIS OF COMMERCIAL PILOCARPINE PREPARATIONS BY HIGH-PERFORMANCE LIQUID CHROMATOGRAPHY
Drake, Michael V.; O'Donnell, James J.; Sandman, Robert P.
Dep. Ophthalmol. Univ. California San Francisco USA US CA 94143
J. Pharm. Sci.; (82) P 358-9; Vol 71; No 3

27-82-0291 (0291)
HPLC DETERMINATION OF D AND L-MOXALACTAM IN HUMAN SERUM AND URINE
Ziemniak, J. A.; Chiarmonte, D. A.; Miner, D. J.; Schentag, J. J.
Sch. Pharm. State Univ. New York Buffalo USA US NY 14209
J. Pharm. Sci.; (82) P 399-402; Vol 71; No 4

27-82-0292 (0292)
AN ISOCRATIC HIGH-PRESSURE LIQUID CHROMATOGRAPHIC DETERMINATION OF NAPROXEN AND DESMETHYLNAPROXEN IN HUMAN PLASMA
Shimek, J. L.; Rao, N. G. S.; Khalil, S. K. Wahba
Coll. Pharm. North Dakota State Univ. Fargo USA US ND 58105
J. Pharm. Sci.; (82) P 436-9; Vol 71; No 4

27-82-0293 (0293)
KINETICS OF CHLORAMBUCIL HYDROLYSIS USING HIGH-PRESSURE LIQUID CHROMATOGRAPHY
Chatterji, Dulal C.; Yeager, Russell L.; Gallelli, Joseph F.
Clin. Cent. NIH Bethesda USA US MD 20205
J. Pharm. Sci.; (82) P 50-4; Vol 71; No 1

27-82-0294 (0294)
DETERMINATION OF RELATED COMPOUNDS IN ASPIRIN BY LIQUID CHROMATOGRAPHY
Pfeiffer, C. D.; Pankey, J. W.
Anal. Lab. Dow Chem. Co. Midland USA US MI 48640
J. Pharm. Sci.; (82) P 511-14; Vol 71; No 5

27-82-0295 (0295)
HIGH-PERFORMANCE LIQUID CHROMATOGRAPHIC ASSAY FOR BUMETANIDE IN PLASMA AND URINE
Smith, David E.
Coll. Pharm. Univ. Michigan Ann Arbor USA US MI 48109
J. Pharm. Sci.; (82) P 520-3; Vol 71; No 5

27-82-0296 (0296)
HIGH-PRESSURE LIQUID CHROMATOGRAPHIC DETERMINATION OF PROMETHAZINE PLASMA LEVELS IN THE DOG AFTER ORAL, INTRAMUSCULAR, AND INTRAVENOUS DOSAGE
Patel, Rajni B.; Welling, Peter G.
Sch. Pharm. Univ. Wisconsin Madison USA US WI 53706
J. Pharm. Sci.; (82) P 529-32; Vol 71; No 5

27-82-0297 (0297)
HIGH-PRESSURE LIQUID CHROMATOGRAPHIC DETERMINATION OF AMITRIPTYLINE AND ITS MAJOR METABOLITES IN HUMAN WHOLE BLOOD
Smith, Gabrielle A.; Schulz, Pierre; Giacomini, Kathleen M.; Blaschke, Terrence F.
Med. Cent. Stanford Univ. Stanford USA US CA 94305
J. Pharm. Sci.; (82) P 581-3; Vol 71; No 5

27-82-0298 (0298)
HIGH-PRESSURE LIQUID CHROMATOGRAPHIC DETERMINATION OF CIMETIDINE IN PLASMA AND URINE
Mihaly, G. W.; Cockbain, S.; Jones, D. B.; Hanson, R. G.; Smallwood, R. A.
Dep. Med. Univ. Melbourne Heidelberg Australia AU 3084
J. Pharm. Sci.; (82) P 590-2; Vol 71; No 5

27-82-0299 (0299)
DETERMINATION OF OCTANOL-WATER EQUIVALENT PARTITION COEFFICIENTS OF INDOLIZINE AND SUBSTITUTED 2-PHENYLINDOLIZINES BY REVERSED-PHASE HIGH-PRESSURE LIQUID CHROMATOGRAPHY AND FRAGMENTATION VALUES
Lins, Claudio L. K.; Block, John H.; Doerge, Robert F.; Barnes, Gerard J.
Sch. Pharm. Oregon State Univ. Corvallis USA US OR 97331
J. Pharm. Sci.; (82) P 614-17; Vol 71; No 6

27-82-0300 (0300)
DETERMINATION OF CYCLOBENZAPRINE IN TABLETS BY HIGH-PERFORMANCE LIQUID CHROMATOGRAPHY
Heinitz, Maxine L.
Food and Drug Adm. Minneapolis USA US MN 55401
J. Pharm. Sci.; (82) P 656-8; Vol 71; No 6

27-82-0301 (0301)
THE MEASUREMENT OF THEOPHYLLINE METABOLISM IN HEPATIC MICROSOMES USING HIGH-PERFORMANCE LIQUID CHROMATOGRAPHY
Hemsworth, Theresa C.; Renton, Kenneth W.
Dep. Pharmacol. Dalhousie Univ. Halifax Can. CA NS B3H 4H7
J. Pharm. Sci.; (82) P 712-13; Vol 71; No 6

27-82-0302 (0302)
REVERSED-PHASE HIGH-PRESSURE LIQUID CHROMATOGRAPHIC DETERMINATION OF SERUM METHOTREXATE AND 7-HYDROXYMETHOTREXATE

Buice, Robert G.; Sidhu, Paramjeet
Cent. Health Sci. Univ. Tennessee Memphis USA US TN 38163
J. Pharm. Sci.; (82) P 74-7; Vol 71; No 1

27-82-0303 (0303)
HIGH-PERFORMANCE LIQUID CHROMATOGRAPHIC ANALYSIS OF DIGOXIN FORMULATIONS
Desta, Belachew; McErlane, K. M.
Fac. Pharm. Sci. Univ. British Columbia Vancouver Can. CA BC V6T 1W5
J. Pharm. Sci.; (82) P 777-80; Vol 71; No 7

27-82-0304 (0304)
HPLC ANALYSIS OF INDOMETHACIN AND ITS IMPURITIES IN CAPSULE AND SUPPOSITORY FORMULATIONS
Kwong, E.; Pillai, G. K.; McErlane, K. M.
Fac. Pharm. Sci. Univ. British Columbia Vancouver Can. CA BC V6T 1W5
J. Pharm. Sci.; (82) P 828-30; Vol 71; No 7

27-82-0305 (0305)
HIGH-PERFORMANCE LIQUID CHROMATOGRAPHIC DETERMINATION OF SUPROFEN IN DRUG SUBSTANCE AND CAPSULES
Lagu, Avinash L.; Young, Richard; McGonigle, Eugene; Lane, Philip A.
Pharm. Anal. Sect. Ortho Pharm. Corp. Raritan USA US NJ 08869
J. Pharm. Sci.; (82) P 85-8; Vol 71; No 1

27-82-0306 (0306)
ANALYSIS OF IODOCHLORHYDROXYQUIN IN CREAM FORMULATIONS AND BULK DRUGS BY HIGH-PERFORMANCE LIQUID CHROMATOGRAPHY
Kubiak, E. J.; Munson, J. W.
Qual. Control Div. Upjohn Co. Kalamazoo USA US MI 49001
J. Pharm. Sci.; (82) P 872-5; Vol 71; No 8

27-82-0307 (0307)
DETERMINATION OF MIANSERIN AND METABOLITES IN PLASMA BY LIQUID CHROMATOGRAPHY WITH ELECTROCHEMICAL DETECTION
Suckow, Raymond F.; Cooper, Thomas B.; Quitkin, Frederic M.; Stewart, Jonathan W.
Anal. Psychopharmacol. Lab. Rockland Res. Inst. Orangeburg USA US NY 10962
J. Pharm. Sci.; (82) P 889-92; Vol 71; No 8

27-82-0308 (0308)
DETERMINATION OF ISOETHARINE IN PLASMA BY REVERSED-PHASE CHROMATOGRAPHY WITH AMPEROMETRIC DETECTION
Park, George B.; Koss, Raymond F.; Utter, Julie; Mayes, Brian A.; Edelson, Jerome
Dep. Drug Metab. Sterling-Winthrop Res. Inst. Rensselaer USA US NY 12144
J. Pharm. Sci.; (82) P 932-4; Vol 71; No 8

27-82-0309 (0309)
HIGH-PERFORMANCE LIQUID CHROMATOGRAPHIC DETERMINATION OF PROGLUMIDE IN PLASMA
Marunaka, Teruyoshi; Minami, Yoshinori; Umeno, Yukihiko; Matsushima, Eiji
Res. Lab. Taiho Pharm. Co., Ltd. Tokushima Japan JP 771-01
J. Pharm. Sci.; (82) P 949-50; Vol 71; No 8

27-82-0310 (0310)
ION-PAIR HIGH-PERFORMANCE LIQUID CHROMATOGRAPHY OF TERBUTALINE AND CATECHOLAMINES WITH AMINOPHYLLINE IN INTRAVENOUS SOLUTIONS
Williams, David A.; Fuvg, Esther Y. Y.; Newton, David W.
Dep. Chem. Massachusetts Coll. Pharm. Allied Health Sci. Boston USA US MA 02115
J. Pharm. Sci.; (82) P 956-8; Vol 71; No 8

27-82-0311 (0311)
STABILITY ASSAY OF ALLANTOIN IN LOTIONS AND CREAMS BY HIGH-PRESSURE LIQUID CHROMATOGRAPHY
Zaidi, Z. R.; Sena, F. J.; Basilio, C. P.
Res. Dev. Lab. Block Drug Co. Inc. Jersey City USA US NJ 07302
J. Pharm. Sci.; (82) P 997-9; Vol 71; No 9

27-82-0312 (0312)
INCOMPATIBILITY OF CEFOTIAM DIHYDROCHLORIDE IN PARENTERAL PREPARATIONS BY HIGH-PERFORMANCE LIQUID CHROMATOGRAPHY
Tanno, Keiki; Ikarashi, Kuniichi; Sasahara, Kazuhisa; Kato, Masami
Sch. Med. Niigata Univ. Niigata Japan JP
Jpn. J. Antibiot.; (82) P 1177-82; Vol 35; No 5; In Japan

27-82-0313 (0313)
QUANTITATIVE DETERMINATION OF DIGITOXIN BY HIGH PERFORMANCE LIQUID CHROMATOGRAPHY
Popov, D. M.; Nikolaev, N. N.; Semenov, V. A.
Vses. Nauchno-Issled. Inst. Fiz. Kul't. Moscow USSR SU
Khim.-Farm. Zh.; (82) P 761-4; Vol 16; No 6; In Russ

27-82-0314 (0314)
A SPECIFIC HPLC METHOD FOR RESIDUAL IMAZALIL AND ITS MAIN METABOLITE T 824
Wynants, J.
Janssen Pharm. Beerse Belg. BE 2340
Meded. Fac. Landbouwwet., Rijksuniv. Gent; (82) P 347-59; Vol 47; No 1

27-82-0315 (0315)
DRUG MONITORING OF 5-FLUOROCYTOSINE IN SERUM BY HIGH-PERFORMANCE LIQUID CHROMATOGRAPHY
Essers, L.; Hantschke, D.; Leisse, K. H.
Inst. Med. Mikrobiol. Essen Fed. Rep. Ger. DE
Mykosen; (82) P 183-8; Vol 25; No 4; In Ger

27-82-0316 (0316)
HIGH-PERFORMANCE LIQUID CHROMATOGRAPHIC METHOD FOR DETERMINATION OF GUAZATINE RESIDUES IN CROPS
Mori, Yoshikazu; Yagura, Yasuji; Kataoka, Yasuki
Chem. Div. Dainippon Ink Chem. Co. Ibaraki Japan JP 314-02
Nippon Noyaku Gakkaishi; (82) P 53-6; Vol 7; No 1; In Japan

27-82-0317 (0317)
QUANTITATIVE CALIFORNIUM-252 PLASMA DESORPTION MASS SPECTROMETRY FOR PHARMACEUTICALS. A NEW APPROACH TO COUPLING LIQUID CHROMATOGRAPHY WITH MASS SPECTROMETRY
Jungclas, Hartmut; Danigel, Harald; Schmidt, Lothar
Kernchem. Philipps-Univ. Marburg Fed. Rep. Ger. DE D-3550
Org. Mass Spectrom.; (82) P 86-90; Vol 17; No 2

27-82-0318 (0318)
HIGH-PERFORMANCE LIQUID CHROMATOGRAPHY ANALYSIS OF SOME BRANDS OF GENTAMICIN USED CLINICALLY
White, L. O.; Reeves, D. S.
Dep. Microbiol. Southmead Hosp. Bristol UK GBBS10 5NB
CONFERENCE; Curr. Chemother. Immunother., Proc. Int. Congr. Chemother., 12th; P 270-1 (82); Vol 1,; Avail.: Periti, Piero; Gialdroni Grassi, Giuliana; Date: 81; Publisher: Am. Soc. Microbiol., Washington, D. C.

27-82-0319 (0319)
HIGH-PRESSURE LIQUID CHROMATOGRAPHY OF PARACETAMOL, ACETYLSALICYLIC ACID, AND

CODEINE PHOSPHATE
Stuber, B.; Mueller, K. H.
Dr. Spirig A. G. Olten Switz. CH CH-4601
Pharm. Acta Helv.; (82) P 181; Vol 57; No 7; In Ger

27-82-0320 (0320)
SEPARATION OF VALTRATE AND ISOVALTRATE BY MEANS OF PREPARATIVE LIQUID CHROMATOGRAPHY
Hazelhoff, B.; Jellema, R.; Grobben, H.; Malingre, T. M.
Lab. Pharmacogn. State Univ. Groningen Neth. NL 9713 AW
Pharm. Weekbl., Sci. Ed.; (82) P 21-4; Vol 4; No 1

27-82-0321 (0321)
CYCLOSPORIN A: CORRELATION BETWEEN HPLC AND RIA SERUM LEVELS
Abisch, E.; Beveridge, T.; Gratwohl, A.; Niederberger, W.; Nussbaumer, K.; Schaub, P.; Speck, B.; Trapp, M.
Biopharm. Dep. Sandoz Ltd. Basel Switz. CH 4002
Pharm. Weekbl., Sci. Ed.; (82) P 84-6; Vol 4; No 3

27-82-0322 (0322)
LIQUID CHROMATOGRAPHIC DETERMINATION OF CHLORDIAZEPOXIDE AND METABOLITES IN PLASMA
Divoll, Marcia; Greenblatt, David J.; Shader, Richard I.
Sch. Med. Tufts Univ. Boston USA US MA 02111
Pharmacology; (82) P 261-6; Vol 24; No 5

27-82-0323 (0323)
TLC- AND HPLC-ANALYSIS OF ELEUTHEROCOCCUS PREPARATIONS
Wagner, H.; Heur, Y. H.; Obermeier, A.; Tittel, G.; Bladt, S.
Inst. Pharm. Biol. Univ. Muenchen Munich Fed. Rep. Ger. DE D-8000/2
Planta Med.; (82) P 193-8; Vol 44; No 4; In Ger

27-82-0324 (0324)
NEW CARDIOACTIVE DRUGS. II. DETECTION AND ISOLATION OF CARDIOTONIC AMINES WITH ION-PAIR HPLC
Wagner, H.; Grevel, J.
Inst. Pharm. Biol. Univ. Muenchen Munich Fed. Rep. Ger. DE
Planta Med.; (82) P 36-40; Vol 44; No 1; In Ger

27-82-0325 (0325)
DETERMINATION OF THE RADIOCHEMICAL PURITY OF HIPPURAN BY HPLC
Pfeiffer, G.; Beranek, I.
Swiss Fed. Inst. React. Res. Wuerenlingen Switz. CH
Prog. Radiopharmacol.; (82) P 27-30; Vol 3; No Proc. Eur. Symp. Radiopharmacol., 3rd

27-82-0326 (0326)
A LIQUID CHROMATOGRAPHIC METHOD FOR QUANTITATING AMITRIPTYLINE IN BRAIN TISSUE
Preskorn, Sheldon H.; Glotzbach, Robin K.
Med. Cent. Univ. Kansas Kansas City USA US KS 66103
Psychopharmacology (Berlin); (82) P 23-4; Vol 78; No 1

27-82-0327 (0327)
A COMPARISON OF DIPHENYLHYDANTOIN METABOLISM IN DIFFERENT TISSUES USING HIGH PERFORMANCE LIQUID CHROMATOGRAPHY
MacKinney, Archie A., Jr.; Shinnick, Fred L.; Gorder, Cherna
Hematol. Lab. William S. Middleton Mem. Veterans Hosp. Madison USA US WI 53705
Res. Commun. Chem. Pathol. Pharmacol.; (82) P 395-411; Vol 36; No 3

27-82-0328 (0328)
REVERSED-PHASE HIGH PERFORMANCE LIQUID CHROMATOGRAPHIC DETERMINATION OF METHADONE IN HUMAN PLASMA
Buice, Robert G.; Sidhu, Paramjeet; Gurley, B. J.; Hamilton, Cindy
Coll. Pharm. Univ. Tennesse Memphis USA US TN 38163
Res. Commun. Subst. Abuse; (82) P 97-107; Vol 3; No 1

27-82-0329 (0329)
SIMULTANEOUS DETERMINATION OF CHLORPROMAZINE AND LEVOMEPROMAZINE IN HUMAN PLASMA BY HIGH-PERFORMANCE LIQUID CHROMATOGRAPHY USING ELECTROCHEMICAL DETECTION
Murakami, Koji; Murakami, Kazuo; Ueno, Tadataka; Hijikata, Junko; Shirasawa, Kimio; Mutoh, Toshitake
Dep. Psychiatry Sagami Hosp. Sagamihara Japan JP
Rinsho Kagaku; (82) P 50-5; Vol 11; No 1; In Japan

27-82-0330 (0330)
DETERMINATION OF ANTIEPILEPTIC DRUGS BY HIGH PERFORMANCE LIQUID CHROMATOGRAPHY
Maeda, Toshie; Miyagawa, Fumio; Nakamura, Maspo; Saito, Mitsuaki
Sch. Med. St. Marianna Univ. Kawasaki Japan JP
Sei Marianna Ika Daigaku Zasshi; (82) P 153-61; Vol 10; No 1; In Japan

27-82-0331 (0331)
CLINICAL PHARMACOKINETIC STUDIES ON THE CONCURRENT USE OF TWO ANTIPSYCHOTIC DRUGS. I. SIMULTANEOUS DETERMINATION OF CHLORPROMAZINE AND LEVOMEPROMAZINE IN HUMAN PLASMA BY HIGH-PERFORMANCE LIQUID CHROMATOGRAPHY USING ELECTROCHEMICAL DETECTION
Ueno, Tadataka
Sch. Med. Showa Univ. Tokyo Japan JP
Showa Igakkai Zasshi; (82) P 35-43; Vol 42; No 1; In Japan

27-82-0332 (0332)
MICRO DETERMINATION OF THEOPHYLLINE BY HIGH-PERFORMANCE LIQUID CHROMATOGRAPHY
Kubo, Hiroaki
Dep. Pharm. Sci. Kitasato Univ. Tokyo Japan JP
Sogo Rinsho; (82) P 43-9; Vol 31; No 1; In Japan

27-82-0333 (0333)
QUANTITATIVE DETERMINATION OF o-(.BETA.-HYDROXYETHYL)QUERCETINS IN URINE BY HIGH-PERFORMANCE LIQUID CHROMATOGRAPHY
Kuhnz, W.; Jung, G.; Voelter, W.
Inst. Org. Chem. Univ. Tuebingen Tuebingen Fed. Rep. Ger. DE D-7400
Stud. Org. Chem.(Amsterdam); (82) P 293-8; Vol 11; No Flavonoids Bioflavonoids

27-82-0334 (0334)
PURIFICATION OF HUMAN LYMPHOKINES BY HIGH PERFORMANCE LIQUID CHROMATOGRAPHY
McEntire, J. E.; Dunn, P.; Papermaster, B. W.; Gehrke, C. W.
Cancer Res. Cent. Columbia USA US MO 65201
CONFERENCE; Hum. Lymphokines: Biol. Immune Response Modif., (Proc. Int. Symp.), 3rd; (82) P 361-77; Avail.: Khan, Amanullah; Hill, Norwood Oakley; Date: 81; Publisher: Academic, New York, N. Y.

27-82-0335 (0335)
LIQUID CHROMATOGRAPHIC DETERMINATION OF QUINIDINE IN SERUM AND URINE
Patel, C. P.
Miles Lab. Elkhart USA US IN 46515
Ther. Drug Monit.; (82) P 213-17; Vol 4; No 2

27-82-0336 (0336)
INTERFERENCES IN A HIGH PRESSURE LIQUID CHROMATOGRAPHIC ASSAY OF THEOPHYLLINE

Leslie, James; Miller, Ann K.
Sch. Pharm. Univ. Maryland Baltimore USA US MD 21201
Ther. Drug Monit.; (82) P 323-4; Vol 4; No 3

27-82-0337 (0337)
SPECIFIC DETERMINATION OF PENTAMETHYLMELAMINE AND ITS DEMETHYLATED METABOLITES IN HUMAN PLASMA BY HIGH PRESSURE LIQUID CHROMATOGRAPHY
Leyland-Jones, Brian R.; Deesen, Patricia E.; Casper, Ephraim S.; Young, Charles W.
Clin. Pharmacol. Lab. Mem. Sloan-Kettering Cancer Cent. New York USA US NY 10021
Ther. Drug. Monit.; (82) P 185-90; Vol 4; No 2

27-82-0338 (0338)
REVERSED-PHASE HPLC OF BENZODIAZEPINES AND ITS APPLICATION IN THE QUALITY CONTROL OF DIAZEPAM
Zhou, Yuanyao; Li, Fang
Wuhan Munic. Inst. Drug Control Wuhan Peop. Rep. China CN
Yaowu Fenxi Zazhi; (82) P 129-34; Vol 2; No 3; In Ch

27-82-0339 (0339)
ANALYSIS OF AMINO ACID DRUGS BY ION-EXCHANGE HPLC. I. SEPARATION AND DETERMINATION OF PROSTALL CAPSULES
Yu, Zong; Kong, Xiangling; Zhou, Jingyuan
Tianjin Inst. Drug Control Tianjin Peop. Rep. China CN
Yaowu Fenxi Zazhi; (82) P 154-7; Vol 2; No 3; In Ch

27-82-0340 (0340)
COMPARISON BETWEEN HIGH-PERFORMANCE LIQUID CHROMATOGRAPHY AND ULTRAVIOLET SPECTROPHOTOMETRY FOR DETERMINING THE CONCENTRATION OF PHENYTOIN IN RABBIT PLASMA
Wang, Wenling; Lou, Yaqing; Duan, Zhongping; Wang, Weiping
Inst. Clin. Pharmacol. Beijing Med. Coll. Beijing Peop. Rep. China CN
Yaoxue Tongbao; (82) P 245-6; Vol 17; No 4; In Ch

27-82-0341 (0341)
DETERMINATION OF PRAZIQUANTEL IN RABBIT PLASMA BY HPLC AND ITS PHARMACOKINETIC PARAMETERS
Xu, Peipei; Ren, Zhonglu; Li, Yongxue; Lu, Minglian; Li, Duan; Xu, Hong; Zhao, Mingxia
Dep. Pharm. Shanghai First Med. Coll. Shanghai Peop. Rep. China CN
Yaoxue Tongbao; (82) P 364; Vol 17; No 6; In Ch

27-82-0342 (0342)
SEPARATION AND ASSAY OF SANTONIN ISOMERS BY HPLC
Liu, Xiulin
Liaoning Provincial Inst. Drug Control Shenyang Peop. Rep. China CN
Yaoxue Xuebao; (82) P 703-7; Vol 17; No 9; In Ch

27-82-0343 (0343)
HIGH-PRESSURE LIQUID CHROMATOGRAPHIC ANALYSIS OF DRUG RESIDUES IN HONEY. II. CHLORAMPHENICOL AND SULFATHIAZOLE
Juergens, Uwe
Landesuntersuchungsamt Gesundheitswesen Nordbayern Erlangen Fed. Rep. Ger. DE D-8520
Z. Lebensm.-Unters. Forsch.; (82) P 208-10; Vol 174; No 3; In Ger

27-82-0344 (0344)
STUDIES ON ANALYSIS OF FEED ADDITIVES BY HIGH-PERFORMANCE LIQUID CHROMATOGRAPHY. I. DETERMINATION OF NICARBAZIN IN CHICKEN MUSCLE AND EGG BY HIGH-PERFOR-

MANCE LIQUID CHROMATOGRAPHY
Hoshino, Youji; Horie, Masakazu; Nose, Norihide
Saitama Pref. Inst. Public Health Urawa Japan JP 338 Shokuhin Eiseigaku Zasshi; (82) P 265-9; Vol 23; No 3; In Japan

CHAPTER 7.28
Vitamins

--- 1982 ---

28-82-0001 (0001)
DETERMINATION, BY HPLC, OF CAROTENOIDS IN PAPRIKA PRODUCTS
Baranvai, Marta; Matus, Zoltan; Szabolcs, Jozsef
Inst. Chem. Univ. Med. Sch. Pecs Hung. HU H-7643
Acta Aliment.; (82) P 309-23; Vol 11; No 3

28-82-0002 (0002)
HIGH PERFORMANCE LIQUID CHROMATOGRAPHY OF FOLATES: IDENTIFICATION OF POLY-.GAMMA.-GLUTAMATE CHAIN LENGTHS OF LABELED AND UNLABELED FOLATES
Shane, Barry
Sch. Hyg. Public Health Johns Hopkins Univ. Baltimore USA US MD 21205
Am. J. Clin. Nutr.; (82) P 599-608; Vol 35; No 3

28-82-0003 (0003)
QUANTITATION OF URINARY NIACIN METABOLITES BY REVERSED-PHASE LIQUID CHROMATOGRAPHY
Carter, Eric G. A.
Dep. Nutr. Sci. Univ. California Berkeley USA US CA 94720
Am. J. Clin. Nutr.; (82) P 926-30; Vol 36; No 5

28-82-0004 (0004)
DETERMINATION OF THREE DIFFERENT POOLS OF REDUCED ONE-CARBON-SUBSTITUTED FOLATES. III. REVERSED-PHASE HIGH-PERFORMANCE LIQUID CHROMATOGRAPHY OF THE AZO DYE DERIVATIVES OF p-AMINOBENZOYLPOLY-.GAMMA.-GLUTAMATES AND ITS APPLICATION TO THE STUDY OF UNLABELED ENDOGENOUS PTEROYLPOLYGLUTAMATES OF RAT LIVER
Eto, Isao; Krumdieck, Carlos L.
Dep. Nutr. Sci. Univ. Alabama Birmingham USA US AL 35294
Anal. Biochem.; (82) P 323-9; Vol 120; No 2

28-82-0005 (0005)
THE MEASUREMENT OF ERYTHROCYTE THIAMIN PYROPHOSPHATE BY HIGH-PERFORMANCE LIQUID CHROMATOGRAPHY
Warnock, L. G.
VA Med. Cent. Nashville USA US TN 37203
Anal. Biochem.; (82) P 394-7; Vol 126; No 2

28-82-0006 (0006)
SEPARATION OF RETINOL, RETINOIC ACID, MANNOSE, DOLICHYL MANNOSYL PHOSPHATE, MANNOSYL RETINYL PHOSPHATE BY HIGH-PRESSURE LIQUID CHROMATOGRAPHY
Kurokawa, Tadashi; De Luca, Luigi M.
Lab. Exp. Pathol. Natl. Cancer Inst. Bethesda USA US MD 20205
Anal. Biochem.; (82) P 428-32; Vol 119; No 2

28-82-0007 (0007)
ANALYSIS OF VITAMIN D VIA HPLC. NEW APPLICATIONS IN PRESENCE OF VITAMIN A AND BHT
Hofsass, H.; Delano, N.; Korsch, B. H.; Grant, A.; Gehrlein, L.; Alicino, N. J.
Harrison USA US NJ 07029
Anal. Chem. Symp. Ser.; (82) P 367-80; Vol 10; No Adv. Steroid Anal.

28-82-0008 (0008)
A RELIABLE SEMIAUTOMATED METHOD FOR THE DETERMINATION OF TOTAL THIAMINE IN WHOLE BLOOD BY THE THIOCHROME METHOD WITH HIGH-PERFORMANCE LIQUID CHROMATOGRAPHY
Schrijver, Jaap; Speek, Andries J.; Klosse, Jan A.; Van Rijn, Herman J. M.; Schreurs, Wil H. P.

Dep. Clin. Biochem. Inst. CIVO-Toxicol. Nutr. Zeist Neth. NL
Ann. Clin. Biochem.; (82) P 52-6; Vol 19; No 1

28-82-0009 (0009)
STEREOCHEMISTRY OF NATURALLY-OCCURRING 25-HYDROXYVITAMIN D3-26,23 LACTONE AS DETERMINED BY RADIOLIGAND BINDING ANALYSIS AND HIGH-PERFORMANCE LIQUID CHROMATOGRAPHY
Horst, R. L.; Reinhardt, T. A.; Williams, D. H.
Natl. Anim. Dis. Cent. U. S. Dep. Agric. Ames USA US IA 50010
Biochem. Biophys. Res. Commun.; (82) P 541-6; Vol 108; No 2

28-82-0010 (0010)
METHODS FOR DETERMINATION OF UBIQUINONE BY HIGH PERFORMANCE LIQUID CHROMATOGRAPHY
Katsui, Goichiro
Eisai Co. Ltd. Tokyo Japan 112
Bitamin; (81) P 305-12; Vol 55; No 7; In Japan

28-82-0011 (0011)
ASSAY METHODS OF VITAMIN A. 3. ASSAY METHODS OF VITAMIN A IN PHARMACEUTICAL PREPARATIONS, FOODS AND FEEDS BY HIGH-PERFORMANCE LIQUID CHROMATOGRAPHY
Kastui, Goichiro
Eisai Co. Ltd. Tokyo Japan JP 112
Bitamin; (82) P 151-6; Vol 56; No 3; In Japan

28-82-0012 (0012)
ASSAY METHODS OF VITAMIN K BY HIGH PERFORMANCE LIQUID CHROMATOGRAPHY
Katsui, Goichro
Eisai Co. Ltd. Tokyo Japan JP 112
Bitamin; (82) P 183-91; Vol 56; No 4; In Japan

28-82-0013 (0013)
METHODS FOR ASSAY OF VITAMIN C. 2. HIGH-PERFORMANCE LIQUID CHROMATOGRAPHIC METHODS
Arakawa, Nobuhiko; Otsuka, Megumi; Kurata, Tadao
Dep. Food Nutr. Ochanomizu Univ. Tokyo Japan JP 112
Bitamin; (82) P 307-14; Vol 56; No 7; In Japan

28-82-0014 (0014)
DETERMINATION OF .ALPHA.- AND .BETA.-CAROTENE IN FRUIT AND VEGETABLES BY HIGH PERFORMANCE LIQUID CHROMATOGRAPHY
Bushway, R. J.; Wilson, A. M.
Dep. Food Sci. Univ. Maine Orono USA US ME 04469
Can. Inst. Food Sci. Technol. J.; (82) P 165-9; Vol 15; No 3

28-82-0015 (0015)
DETERMINATION OF ASCORBIC ACID IN WHEAT FLOURS, BREAD DOUGH CONDITIONERS, AND COMMERCIAL VITAMIN C TABLETS BY HIGH-PERFORMANCE LIQUID CHROMATOGRAPHY
Lookhart, G. L.; Hall, S. B.; Finney, K. F.
Agric. Res. Serv. U. S. Grain Market. Res. Lab. Manhattan USA US KS 66502
Cereal Chem.; (82) P 69-71; Vol 59; No 1

28-82-0016 (0016)
STUDIES ON REDUCED AND OXIDIZED UBIQUINONES. I. SIMULTANEOUS DETERMINATION OF REDUCED AND OXIDIZED UBIQUINONES IN TISSUES AND MITOCHONDRIA BY HIGH-PERFORMANCE LIQUID CHROMATOGRAPHY
Ikenoya, Satoru; Takada, Masahiro; Yuzuriha, Teruaki; Abe, Kouichi ; Katayama, Kouichi
Res. Lab. Eisai Co., Ltd. Tokyo Japan 112
Chem. Pharm. Bull.; (81) P 158-64; Vol 29; No 1

28-82-0017 (0017)
ISOLATION AND CHARACTERIZATION OF A VITAMIN D3 COMPOUND FROM A PLANT BY LIQUID CHROMATOGRAPHY
Rambeck, Walter A.; Zucker, H.
Vet. Fac. Univ. Munich Munich Fed. Rep. Ger. DE
Chromatogr. Sci.; (82) P 319-26; Vol 20; No Biol./Biomed. Appl. Liq. Chromatogr. 4

28-82-0018 (0018)
LC DETERMINATION OF VITAMINS A AND D IN FOODS AND ANIMAL FEEDSTUFFS
Wiggins, R. A.; Zai, E. S.; Lumley, I.
Lab. Gov. Chem. Dep. Ind. London UK GB
Chromatogr. Sci.; (82) P 327-41; Vol 20; No Biol./Biomed. Appl. Liq. Chromatogr. 4

28-82-0019 (0019)
IMPROVED DIRECT DETERMINATION OF ALPHA- AND GAMMA-TOCOPHEROLS IN PLASMA AND PLATELETS BY LIQUID CHROMATOGRAPHY, WITH FLUORESCENCE DETECTION
Lehmann, Joanna; Martin, Hilda L.
Lipid Nutr. Lab. Beltsville Hum. Nutr. Res. Cent. Beltsville USA US MD 20705
Clin. Chem. (Winston-Salem, N. C.); (82) P 1784-7; Vol 28; No 8

28-82-0020 (0020)
ASCORBIC ACID IN LYMPHOCYTES: CELL PREPARATION AND LIQUID-CHROMATOGRAPHIC ASSAY
Lee, Winnie; Hamernyik, Peggy; Hutchinson, Martha; Raisys, Vidmantas A.; Labbe, Robert F.
Dep. Lab. Med. Univ. Washington Seattle USA US WA 98195
Clin. Chem. (Winston-Salem, N. C.); (82) P 2165-9; Vol 28; No 10

28-82-0021 (0021)
LIQUID-CHROMATOGRAPHIC DETERMINATION OF THE TOTAL THIAMIN CONTENT OF BLOOD
Kimura, Mieko; Fujita, Tomio; Itokawa, Yoshinori
Fac. Med. Kyoto Univ. Kyoto Japan JP 606
Clin. Chem. (Winston-Salem, N. C.); (82) P 29-31; Vol 28; No 1

28-82-0022 (0022)
LIQUID-CHROMATOGRAPHIC ASSAY FOR RETINOL (VITAMIN A) AND RETINOL ANALOGS IN THERAPEUTIC TRIALS
McClean, Sharon W.; Ruddel, Mark E.; Gross, Earl G.; DeGiovanna, John J.; Peck, Gary L.
Clin. Chem. Serv. NIH Bethesda USA US MD 20205
Clin. Chem. (Winston-Salem, N. C.); (82) P 693-6; Vol 28; No 4, Pt. 1

28-82-0023 (0023)
A RAPID AND SIMPLE METHOD FOR THE MEASUREMENT OF PLASMA 25-HYDROXYVITAMIN D2 AND 25-HYDROXYVITAMIN D3 USING SEP-PAK C18 CARTRIDGES AND A SINGLE HIGH-PERFORMANCE LIQUID CHROMATOGRAPHIC STEP
Turnbull, Hazel; Trafford, D. J. H.; Makin, H. L. J.
Dep. Chem. Pathol. London Hosp. Med. Coll. London UK GB E1 2AD
Clin. Chim. Acta; (82) P 65-76; Vol 120; No 1

28-82-0024 (0024)
DETERMINATION OF FAT-SOLUBLE VITAMINS BY HPLC
Rougereau, Andre; Guiller, Alain; Gore, Jacques; Person, Odile
Inst. Nutr. Tours Fr. FR 37000
CONFERENCE; Recent Dev. Food Anal., Proc. Eur. Conf. Food Chem., 1st; (82) P 33-40; Avail.: Baltes, Werner; Czedik-Eysenberg, Peter B.; Pfannhauser, Werner; Date: 81; Publisher: Verlag Chem., Weinheim, Fed. Rep. Ger.

28-82-0025 (0025)
ANALYSIS OF VITAMINS IN FOOD BY HPLC
Hedlund, B.
Swed. Food Inst. SIK Goeteborg Swed. SE S-400 23
CONFERENCE; Recent Dev. Food Anal., Proc. Eur. Conf. Food Chem., 1st; (82) P 27-32; Avail.: Baltes, Werner; Czedik-Eysenberg, Peter B.; Pfannhauser, Werner; Date: 81; Publisher: Verlag Chem., Weinheim, Fed. Rep. Ger.

28-82-0026 (0026)
STUDIES ON THE OPTIMIZATION OF THE SEPARATION OF WATER-SOLUBLE VITAMINS BY AMINE-BOUND PHASE HPLC
Calcagno, C.; Lanteri, S.
Ist. Sci. Farm. Univ. Genova Genoa Italy IT
Farmaco, Ed. Prat.; (82) P 152-9; Vol 37; No 5; In Ital

28-82-0027 (0027)
QUANTITATIVE DETERMINATION OF VITAMINS D2 AND D3 IN MARGARINE BY HPLC
Mueller-Mulot, W.; Rohrer, G.
Kontroll-Abt. Hoffmann-La Roche A.-G. Grenzach-Wyhlen Fed. Rep. Ger. DE 7889/1
Fette, Seifen, Anstrichm.; (82) P 354-8; Vol 84; No 9; In Ger

28-82-0028 (0028)
DETERMINATION OF REDUCED BIOPTERINS BY HIGH PRESSURE LIQUID CHROMATOGRAPHY AND SUBSEQUENT ELECTROCHEMICAL DETECTION
Braeutigam, Matthias; Dreesen, Rainer; Herken, Hans
Pharmakol. Inst. Freien Univ. Berlin Fed. Rep. Ger. DE
Hoppe-Seyler's Z. Physiol. Chem.; (82) P 341-3; Vol 363; No 3

28-82-0029 (0029)
A HIGH-PERFORMANCE LIQUID CHROMATOGRAPHY METHOD FOR DETERMINING ASCORBIC ACID CONTENT OF FRESH FRUITS AND VEGETABLES
Watada, Alley E.
Hortic. Crops Qual. Lab. Dep. Agric. Beltsville USA US MD 20705
HortScience; (82) P 334-5; Vol 17; No 3, Sect. 1

28-82-0030 (0030)
DETERMINATION OF VITAMIN D AND RETINOID ACTIVITY IN EGGS BY HPLC
Sivell, Lorna M.; Wenlock, R. W.; Jackson, Patricia A.
Food Sci. Div. Minist. Agric., Fish. Food London UK GB
Hum. Nutr.: Appl. Nutr.; (82) P 430-7; Vol 36A; No 6

28-82-0031 (0031)
ASCORBIC ACID IN MEAT PRODUCTS. ANALYSIS BY REVERSED-PHASE ION-PAIR HIGH-PRESSURE LIQUID CHROMATOGRAPHY
Parolari, Giovanni
Stazione Sper. Ind. Conserve Alimentari Parma Italy IT
Ind. Conserve; (82) P 19-22; Vol 57; No 1; In Ital

28-82-0032 (0032)
SEMIAUTOMATED FLUOROMETRIC DETERMINATION OF PYRIDOXAL-5'-PHOSPHATE (VITAMIN B6) IN WHOLE BLOOD BY HIGH-PERFORMANCE LIQUID CHROMATOGRAPHY (HPLC)
Schrijver, Jaap; Speek, Andries J.; Schreurs, Wil H. P.
Inst. CIVO-Toxicol. Nutr. TNO Zeist Neth. NL
Int. J. Vitam. Nutr. Res.; (81) P 216-22; Vol 51; No 3

28-82-0033 (0033)
HIGH-PERFORMANCE LIQUID CHROMATOGRAPHIC DETERMINATION OF ASCORBIC ACID IN FOOD: COMPARISON WITH OTHER METHODS
Floridi, A.; Coli, R.; Fidanza, A. Alberti; Bourgeois, C. F.; Wiggins, R. A.
Ist. Sci. Aliment. Univ. Perugia Perugia Italy IT

Int. J. Vitam. Nutr. Res.; (82) P 193-6; Vol 52; No 2

28-82-0034 (0034)
A METHOD FOR THE DETERMINATION OF VITAMIN K3 IN PREMIXES AND ANIMAL FEEDS WITH THE AID OF HIGH PERFORMANCE LIQUID CHROMATOGRAPHY
Manz, U.; Maurer, R.
Dep. Vitamin Nutr. Res. Cent. Res. Units F. Hoffmann-La Roche & Co., Ltd. Basel Switz. CH
Int. J. Vitam. Nutr. Res.; (82) P 248-52; Vol 52; No 3

28-82-0035 (0035)
ASCORBIC ACID CONTENT OF SOME TROPICAL FRUIT PRODUCTS DETERMINED BY HIGH-PERFORMANCE LIQUID CHROMATOGRAPHY
Shaw, Philip E.; Wilson, Charles W., III
U. S. Citrus Subtrop. Prod. Lab. Winter Haven USA US FL 33880
J. Agric. Food Chem.; (82) P 394-6; Vol 30; No 2

28-82-0036 (0036)
ANALYSIS OF FAT-SOLUBLE VITAMINS. XXVI. HIGH PERFORMANCE LIQUID CHROMATOGRAPHIC DETERMINATION OF VITAMIN D IN FORTIFIED MILK AND MILKPOWDER
Borsje, Ben; De Vries, Ellen J.; Zeeman, Jakob; Mulder, Frits J.
Duphar B. V., Res. Lab. Weesp Neth. NL
J. Assoc. Off. Anal. Chem.; (82) P 1225-7; Vol 65; No 5

28-82-0037 (0037)
SIMULTANEOUS DETERMINATION OF VITAMINS A AND D IN DOSAGE FORMS BY HIGH-PRESSURE LIQUID CHROMATOGRAPHY
Santoro, Maria Ines R. M.; Magalhaes, Joao F.; Hackmann, Erika R. M.
Fac. Cienc. Farm. Univ. Sao Paulo Sao Paulo Brazil BR
J. Assoc. Off. Anal. Chem.; (82) P 619-23; Vol 65; No 3

28-82-0038 (0038)
HIGH-PERFORMANCE LIQUID CHROMATOGRAPHIC DETERMINATION OF VITAMIN D IN FORTIFIED MILKS, MARGARINE, AND INFANT FORMULAS
Thompson, James N.; Hatina, George; Maxwell, William B.; Duval, Suzanne
Nutr. Res. Lab. Health and Welfare Canada Ottawa Can. CA ON K1A 0L2
J. Assoc. Off. Anal. Chem.; (82) P 624-31; Vol 65; No 3

28-82-0039 (0039)
REVERSE PHASE LIQUID CHROMATOGRAPHIC DETERMINATION OF VITAMIN D2 AND D3 IN MILK
Muniz, Juan F.; Wehr, C. Timothy; Wehr, H. Michael
Lab. Serv. Oregon Dep. Agric. Salem USA US OR 97310
J. Assoc. Off. Anal. Chem.; (82) P 791-7; Vol 65; No 4

28-82-0040 (0040)
APPLICATION OF GEL PERMEATION CHROMATOGRAPHY AND NONAQUEOUS REVERSE PHASE CHROMATOGRAPHY TO HIGH-PERFORMANCE LIQUID CHROMATOGRAPHIC DETERMINATION OF RETINYL PALMITATE AND .ALPHA.-TOCOPHERYL ACETATE IN INFANT FORMULAS
Landen, William O., Jr.
FDA Atlanta USA US GA 30309
J. Assoc. Off. Anal. Chem.; (82) P 810-16; Vol 65; No 4

28-82-0041 (0041)
HIGH PERFORMANCE LIQUID CHROMATOGRAPHY OF VITAMIN A: A QUANTITATIVE DETERMINATION
Bohman, Ove; Engdahl, Kjell Aake; Johnsson, Haakan
Inst. Chem. Univ. Uppsala Uppsala Swed. SE S-751 21
J. Chem. Educ.; (82) P 251-2; Vol 59; No 3

28-82-0042 (0042)
SEPARATION OF THIAMIN AND ITS COMMON DEGRADATION AND OXIDATION PRODUCTS BY HIGH-PERFORMANCE LIQUID CHROMATOGRAPHY
Panijpan, Bhinyo; Kimura, Mieko; Itokawa, Yoshinori
Fac. Med., Kyoto Univ., Kyoto Japan
J. Chromatogr. (82) P 144-7; Vol 245; No 1

28-82-0043 (0043)
HIGH-PERFORMANCE LIQUID CHROMATOGRAPHY OF FAT-SOLUBLE VITAMINS. SEPARATION AND IDENTIFICATION OF VITAMINS D2 AND D3 AND THEIR ISOMERS IN FOOD SAMPLES IN THE PRESENCE OF VITAMIN A, VITAMIN E AND CAROTENE
Zonta, Fabio; Stancher, Bruno; Bielawny, Jan
Ist. Merceol. Univ. Trieste Trieste Italy IT 34127
J. Chromatogr.; (82) P 105-12; Vol 246; No 1

28-82-0044 (0044)
SEPARATION AND DETERMINATION OF THIAMIN AND ITS PHOSPHATE ESTERS BY REVERSED-PHASE HIGH-PERFORMANCE LIQUID CHROMATOGRAPHY
Kimura, Mieko; Panijpan, Bhinyo; Itokawa, Yoshinori
Fac. Med. Kyoto Univ. Kyoto Japan JP
J. Chromatogr.; (82) P 141-3; Vol 245; No 1

28-82-0045 (0045)
TWO IMPROVED HIGH-PERFORMANCE LIQUID CHROMATOGRAPHIC SEPARATIONS OF BIOLOGICALLY SIGNIFICANT FORMS OF FOLATE
Reingold, Ruth N.; Picciano, Mary Frances
Dep. Foods Nutr. Univ. Illinois Urbana USA US IL 61801
J. Chromatogr.; (82) P 171-9; Vol 234; No 1

28-82-0046 (0046)
QUANTITATION OF PHYSIOLOGICAL .ALPHA.-TOCOPHEROL, METABOLITES, AND RELATED COMPOUNDS BY REVERSED-PHASE HIGH-PERFORMANCE LIQUID CHROMATOGRAPHY
Howell, Scott K.; Wang, Yeu Ming
M.D. Anderson Hosp. Tumor Inst. Univ. Texas System Cancer Cent. Houston USA US TX 77030
J. Chromatogr.; (82) P 174-80; Vol 227; No 1

28-82-0047 (0047)
SEPARATION OF VITAMIN B6 COMPOUNDS BY PAIRED-ION HIGH-PERFORMANCE LIQUID CHROMATOGRAPHY
Tryfiates, George P.; Sattsangi, Saroj
Sch. Med. West Virginia Univ. Morgantown USA US WV 26506
J. Chromatogr.; (82) P 181-6; Vol 227; No 1

28-82-0048 (0048)
HIGH-PERFORMANCE LIQUID CHROMATOGRAPHIC DETERMINATION OF CAROTENE AND VITAMIN A AND ITS GEOMETRIC ISOMERS IN FOODS. APPLICATIONS TO CHEESE ANALYSIS
Stancher, Bruno; Zonta, Fabio
Ist. Merceol. Univ. Trieste Trieste Italy IT 34127
J. Chromatogr.; (82) P 217-25; Vol 238; No 1

28-82-0049 (0049)
COMPARISON BETWEEN STRAIGHT AND REVERSED PHASES IN THE HIGH-PERFORMANCE LIQUID CHROMATOGRAPHIC FRACTIONATION OF RETINOL ISOMERS
Stancher, Bruno; Zonta, Fabio
Ist. Merceol. Univ. Trieste Trieste Italy IT 34127
J. Chromatogr.; (82) P 244-8; Vol 234; No 1

28-82-0050 (0050)
SIMULTANEOUS DETERMINATION OF ASCORBIC ACID AND DEHYDROASCORBIC ACID BY

REVERSED-PHASE ION-PAIR HIGH-PERFORMANCE LIQUID CHROMATOGRAPHY WITH PRECOLUMN DERIVATIZATION
Keating, Ross W.; Haddad, Paul R.
Dep. Anal. Chem. Univ. New South Wales Kensington Australia AU 2033
J. Chromatogr.; (82) P 249-55; Vol 245; No 2

28-82-0051 (0051)
DETERMINATION OF THE B2 VITAMER FLAVIN-ADENINE DINUCLEOTIDE IN WHOLE BLOOD BY HIGH-PERFORMANCE LIQUID CHROMATOGRAPHY WITH FLUOROMETRIC DETECTION
Speek, A. J.; Van Schaik, F.; Schrijver, J.; Schreurs, W. H. P.
Dep. Clin. Biochem. TNO Zeist Neth. NL 3700 AJ
J. Chromatogr.; (82) P 311-16; Vol 228

28-82-0052 (0052)
DIFFERENTIAL DETERMINATION OF L-ASCORBIC ACID AND D-ISOASCORBIC ACID BY REVERSED-PHASE HIGH-PERFORMANCE LIQUID CHROMATOGRAPHY WITH ELECTROCHEMICAL DETECTION
Tsao, C. S.; Salimi, S. L.
Linus Pauling Inst. Sci. Med. Palo Alto USA US CA 94306
J. Chromatogr.; (82) P 355-8; Vol 245; No 3

28-82-0053 (0053)
SEPARATION AND DETERMINATION OF CIS/TRANS-.BETA.-CAROTENES BY HIGH-PERFORMANCE LIQUID CHROMATOGRAPHY
Tsukida, Kiyoshi; Saiki, Kayoko; Takii, Tomoko; Koyama, Yasushi
Kobe Women's Coll. Pharm. Kobe Japan JP 658
J. Chromatogr.; (82) P 359-64; Vol 245; No 3

28-82-0054 (0054)
THIAMIN ANALYSIS AND SEPARATION OF THIAMIN PHOSPHATE ESTERS BY HIGH-PERFORMANCE LIQUID CHROMATOGRAPHY
Hilker, Doris M.; Clifford, Andrew J.
Dep. Nutr. Univ. California Davis USA US CA 95616
J. Chromatogr.; (82) P 433-8; Vol 231; No 2

28-82-0055 (0055)
MEASUREMENT OF VITAMIN A AND VITAMIN E IN HUMAN SERUM BY HIGH-PERFORMANCE LIQUID CHROMATOGRAPHY
Driskell, William J.; Neese, Jane W.; Bryant, Clyde C.; Bashor, Mark M.
Clin. Chem. Div. U.S. Dep. Health Human Serv. Atlanta USA US GA 30333
J. Chromatogr.; (82) P 439-44; Vol 231; No 2

28-82-0056 (0056)
SEPARATION OF CAROTENOIDS BY HPLC. PART 2. SEPARATION OF CROCETIN GLYCOSYL ESTERS BY HIGH-PERFORMANCE LIQUID CHROMATOGRAPHY
Pfander, Hanspeter; Rychener, Martin
Inst. Org. Chem. Univ. Berne Bern Switz. CH CH-3012
J. Chromatogr.; (82) P 443-7; Vol 234; No 2

28-82-0057 (0057)
DETERMINATION OF VITAMIN D3 IN COD-LIVER OIL BY HIGH-PERFORMANCE LIQUID CHROMATOGRAPHY
Pask-Hughes, R. A.; Calam, D. H.
Natl. Inst. Biol. Stand. Control London UK GB NW3 6RB
J. Chromatogr.; (82) P 95-104; Vol 246; No 1

28-82-0058 (0058)
ION-PAIR HIGH-PRESSURE LIQUID CHROMATOGRAPHY OF CIS-TRANS ISOMERS OF RETINOIC ACID IN TISSUES OF VITAMIN A-SUFFICIENT RATS
Sundaresan, P. Ramnathan; Bhat, Pangala V.

Div. Toxicol. FDA Washington USA US DC 20204
J. Lipid Res.; (82) P 448-55; Vol 23; No 3

28-82-0059 (0059)
IMPROVED HIGH-PRESSURE LIQUID-CHROMATOGRAPHIC ASSAY OF SERUM 45-HYDROXY-CHOLECALCIFEROL AND 25-HYDROXYERGOCALCIFEROL AFTER REVERSE-PHASE SEP-PAK C18 CARTRIDGE PREPARATION OF SAMPLE
Kohl, E. Alan; Schaefer, Phillip C.
Health Sci. Cent. Univ. Texas San Antonio USA US TX 78284
J. Liq. Chromatogr.; (81) P 2023-37; Vol 4; No 11

28-82-0060 (0060)
HIGH PERFORMANCE LIQUID CHROMATOGRAPHIC METHODS FOR THE QUANTITATION OF VITAMIN A PALMITATE IN LIQUID MULTIVITAMIN FORMULATIONS
Van Antwerp, J.; Lepore, J.
Qual. Control Dep. Hoffmann-La Roche Inc. Nutley USA US NJ 07110
J. Liq. Chromatogr.; (82) P 571-84; Vol 5; No 3

28-82-0061 (0061)
THE DETERMINATION OF FOLIC ACID (PTEROYLMONOGLUTAMIC ACID) IN FORTIFIED PRODUCTS BY REVERSED-PHASE HIGH-PRESSURE LIQUID CHROMATOGRAPHY
Hoppner, K.; Lampi, B.
Nutr. Res. Lab. Health Prot. Branch Ottawa Can. CA ON
J. Liq. Chromatogr.; (82) P 953-66; Vol 5; No 5

28-82-0062 (0062)
THE CONTENT OF PHYLLOQUINONE (VITAMIN K1) IN HUMAN MILK, COWS' MILK AND INFANT FORMULA FOODS DETERMINED BY HIGH-PERFORMANCE LIQUID CHROMATOGRAPHY
Haroon, Yacoob; Shearer, Martin J.; Rahim, Seema; Gunn, William G. ; McEnery, Gerald; Barkhan, Percy
Dep. Haematol. Guy's Hosp. London UK GB SE1 9RT
J. Nutr.; (82) P 1105-17; Vol 112; No 6

28-82-0063 (0063)
HPLC DETERMINATION OF RETINOIDS IN THE LIVER OF KNOWN STRAIN MICE
Smirnov, A. Yu.; Sergeev, A. V.; Shevchenko, V. E.
Vses. Onkol. Nauchn. Tsentr Moscow USSR SU
Khim.-Farm. Zh.; (82) P 913-17; Vol 16; No 8; In Russ

28-82-0064 (0064)
DETERMINATION OF VITAMIN D IN MILK BY HIGH-PRESSURE LIQUID CHROMATOGRAPHY
Mouillet, L.; Luquet, F. M.; Gagnepain, M. F.; Sorgue, Y.
ISHA Longjumeau Fr. FR 91163
Lait ; (82) P 44-54; Vol 62; No 611-612; In Fr

28-82-0065 (0065)
HIGH-PERFORMANCE LIQUID CHROMATOGRAPHIC SEPARATION IN SHORT COLUMNS OF INDIVIDUAL TOCOPHEROLS AND TOCOTRIENOLS IN EDIBLE FATS AND OILS
Gertz, Christian; Herrmann, Karl
Chem. Untersuchungsamt Stadt Hagen Hagen Fed. Rep. Ger. DE 5800/1
Lebensmittelchem. Gerichtl. Chem.; (82) P 53-8; Vol 36; No 3; In Ger

28-82-0066 (0066)
A MICROMETHOD FOR THE DETERMINATION OF TOCOPHEROLS IN VEGETABLE OILS USING HIGH PERFORMANCE LIQUID CHROMATOGRAPHY
Koyama, Hiroshi; Kumazaki, Ikumi; Matsuhisa, Tsugio
Coll. Agric. Meijo Univ. Nagoya Japan JP
Meijo Daigaku Nogakubu Gakujutsu Hokoku; (82) P 25-30; Vol 18

28-82-0067 (0067)
APPLICATION OF HIGH PERFORMANCE LIQUID CHROMATOGRAPHY TO THE ANALYSIS OF CAROTENOID PIGMENTS
Nakazoe, Junichi
Tokai Reg. Fish. Res. Lab. Tokyo Japan JP 104
Nippon Suisan Gakkaishi; (82) P 1007; Vol 48; No 7

28-82-0068 (0068)
AN HPLC METHOD FOR THE SEPARATION AND QUANTITATION OF WATER-SOLUBLE VITAMINS IN VITAMIN-MINERAL FORMULATIONS
Jenkins, Charles
USA US
Pharm. Technol.; (82) P 53-6, 59, 61, 62-5; Vol 6; No 3

28-82-0069 (0069)
EVIDENCE FOR A VITAMIN D-RELATED CONTAMINANT IN THE ASSAY BY HPLC OF 25-HYDROXYCHOLECALCIFEROL IN SHEEP SERUM
Smith, B. S. W.; Wright, H.; Slater, J. S.; Harkins, D. C.
Moredun Res. Inst. Anim. Dis. Res. Assoc. Edinburgh UK GB
Proc. Workshop Vitam. D; (82) P 801-3; Vol 5th; No Vitam. D: Chem., Biochem. Clin. Endocrinol. Calcium Metab.

28-82-0070 (0070)
SIMPLIFIED METHOD FOR THE DETERMINATION OF 1,25-DIHYDROXYVITAMIN D USING AUTOMATED HIGH PRESSURE LIQUID CHROMATOGRAPHY: APPLICATION TO THE DIFFERENTIAL DIAGNOSIS OF HYPERCALCEMIA
Endres, D.; Lu, J.; Mueller, J.; Adams, J.; Holick, M.; Broughton, A.
Nichols Inst. San Juan Capistrano USA US CA 92675
Proc. Workshop Vitam. D; (82) P 813-15; Vol 5th; No Vitam. D: Chem., Biochem. Clin. Endocrinol. Calcium Metab.

28-82-0071 (0071)
A HIGH-PERFORMANCE LIQUID CHROMATOGRAPHIC METHOD FOR SIMULTANEOUS DETERMINATION OF 25-HYDROXYVITAMIN D2 AND 25-HYDROXYVITAMIN D3 IN HUMAN PLASMA
Kobayashi, T.; Okano, T.; Shida, S.; Nakao, H.; Kuroda, E.; Kodama, S.; Matsuo, T.
Kobe Women's Coll. Pharm. Kobe Japan JP 658
Proc. Workshop Vitam. D; (82) P 817-19; Vol 5th; No Vitam. D: Chem., Biochem. Clin. Endocrinol. Calcium Metab.

28-82-0072 (0072)
RECENT ASPECTS OF HPLC OF ISOMERS OF RETINOL
Zonta, F.
Ist. Merceol. Univ. Trieste Trieste Italy IT
Rass. Chim.; (81) P 203-8; Vol 33; No 4; In Ital

28-82-0073 (0073)
SYSTEMATIC SEPARATION OF CAROTENOID PIGMENTS FROM CITRUS FLESH BY HIGH-PERFORMANCE LIQUID CHROMATOGRAPHY
Satoh, Hisao; Komai, Toshiko; Kitahara, Keiichi; Aoyagi, Reiko; Honda, Teruo; Noguchi, Fuminori
Dep. Chem. Tokyo Med. Coll. Tokyo Japan JP
Tokyo Ika Daigaku Kiyo; (82) P 13-21; Vol 8; In Japan

28-82-0074 (0074)
DETERMINATION OF TOCOPHEROLS IN RICE BRAN OILS BY HIGH PERFORMANCE LIQUID CHROMATOGRAPHY
Tanabe, Keizo; Yamaoka, Masakazu; Tanaka, Akio; Kato, Akio; Amemiya, Junko
Natl. Chem. Lab. Ind. Ibaraki Japan JP
Yukagaku; (82) P 205-8; Vol 31; No 4; In Japan

INDEX

INDEX

INDEX CHAPTER 1

INDEX CHAPTER 2

INDEX

INDEX

INDEX CHAPTER 3

INDEX

INDEX CHAPTER 4

INDEX CHAPTER 5.1

INDEX CHAPTER 5.2

INDEX

INDEX CHAPTER 5.3

INDEX CHAPTER 5.4

INDEX

INDEX CHAPTER 6

INDEX

INDEX CHAPTER 7

INDEX CHAPTER 8

INDEX

INDEX

INDEX

INDEX

INDEX

INDEX CHAPTER 9

INDEX

INDEX CHAPTER 10

INDEX CHAPTER 11.1

INDEX

INDEX CHAPTER 11.2

INDEX CHAPTER 12

INDEX

INDEX CHAPTER 13

INDEX CHAPTER 14

INDEX

INDEX

INDEX CHAPTER 15

INDEX

INDEX CHAPTER 16

INDEX

INDEX CHAPTER 17

INDEX

INDEX CHAPTER 18

INDEX

INDEX CHAPTER 19

INDEX CHAPTER 20

INDEX CHAPTER 21

INDEX

INDEX CHAPTER 22

INDEX CHAPTER 23

INDEX

INDEX CHAPTER 24

INDEX CHAPTER 25

INDEX

INDEX CHAPTER 26

INDEX

INDEX

INDEX

INDEX

INDEX CHAPTER 27.1

INDEX CHAPTER 27.2

INDEX

INDEX CHAPTER 27.3

INDEX

INDEX CHAPTER 27.4

INDEX

INDEX

INDEX

INDEX

INDEX

INDEX

INDEX CHAPTER 28

INDEX

INDEX